THE ROUTLEDGE HANDBOOK OF THE POLITICAL ECONOMY OF THE ENVIRONMENT

Featuring a stellar international cast list of leading and cutting-edge scholars, *The Routledge Handbook of the Political Economy of the Environment* presents the state of the art of the discipline that considers ecological issues and crises from a political economy perspective. This collective volume sheds new light on the effect of economic and power inequality on environmental dynamics and, conversely, on the economic and social impact of environmental dynamics.

The chapters gathered in this handbook make four original contributions to the field of political economy of the environment. First, they revisit essential concepts and methods of environmental economics in light of their political economy. Second, they introduce readers to recent theoretical and empirical advances in key issues of political economy of the environment with a special focus on the relationship between inequality and environmental degradation, a nexus that has dramatically come into focus with the COVID crisis. Third, the authors of this handbook open the field to its critical global and regional dimensions: global issues, such as the environmental justice movement and inequality and climate change as well as regional issues such as agriculture systems, air pollution, natural resources appropriation and urban sustainability. Fourth and finally, the work shows how novel analysis can translate into new forms of public policy that require institutional reform and new policy tools. Ecosystems preservation, international climate negotiations and climate mitigation policies all have a strong distributional dimension that chapters point to. Pressing environmental policy such as carbon pricing and low-carbon and energy transitions entail numerous social issues that also need to be accounted for with new analytical and technological tools.

This handbook will be an invaluable reference, research and teaching tool for anyone interested in political economy approaches to environmental issues and ecological crises.

Éloi Laurent is Senior Economist at OFCE/Sciences Po, France, Professor at the School of Management and Innovation at Sciences Po and Ponts ParisTech and a visiting professor at Stanford University, USA.

Klara Zwickl is Assistant Professor at the WU Vienna University of Economics and Business, Austria.

'Tackling the unprecedented challenges of the 21st century will depend to a great extent on the social sciences we use to understand them and to suggest appropriate interventions. Yet the pre-eminent social science, economics, remains largely blind and uninterested regarding two of the most pressing challenges: the environment and inequality. This landmark volume tackles both of these gaps together and head on, introducing readers to a growing body of exciting work that aims to revolutionize economics, and thereby to tackle the grievous and daunting injustices of our age. A must-read for anyone interested in a deeper reconstruction of the economic ideas that have shaped our current crisis and how they can be transformed.'

– ***David Tyfield***, *Professor of Sustainable Transitions and Political Economy, Lancaster University, UK*

THE ROUTLEDGE HANDBOOK OF THE POLITICAL ECONOMY OF THE ENVIRONMENT

Edited by Éloi Laurent and Klara Zwickl

LONDON AND NEW YORK

First published 2022
by Routledge
2 Park Square, Milton Park, Abingdon, Oxon OX14 4RN

and by Routledge
605 Third Avenue, New York, NY 10158

Routledge is an imprint of the Taylor & Francis Group, an informa business

British Library Cataloguing-in-Publication Data
A catalogue record for this book is available from the British Library

Library of Congress Cataloging-in-Publication Data
Names: Laurent, Éloi, editor. | Zwickl, Klara, editor.
Title: The Routledge handbook of the political economy of the environment / edited by Éloi Laurent and Klara Zwickl.
Description: Abingdon, Oxon ; New York, NY : Routledge, 2022. | Series: Routledge international handbooks | Includes bibliographical references and index.
Identifiers: LCCN 2021020649 (print) | LCCN 2021020650 (ebook) | ISBN 9780367410704 (hardback) | ISBN 9781032058603 (paperback) | ISBN 9780367814533 (ebook)
Subjects: LCSH: Environmental economics.
Classification: LCC HC79.E5 R6749 2022 (print) | LCC HC79.E5 (ebook) | DDC 333.7—dc23
LC record available at https://lccn.loc.gov/2021020649
LC ebook record available at https://lccn.loc.gov/2021020650

ISBN: 978-0-367-41070-4 (hbk)
ISBN: 978-1-032-05860-3 (pbk)
ISBN: 978-0-367-81453-3 (ebk)

DOI: 10.4324/9780367814533

Typeset in Bembo
by Apex CoVantage, LLC

CONTENTS

CONTRIBUTORS

Michael Ash is Professor of Economics and Public Policy at the University of Massachusetts Amherst and co-directs the Corporate Toxics Information Project of the Political Economy Research Institute. Ash was staff labor economist for the Council of Economic Advisers.

Michel Bourban is Postdoctoral Researcher in the Department of Politics and International Studies at the University of Warwick, UK.

James K. Boyce is Professor Emeritus of Economics at the University of Massachusetts Amherst. His books include *Economics for People and the Planet* (2019), *Economics, the Environment and Our Common Wealth* (2013) and *The Political Economy of the Environment* (2002).

Juan-Camilo Cardenas is Professor of Economics at the Universidad de Los Andes (Colombia) and Research Associate at the Center of the Sustainable Development Goals for Latin America (CODS); he is also Adjunct Professor and Affiliated Faculty at the University of Massachusetts Amherst and the Ostrom Workshop at Indiana University respectively. He holds a Ph.D. in resource economics from the University of Massachusetts Amherst and did his post-doctorate at the Workshop in Political Theory and Policy Analysis (Indiana University) under the mentorship of Elinor Ostrom.

Julien Caudeville is Researcher at the French Institute of Industrial Environment and Risks (INERIS), expert in risk assessment and spatial statistics. His field of research deals with environmental health inequalities and environmental exposure modeling.

Ian M. Cook is Research Fellow at the Central European University (CEU). An anthropologist with a regional focus on south India, he works primarily on cities, podcasting, environmental justice, access to education and doing academia differently.

Patrick Criqui is Senior Researcher, Emeritus, at CNRS, Université Grenoble Alpes. He has worked on climate change policy modelling for different EU Research Framework Programmes. He participated in the Deep Decarbonization Pathways Project before COP-21 and now deals with low emission development strategies.

Céline Guivarch is Researcher at the Ecole Nationale des Ponts et Chaussées and the Centre International de Recherche sur l'Environnement et le Développement (CIRED).

Ruben Haalebos is Data Scientist in the London Stock Exchange Group, working in the Sustainable Investment department on subjects at the crossroads between environment, sovereign and corporates.

Stéphane Hallegatte is the Lead Economist of the World Bank Climate Change Group. He is the author of dozens of articles published in international journals and of several books and World Bank reports on green growth, sustainable development, and the economics of natural disasters and climate change.

Alain Karsenty is Senior Researcher at Cirad (Agricultural Research for Development), Montpellier, France. His research and expertise area covers the economic instruments for the environment. He has an extensive knowledge of land tenure, concessions and forest policies in developing countries.

Fridolin Krausmann is Professor at the Institute of Social Ecology Vienna (SEC), University of Natural Resources and Life Sciences, Vienna (BOKU), where he studies long-term changes of resource use in local systems, in national economies and at the global scale.

Éloi Laurent is Senior Economist at OFCE/Sciences Po, France, Professor at the School of Management and Innovation at Sciences Po and Ponts ParisTech and a visiting professor at Stanford University, USA.

Paul Malliet is Senior Economist at the OFCE, working on macroeconomic modelling of climate and energy transition policies.

Antoine Missemer is Researcher in economics and history of ideas at CNRS and member of CIRED – International Research Centre on Environment and Development, Paris, France.

James C. Murombedzi is a political economist whose research interests include climate change policy and governance, environment and development, and land and natural resources policy.

David N. Pellow is the Dehlsen Chair and Professor of Environmental Studies and Director of the Global Environmental Justice Project at the University of California, Santa Barbara. His teaching, research and activism focus on environmental justice in the U.S. and globally.

Richard Puchalsky (M.S. Astronomy, University of Maryland, 1990) is a freelance data librarian who facilitates public access to environmental and financial databases. Projects include PERI's Toxic 100 and Greenhouse 100, and Good Jobs First's Subsidy Tracker and Violation Tracker.

Alfredo R. M. Rosete is Assistant Professor of Economics at Central Connecticut State University. His primary research areas are agriculture and development economics, immigration, and political economy. He has published articles in the *Journal of Globalization and Development*, the *Review of Radical Political Economy* and the *Journal of Rural Studies*.

Anke Schaffartzik is Senior Scientist at the Institute of Social Ecology Vienna (SEC), University of Natural Resources and Life Sciences, Vienna (BOKU), where she collaboratively investigates linkages between resource use and socio-economic development across levels of spatial scale.

Diane M. Sicotte is Associate Professor of Sociology at Drexel University in Philadelphia, PA, USA, where she researches environmental inequality, plastics and society, and labor unions in the fossil fuel economy.

Tamara Steger is Associate Professor in the Department of Environmental Sciences and Policy at Central European University in Vienna, Austria.

Simon Sturn is Researcher at the WU Vienna University of Economics and Business. He works on the empirical analysis of environmental, labor and macroeconomic policies. His work has been published in such journals as *Applied Economics*, *Economic Modelling*, *Industrial & Labor Relations Review* and *Oxford Economic Papers*.

Nicolas Taconet is Researcher at the Ecole Nationale des Ponts et Chaussées and the Centre International de Recherche sur l'Environnement et le Développement (CIRED).

Henri Waisman leads the Deep Decarbonization Pathways initiative at Institut du Développement Durable et des Relations Internationales (IDDRI, France) and is an author of the IPCC report "Global Warming of 1.5°C". His work focuses on long-term low emissions development strategies.

Brian Walsh is Senior Economist in the World Bank Climate Change Group, specializing in predictive analytics for risk management and poverty reduction.

Klara Zwickl is Assistant Professor at the WU Vienna University of Economics and Business, Austria. She works on environmental inequality and distributional effects of environmental policies. She has published in journals such as *Ecological Economics*, *The Energy Journal* and *Empirical Economics*.

1
INTRODUCTION
Political economy of the environment in the century of ecological crises

Éloi Laurent and Klara Zwickl

What is political economy of the environment?

Before the 20th century, economics was not understood as a distinct discipline from political science. In fact, the most influential economic theorists of the late 18th and 19th century – including Adam Smith, David Ricardo, John Stuart Mill, and Karl Marx – considered themselves as political economists, emphasizing that, because economic and political processes and outcomes are closely connected, they should be studied jointly. The term "political economy" is said to have originated in France in the 17th century, a time and place where the role of the state in shaping national economies was decisive. The first book titled *Traité d'économie politique* was published by French mercantilist Antoine de Montchrétien in 1615, though some evidence points towards an even earlier use of the expression (Groenewegen, 1991).

Until the late 19th century, the two terms – economics and political economy – were largely used interchangeably. In his influential book *Principles of Economics*, published in 1890, Alfred Marshall provided a new definition of economics, separating it from other social sciences and from political processes. More specifically, Marshall (1920) introduced a theory of consumption and production based on the concepts of supply and demand and constraint optimization of households and firms. Furthermore, Marshall's contribution served as a starting point for neoclassical welfare theory founded on the concepts of consumer and producer surplus. Society is considered as the sum of individuals, who maximize their utility subject to a budget constraint, the only power considered being purchasing power, which determines the household's budget constraint. Political processes and distributional considerations, such as inequality in income, wealth, or influence, are no longer considered relevant for this new branch of economics.

With the development of neoclassical economics in the decades following Marshall's principles, political, distributional and institutional factors were increasingly neglected in economic theory. In 1932, Lionel Robbins provided a general definition of economics that is still used in many textbooks today that reflected this neglect: "the science which studies human behavior as a relationship between ends and scarce means which have alternative uses" (Robbins, 1972: 15). While the optimal use of scarce resources became the central focus of economic analysis, determining which groups in society were able to access these resources and at what costs to others or to the environment was gradually overlooked. Similarly, power disparities, which shape the allocation and distribution of scarce resources, were often omitted. They still are. In fact, in

DOI: 10.4324/9780367814533-1

purely competitive markets, power dynamics become irrelevant, or to quote Paul Samuelson, "in a perfectly competitive market it really doesn't matter who hires who", whether capital hires labor or labor capital (Samuelson, 1957: 894).

While the perfectly competitive market serves as the baseline for neoclassical economic models, several market failures have been identified by economists, including information asymmetries, externalities, market power, principle-agent problems, and public goods. Over the last decades, a large body of theoretical and empirical literature in various fields, such as behavioral economics, environmental economics, labor economics, and public economics, has presented compelling evidence for their existence and magnitude. In fact, these market failures are increasingly viewed as of such significance that they have come to describe the baseline, not its exceptions.

Modern behavioral microeconomic theory emphasizes that market failures arise because of incomplete contracts (Bowles, 2004). In some cases, such as in labor or credit markets, information asymmetries and principal-agent problems make it difficult to set up complete and enforceable contracts. For example, while employers can hire workers, due to a lack of information and principal-agent problems, they cannot exactly determine their work effort. Public goods, such as climate change mitigation, are underprovided because no single country can decide on the level of total global carbon emissions and therefore faces strong incentives to freeride on others.

When contracts are incomplete, it is power and norms that drive economic outcomes (Bowles, 2004). For example, the distribution of power between plant operators and local residents will decide on the magnitude of externalities resulting from industrial pollution in the neighborhood. In many cases, property rights can be reallocated so that negative externalities are internalized and public goods are no longer underprovided. Furthermore, plant operators can be mandated to install pollution mitigation equipment to improve local air quality. Another illustration is the "Tragedy of the Commons", in which elusive or non-existant property rights lead to an overexploitation of a local environmental good. If an appropriate institutional structure can be set up, access to the good can be regulated and the resource can be preserved. Again, the distribution of power will decide if and how property rights are altered: the local environmental good either could be privatized or publicized and resource use determined by its owner or could be turned into a common where parties decide on its access and governance structure. The property rights of global climate could also be changed to solve the public goods problem, most obviously through a binding international agreement, but also, as the next best alternative, through measures such as carbon border adjustment, which as a first step would prevent carbon leakage to countries with no climate policy. Which policy measures will be taken is decided by power disparities within and between countries.

Because power is so central in determining not only political, economic, and social but also environmental outcomes, political economy is experiencing a revival. Over the last decades, numerous publications on "the political economy of" different subjects have emerged aiming at including power and political processes back into economic analysis. This handbook aims at contributing to this revival.

The notion of *political economy of the environment* was largely shaped by James K. Boyce's landmark book (Boyce, 2002). The novelty of his approach is to investigate both current environmental challenges (such as climate change or air pollution) and current economic issues (such as development and globalization) through the lens of power inequality. He identifies the winners and losers from environmental degrading activities, as well as the dynamics between the two, by revealing how the winners are able to pursue their activities and what the consequences are for the losers. He distinguishes between five different dimensions of power: purchasing power, decision power, agenda power, value power, and event power. Purchasing power is defined, as in most economics textbooks, as the ability to purchase goods and services produced in the economy. Decision power refers to the ability to decide on more or fewer environmental regulations. The distribution of agenda power refers to the capacity to determine which topics receive attention by politicians and the media, while value power affects how people's values and beliefs are shaped and shifted. Finally, event power refers to the ability to influence the events under which people then have to make decisions (Boyce, 2002: 8f, see Boyce 2019a for an updated synthesis on those themes and contemporary issues).

We define the political economy of the environment along these "Boycian" lines – as the field that analyzes the economic and social impacts of environmental degradation, the uneven distribution of environmental resources, benefits and damage, the social dynamics behind environmental outcomes, and the determination and implementation of environmental policy in the context of incomplete and imperfect information in economies where power inequalities abound. Political economy of the environment is thus concerned on the one hand with the effect of economic and power inequality on environmental dynamics, investigating how higher wealth and power disparities affect the distribution of environmental costs and benefits, as well as overall environmental outcomes. On the other hand, it examines the economic and social impact of environmental dynamics, analyzing differential vulnerability to environmental hazards and degradation along a plurality of justice criteria, both within and between countries. These topics are emerging as critical concerns in the current "decisive decade" of environmental policy, where the quest for sustainability and social justice are intertwined and interdependent.

Economic, social, and environmental inequality and their interactions are the core drivers and outcomes of the political economy of the environment and will therefore logically be the key focus of this volume, while its general framework is an integrated view of inequality and the environment, a view that standard economics has yet to acknowledge.

Inequality and the environment: from blind spots to linkages

In his *General Theory*, a book that laid the foundations of modern macroeconomic analysis and revolutionized economic policy, John Maynard Keynes made no mystery of his attachment to equality: "The outstanding faults of the economic society in which we live are its failure to provide for full employment and its arbitrary and inequitable distribution of wealth and incomes" (Keynes, 1936: 331). In making this forceful argument, Keynes found himself in a rare agreement with Arthur Cecil Pigou, who a few years earlier had highlighted the centrality of injustice in economic thinking: "Wonder, Carlyle declared, is the beginning of philosophy. It is not wonder, but rather the social enthusiasm which revolts from the sordidness of mean streets and the joylessness of withered lives, that is the beginning of economic science" (Pigou, 1920). Pigou and Keynes, together, echoed David Ricardo or John Stuart Mill, who, along with scores of their colleagues, thought that inequality was the key issue of economic analysis. Yet they were at odds with the evolution of their discipline, which, as aforementioned, was shifting away from distributional considerations, dis-embedding itself from its ethical cradle.

The economics of inequality has made a noted comeback in the last 15 years, which stands in contrast with its eclipse from academic and policy debates between the late 1970s and early 2000s. No text better embodies this comeback of inequality economics than Thomas Piketty's *Capital in the 21st Century*, which carefully and powerfully documented the contemporary rise in income and wealth inequality while achieving academic recognition and attaining global fame (Piketty, 2014). And yet, it is striking that for a book first published in French in 2013 by an economist in his early 40s, fewer than 1% of *Capital*'s pages are devoted to environmental issues. Piketty, who has probably contributed to educating several generations of students, academics, and policymakers on the reality of inequality, appears largely blind himself to ecological crises in plain sight (the follow up to *Capital in the 21st Century*, *Capital and Ideology*, published in 2020, has the same proportion of pages devoted to environmental issues, Piketty, 2020).

To put it simply, it seems that economics has finally opened its eyes to inequality only to close them to environmental challenges. And yet environmental economics is at least 150 years old and rich of at least three ages that have followed each other since Stanley Jevons published his founding work on the economics of energy in 1865[1]: resource economics, externality economics, and sustainability economics (Laurent, 2020). We have therefore gone through three ages of environmental economics for a century and a half, eras punctuated by countless publications that have not cancelled each other out in succession but juxtaposed, superimposed, and often mutually enriched one other to shed light on the major ecological issues of our time. And yet, at the dawn of the decisive decade to preserve the biosphere, economics still largely disregard the environment.

The vast majority of professional economists ignore environmental issues, in the double sense of unfamiliarity and indifference. When they do care, it is usually to downplay their impact and to suggest remedies that might worsen environmental damage, such as accelerating economic growth or monetizing ecosystem services. There is no doubt that there are thousands of economists around the world, some writing in this handbook, who are genuinely concerned with ecological problems and are working constructively to understand and remedy them (whether in economics departments, think tanks, or organizations such as the Intergovernmental Panel on Climate Change [IPCC] or the Intergovernmental Science-Policy Platform on Biodiversity and Ecosystem Services [IPBES]). However, it also clear that they represent a small minority in the vast field of economic research and decision-making (a meager 4% of professional economists registered on the Ideas/RePEc database), even while humanity enters the third and critical decade of the "century of the environment" (according to the words of Harvard natural scientist Edward Wilson).

The 2018 Sveriges Riksbank Prize in Economic Sciences in Memory of Alfred Nobel, jointly awarded to William Nordhaus and Paul Romer, somewhat paradoxically illustrates this point (Laurent, 2020). Before the 2018 award, Elinor Ostrom was the only recipient to have been honored indirectly for her contribution to environmental economics ("for her analysis of economic governance, especially the commons"): a single winner, for 51 prizes awarded to 84 individuals during half a century, a century during which environmental issues (such as air pollution, climate change, degradation of ecosystems, and the destruction of biodiversity, just to name a few) have literally jumped out in the eyes of public opinion.

The 2018 award was therefore welcome in principle. William Nordhaus, widely regarded by the profession as one of the world's two most influential climate economists,[2] was honored "for integrating climate change into long-run macroeconomic analysis". However, this "integration" turns out to be seriously flawed. Boyce (2019b) demonstrates that the increase in global average temperature that would accompany the "optimal" price of carbon recommended by the Nordhaus model (known as "DICE") is 3.5 °C by 2100 and continues to rise afterwards. The DICE model therefore recommends a temperature that is twice that of the scientific consensus patiently developed over three decades, with this recommendation being made based on a very fragile methodology, explicitly questioned by the IPCC.[3] Is the unfortunate episode of the 2018 economics prize isolated or does it reflect a deeper problem? In an attempt to shed light on this question, we can first turn to the study recently published by Andrew Oswald and Nicholas Stern (Oswald and Stern, 2019) aimed at assessing the place of environmental issues in academic publications in economics. Out of 77,000 articles published in the 10 most influential journals in the discipline, exactly 57 were devoted to climate change, that is, less than 0.1%. According to another account, it appears that out of 44,000 articles published since 2000 in 50 leading journals, 11 were devoted to the decline in biodiversity, again in the order of 0.1% (Goodall and Oswald, 2019). We can then combine these bibliometric data (which relate to the volume of publications) with other indicators reflecting their degree of recognition or disciplinary impact.

Of the 20 articles considered in 2011 as the most important in a century of existence of the *American Economic Review* by eminent representatives of the discipline, none deals with environmental issues (Arrow et al., 2011). Of the 100 most cited economists listed by Ideas/RePEc, not one is an environmental economist. Of the 100 most cited works listed by Ideas/RePEc, not one deals with environmental economics. Out of the 70 most cited articles in the five most influential academic journals in economics over the period 1991–2015 (i.e., 1% of articles), none deals with environmental issues (Linnemer and Visser, 2016). As has been said, around 2% of the Bank of Sweden's economics prizes were awarded to environmental economists. Similarly, the John Bates Clark Medal, considered the most prestigious recognition just after that of the Bank of Sweden, was awarded to only one environmental economist out of 41 recipients: Kenneth E. Boulding (recognized for his rich and diverse work 60 years ago), again around 2%.

This lack of interest of contemporary economics in environmental issues is all the more damaging given that the ecological transition is now a social science issue (i.e. a social-ecological transition), whereas the hard sciences have largely worked to reveal the extent and urgency of ecological crises. We now need to quickly change attitudes and behaviors to prevent human well-being from self-destructing over the coming decades, an urgency made clear in 2020, a year where ecological shocks, from COVID-19 to climate, have visibly threatened humanity. In other words, it is social sciences, including economics, that hold the key to the problems that the hard sciences have revealed because they are the disciplines of human change. Yet, the vast majority of economists are still missing in action when they are needed most.

Although it is clearly overambitious, we believe that there are ways to turn this unfortunate situation around and we hope that this handbook can contribute to this positive turn.

First, it might be useful to admit that the rules of the human household (such as economic growth) cannot be imposed on the laws of the great natural household (such as climate). In other words, there is, in reality, no economy without environment, whereas, as we have just seen, there is today so little focus on environment in economics.

Then, it might be useful to articulate social issues and ecological challenges to better underline the complementary relationship between human well-being and the preservation of the biosphere. In other words, it might be useful to work on the link between social and natural systems, combining social justice and environmental sustainability to show that it is socially beneficial to mitigate our ecological crises and ecologically beneficial to mitigate our social crises.

It is therefore important to develop a new approach to economics, more lucid about its failures and internal limits, more open to understanding other forms of knowledge, and therefore more socially useful. In short, a sustainable political economy calibrated for the 21st century, bounded upstream by biophysics and downstream by ethics (Laurent, 2020).

This revolution also means reforming environmental economics itself, which is still too focused on mainstreaming ecological crises for decision makers relying on the neoclassical economics framework and toolbox: markets, prices, equilibria. The current destruction of the biosphere, coupled with its social causes and consequences, actually offer a chance to go back to the key question of economic analysis, the one dear to Ricardo, Mill, Keynes, and Pigou and so many others: justice, which is a response to power.

In short, economics should pay much closer attention to environmental challenges, and environmental economics could benefit from a more careful consideration of the role of inequality. It could then play an essential role in advancing the "just transition". This is precisely what this handbook is about.

The purposes, structure, and substance of this handbook

By gathering work from leading and cutting-edge scholars, this handbook intends to make four original contributions to the field of political economy of the environment.

First, it aims at revisiting essential concepts and methods of environmental economics in the light of their political economy. This introduction has just started a discussion on the past and future of the discipline in a critical time for ecological crises. In chapter 2, "Political Economy of the Environment: A Look Behind and Ahead", James K. Boyce goes further in framing the discussion of the relationship between inequality and the environment. He explores evidence indicating that power disparities affect not only the distribution of environmental costs, but also their magnitude. The latter can be explained by the fact that in societies with larger income, wealth, and power disparities, it becomes easier for the beneficiaries of pollution to impose the costs upon others. Boyce then discusses four normative criteria used in the field of political economy of the environment to decide on environmental policy objectives: efficiency, safety, sustainability, and justice. While neoclassical economists are only concerned with efficiency, the other three criteria can generate quite different policy implications, and moreover can supplement and reinforce each other.

Second, this handbook aims at introducing readers to recent theoretical and empirical advances in key issues of political economy of the environment with a special focus on the relationship between inequality and environmental degradation (Part 1). In chapter 3, "The Sustainability-Justice Nexus", Éloi Laurent focuses on the links between justice and sustainability and more precisely between the inequality and ecological crises. The chapter briefly reviews the available empirical evidence on inequality and ecological crises and details the transmission channels of the inequality crisis to environmental degradation. He finally provides a novel typology of environmental justice, distinguishing between procedural, recognitive, and distributive justice, and illustrates them with respect to different environmental hazards, including air pollution, noise and chemical pollution, and exposure to socio-ecological disasters.

In chapter 4, "A Socio-Metabolic Perspective on (Material) Growth and Inequality", Anke Schaffartzik and Fridolin Krausmann analyze disparities in global material extraction, which they consider both the consequences of as well as the prerequisite for globally unsustainable patterns of economic growth. They present an ecological socio-metabolic perspective, according to which society cannot only be understood in terms of social relations, but it is also dependent upon material and energy inputs, as well as output of wastes and emissions. After presenting the key concepts of social metabolism, they analyze its changes over time and derive patterns of metabolic inequality as one key aspect of international socio-ecological inequality.

In chapter 5, "The History of Environmental and Energy Economics through the Lens of Political Economy", Antoine Missemer highlights the value of mobilizing a political economy perspective when writing the history of environmental and energy economics. Through the examples of the emergence of the concept of natural capital in the 1900s and of the first experiments to measure energy-growth decoupling in the 1920s, his chapter shows how political economy is able to renew our understanding of past paradigms, ideas, and theories related to natural resources and the environment. Not only can political economy provide insights on the context and motives surrounding the emergence of concepts and models, it may also lead the historian to reconsider his or her research questions as he or she discovers that energy and environmental issues conceal true distributional, social issues. In other words, theoretical and conceptual contributions apparently limited to the sole field of environmental and energy economics can be illuminated in a new light through the lens of political economy.

Third, with this handbook, we want to open the field to its critical global and regional dimensions. Global issues, such as the environmental justice movement and inequality and climate change, as well as regional issues, such as agriculture systems, air pollution, natural resources appropriation, and urban sustainability, are presented.

In chapter 6, "Global Environmental and Climate Justice Movements", David N. Pellow gives an overview of stylized facts and trends in global environmental inequalities and environmental conflicts. He presents the critical environmental justice framework, which he co-developed, addressing some shortcomings in existing environmental justice studies, including the importance of understanding multiple dimensions of inequality, as well as their intersections; the need to understand environmental disparities at multiple spatial scales; and a critical reflection of the role of state power in shaping and reinforcing environmental inequality. He then applies this critical environmental justice framework to climate justice movements.

In chapter 7, "Global Inequalities and Climate Change", Céline Guivarch and Nicolas Taconet synthesize recent works on the links between climate change and inequality to show how climate change impacts and mitigation affect inequalities, both between countries and between individuals. They first analyze inequalities in exposure and vulnerability to climate change and then study inequality in the contribution to greenhouse gas emissions between countries and individuals. Finally, they show how inequality can shed light on the fairness of actions to fight climate change.

In chapter 8, "Natural Disasters, Poverty and Inequality: New Metrics for Fairer Policies", Stéphane Hallegatte and Brian Walsh develop a framework to reassess the well-being losses associated with natural disasters. They first present stylized facts on the impacts of natural disasters on poverty and show how poverty conversely exacerbates natural disaster impacts, and then they develop a new disaster risk management strategy, which considers the fact that poor people are not only disproportionately affected by natural hazards, but also lose more in relative terms when they are affected. While previous disaster risk assessments have focused on asset losses, including three components – hazard, exposure, and vulnerability – the authors instead focus on well-being losses and include a fourth component, socioeconomic resilience. They illustrate the importance of the latter by presenting some results from a multi-metric assessment of disaster risk in the Philippines, finding that different measures of disaster risks – asset declines, poverty increases, well-being declines, and effects on socioeconomic resilience – suggest different priorities for policy interventions.

In chapter 9, "Contracts and Dispossession: Agribusiness Venture Agreements in the Philippines", Alfredo R.M. Rosete investigates under which conditions voluntary agribusiness partnerships between smallholders and agribusiness firms in the Davao region of the Philippines can lead to the loss of smallholders right to their land. He first gives an overview of agribusiness partnerships in the context of land reform in the Philippines and presents his theoretical framework, in which property rights are understood as a set of abilities over an asset. He argues that rather than the types of contracts, it is the configuration of rights in a contract that will determine whether smallholders are threatened by dispossession. He then presents some insights from his fieldwork in different locations of the Davao region and compares cases where smallholders were able to maintain effective control over their land to cases that resulted in dispossession of their land.

In Chapter 10, "Natural Resources, Climate Change and Inequality in Africa", James C. Murombedzi shows that the use and governance of natural resources are among the most central of issues for the daily lives of the majority of Africans: patterns of rural resource use are fundamental to rural and national economies, as well as to local and global concerns about sustainability. Resource degradation through unsustainable use patterns as well as climate change have resulted in a global environmental and developmental emergency whose impacts include increasing concentration of wealth and growing poverty and inequality. He argues that the use and control of natural resources has historically generated inequality in Africa and that climate change and the responses to it have exacerbated these inequalities. The dominance of the market, representing corporate interests over social and environmental interests, is clearly socially, economically, and environmentally unsustainable. Instead of the current production system that emphasizes market mechanisms to allocate the costs and benefits of nature, Murombedzi calls for a social structure of accumulation that places economic justice over profit and, more practically, institutes an inclusive, sustainable model for growth.

In chapter 11, "From Western Pennsylvania to the World: Environmental Injustice and the Ethane-to-Plastics Global Production Network", Diane M. Sicotte examines environmental injustice produced in one strand of the global production network that manufactures plastics using ethane, a gas liquid found in natural gas. Record quantities of ethane are currently being produced in the ethane-rich Marcellus Shale region in Western Pennsylvania. Low gas prices and debt financing have increased incentives for the petrochemicals industry to invest in infrastructures that enable cheap ethane to be transformed into profitable plastic. Ethane is produced from fracked natural gas, and fracking has generated both distributional and procedural injustice in communities near natural gas wellheads. Environmental injustice is also generated at the sites where high-pressure ethane transmission pipelines are currently being built, at LNG export terminals where ethane is exported, and at the fence line of the petrochemical plants where ethane is transformed into plastic resins. More than a third of plastics are made into packaging and other disposable items, and the industry intends to increase production of disposable plastic items. This plan rests upon the fiction that increased recycling of plastics, and not source reduction, is the solution to widespread pollution with plastic garbage. In the US, plastic waste is either disposed of in marginalized communities or exported to poorer nations, causing global threats to health and safety as well as distributional injustice. Examining this global production network illustrates the political economic context behind the increasing production and use of unsustainable products and how the profitability of unsustainable practices is supported by negative externalities visited upon less powerful groups of people.

In chapter 12, "Latin America Caught between Inequality and Natural Capital Degradation: A View from Macro and Micro Data", Juan-Camilo Cardenas explores the relationship between inequality and the possibilities of environmental sustainability in the Latin American region. Given that the region is both rich in natural resources and at the same time ridden with environmental conflicts and that it suffers from deep problems with respect to inequality and social exclusion, the chapter explores how these extremes offer some extra challenges for the region to create an inclusive and sustainable path for the future. Data from a recent survey conducted in seven countries are also presented, which offer some insights into the interactions between prosocial preferences and pro-environmental attitudes and behaviors and how to tackle the double challenge of stopping environmental degradation while reducing inequality.

In chapter 13, "Air Quality Co-benefits of Climate Mitigation in the European Union", Klara Zwickl and Simon Sturn assess the role of positive spillovers of climate mitigation on public health through air quality improvements in the EU. Since carbon emissions simultaneously release global greenhouse gases and local co-pollutants, carbon mitigation can provide substantial air quality improvements. These so-called air quality co-benefits have two implications for EU climate policy: first, considering co-pollutant damages along with those from greenhouse gases raise the total damages from fossil fuel combustion, this consequently calls for a higher carbon price. In fact, if air quality co-benefits are large, carbon mitigation can be justified independently of its climate goals. Second, co-pollutant damages have been found to vary across sectors and space, suggesting more stringent climate policies in sectors and locations with high co-benefits.

In chapter 14, "Designing Urban Sustainability: Environmental Justice in EU-funded Projects", Ian M. Cook and Tamara Steger analyze how European Union–funded projects can support environmental justice in urban sustainability initiatives. They collect data from EU-funded projects focusing on urban sustainability and/or social justice and provide examples of how some projects have promoted approaches to inclusivity through enabling new networks and the co-construction of knowledge across different groups in society, including civil society institutions working with deprived urban communities. They also address some shortcomings in achieving urban environmental justice goals through EU-funded projects, namely the fact that many EU projects only last a few years, while the reduction of structural inequalities requires persistent, long-term measures.

Fourth and finally, this handbook wants to show how novel analysis can translate into new forms of public policy that require institutional reform and new policy tools (Part 2). Ecosystems preservation, international climate negotiations, and climate mitigation policies all have a strong distributional dimension that chapters in this part point to. Pressing environmental policy such as carbon pricing and low-carbon and energy transitions entail numerous social issues that also need to be accounted for with new analytical and technological tools.

In chapter 15, "From the Welfare State to the Social-Ecological State", Éloi Laurent argues for a necessary and possible metamorphosis from the welfare state of the 20th century into a social-ecological state freed from growth and aiming at full health calibrated for the 21st century. The chapter first retraces the genealogy of the social-ecological state to then clarify its philosophy, its perceived need of economic growth, its various functions and, as an illustration of its necessary generalization to all levels of governance, its application to urban policy.

In chapter 16, "Promoting Justice in Global Climate Policies", Michel Bourban shows that global inequalities raised by climate change lead to multiple injustices and adopts a climate justice perspective to assess the degree of (un)fairness of current mitigation and adaptation policies and to propose institutional reforms that would make them fairer. The chapter explains in what sense political economists and normative political theorists could join efforts to promote climate justice, despite the historic tensions between their two fields. It then details mitigation policies and explains why current nationally determined contributions (NDCs) are unfair. It also proposes a roadmap for a just energy transition that would implement a price-signal approach complementing the pledge-and-review approach that dominates climate negotiations. The chapter then moves to adaptation policies and highlights that adaptation finance transfers to vulnerable countries are currently insufficient. It also supports the promotion of democratization processes in vulnerable societies to avoid a trade-off between two influential criteria guiding the allocation of adaptation finance: vulnerability and good governance. The conclusion sketches a research agenda for future interdisciplinary research on climate justice.

In chapter 17, "Carbon Pricing and Climate Justice", James K. Boyce argues that there is only one way to guarantee that we keep fossil fuels in the ground to the extent needed to meet the 1.5–2 °C warming target of the Paris Agreement: imposing hard limits on the amount of fossil carbon allowed to enter our economies and thence the earth's atmosphere. But, he states, this would be almost certain to raise fuel prices, perhaps quite substantially. A robust and effective carbon price – implemented by means of a carbon tax, a carbon cap, or a combination of the two – will indeed have major distributional impacts that must be addressed for reasons of both economic equity and political sustainability. He proposes that carbon dividends are one way to do so. Carbon pricing, he argues, should be implemented as a complement, not an alternative, to other policies to advance the clean energy transition, including public investment and smart regulations.

In chapter 18, "Political Economy of Border Carbon Adjustment", Paul Malliet and Ruben Haalebos observe that if the issue of climate change should lead the international community to act jointly in order to tackle it, diverging levels of effort take place in reality. While pricing carbon is recognized as the cornerstone economic policy to shift emissions efficiently and has been implemented unilaterally in several jurisdictions, this desynchronization between world countries in the implementation of such policies can lead to pervasive effects, such as carbon leakage, which can be countered by taxing imported emissions through border carbon adjustment. This leads to redistributive effects among final consumers that are different from those induced by domestic carbon pricing schemes and that can, when associated to a revenues recycling scheme, lower the burden of carbon pricing policy on households and eventually limit the share of net contributors to a carbon pricing policy.

In chapter 19, "Political Economy of Forest Protection", Alain Karsenty starts by observing that converting forest ecosystems to other land uses entails major negative consequences for the climate, biodiversity, and human well-being. He observes that several international initiatives have emerged to address these issues but that, so far, none of these initiatives has succeeded in curbing deforestation and the conversion of natural ecosystems to artificialized areas. Diagnoses of the "forest crisis" are generally correct, but they often overlook major political economy issues, such as the fact that governments are not benevolent institutions acting for the common welfare of their people and that urban elites have little interest in the fate of forest-dependent people, who are not that numerous and are often voiceless. Endeavors to tackle the forest crisis without questioning the unabated global demand for biomass, energy and agricultural land, and the rules of international trade seem illusory. Results-based payments have to be rethought, he argues, without tying one's hands with an automatic payment procedure based on an unverifiable level/reference scenario. The only meaningful criterion being the coherence of public policies that potentially have impacts on forests, he pleads for an agenda merging food security and forest protection.

In chapter 20, "Informing the Political Economy of Energy and Climate Transitions: Modelling Tools, Pathways Design Frameworks and Analytical Challenges", Patrick Criqui and Henri Waisman show that measures adopted to reduce greenhouse gas emissions will have major impacts on human activities, with heterogeneous effects according to national circumstances and differential impacts on various sectors, activities, and categories of households. Climate mitigation is therefore not a question of pure economics but of political economy, involving the resolution of conflicts of interest. They argue that a careful attention to redistributive effects is required when designing low emission development strategies and selecting policy instruments and measures. Their chapter starts by analyzing the emergence of research communities and models investigating the economic and social dimensions of energy and climate policies and discusses the use of economic models in climate debates and negotiations. The chapter further highlights the strengths and limitations of conventional economic modelling approaches to argue how a new modelling paradigm, based on national decarbonization scenarios, progressively emerged in the lead-up to the Paris Agreement, and discusses how this new paradigm is relevant to inform the political economy dimensions of climate action. The chapter finally presents key insights from recent literature investigating the major implications of the carbon neutrality objective as codified in the Paris Agreement. According to the authors, the magnitude of these impacts calls for new approaches and new solutions and opens new research avenues for the economic analysis of climate policies.

In chapter 21, "Diagnostics and Policy Tools to Measure and Mitigate Environmental Health Inequalities", Julien Caudeville defines environmental health inequalities (EHIs) as health hazards disproportionately or unfairly distributed among the most vulnerable social groups or territories, which are generally the most discriminated, poor populations and minorities affected by environmental risks. However, he notes, constructing methods and tools to help orient public policies in order to reduce territorialized EHI requires the evaluation of phenomena not always easy to apprehend and the reliability and representativeness of information that usually demand statistical processing. He goes on to present the European institutional and scientific contexts in which EHI characterization operates. The connection between the environment and health in public policies in Europe is reviewed to address the need of integrating data from environmental health tracking information systems, as means of characterizing the exposure of populations living in a territory. To enhance mitigation of EHI and prevention of health inequalities in the scientific exposome concept emergence context, an environmental health methodological framework is presented using different examples to identify a common taxonomy for conceptualizing and operationalizing environmental exposures as an important step towards articulating a science of environmental health disparities.

In chapter 22, "Building on the Right to Know: Data Interlinkage and Information Intermediation for Environmental and Corporate Regulation", Richard Puchalsky, Michael Ash, and James K. Boyce give a detailed overview of how environmental inequalities in air pollution exposure and other environmental risks can be empirically assessed in the United States, using facility-level data collected by the US Environmental Protection Agency (EPA), including the Toxic Release Inventory, as well as the Risk Screening Environmental Indicators, and the Greenhouse Gas Reporting Program. They give a detailed explanation of how these data are collected and which steps are necessary to use them for empirical environmental justice analyses. They also present some helpful methods to link different databases, a task not performed by the EPA, and present some results on the top toxic air polluting companies as well as the top greenhouse gas emitting companies in the US. Finally, they discuss the use of these data in academic research and for public intermediation.

In the handbook's conclusion, "New Frontiers in the Political Economy of the Environment", we put forward the notion of social-ecological frontiers and outline four new frontiers which are currently emerging in the field political economy of the environment, which are not sufficiently covered in this volume. These include first, the emerging contradiction between the reality of digital transition and the necessity of ecological transition; second, socio-ecological analysis for urban environmental justice; third, the development of a broad framework to understand the distributional effects of environmental policies; and fourth, a conceptual understanding of public views and attitudes towards environmental conservation.

Notes

1 More precisely on the economic consequences for the British economy of the supposed depletion of coal, coal which was in 2020 the second largest source of energy in the world and the first energy source of greenhouse gases.
2 Along with Martin Weitzman, who ended his life in August 2019.
3 IPCC, Climate Change 2014. Synthesis Report, 2014, p. 79.

Bibliography

Arrow, K. J., Bernheim, D. B., Feldstein, M. S., McFadden, D. L., Poterba, J. M. & Solow, R. M. 2011. "100 Years of the American Economic Review: The Top 20 Articles". *American Economic Review*, 101, 1–8.

Bowles, S. 2004. *Microeconomics: Behavior, Institutions, and Evolution*, Princeton, Princeton University Press.

Boyce, J. K. 2002. *The Political Economy of the Environment*, Cheltenham, Edward Elgar.
Boyce, J. K. 2019a. *Economics for People and the Planet: Inequality in the Era of Climate Change*, London, Anthem Press.
Boyce, J. K. 2019b. *The Case for Carbon Dividends*, Medford, Polity Press.
Goodall, A. H. & Oswald, A. J. 2019. "Researchers Obsessed with FT Journals List Are Failing to Tackle Today's Problems". *Financial Times*, May 8.
Groenewegen, P. 1991. "'Political Economy' and 'Economics'". *In:* Eatwell, J., Milgate, M. & Newman, P. (eds.) *The World of Economics*, London: Palgrave Macmillan.
Keynes, J. M. 1936. *The General Theory of Employment, Interest and Money*, London, Macmillan and Co.
Laurent, É. 2020. *The New Environmental Economics: Sustainability and Justice*, Cambridge, Polity Press.
Linnemer, L. & Visser, M. 2016. *The Most Cited Articles from the Top-5 Journals (1991–2015).* CESifo Working Paper Series No. 5999. https://papers.ssrn.com/sol3/papers.cfm?abstract_id=2821483.
Marshall, A. 1920. "The Substance of Economics". *In: Principles of Economics*, 8th ed., London, Palgrave Macmillan.
Oswald, A. J. & Stern, N. 2019. "Why Does the Economics of Climate Change Matter so Much, and Why Has the Engagement of Economists Been so Weak?". *Royal Economic Society Newsletter.*
Pigou, A. C. 1920. *The Economics of Welfare*, London, Palgrave Macmillan.
Piketty, T. 2014. *Capital in the Twenty-First Century*, Translated by Arthur Goldhammer, Cambridge, MA, Belknap Press of Harvard University Press.
Piketty, T. 2020. *Capital and Ideology*, Translated by Arthur Goldhammer, Cambridge, MA, Belknap Press of Harvard University Press.
Robbins, L. 1972. *An Essay on the Nature and Significance of Economic Science*, London, Palgrave Macmillan.
Samuelson, P. A. 1957. "Wages and Interest: A Modern Dissection of Marxian Economic Models". *The American Economic Review*, 47, 884–912.

2

POLITICAL ECONOMY OF THE ENVIRONMENT

A look back and ahead

James K. Boyce

Environmental economics extends the purview of economic inquiry beyond items that carry price tags in markets – the goods and services that count in measuring national income – to include non-marketed attributes of our natural environment such as clean air, clean water, biodiversity, and global climate stability. This is founded on growing recognition of the environment's crucial role as a source for raw materials and as a sink for the disposal of wastes generated in economic activities.

If economics is defined as being concerned with the allocation of scarce resources among competing ends – a common definition found in textbooks – then environmental economics widens these competing ends to encompass the protection of natural resources and environmental quality.

Political economy analyzes the allocation of scarce resources not only among competing ends but also among competing individuals, groups, and classes. The political economy of the environment extends the purview of environmental economics beyond the allocation of scarce resources among competing market and non-market ends to their allocation among competing people.

In analyzing environmental degradation, the political economy of the environment poses three basic questions:

- Who wins? Who benefits from economic activities that degrade the environment? If no one benefits (or at least thinks they do), these activities would not occur.
- Who loses? Who is harmed by environmentally degrading activities? If no one is harmed in current or future generations, these would not matter from the standpoint of human well-being.
- Who decides? Why can the beneficiaries of these activities impose environmental costs on the people who are harmed by them? Who chooses the ends and means in environmental policy and practice?

This analytical framework has both a positive agenda and a normative agenda. The aim of positive analysis is to describe what happens and why.[1] The aim of normative analysis is to prescribe what should happen. In both respects, as this chapter discusses, the political economy of the environment departs from neoclassical economics.

DOI: 10.4324/9780367814533-2

Inequality and the environment: positive issues

There are three possible reasons why those who benefit from environmentally degrading activities are able to impose environmental costs on others (Boyce 1994, 2007). One possibility is that those who are harmed are future generations who are not here to defend themselves, whereas the winners are here today. The second is imperfect information: those who are bear the costs may be unaware of the harm or unaware of its causes. The third possibility is inequality in the distribution of wealth and power: those who bear the costs do not have sufficient purchasing power or political influence to prevail in social contests over use and abuse of the environment.

In the first case, addressing environmental degradation requires an ethic of intergenerational responsibility on the part of those of us who are alive today. In the second, the remedy is wider access to environmental information, and in particular right-to-know laws that protect the public's right to information about environmental harms and who is responsible for them.[2] In the third case, a solution requires the redistribution of power.

Power and social decisions

Both purchasing power and political power are implicated in environmental decisions. Purchasing power underpins the monetary valuation of environmental harms in cost-benefit analysis, just as it underpins consumer demand in actually existing markets for goods and services. In cost-benefit analysis, and in markets, each dollar – not each person – counts equally. Costs and benefits that go to people with more dollars receive greater weight than if they go to people with less.

Political power matters, too. Decision makers do not necessarily attach the same importance to all benefits and costs as measured by cost-benefit analysts. When the people who are harmed have no political power, costs imposed upon them can be simply ignored. This is not merely a hypothetical possibility. It was illustrated in 2017 by the U.S. Environmental Protection Agency's decision to assign *zero* value to climate-change impacts outside the United States in mounting a cost-benefit case to repeal an Obama-era policy that would have curbed carbon emissions from power plants (Mooney 2017; Boyce 2018). But even among those who are not excluded entirely from the political process, power often is distributed quite unequally.

Both sorts of power – purchasing power and political power – tend to be correlated. Those with more wealth typically wield more political clout, and vice versa. The joint effect can be described by a power-weighted social decision rule, in which environmental outcomes are shaped by inequality in the distribution of wealth and influence (Boyce 1994).

Two predictions follow. The first is that the distribution of environmental costs will not be random. Instead, risks and harm will be inflicted disproportionately on those with less economic wealth and less political power. The second is that wider inequalities will tend to result in higher levels of environmental degradation. Both propositions – one on the direction of environmental costs, the other on their magnitude – have been supported by the growing body of research on the political economy of the environment carried out in the past quarter century.

Inequality and the direction of environmental harm

In the United States, environmental justice researchers have documented systematic disparities in exposure to hazards along the social fault lines of race, ethnicity, and class. African-Americans, Latinos, and low-income communities are more likely to have hazardous facilities sited in their midst and more likely to face disproportionate exposure to pollution.

One of the earliest studies, by sociologist Robert Bullard (1983), examined the distribution of hazardous waste sites in Houston, Texas, revealing that they were located primarily in African-American neighborhoods. Subsequent research has found similar patterns in many parts of the country. Race and ethnicity are strong correlates of proximity and exposure, even after controlling for neighborhood income; indeed, these are often a stronger predictor than income (Bullard et al. 2008; Zwickl et al. 2014; Mohai and Saha 2015a).

Researchers have investigated the direction of causality that underlies these correlations. Are hazardous facilities sited from the outset in communities with less wealth and power, or do post-siting demographic changes explain the pattern, as wealthier residents move out, property values decline, and poorer people move in? Time-series data on hazardous facilities are not readily available, so few studies have explored this question directly, but those that have done so have found compelling evidence of disparities in the initial siting decisions, as well as some evidence of post-siting changes in the demography of nearby neighborhoods (Mohai and Saha 2015b; Pastor et al. 2001).

Researchers also have begun to explore the economic and health consequences of these environmental disparities. Disproportionate pollution exposure has adverse effects on children in particular, resulting in higher rates of infant mortality, lower birth weights, a higher incidence of neurodevelopmental disabilities, more frequent and intense asthma attacks, and lower school test scores. And among adults, pollution exposure is linked to lost workdays due to illness and the need to care for sick children.[3] These effects exacerbate the vulnerabilities that make some communities more susceptible to environmental harm in the first place.

Environmental inequalities can be found across the world, not only in the United States. In England and the Netherlands, for example, poorer and more non-white neighborhoods have higher air concentrations of particulate matter and nitrogen oxides (Fecht et al. 2015). In Delhi, India, a mega-city whose residents breathe some the world's dirtiest air, not all are equally exposed: the poor live in more polluted neighborhoods, they cannot afford air conditioning or air purifiers, and they spend more time working outdoors where pollution levels are higher, and at the same time they receive fewer benefits from power generation, transportation, and other activities that cause the pollution (Garg 2011; Foster and Kumar 2011; Kathuria and Khan 2016).

Although most research on environmental justice has focused on race, ethnicity, and income, power disparities in other dimensions may have environmental consequences, too. In some cases, for instance, particularly activities involving resource extraction or solid waste disposal, rural areas may suffer disproportionate environmental harm compared to urban areas (Kelly-Reif and Wing 2016).

To take another example, gender-based inequalities may translate into disparate environmental harms inflicted on women. The prime example, perhaps, is the exposure of women to indoor air pollution – a leading cause of premature mortality worldwide – in places where solid fuels such as wood, crop residues, and dung are used for cooking, notably south Asia and sub-Saharan Africa.[4]

The impacts of power disparities can operate across national borders, too, displacing environmental harm originating in high-income countries onto vulnerable communities in low-income countries. In a 1992 memorandum, Lawrence Summers, then chief economist at the World Bank, wrote, 'the economic logic of dumping a load of toxic waste in the lowest-wage country is impeccable.'[5] All too often environmental practice follows this script, as millions of tons of toxic waste are shipped each year from advanced industrialized countries of the global North to Africa, Asia, and Latin America (Kellenberg 2015).

Inequality and the magnitude of environmental degradation

The impact of inequality on the total magnitude of environmental degradation has received somewhat less attention from researchers, in part because quantitative analysis has been hindered by a paucity of the necessary data. Year-to-year variations in inequality and environmental quality are likely to be small, and the environmental impacts of inequality are likely to operate on a multi-year time frame, features that render time-series analysis problematic. Cross-sectional analysis, meanwhile, is complicated by issues of choosing the appropriate spatial scale and by the need to control for a large number of potentially confounding variables. Notwithstanding these difficulties, the topic has received growing attention.

Before turning to the evidence, it is useful to consider why one might expect greater inequality to lead to more environmental harm. One reason has already been discussed: the concentration of environmental costs at the lower end of the wealth-and-power spectrum. The wider the extent of inequality, the less weight these costs receive both in the economic scales of cost-benefit analysis and in the political calculations of public-sector and private-sector decision makers.

The second reason is the converse of the first: the benefits from environmentally degrading activities tend to be concentrated at the upper end. The externalization of environmental costs leads to lower production costs, generating benefits in the form of higher profits for the firm's shareholders, higher compensation for its executives, lower prices to consumers of its products, or a combination of these.[6] In general, shareholders and executives occupy relatively high rungs on the wealth-and-power spectrum. Insofar as the benefits of cost externalization are passed along to consumers, they accrue in proportion to consumption, benefiting those with the most purchasing power. The wider the extent of inequality, the more weight these benefits receive in cost-benefit analysis and in the eyes of decision makers.

Of course, many affluent individuals prefer to live in a clean and safe environment. To a considerable extent, however, environmental quality is an impure public good in that while not entirely private, it also is not equally available (or unavailable) to everyone. Relatively wealthy and powerful people can afford to live in neighborhoods with cleaner air. They also can afford bottled water, air conditioners, and air purifiers. In the event of illness caused by pollution exposure, they can obtain better medical care. At the same time, they can more effectively prevent the siting of environmental hazards in their own neighborhoods. To be sure, they may not escape the consequences of environmental degradation altogether, but in their private calculations they balance a relatively small share of the costs against a relatively large share of the benefits.

In sum, one can expect greater inequality to lead to more environmental degradation by making it politically easier, as well as more 'efficient' by the canons of neoclassical economics, for those who benefit from it to impose the costs upon others.

Cross-national data on several dimensions of environmental quality became available to researchers in the early 1990s. One of the first questions that economists used these data to address was the relationship between environmental degradation and per capita income. In a well-known study, Grossman and Krueger (1995) analyzed several indicators of air and water quality and found that pollution tended to rise with per capita incomes up to a turning point, in the neighborhood of $5,000, after which environmental quality improves. The result was an inverted U-shaped relationship between per capita income and environmental degradation that resembles the curve postulated by Kuznets (1955) on the relationship between per capita income and income inequality. The new relationship became known as the 'environmental Kuznets curve' (EKC).

The EKC appeared to offer an escape from the bleak idea that economic growth is incompatible with environmental protection. Maybe there are no environmental limits to growth, after all. Maybe humans are not, as a prominent environmental historian once declared, a 'cancerous' species that 'endangers the larger whole' (Nash 2001, p. 386). A spirited debate ensued between some who saw economic growth as the solution to environmental ills and others who instead saw it as the root disease.

Few noticed that Grossman and Krueger also reported that, in a number of cases, further growth in per capita income led to a second turning point after which pollution again began to rise – a result that would seem to bring little comfort to the growth-as-cure school of thought. Moreover, Grossman and Krueger cautioned, 'there is nothing at all inevitable about the relationships that have been observed in the past' (p. 372).

In a follow-up paper, Grossman and Krueger (1996) observed that policy responses driven by 'vigilance and advocacy' on the part of the public are likely to be the main explanation for improvements in environmental quality. This suggests that the similarity between the EKC and the original Kuznets curve may not be mere coincidence. If, as Kuznets suggested, there is a turning point after which inequality falls as per capita income rises, then parallel improvements in environmental quality may be driven not by per capita income itself but instead by less inequality.

When proxies for inequality in the distribution of wealth and power were added as possible determinants of cross-country variations in environmental quality, the results supported the hypothesis that they are inversely related. Indeed, controlling for proxy variables such as political rights and civil liberties in many cases caused the EKC relationship between pollution and per capita income to weaken or disappear (Torras and Boyce 1998; Harbaugh et al. 2002; Neumayer 2002; Farzin and Bond 2006).

Today more cross-national evidence has become available. Researchers have found that greater inequality is associated with worse environmental performance not only in terms of air and water pollution, but also in other respects. The proportion of plants and animals threatened with extirpation or extinction is higher in countries with more unequal income distributions (Mikkelson et al. 2007; Holland et al. 2010). Rates of deforestation are higher in countries with higher levels of corruption, a variable that can be interpreted as both a cause and a consequence of inequality (Koyunco and Yilmaz 2009). In upper-income countries, private patents on environmental innovations and public expenditure on environmental research and development both are lower in countries with wider income inequality (Vona and Patriarca 2011).

The evidence for adverse environmental effects of inequality generally is strongest for variables that have immediate impacts on human health, including air and water pollution, as one might expect (Cushing et al. 2015). For environmental impacts that are widely dispersed across time and space, the evidence is more mixed. Recent studies nevertheless have reported evidence of an inverse relationship between inequality and carbon dioxide emissions (Knight et al. 2017; McGee and Greiner 2018). Part of the explanation may be that fossil fuel combustion also generates conventional air pollutants, such as sulfur dioxide and nitrogen oxides, that trigger public demands for emission reductions.

Interstate studies in the U.S. have also found evidence that inequality adversely affects environmental outcomes. States with more unequal distributions of power tend to have weaker environmental policies, leading to greater environmental stress and worse public health outcomes (Boyce et al. 1999). Interstate differences in inequality also have been found to be correlated with carbon dioxide emissions (Jorgenson et al. 2017).

Taking metropolitan areas as the unit of observation, Morello-Frosch and Jesdale (2006) found that U.S. cities with more residential segregation by race and ethnicity tend to have higher cancer risks from air pollution for all population groups. Similarly, Ash et al. (2013) found that in metropolitan areas that rank highest in terms of racial and ethnic disparities in industrial air pollution exposure, average exposure levels are higher for Anglo whites, too, implying that that environmental justice can be 'good for white folks.'

The implication of all these studies is that protecting the environment and reducing inequality can and should be complementary goals. With lower levels of inequality, the public is better able protect the air, water, and natural resources on which human well-being depends.

Values and the environment: normative issues

Policy prescriptions invariably rest on normative criteria, the explicit or implicit ethical principles by which we assess alternative courses of action and states of the world as better or worse. Neoclassical economics invokes one overriding criterion for this purpose – efficiency – and neoclassical environmental economists have invested a great deal of time and effort in trying to operationalize this for policymaking purposes. Political economists often invoke other criteria, including safety, sustainability, and justice. How best to operationalize these, and how to combine them, are key issues yet to be fully resolved.

Efficiency

The term 'efficiency,' as deployed in neoclassical economics, refers to something more than cost-effectiveness. In everyday speech, these notions are often used as synonyms. When we speak, for example, of the most efficient way to travel from point A to point B, we are really talking about cost-effectiveness, the lowest-cost means to achieve this end. But when neoclassical economists speak of efficiency, they are not only referring to decisions about the means, but also how to choose the ends themselves, asking for example whether it is desirable to travel from A to B at all.

Cost-effectiveness can be applied to the pursuit of ends chosen on the basis of any of the aforementioned criteria. For example, policymakers may use a safety criterion to decide upon air quality standards and then try to choose the most cost-effective ways of attaining the safety objective. In invoking efficiency to choose the standards themselves, neoclassical economics goes considerably further, requiring the policymaker to put a monetary value on protecting public health and saving human lives and to weigh this against the costs of doing so in order to decide on the 'efficient' level of clean air.

In theory, neoclassical efficiency is based on a seemingly noncontroversial idea: 'Pareto optimality,' the proposition that an optimal state of the world is one where no individual can be made better off without making someone else worse off.[7] Because it is silent when it comes to how the economic pie should be distributed, innumerable outcomes could qualify as Pareto optimal. Even if saving the life of an impoverished child at the cost of one dollar to a millionaire, strict Pareto optimality offers no grounds for advocating it, because the millionaire would be made fractionally worse off. Efficiency in this sense of the term amounts to saying that twenty-dollar bills should not be left lying on the ground. As a basis for policy making it has little cutting power, since just about any policy, even one that makes very many people very much better off, will make someone at least somewhat worse off.

To escape from this prescriptive cul-de-sac and arrive at a more practical basis for its policy prescriptions, neoclassical economics replaces strict Pareto optimality with a more flexible criterion, that of a 'potential Pareto improvement.' One state of the world can now be judged preferable to another one if those who are made better off could, in theory, compensate those who are made worse off and still come out ahead. Whether compensation is really paid is shrugged off as a distributional issue that is extraneous to making a policy prescription based on efficiency. By this sleight of hand, the policy goal becomes simply the biggest economic pie, its size being measured by its monetary value, regardless of how the pie is sliced. In macroeconomics, this translates into maximizing GDP. In microeconomics, it translates into maximizing net benefits, calculated by the tools of cost-benefit analysis.

Economists have devised a number of quasi-ingenious methods to assign monetary values to things without a market price tag, from the value of a statistical life (meaning the value of avoiding a risk of premature death) to the value of endangered species, clean air, and climate stability. Mostly these methods rest on willingness to pay: how much people in a given population would be willing to pay to reduce their risk of premature death, save the whales, and so on. Just as in real markets, individual preferences count insofar as they are backed by the ability to pay. In markets for food, hunger generates effective demand only if it is backed by purchasing power. So, too, in the shadow markets of cost-benefit analysis, the value of a clean and safe environment rests not only on what people desire but also on what they can pay for it.[8]

The result is encapsulated in the aforementioned memorandum by World Bank chief economist Lawrence Summers, maintaining that toxic waste should be dumped in the country with the lowest wages. 'The arguments against all of these proposals for more pollution in LDCs [less developed countries],' Summers concluded, citing 'intrinsic rights to certain goods' and 'moral reasons' as examples of such arguments, 'could be turned around and used more or less effectively against every Bank proposal for liberalization.' Or, one might add, against any policy prescription based exclusively on the normative criterion of neoclassical efficiency.

Safety

Existing environmental laws and policies often rest on a quite different normative foundation: safety. In the United States, for example, the Clean Air Act directs the Environmental Protection Agency to establish air quality standards for 'the protection of public health and welfare' while 'allowing an adequate margin of safety' – not to decide on standards by weighing the benefits of protecting public health against its costs.[9] In such a world, economists play a more modest role. They can recommend how to pursue the objective most cost-effectively, but it is not their job to decide on the objective itself.

Safety is generally a matter of degree, so there is often some arbitrariness in deciding what qualifies as 'safe.' In practice, environmental policymakers often follow a rule of thumb, such as defining the acceptable risk from pollution as adverse health impacts on one in 10,000 people, or one in 100,000, in a given year (Kutlar Joss et al. 2017; Hunter and Fewtrell 2001). Similarly, in international climate policy, the Paris Agreement's goal of holding the increase in the global average temperature to well below 2 °C above pre-industrial levels and pursuing efforts to limit the temperature increase to 1.5 °C is based on scientific assessments as to what is safe, rather than judgments by neoclassical economists as to what is efficient.[10]

The ethical underpinning for the safety criterion is the principle that everyone has the right to live in a clean and safe environment. In many countries, this right is enshrined in the most fundamental of legal documents, the national constitution. The post-apartheid Constitution of the Republic of South Africa mandates, for example, that 'every person shall have the right to an environment that is not detrimental to his or her health or well-being.'[11] Insofar as rights are held equally by all, the safety criterion provides a far more egalitarian basis for environmental policy than willingness to pay.

The economics of implementing the safety criterion are relatively straightforward. All that is required is an assessment of the costs of alternative means of meeting the standard, as opposed to the calculation and comparison of the benefits and costs of a wide range of possibilities.

One conceptual issue that is worth considering, however, is the difference between saying that each individual enjoys an equal right to risk mitigation and saying that each statistical life counts equally. In the latter case, the same level of risk to an individual – for example, from air pollution – would carry more weight in densely populated areas than in sparsely populated areas simply because more people are impacted in the former. In other words, paraphrasing Summers, by this logic a load of toxic waste should be dumped in the location with the lowest population density. To be sure, few would advocate siting a nuclear waste dump in proximity to a major population center. But from the perspective of individual rights, what is deemed safe should not vary depending on whether one lives in the city or the countryside.

Sustainability

The ethical underpinning for the sustainability criterion is intergenerational equity. Often this is translated into the goal of ensuring that the well-being of future generations is no less than that of the present generation. The Brundtland Commission in 1987 expressed this idea in terms of human needs: sustainable development 'meets the needs of the present without compromising the ability of future generations to meet their own needs' (World Commission on Environment and Development 1987, p. 8). Alternatively, sustainability is sometimes defined in terms of a nondecreasing stock of natural capital or of total natural and human-made capital (called 'strong' and 'weak' sustainability, respectively).

The sustainability criterion departs markedly from neoclassical efficiency, where the well-being of future generations is handled by discounting future costs and benefits to obtain their 'present values.' With a fairly modest discount rate of 4 percent, for example, a $100 million cost (in today's dollars) to be incurred 100 years from now is valued at only $2 million today. In other words, it would be inefficient for the present generation to spend more than $2 million in order to avoid this cost on behalf of future generations.

Private firms often use discounted cash flow analysis to make investment decisions, since money has 'time value' by virtue of its potential earning capacity. Individuals also exhibit 'time preference' in their decisions, valuing a dollar today more than the same dollar a year or more hence. Inequalities of wealth and power may increase the discount rates used in private decisions, further devaluing the well-being of future generations. Among the very poor, the imperatives of day-to-day survival may become so pressing as to overshadow concerns about tomorrow. Among the very rich, fear that popular discontent will one day dislodge them from their privileged positions may encourage a cut-and-run strategy for natural resource management, exemplified by the rapacious deforestation across much of Southeast Asia in the 1960s and 1970s under the rule of dictators like Marcos in the Philippines (Broad and Cavanagh 1993; Boyce 1993).

Neoclassical cost-benefit analysis elevates discounting from a private calculus into an ethical principle for public policy decisions that will impact future generations. The effect of discount rates is to count their well-being for less – often stunningly less – than our own. One rationale proffered for this seemingly callous stance is the belief that human well-being is on an upward escalator that inexorably rises over time. Citing a forecast that global per capita income will grow from about $10,000 today to roughly $130,000 (in today's dollars) in the next two centuries, climate economist (and future Nobel laureate) William Nordhaus argued, for example, that 'while there are plausible reasons to act quickly on climate change, the need to redistribute to a wealthy future does not seem to be one of them' (Nordhaus 2008). Yet one might think that climate change itself would be enough to cast a rather large shadow over the comforting assumption of a dramatically wealthier future for humankind.

In effect, the sustainability criterion imposes a constraint on decision makers today. Efforts to translate this into an operational criterion pose several questions, however. What, precisely, is to be sustained? How should it be measured? Is human-made capital, for example, a good substitute for natural capital? Even if we adopt a stringent constraint such as maintaining the stock of natural capital, how do we combine diverse resources like clean air, clean water, minerals, and biodiversity into one measure? Instead of trying to come up with a single metric, should we measure sustainability as a multivariable vector?[12] Why should we take today's levels as a benchmark? If human well-being, or the stock of capital, grows or declines over time, does the threshold for sustainability rise or fall with them? These practical issues may be no more (or less) insuperable than the monetary valuations required to operationalize the neoclassical efficiency criterion, but to date they have received relatively little attention.

Justice

Justice is often regarded as a central normative goal in the political economy of the environment. The distribution of environmental costs and benefits is important not only because of what it tells us about how the world works, but also because justice is a compelling end in itself.

Whereas sustainability addresses intergenerational equity, justice addresses intragenerational equity. While neoclassical efficiency focuses on the size of the pie, justice focuses on how it is sliced. Whereas the safety criterion aims to protect public health, justice seeks to ensure that environmental health – whatever its level – is distributed fairly across the population.

Environmental justice most often refers to equity across subgroups of the population defined on the basis of race, ethnicity, income, gender, or other attributes. As discussed earlier, a large body of evidence has found systematic environmental disparities to exist, with disproportionate costs imposed on certain racial and ethnic groups, on low-income communities, and in some cases on women.

An alternative approach is to rank the whole population by the environmental attribute in question – exposure to air pollution, for example – and compute a distributional measure such as the Gini coefficient to assess the extent of disparity. This vertical measure of inequality has been applied to environmental quality much less often than horizontal (intergroup) measures, but it, too, may be regarded as salient to environmental justice.[13] Rather than relying on a single measure of justice, an alternative approach could be to treat it as a vector of variables encompassing both horizontal and vertical equity.[14]

As a normative goal, justice requires the reduction or elimination of environmental disparities. In principle, this could be achieved either by reducing pollution and resource depletion in overburdened communities or by increasing them in less burdened communities. The latter possibility has led some critics to accuse environmental justice advocates of 'Nimbyism,' the 'not-in-my-back-yard' ethic that contributed to the environmental disparities in the first place. In response, proponents have countered that their ultimate goal is 'Not in anybody's back yard,' a formulation close to the safety criterion.

In implementing the justice criterion, two additional issues warrant mention. The first is how to aggregate across diverse dimensions of environmental quality. There is an important difference, for example, between a situation where one type of pollution is concentrated in one community and another type in another community, versus a scenario in which both are concentrated in the same community. The theoretical and empirical literature on environmental justice suggests that the latter situation is quite common, but the extent to which different environmental impacts offset each other across communities, as opposed to being additional or perhaps even multiplicative, deserves more attention.

The second issue involves spatial scale. Two adjacent locations each many have equitable distributions of environmental costs within them, but a highly inequitable distribution between them. This means that if combined into a single spatial unit – as we move, for example, from a subnational to the national scale – the measured extent of environmental inequality may change rather dramatically. This is particularly relevant to environmental justice on a global scale. If highly polluting production processes are shifted offshore from North America to Asia, or from western Europe to eastern Europe, for example, this could diminish environmental disparities within countries while exacerbating them internationally.[15]

Multiple criteria and incomplete orderings

The four criteria discussed previously – efficiency, safety, sustainability, and justice – offer distinct normative bases for evaluating outcomes and prescribing policies. In some cases they will lead to divergent conclusions, but in others they may lead to the same results.

Often, in fact, there may be a substantial degree of compatibility among safety, sustainability, and justice, the alternatives to neoclassical efficiency that are favored by political economists.[16] Higher levels of environmental degradation that are linked to wider disparities of wealth and power are likely to contradict all three normative goals. And at least in cases where these outcomes reflect disparities in political power, rather than simply disparities in purchasing power, they may contradict neoclassical efficiency, too.[17]

Multiple-criteria decision analysis offers an alternative to relying solely on one criterion or another. When rankings across alternative outcomes coincide across all criteria, decision-making is relatively easy. In cases where they diverge, the result is an incomplete ordering.[18] Rather than sweeping these different conclusions under the rug by relying on one criterion alone or by collapsing multiple criteria into a single metric, the best course of action may be to acknowledge this reality and deliberate as to the best course of action accordingly.

Concluding remarks

The political economy of the environment aims to deepen our understanding of the interplay among the economy, the environment, and human well-being. In contrast to neoclassical environmental economics, it pays attention not only to the net magnitude of costs and benefits but also to their distribution.

In the realm of positive analysis – descriptions of how the world works – this means exploring the multiple ways in which the distribution of wealth and power affects environmental outcomes. The political economy of the environment posits that our relationships with nature are tied intimately to our relationships with each other.

Research has demonstrated that the costs of environmental degradation do not fall randomly across the population. 'Negative externalities,' as these are called in neoclassical economics, are not impersonal side effects of economic activities. Instead, their dispersion maps that of purchasing power and political power. More research is needed to better understand the dynamics behind this and the reasons for variations in the patterns and extent of disparities across time and space.

Research also has supported the hypothesis that inequalities affect the overall magnitude of environmental degradation, as well as the distribution of the resulting costs and benefits. This would imply that the goals of protecting the environment and working for a more equitable distribution of wealth and power are complementary. Again, more research is needed to better understand the nature and strength of these effects across the multiple dimensions of environmental quality.

In the realm of normative analysis – prescriptions for how the world should work – political economists advocate a wider range of criteria for decision-making than relying solely on neoclassical efficiency, defined as the maximization of net benefits regardless of their distribution. This does not mean that political economists regard the overall magnitude of net benefits as unimportant, but simply that they do not regard this as the only gauge by which outcomes should be measured and compared. Nor does it mean that political economists are unwilling to consider cost-effectiveness in deciding on the means to pursue environmental ends, however the ends are chosen.

Safety, sustainability, and justice are the alternative criteria that political economists invoke for evaluating environmental outcomes and recommending policies. More research is needed to operationalize these fully for policymaking purposes. And more research is needed to explore how multiple criteria can be brought to bear on decision-making processes.

In sum, the political economy of the environment deals with some of the most urgent questions of our time, yet as a field of inquiry and research it is still at a fairly early stage of development. There is ample room for important work to be done.

Notes

1 As Amartya Sen (1980) has pointed out, even "positive" analysis has a normative component in that the values of the analyst affect choices as to what to describe.
2 See, for example, chapter 22 by Puchalsky et al. in this volume.
3 For a brief review of relevant literature, see Boyce et al. (2016).
4 See, for example, Okello et al. (2018). Austin and Meija (2017) find that the ratio of female to male premature deaths from indoor air pollution is inversely related to indicators of women's status.
5 'Let Them Eat Pollution,' *The Economist,* 8 February 1992. The economic logic invoked here is the neoclassical efficiency criterion as implemented in cost-benefit analysis. 'The measurement of the costs of health-impairing pollution depends on the forgone earnings from increased morbidity and mortality,' Summers argued. 'From this point of view a given amount of health-impairing pollution should be done in the country with the lowest cost, which will be the country with the lowest wages.' See later discussion.
6 The partitioning of the internal benefits of environmental cost externalization (and, conversely, the costs of pollution taxes and regulatory compliance) across shareholders, executives, and consumers has received remarkably little attention from empirical researchers. Theoretical models often assume full pass-through to consumers, an assumption that seems incongruent with widespread corporate opposition to environmental policies.
7 Controversy can arise, however, when adherence to the Pareto criterion violates other norms, such as liberty. See, for example, Sen (1987).
8 It is sometimes claimed that in focusing only on the size of the economic pie, neoclassical efficiency is neutral regarding how pie is distributed. This is not strictly true. The prices used to measure the size of the pie reflect the distribution of purchasing power. If, for example, income were reallocated from rich to poor, demand for rice and beans would go up, and demand for champagne and caviar would go down, changing their prices and thereby altering the 'efficient' composition of output.
9 42 U.S. Code § 7409 – National primary and secondary ambient air quality standards, section (b)(1).
10 For discussion, see Schleussner et al. (2016). For a comparison of very different carbon price recommendations based on the criteria of safety and neoclassical efficiency, see chapter 17 in this volume.
11 For more examples, see Popovic (1996).
12 The vector approach to sustainability assessment is suggested by Pearce et al. (1990).
13 For discussion and comparisons of vertical and horizontal measures, see Boyce et al. (2016).
14 In this approach, environmental justice could be defined in terms of an $n + 1$ dimensional vector, where n = the number of horizontal differentiations on the basis of race, ethnicity, gender, region, or other attributes, with one measure of vertical inequality added.

15 Studies of pollution offshoring have reached mixed conclusions; see, for example, Li and Zhou (2017), Cherniwchan et al. (2017), and Brunel (2017).
16 For further discussion of the mutually reinforcing links between sustainability and justice, see Laurent (2019).
17 For discussion of the relationship between environmental injustice and efficiency, see Glasgow (2005). For a discussion of trade-offs and compatibilities across criteria applied to urban development, see Kremer et al. (2019).
18 See Sen (2004) for a discussion of alternative approaches to incompleteness.

References

Ash, M. et al. 2013. Is environmental justice good for white folks? *Social Science Quarterly* 94, 616–636.

Austin, K.F. and M.T. Meija. 2017. Household air pollution as a silent killer: Women's status and solid fuel use in developing nations. *Population and Environment* 39, 1–25.

Boyce, J.K. 1993. *The Philippines: The Political Economy of Growth and Impoverishment in the Marcos Era*. London: Palgrave Macmillan.

Boyce, J.K. 1994. Inequality as a cause of environmental degradation. *Ecological Economics* 11, 169–178.

Boyce, J.K. 2007. Inequality and environmental protection. In Jean-Marie Baland, Pranab Bardhan, and Samuel Bowles, eds., *Inequality, Collective Action, and Environmental Sustainability*. Princeton: Princeton University Press, 314–348.

Boyce, J.K. 2018. Carbon pricing: Effectiveness and equity. *Ecological Economics* 150, 52–61.

Boyce, J.K. et al. 1999. Power distribution, the environment, and public health: A state-level analysis. *Ecological Economics* 29, 127–140.

Boyce, J.K. et al. 2016. Measuring environmental inequality. *Ecological Economics* 124, 114–123.

Broad, R. and J. Cavanagh. 1993. *Plundering Paradise: The Struggle for the Environment in the Philippines*. Berkeley: University of California Press.

Brunel, C. 2017. Pollution offshoring and emission reductions in EU and US manufacturing. *Environmental and Resource Economics* 68, 621–641.

Bullard, R.D. 1983. Solid waste sites and the black Houston community. *Sociological Inquiry* 53, 273–288.

Bullard, R.D. et al. 2008. Toxic wastes and race at twenty: Why race still matters after all of these years. *Environmental Law* 38, 371–411.

Cherniwchan, J. et al. 2017. Trade and the environment: New methods, measurements, and results. *Annual Review of Economics* 9, 59–85.

Cushing, L. et al. 2015. The haves, the have-nots, and the health of everyone: The relationship between social inequality and environmental quality. *Annual Review of Public Health* 36, 193–209.

Farzin, Y.H. and C.A. Bond. 2006. Democracy and environmental quality. *Journal of Development Economics* 81, 213–235.

Fecht, D. et al. 2015. Associations between air pollution and socioeconomic characteristics, ethnicity and age profile of neighbourhoods in England and the Netherlands. *Environmental Pollution* 198, 201–210.

Foster, A. and N. Kumar. 2011. Health effects of air quality regulations in Delhi, India. *Atmospheric Environment* 45, 1675–1683.

Garg, A. 2011. Pro-equity effects of ancillary benefits of climate change policies: A case study of human health impacts of outdoor air pollution in New Delhi. *World Development* 39, 1002–1025.

Glasgow, J. 2005. Not in anybody's backyard the non-distributive problem with environmental justice. *Buffalo Environmental Law Journal* 13, 69–123.

Grossman, G.M. and A.B. Krueger. 1995. Economic growth and the environment. *Quarterly Journal of Economics* 110(2), 353–377.

Grossman, G.M. and A.B. Krueger. 1996. The inverted-U: What does it mean? *Environment and Development Economics* 1, 119–122.

Harbaugh, W.T. et al. 2002. Reexamining the empirical evidence for an environmental Kuznets curve. *Review of Economics & Statistics* 84, 541–551.

Holland, T.G. et al. 2010. A cross-national analysis of how economic inequality predicts biodiversity loss. *Conservation Biology* 23, 1304–1313.

Hunter, P. and L. Fewtrell. 2001. Acceptable risk. In World Health Organization, ed., *Water Quality: Guidelines, Standards and Health*. London: IWA Publishing, 207–227.

Jorgenson, A. et al. 2017. Income inequality and carbon emissions in the United States: A state-level analysis, 1997–2012. *Ecological Economics* 134, 40–48.
Kathuria, V. and N.A. Khan. 2016. Vulnerability to air pollution: Is there any inequity in exposure? *Economic and Political Weekly*, 28 July, 3158–3165.
Kellenberg, D. 2015. The economics of the international trade of waste. *Annual Review of Resource Economics* 7, 109–125.
Kelly-Reif, K. and S. Wing. 2016. Urban-rural exploitation: An underappreciated dimension of environmental injustice. *Journal of Rural Studies* 47, 350–358.
Knight, K. et al. 2017. Wealth inequality and carbon emissions in high-income countries. *Social Currents* 4, 403–412.
Koyunco, C. and R. Yilmaz. 2009. The impact of corruption on deforestation: Cross-country evidence. *Journal of Developing Areas* 42, 213–222.
Kremer, P. et al. 2019. The future of urban sustainability: Smart, efficient, green or just? *Sustainable Cities and Society* 51, 101761.
Kutlar Joss, M. et al. 2017. Time to harmonize national ambient air quality standards. *International Journal of Public Health* 62, 453–462.
Kuznets, S. 1955. Economic growth and income inequality. *American Economic Review* 45, 1–28.
Laurent, E. 2019. *The New Environmental Economics: Sustainability and Justice.* London: Polity Press.
Li, X. and Y.M. Zhou. 2017. Offshoring pollution while offshoring production? *Strategic Management Journal* 38, 2310–2329.
McGee, J.A. and P.T. Greiner. 2018. Can reducing oncome inequality decouple economic growth from CO_2 emissions? *Socius* 4, 1–11.
Mikkelson, G. M. et al. 2007. Economic inequality predicts biodiversity loss. *PLoS One* 5, May.
Mohai, P. and R. Saha. 2015a. Which came first, people or pollution? Assessing the disparate siting and post-siting demographic change hypothesis of environmental injustice. *Environmental Research Letters* 10, 115008.
Mohai, P. and R. Saha. 2015b. Which came first, people or pollution? A review of theory and evidence from longitudinal environmental justice studies. *Environmental Research Letters* 10, 125011.
Mooney, C. 2017. New EPA document reveals sharply lower estimate of the cost of climate change. *Washington Post*, 11 October.
Morello-Frosch, R. and B.M. Jesdale. 2006. Separate and unequal: Residential segregation and estimated cancer risks associated with ambient air toxics in U.S. metropolitan areas. *Environmental Health Perspectives* 114, 368–393.
Nash, R.F. 2001. *Wilderness and the American Mind*, 4th edn. New Haven: Yale University Press.
Neumayer, E. 2002. Do democracies exhibit stronger international environmental commitment? *Journal of Peace Research* 39, 139–164.
Nordhaus, W. 2008. 'The question of global warming': An exchange. *New York Review of Books*, 25 September.
Okello, G. et al. 2018. Women and girls in resource poor countries experience much greater exposure to household air pollutants than men: Results from Uganda and Ethiopia. *Environment International* 119, 429–437.
Pastor, M. et al. 2001. Which came first? Toxic facilities, minority move-in, and environmental justice. *Urban Affairs Review* 23, 1–21.
Pearce, D. et al. 1990. *Sustainable Development.* Cheltenham: Edward Elgar.
Popovic, N.A.F. 1996. In pursuit of environmental human rights. *Columbia Human Rights Law Review* 27, 487–620.
Schleussner, C.F. et al. 2016. Science and policy characteristics of the Paris Agreement temperature goal. *Nature Climate Change* 6, 827–835.
Sen, A.K. 1980. Description as choice. *Oxford Economic Papers* 32, 353–369.
Sen, A.K. 1987. *On Ethics and Economics.* Oxford: Blackwell.
Sen, A.K. 2004. Incompleteness and reasoned choice. *Synthese* 140, 43–59.
Torras, M. and J.K. Boyce. 1998. Income, inequality, and pollution: A reassessment of the environmental Kuznets curve. *Ecological Economics* 25, 147–160.
Vona, F. and F. Patriarca. 2011. Income inequality and the development of environmental technologies. *Ecological Economics* 70, 2201–2213.
World Commission on Environment and Development. 1987. *Our Common Future.* Oxford: Oxford University Press.
Zwickl, K. et al. 2014. Regional variation in environmental quality: Industrial air toxics exposure in U.S. cities. *Ecological Economics* 107, 494–509.

PART 1

Inequality and the environment

A theoretical, empirical and historical framework

3

THE SUSTAINABILITY-JUSTICE NEXUS[1]

Éloi Laurent

Introduction: the social-ecological approach

"As a system approaches its ecological limits, inequality only increases": with these words written more than 30 years ago, the Brundtland Commission (United Nations World Commission on Environment and Development 1987) sealed the profound intertwining of unsustainability and inequality. By linking justice and sustainability in a "sustainability-justice nexus" (Agyeman et al. 2002)[2], a number of scholars have echoed this linkage in recent years, arguing that our societies will be more just if they are more sustainable and more sustainable if they are more just (Laurent 2011a, 2020; Dasgupta and Ramanathan 2014; Motesharrei et al. 2014; Gough 2017; Chancel 2020). Bridging the challenge of sustainability and the issue of justice inevitably leads to the need to think about social and ecological problems and policies together. When this is done, it appears that it makes environmental sense to mitigate our social crisis and social sense to mitigate our environmental crises. This is the basic statement of the social-ecological approach and the focus of this chapter.

The social-ecological approach (Laurent 2011a, 2020) considers the reciprocal relationship between its two dimensions, demonstrating how social logics determine environmental degradation and crises, and in turn exploring the social consequences of this degradation of the human environment. On the matter of inequality, the social-ecological approach is a two-way street: social inequalities feed ecological crises while ecological crises in turn aggravate social inequalities.

The first arrow of causality, which runs from inequality to environmental degradation, can be labelled "integrative social-ecology," as it shows that the gap between the rich and the poor and the interaction of the two groups lead to increased environmental degradation and accelerated ecological crises that affect every member of a given community (e.g. greater inequality leads to a lesser adaptation capacity of groups to climate change).

The reciprocal arrow of causality, that goes from ecological crises to social injustice, can be labelled "differential social-ecology," as it shows that the social impact of ecological crises is not the same for different individuals and groups, given their socioeconomic status (the most vulnerable socially appear to be "ecological sentinels" in the sense that they are first and foremost affected by current ecological crises, such as low-income groups in flooding regions, for instance in the US state of Louisiana).

DOI: 10.4324/9780367814533-4

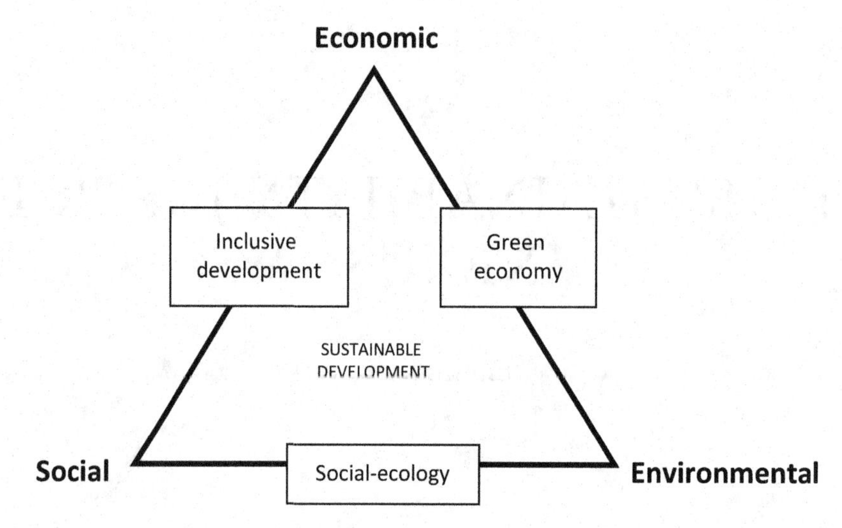

Figure 3.1 The three linkages of sustainable development

Source: Own illustration.

Environmental risk is certainly a collective and global horizon, but it is socially differentiated. Who is responsible for what and who bears the consequences? Such is the basic question of the social-ecological approach, and it is a response to a blind spot in sustainable development studies.

Indeed, as sustainable development assembles three dimensions (economic, social, and ecological) and while the economic-social and economic-ecological links have been explored in great detail in recent years (resulting in, respectively, the "inclusive development" and "green economy" paradigms, see Figure 3.1), the social-ecological link is more obscure, although it is becoming the focus of growing academic and policy attention. This chapter intends to shed some light on what appears to be a missing link in sustainable development.

The first section briefly reviews the available empirical evidence on the inequality and ecological crises; the second section reviews the transmission channels of the inequality crisis to environmental degradations; the final section reviews the different types of social inequality resulting from environmental degradations and crises.

Tacking stock of our twin crises[3]

Empirically, it is hardly debatable that intranational inequality on the one hand and environmental degradation and natural resource consumption on the other have been going up simultaneously in the last three decades.

Science and data with respect to climate change, biodiversity destruction, ecosystem degradation, and natural resource consumption, ever more precise and robust, are now widely consensual: climate is changing rapidly and in some places dramatically, ecosystems and nature's contributions to human well-being have been substantially degraded, biodiversity has been considerably eroded, and natural resource consumption is at an all-time high.[4]

Regarding inequality dynamics, two fundamental results have been established recently. First, Piketty and his co-authors have documented the rise of domestic inequality (or "within countries" inequality): in the last three decades, income and wealth inequality has grown in all regions of the world, albeit at different speeds and from different initial situations (Alvaredo et al. 2018). Second, Milanovic (2016) has shown that if domestic inequality has increased, international inequality (or "between countries" inequality) has on the contrary started to decrease since 2000 but at a very slow pace: the Gini index of unweighted international inequality stands today close to 0.55. In short, the world is more equal than 30 years ago (but still at a very high level of inequality), but countries are a bit more unequal. The most remarkable evolution is the fact that weighted international inequality has started to decrease under the influence of the development of Global Asia, especially China and India. It remains that when all trends are conflated to measure inequality between global citizens (considering all people in the world regardless of where they live), inequality has in fact increased: the top 1% income share has grown significantly while the bottom 50% has stagnated (Alvaredo et al. 2018), and the world's Gini index stands close to 0.63.

The twin crises of inequality and damage to the biosphere are thus an undeniable empirical reality that demands to be analytically explained and remediated through policy.

According to Dobson (1999), there are three ways to relate justice principles and environmental issues: by considering that environmental quality (goods and "bads") is unevenly distributed among the members of a society; by making justice into a condition of possibility of sustainability; and, finally, by doing justice to the environment or by doing justice in the name of the environment (that is, by recognizing a right to nature and creatures who inhabit it in one form or another).

When considering the second option, Dobson denies that inequalities contribute to environmental unsustainability. He recognizes that environmental amenities and damages are distributed according to the level of income, but according to him, the fact that greater social justice (for example, through a more even distribution of environmental ills) improves environmental sustainability has not received empirical evidence. On the contrary, Dobson argues, there are many cases in which social justice and sustainability will contradict one another: improving the distribution of income and power could aggravate and not correct environmental unsustainability. This section argues against this view and highlights five channels through which inequality in fact harms sustainability.

Let's first consider the micro-ecological level, that is, the behavior of rich and poor in isolation. With respect to the rich, Veblen showed that the middle class's desire to imitate the lifestyles of the upper class can lead to a cultural epidemic of environmental degradation. Veblen called this phenomenon "conspicuous consumption," and the bigger cars, larger houses, more luxurious goods, and so on that the rich buy and the middle class desire have a heavy environmental toll. With regard to the poor, Indira Gandhi explained in her speech at the first international environmental summit in Stockholm in 1972 that "poverty and need are the biggest polluters." In the developing world, poverty is indeed leading to unsustainable environmental degradation, such as the dramatic depletion of forest cover in Haiti and Madagascar, the product of a losing trade-off between present and future welfare (see UNEP 2010; Barrett et al. 2011). Since the wealth of the world's poor lies in natural capital because of lack of access to other forms of capital, the depletion of such natural resources leads to further impoverishment. The eradication of poverty, thus, is not only a social benefit but also an environmental one, provided that it takes the form not of a game of consumerist catch-up but of a redefinition of comprehensive wealth, its components, and its indicators (World Bank 2009).

On the macro-ecological level – where the interaction of rich and poor and its environmental outcome is considered – it can be shown that a political economy process lies behind environmental degradation (Boyce 1994, 2002). "Winners" of environmental degradation are able to impose the costs onto the losers because the losers are either not yet born, ignorant of the consequences of the degradation, or lacking in the power to limit them. Five macro-ecological channels through which rich and poor interact in environmental degradation, crises, and policies stand out in particular.

How inequality pollutes the planet[5]

Inequality increases the need for environmentally harmful and socially unnecessary economic growth

Inequality inflates the need for economic growth. If wealth accumulation in a given country is increasingly captured by a small fraction of the population, as it has been the case in many OECD (Organisation for Economic Co-operation and Development), emerging, and developing countries in the last three decades, the rest of the population will need to compensate with additional economic development. Krugman (2002) summed this up well: "Here's a radical thought: if the rich get more, that leaves less for everyone else." Since virtually no country in the world has managed to decouple (in absolute or net terms) economic growth from its negative environmental impact (Parrique et al. 2019), for example carbon emissions or waste, more economic growth currently means more of such "bads," whether locally or globally.[6] In the United States, between 1993 and 2011, 1% of the population managed to capture 75% of economic growth. A more even distribution of income (i.e. a growth of income of 2% for the top 10% and bottom 90% of the income distribution alike) would have reduced the total growth necessary to meet the needs of the vast majority of Americans and led to a small decline in CO_2 emissions (Laurent 2020). A more comprehensive empirical assessment has shown that if the US were to reduce inequality to the level of Sweden (halving their Gini index), emissions would go up by only 1.5%, suggesting that a very substantial but lower reduction in inequality would in fact decrease emissions (Laurent 2013).

But the equalization of economic conditions could, in fact, increase the ecological challenge since the marginal increase of environmental degradation is higher at the bottom of the income distribution than at the top.

By definition, there are two ways to reduce inequality: from the bottom up and from the top down. Reducing the income of the richest segments of the world's population (the 10% that emits roughly 50% of CO_2) via adequate taxation will logically result in important cuts in emissions. Second, assuming higher emission intensity of consumption per expenditure by poorer households (based on the classical Keynesian argument of marginal propensity to consume) omits two important facts. The first one is that "luxuries" yield much more carbon emissions than "necessities" do (to borrow the very relevant distinction introduced by Gough 2017). This is the case in any given developed or developing country (as we have seen for emissions in the UK); this is also the case internationally (Otto et al. 2019): a typical "super-rich" household of two people produces a carbon footprint of 129.3 tCO2e per year, with motor vehicle use generating 9.6 tCO2e per year, household energy emitting 18.9 tCO2e per year, secondary consumption 34.3 tCO2e per year, and 66.5 tCO2e per year generated by the leading emission contributor: air travel). Second, savings as well as consumption result in environmental degradation given the short-termism and inclination toward investments in fossil fuels of prominent financial institutions fueled by the global savings surplus (such as Goldman Sachs).

Moreover, such conclusions assume that the reduction of inequality would entail spreading the lifestyles, wasteful consumption, and ecological footprint of the richest. If so, then the ecological pressure would indeed become unbearable: ecological footprint data clearly show that high-income countries drive the global "ecological deficit" (WWF 2018). But an alternative view holds that shifting from captured development to shared development while redefining development itself can in fact create the necessary room for sustainable social progress (Laurent 2018).

Inequality increases the ecological irresponsibility of the richest, within each country and among nations

Widening inequality exacerbates the fundamental tendency of capitalist enterprises to maximize profits by externalizing cost and turning socially deprived areas into "pollution havens" within countries and across their borders. The financialization of the economy (Epstein 2016) over the past three decades has exacerbated this tendency by shortening time horizons and increasing indifference to unsustainable natural assets management. As the gap between rich and poor grows, governments and businesses find it easier to transfer the environmental damage of the activities of the rich to the neighborhoods of the poor. Income and power inequality, which tends to dissociate polluters from payers, thus act as a disincentive for ecological responsibility or as an accelerator of ecological irresponsibility (Princen 1997).

On the consumption side, the richest consumers present a paradox. They declare in surveys that they care more about the environment than the poor do, and they are indeed, according to the same surveys, more likely to adopt the best environmental practices or to favor more ambitious environmental policy (Laurent 2020). However, at the same time, they pollute more than the poor do in absolute terms because of their higher incomes and more expensive lifestyles. They are also more able to protect themselves from the negative impacts of their behavior as they become richer.

Widening inequality therefore increases not only the demand for a better environment among the richest but also their ability to acquire this good at a lower cost by transferring all corresponding environmental damages to the poorest. For example, in Spain, water has increasingly been diverted from small agricultural enterprises to large coastal tourist facilities. Wealthy tourists enjoy water as a natural amenity and are able to transfer the cost of its abduction and stress to growingly impoverished farmers who now face structural droughts (Sinha et al. 2020).

On the production side, a company faces two essential options to reduce the environmental cost of its production. On the one hand, it can try to adopt the best available technology and to reduce the environmentally harmful impact of its production, a decision that can entail a high economic cost in the short run. On the other hand, it can seek to minimize the economic cost of the social compensation public authorities might demand from it. Income and power inequality will lead the company to relocate to a socially deprived area where people have low incomes and weak political mobilization capacities. The residents of that area would be, presumably, less willing to pay for environmental quality and therefore would demand lesser compensation for environmental damage. Likewise, the feeble political capability of the residents would limit the risk of the emergence of collective action to resist the damaging production.

These dynamics also apply internationally and explain why inequality between countries can result in tragic but avoidable environmental disasters like the chemical pollution in Bhopal in December 1984 or the current degradation of the Niger Delta. Climate change is another case in point: Western societies are less likely to reduce their greenhouse gas emissions because they have little economic incentive to do so as long as they are able to adapt to the most devastating effects of climate change. The reverse is of course true for low-income countries, which contribute little to global emissions but will pay the highest human price for the coming destructive climate. The most striking example of this global injustice may be Africa. The continent accounts for less than 3% of global emissions, but water stress in Africa due to climate change could threaten the well-being of up to 600 million people in the coming decades.

These mechanisms could also account for the striking disparity in biodiversity preservation around the world, as measured by the World Wildlife Fund's Living Planet Index. The decline of the index has been uneven. From 1990 to 2008, the index increased in developed countries by 7%, but it plummeted by 31% in middle-income countries and by 60% in low-income countries. According to the WWF, geographic factors explain only a fraction of the difference. International inequality likely plays an important role, for richer countries are able to preserve their biodiversity while simultaneously exploiting that of countries rich in natural capital but poor in income. For this very reason, evaluations of the ecological impact of a region like the EU, which imports much of its energy and raw materials, should take into account the damage done outside the region, in the original source of production and extraction (Laurent 2020).

Inequality, which affects the health of individuals and groups, diminishes the social-ecological resilience of communities and societies and weakens their collective ability to adapt to accelerating environmental change

A substantial body of research, initiated by Richard Wilkinson and Michael Marmot, has confirmed the negative impact of social inequality on physical and mental health at the local and national level (via stress, violence, less access to health care, etc.).[7] Inequality also acts as an underlying driver of many diseases perceived as natural or biological in the developing world. Paul Farmer, for instance, has asserted, "inequality itself constitutes our modern plague" (see studies from the World Health Organization [WHO] on "preventable burden" of diseases, especially Prüss-Üstün et al. 2016). Myriad governmental and international institutions have already begun to embrace this avenue of research in crafting policy agendas (the WHO, to name only one).

The concepts of social-ecological resilience and vulnerability are in fact now common in the discourse of environmental science. Environmental scientists have begun to describe vulnerability to "natural" disasters as a function of exposure and sensitivity to a given shock, on the one hand, and adaptive capacity and resilience, on the other. Considered within this framework, inequality increases exposure and sensitivity and weakens adaptive capacity and resilience: it acts as a multiplier of the social damage caused by environmental shocks for developed and developing countries alike (as was shown with the COVID-19 pandemic human impact on unequal societies such as the United States).

Inequality hinders collective action aimed at preserving natural resources

According to the "logic of collective action" (the classic theoretical framework formulated by Mancur Olson), a small group of wealthy individuals, convinced that they are the ones who will receive the greatest benefit from environmental protection, would be ready to pay the high cost of ambitious environmental policies. The few (richest), the argument goes, have a logistic comparative advantage over the many (poor). Accordingly, a larger group of people, with more heterogeneous revenues, would not be able to find ways to effectively organize to protect the environment.

This line of reasoning, which suggests that inequality is actually favorable to the preservation of natural resources, has been proven wrong both theoretically and empirically (Baland

and Platteau 1997; Klooster 2000). A number of studies have shown that inequality is, in fact, adverse to the sustainable management of common resources as it disrupts, demoralizes, and disorganizes human communities (see for instance Andersson and Agrawal 2011). The work of Elinor Ostrom in particular demonstrated that institutions that allow communities to preserve resources essential to their long-term well-being are based on principles of reciprocity and fairness, the very opposite of inequality. Adding to the evidence, Ostrom (2010) links equality and the ability of communities around the world to organize efficiently in order to exploit sustainably natural resources and to resist ecological shocks such as climate change.

Her critics, however, make one important point: the difficulty of extrapolating from a purely local context.

In order to account for scale, an analysis of the negative impacts of inequality on environmental decision-making must look toward national and international examples as well. The contemporary United States provides a useful illustration in this respect. Since the 1980s, the US has retreated from the ecological world stage, gradually transferring its prior role of global environmental leader to the European Union. Rapidly increasing income inequality and the corresponding political repercussions might provide an illuminating explanation for this turn of events.

Environmental policy making requires a broad consensus transcending party boundaries, and the simultaneous rise of income inequality and political polarization (understood as growing distance between parties) has reduced the possibility of such bipartisan cooperation. It is now almost impossible in the US to enact ambitious legislation of the caliber of that passed in the 1970s, which later became a model for other nations. While the EPA was formed in 1970, at the beginning of the golden decade for environmental legislation, it is now much more difficult even to confirm a director for the agency. The EPA is also, internally, the subject of political pressure motivated by industrial lobbying, especially from fossil fuels companies that have been empowered by growing economic inequality.

As studies have identified a correlation between income inequality and political polarization in the US, we can think of environmental policy as one of the many policy casualties of the "dance" between these two trends (McCarty et al. 2006). Political polarization and economic inequality both deepened over the past decade. Correspondingly, inertia in the face of environmental degradation has worsened, with the devastation of the Appalachian region and the sabotage of climate negotiations. In this latter case, as with other domestic and global environmental challenges, polarization is combined with an overall shift to the right of the political spectrum, so that the status quo caused by polarization results in a more pro-business and anti-environmental policy.

This polarization dynamic at the local and national level replicates itself on the global scene. Recent research, for instance, has shown that "support is higher for global climate agreements that distribute costs according to prominent fairness principles" (Bechtel and Scheve 2013). Equality and fairness among parties to international environmental negotiations appears to be a key feature of successful global ecological governance (like the Montreal Protocol on ozone layer–depleting substances). On the contrary, inequality in the negotiation process (procedural inequality) and/or distribution of costs (distributive inequality) among nation-states can alter the progress of ecological sovereignty pooling, as with United Nations Framework Convention on Climate Change (UNFCCC) conferences.

Finally, a recent study goes a step further by arguing that inequality could play a key role in bringing about a global ecological collapse. The study investigates the possibility of civilizational collapse, drawing on a rich literature and relying on a new model named "HANDY" (human and nature dynamical), whose particularity is to add a social stratification variable to already existing features of earth models. Humans, in the model, are divided between "Elites" and "Commoners" and their consumption of natural resources is differentiated according to their economic and political power (Motesharrei et al. 2014).

The model's key insight is that ecological collapse can not only come about because of "the stretching of resources due to the strain placed on the ecological carrying capacity" but also due to "the economic stratification of society into Elites [rich] and Masses (or 'Commoners') [poor]." The grim conclusion of the authors regarding one of their key scenarios goes as follows: "the Elites eventually consume too much, resulting in a famine among Commoners that eventually causes the collapse of society." Yet, Motesharrei et al. (2014) also show that this seemingly irresistible collapse by inequality can be prevented through a reduction of current levels of social stratification, a more equal distribution in the consumption of natural resources, and a higher efficiency in this consumption (although technological progress alone cannot, in the model, prevent the eventual collapse).

Inequality reduces the political acceptability of environmental preoccupations and the ability to offset the potential socially regressive effects of environmental policies

Surveys on the political economy of environmental policies have shown that people generally view such policies as socially regressive, which they can, in fact, be (Serret and Johnstone 2006). Growing relative and absolute inequality can thus translate into a reduced acceptability of short-term social (real or perceived) "sacrifices" for long-term (social-ecological) benefits. The failure of France to adopt a carbon tax in 2009/2010 illustrates this argument (Laurent 2020). The socially regressive effect of the tax was obvious, as the bottom 20% of French households spend 2.5 times as much of their income on energy as the top 20% of households do (Laurent 2020). Unsurprisingly, polls reported that as much as 66% of the French population opposed the carbon tax, mostly on economic grounds, with a sharp division between lower-income and higher-income social categories. The government eventually decided to abandon the project in March 2010 after a grueling political defeat amidst rising unemployment and poverty in the context of the "great recession." In 2018, this time facing a full-blown social revolt in a context of severe tax injustice (the so-called "Yellow vests" unrest), the French government decided again to suspend the carbon tax (Berry and Laurent 2019).

The public budget constraints produced by growing inequality, which translates at the macroeconomic level into lower aggregate demand and lower tax revenues, further exacerbate the problem of political acceptability. Inequality makes it more complex and costly, if not impossible, to implement effective compensation mechanisms to counteract possible regressive effects of certain environmental policies, because there are too many people to compensate with too little resource (Nordic countries have been able to successfully implement carbon taxation precisely because they have very low income inequality levels, dynamic economies, and efficient welfare states which foster social consensus). However, social compensation for policies like carbon taxes is a key factor to their political acceptability and even their economic efficiency. In fact, all countries and localities that have adopted carbon taxes over the last two decades have also adopted compensation mechanisms for households and firms that overcame the initial resistance from citizens and businesses (such as in Nordic countries and the Canadian province of British Columbia see Boyce in chapter 17).

The rise of environmental inequality[8]

While the impact of inequality on environmental crises that has been detailed in the previous section may be harder to grasp, the reverse relation is easier to understand and to explain. Environmental conditions determine well-being, most prominently through health-related factors. Therefore, environmental degradation leads to significant and socially differentiated well-being impact.

Recent work by the World Bank shows that extreme climate shocks disproportionately affect the world's poorest and threaten to tip hundreds of millions of hungry people into poverty (see Hallegatte and Walsh in chapter 8). But climate change is just as much a challenge for solidarity in European countries, as European Environment Agency data show: fluvial and coastal floods have affected millions of people in Europe in the last decade; health consequences include injuries, infections, exposure to chemical hazards, and mental health impacts; and heat waves have become more frequent and intense, causing tens of thousands of premature deaths in Europe.

To understand why environmental inequalities may be unjust, one must adopt an explicit theory of justice. Many conceptions of justice co-exist and determine different streams of environmental justice. One of them consists in embracing the capability-building and human development framework developed by Amartya Sen. In essence, the capability approach recommends that well-being be assessed beyond material conditions and also reflect the quality of life of a given person. Among the determinants of quality of life, environmental conditions appear to be of great and growing importance.[9]

Based on Sen's analytical framework, one can define an environmental inequality as a situation that results in an injustice or is unjust if the well-being and capabilities of a particular population are disproportionately affected by its environmental conditions of existence (Laurent 2020). The environmental conditions of existence consist of, negatively, exposure to pollution and risks, and, positively, access to amenities and natural resources (water, air, food). The particular character of the population in question can be defined according to different criteria: social, demographic, territorial, and so on.

Environmental justice therefore can be said to aim at identifying, measuring, and correcting environmental inequalities that result in social injustice. It implies the adoption of an effective arsenal of public policies grounded on scientific research. Yet, one should be clear that environmental justice does not imply that environmental conditions must be equal for all citizens or groups, but that they should not disproportionately affect their well-being and capabilities with respect to the rest of the population.

Different categories of environmental inequality exist and can be broken down to be properly identified and possibly addressed and mitigated.

In a first typology, one can distinguish three forms of environmental inequalities according to their generating factor:

- Inequalities in exposure, sensitivity, and access: this category refers to the unequal distribution of the quality of the environment between individuals and groups. This quality can be negative (exposure to harmful environmental impacts such as urban air pollution) or positive (access to environmental amenities such as green spaces but also water or energy considered from the perspective of their quality or price). In this category of inequalities are included social vulnerability to social-ecological risks (Seveso sites, heat waves, floods, etc.), the risk of cumulative effect of social and environmental inequalities (the educational difficulties of children in the American city of Flint, Michigan, who are exposed to heavy

lead water pollution), and the risk of longer-term social consequences of environmental inequalities (such as the effect on education or long-term income of the prenatal or perinatal exposure to urban air pollution).
- Distributive inequalities of environmental policies: the unequal effect of environmental policies according to social category, in particular the unequal distribution of the effects of tax or regulatory policies between individuals and groups, according to their place in the social category income scale (vertical inequality) and their location in social space (horizontal inequality). The differential impact of carbon taxes, which are also energy taxes, depending on income level and place of residence, for example, falls within this category of environmental inequalities.
- Inequality in participation in environmental public policies: the unequal access to the definition of environmental policies according to social and political status, policies that nevertheless partly determine the environmental conditions of individuals and groups. A well-known example of this type of environmental inequality is the lack of consultation with local populations on the choice of sites where toxic equipment such as incinerators are installed. Environmental inequality with respect to involvement in policy making means that individuals and groups with more resources have more access to environmental policy making on a local, national, or global level (simply because, for instance, they are informed of hearings or townhalls and can afford to attend and participate).

In order to include in the analysis the critically important issue of unequal impacts of individuals and groups on environmental degradation,[10] a simplified typology of environmental inequalities regarding generating factor consists in dividing them into two categories: the inequality impact *of* individuals and groups on environmental damage and definition of environmental policies and the inequality impact *on* individuals and groups of policies and environmental damage.

A second typology of environmental inequalities consists in considering their inequality vector (air pollution, environmental pollution, access to natural resources, exposure and sensitivity to social-ecological disasters, etc.). Finally, a third typology looks at criteria of inequality: according to age (exposure to heat waves of isolated elderly people), socioeconomic level (living on the ground floor in the event of a flood or under the roofs in the event of a heat wave), the quality of housing (indoor air pollution hits the poorest through insalubrity), the neighborhood (children of poor families in French cities of Marseille or Lille are more exposed to fine particle pollution and therefore its lasting social consequences), and the locality (coastal areas for storms, urban areas deprived of vegetation for heat waves).

We can thus distinguish three typologies of environmental inequalities: the first according to the event generating the inequality (or generating factor), the second according to the inequality vector, and the third according to the inequality criterion. Table 3.1 summarizes this framework.

Using this typology, we can determine that the environmental inequality experienced by a Parisian child living near dense traffic during a spike of pollution due to particulate matter 2.5 (PM 2.5) is an inequality of exposure whose vector is air pollution and criteria are age, neighborhood, and locality (at play with possible others such as race and income level). When it comes to measuring environmental inequality, multiple measures rather than single indicators are needed (see Boyce et al. 2016). We can review different types of environmental inequality according to the criteria presented in Table 3.1. Mitigating these various inequalities should be at the heart of the "just transition" (see chapter 15).

Table 3.1 A unified typology of environmental inequality

Philosophical approach	*Generating factor*	*Inequality vector*	*Inequality criterion*	*Example of environmental inequality*
Procedural justice Recognitive justice	Impact of individuals and groups on environmental policies	Exclusion from public decision-making procedures		Non-participation in the decision to install a toxic site (for example a chemical plant) in the city of residence
	Impact of environmental policies on individuals and groups	Taxation, regulatory policies, information/awareness		Vertical and horizontal income inequalities caused by carbon taxation
Distributive justice	Exposure/sensitivity to damage and access to resources	Pollution, access to natural resources and environmental amenities	Age, gender, socioeconomic level (income, health, education, etc.), spatial location, nationality, ethnic characteristics, etc.	Unequal exposure and sensitivity to fine particle pollution in urban areas
	Impact of individuals and groups on nuisance and damage	Emissions of local and global pollution, consumption of natural resources		Carbon footprint of households in the top income deciles

Source: Author.

Air pollution

According to official WHO estimates, ambient air pollution is responsible for 4.2 million deaths per year worldwide, while indoor air pollution is responsible for 3.8 million deaths per year.

A study published in March 2019 (Lelieveld et al. 2019) estimated that outdoor pollution alone could cause up to 8.8 million additional deaths worldwide due to the underestimated health damage from fine particles and other nanoparticles that not only degrade the respiratory system but also the neurological system.

While air quality is a major determinant of quality of life for Europeans (European Commission 2017), air pollution is the greatest risk they run in terms of environmental health: the aforementioned March 2019 study estimates that the annual rate of excess mortality due to ambient air pollution in Europe would be 790,000 and 659,000 in the EU 28 (leading to a reduction in average life expectancy of 2.2 years). About 80% of heart disease and stroke cases, as well as a similar percentage of lung cancer, are linked to air pollution. Air pollution is also associated with adverse effects on fertility, pregnancy, newborns, and children (Science for Environment Policy 2018).

In France, fine particle pollution causes more than 48,000 (preventable) deaths each year, or about 8% of all deaths, as much as alcohol-related mortality, corresponding to a loss of average life expectancy at 30 years of 9 months. If we add the health impact of two other major air pollutants (ozone and nitrogen dioxide), air pollution is responsible for 58,000 premature deaths, or about 10% of all deaths in France. A recent European study on the health impact of fine particle pollution in France reveals that if WHO standards were met, life expectancy at age 30 could increase from 3.6 to 7.5 months according to the French city studied (Pascal et al. 2013).

Inequality in environmental pollution is evident internationally: 97% of cities in low- and middle-income countries with more than 100,000 inhabitants do not comply with WHO guidelines. But even in wealthy developed countries, while air quality has improved markedly over the past decades, exposure to pollution remains far too high and uneven.

It is estimated that around 20% of Europeans are exposed to dangerous particles in the air they breathe (PM10).

The aforementioned study on air pollution in French cities in France (Pascal et al. 2013) also reveals the extent of territorial inequality linked to this exposure to air pollution: the impact on health varies considerably between urban areas (by a factor of two between Toulouse, the least polluted city studied, and Marseille, the most polluted) and within the urban areas themselves. Proximity to road traffic thus considerably increases morbidity due to atmospheric pollution (near roads with heavy traffic, the study revealed a 15% to 30% increase in new cases of asthma in children and chronic respiratory and cardiovascular pathologies prevalent in adults 65 years of age and over).

Thus, the overall impact of air quality on health makes it possible to highlight territorial inequality and finally the impact of pollution on the most vulnerable social groups living in urban areas. At the bottom of this chain, social injustice is compounded by the fact that air pollution can have lasting effects on children's abilities throughout their lives (Currie 2011). Likewise, current research in toxicology emphasizes the impact of the prenatal and perinatal environment on the biological and social development of children.

The issue of exposure is aggravated by that of sensitivity: a French study (Deguen et al. 2015) shows that even if the rich and poor neighborhoods of Paris are exposed to air pollution, the poorest inhabitants are three times more likely to die from severe pollution than the richest because of poorer health and less access to health care.

Risk, noise, and chemical pollution

With regard to chemical pollution of the environment, a first issue concerns the fairness of the distribution of hazardous or toxic sites (the health harms caused by these facilities cannot be proven, since it is their harmful nature that justifies their classification as toxic sites). Recent studies show that environmental exposure is far from being socially homogenous in the US (see, for example, Mohai and Saha 2015), China (see Liu 2012), or Europe (European Environment Agency 2019).

Noise, considered by many experts to be the second biggest environmental risk behind air pollution in terms of its health impact (measured in lost years of disability-adjusted life) should be treated as a form of environmental pollution. A new report by the European Environmental Agency or EEA (European Environmental Agency 2019), reviewing a number of environmental inequalities faced by European citizens (related to air pollution or exposure to extreme temperatures), documents the importance and unequal distribution of noise pollution in the EU. The EEA estimates that environmental noise causes at least 16,600 cases of premature death in Europe each year, with almost 32 million adults annoyed by it and a further 13 million suffering from sleep disturbance. In addition, cities with poorer populations have higher noise levels. The relationship between social inequalities and exposure to noise was highlighted by a study published in early 2013 by the Regional Health Agency of the Ile-de-France region on the Paris major airport hubs (Laurent 2020). The results reveal that the share of population exposed increases with the level of socioeconomic disadvantage and that districts with a significant proportion of those exposed are those of the most disadvantaged. Other studies on noise, conducted for example in the Marseille region, arrive at less clear-cut conclusions and show in particular that it is rather the intermediate social groups that are most vulnerable to noise.

Chemical pollution is also unevenly distributed within nations as a growing body of research in France has shown in recent years. The PLAINE model built by INERIS allows, for example, the mapping of the presence of nickel, cadmium, chromium, and lead in certain parts of the country (see chapter 21). The results for the Nord-Pas-de-Calais region for cadmium document that two areas find themselves overexposed (Metaleurop and the periphery of the Lille metropolitan area). This issue of chemical pollution and overexposure of certain populations is related to the proliferation of "environmental cancer," that is to say, cancers attributable to environmental factors, which are now estimated at around 10% of all cancers in France.[11]

The occupational dimension of environmental inequalities also becomes more and more transparent. For the first time in 2011, the number of deaths from occupational diseases exceeded the number of deaths by accident in France. A considerable difference in life expectancy can be found between occupational groups (seven years between managers and workers and six years between managers and employees), with a gap that has increased rather than shrunk in the last 30 years (according to data from the French statistical agency Institut national de la statistique et des études économiques or INSEE).

At a more detailed level, exposure to endocrine disruptors (chemicals that may interfere with the body's hormonal system) is not homogeneous among occupations: industry, agriculture, cleaning, and plastic sectors exhibit the greatest degree of exposure. As in the case of particulate matter pollution, prenatal and perinatal exposure to such pollutants may have lasting adverse consequences. For instance, some studies link exposure to arsenic in utero and increased infant mortality, low birth weight, and reduced resistance to childhood infections (Farzan et al. 2013). It is this type of study that lead to the ban of bisphenol A in France, but much remains to be done on the many other endocrine disruptors.

Exposure to social-ecological disasters

Exposure to so-called natural hazards constitutes a major source of social inequality that is expected to worsen over the coming decades as ecological crises such as climate change become more severe. To put it in the phraseology of the United Nations (disaster risk reduction or DRR), "There is no such thing as a 'natural' disaster, only natural hazards": the impact of a given disaster depends on the choices we make for our lives and our environment. Every decision and every action makes us more vulnerable or more resilient.[12]

There are two possible ways to look at natural risks. The first hypothesizes that "natural" disasters occur randomly and that humans can hardly do anything about them (that is the etymology of the word "dis-aster," which essentially points to bad luck or adverse fate). The second way is to think that human responsibility lies at the heart of these events, which rather deserve the name of "catastrophes," which etymologically orients towards the idea of a happy or unhappy ending depending on human behavior. Those two worldviews have been respectively defended by French philosophers Voltaire and Rousseau during the controversy on the causes of the Lisbon earthquake in 1755 (Leigh 1967).

According to the Centre for Research on the Epidemiology of Disasters (CRED), over the 1998–2017 period, 90% of 7,255 listed natural disasters have been linked to climatic factors (rainfall, droughts, storms, etc.), with floods and storms representing 70% of the total. In 2017, 335 natural disasters affected over 95.6 million people, killing an additional 9,697 and costing a total of US $335 billion. But this burden was not shared equally, as Asia seemed to be the most vulnerable continent for floods and storms, with 44% of all disaster events, 58% of the total deaths, and 70% of the total people affected.[13]

The EM-DAT database maintained by the CRED distinguishes between two generic categories for disasters: natural and technological. The natural disaster category is divided into five subgroups, which in turn cover 15 disaster types and more than 30 subtypes. The technological disaster category is divided into three subgroups, which in turn cover 15 disaster types.

This distinction between natural and human disasters is of course necessary and based on a completely understandable logic: the industrial accident that occurred in the Total refinery in Toulouse in September 2001 is not equivalent to the tragic earthquake that devastated Haiti in January 2010. But for the contemporary ecological crises (climate, biodiversity, ecosystems) having a human origin, the resulting disasters (floods, droughts, fires, etc.) can hardly be considered as natural.[14] Even more importantly, existing empirical studies clearly show that major contemporary ecological crises (climate change, destruction of biodiversity, degradation of ecosystems) do not have the same social impact around the world: everywhere they reveal social inequalities (that was the case when Hurricane Katrina hit the city of New Orleans in 2005, hardly affecting high-ground rich districts) and worsen them (many African-Americans were not able to recover from the disaster and had to leave the city).[15] The role of social capital for instance is crucial in social-ecological disasters.

In other words, current developments give increasing weight to Rousseau's view: social factors do play a crucial role in "natural" disasters, which are more appropriately "social-ecological" because their causes and impacts are more and more the results of actions taken by human societies.

An expression circulates in the humanitarian community that reflects the reality that "natural" disasters rarely result in fatality: "earthquakes do not kill people, buildings do." Researchers have tried to give substance to this idea by measuring since 1900 the intensity of the earthquakes that hit the planet, taking into account the population density of the regions affected and the level of wealth of their inhabitants (Hatzfeld et al. 2009). There have been 2.5 million deaths since 1900 due to earthquakes, half of them in China and 200,000 in Iran. The authors first attempt to quantify the annual risk in the different countries of death by earthquake (per million inhabitants). While this figure reaches 92 in Armenia, 41 in Turkmenistan, and 29 in Iran, it is only 0.6 in California and 0.008 in France.

But the study goes further in the analysis of data: taking into account the number of earthquakes of magnitude greater than 6, often considered as the destruction threshold, they obtain a ratio of 2,300 deaths/million/magnitude 6 in Armenia, 1,300 in Turkmenistan, and 300 in Tajikistan, but only two in California and 0.8 in France (Hatzfeld et al. 2009). Countries like Japan have learned throughout the 20th century to literally immune themselves from earthquakes' human impact, but not all countries have had the means to do so, an obvious observation when one considers the consequence of the two similarly powerful earthquakes that devastated Haiti in 2010 and barely affected Japan in 2011.

How to explain this striking difference? Rich countries have simply "learned to protect themselves from earthquakes": there is a decrease in the number of victims over time. This protection, which for some territories borders on seismic immunity, has been acquired through institutional progress.[16]

At the heart of the question of so-called natural disasters lies the issue of international and intranational inequalities. One can thus make two points that both call for action: human impact exacerbates natural disasters and makes some of them more frequent, and much of the damage from all natural disasters occurs because of insufficient and unsustainable planning and a lack of foresight (e.g. the devastation associated with Typhoon Hayian in November 2013). Local and national policymakers must thus anticipate announced and virtually certain future disasters – especially heat waves and floods in rich countries and severe hurricanes in poor countries – if they wish to spare their citizens implacable future injustice. In particular, the role played by structural environmental inequalities but also the lack of social capital in certain communities exposed to social-ecological disasters such as heat waves or hurricanes warrants deeper analysis. For example, minorities face more exposure to the risks connected to urban heat island effect because their neighborhoods often lack tree cover or contain too many impervious surfaces, such as asphalt and concrete.[17]

Conclusion: toward a social-ecological policy

This chapter has attempted to show that the sustainability-justice nexus is a two-way street: inequality exacerbates ecological crises, which in turn aggravate inequality. There are encouraging signs that this nexus is gaining momentum in policymaking circles. At the global level, the United Nations Development Programme (UNDP) and the United Nations Environment Programme (UNEP) have jointly launched the Poverty-Environment Initiative (PEI) in 2005 to help countries integrate poverty-environment objectives into their development plans and policies.

Conversely, it is more and more accepted that environmental inequalities can produce lasting and severe damage to the socially disadvantaged, perpetuating and exacerbating injustice. Studies on the effects of air pollution in Los Angeles have shown how exposure to atmospheric pollution affects school performance through the impact of respiratory diseases developed by exposed children. It has also been shown that children from poor families are more likely to be born with poor health because of the polluted environment experienced by their mothers during pregnancy (Currie 2011).[18] This, in turn, results in poor educational attainment and eventually lower income and lower social status. The question for decision makers interested in social justice is thus clear: how should the reality of environmental inequality inform public policy? The European Environment Agency, for the first time in 2018 (EEA 2018), has proposed an inventory of these inequalities in the European Union, recognizing that better harmonization of social and environmental policies and better local action are necessary in order to successfully address environmental justice issues.

Indeed, because of the growingly intertwined nature of theses twin crises, only a social-ecological policy would be sufficient to address the sustainability-inequality nexus (see Laurent 2020 and chapter 15).

Notes

1 This chapter draws from several papers and books written in the last 10 years.
2 The authors mentioned the existence of a nexus between "sustainability, environmental justice and equity."
3 This section is adapted from Laurent 2020.
4 See recent reports from the IPCC, IPBES, and UN Resource Panel.
5 This section is adapted from Laurent 2015.
6 Absolute decoupling of GDP growth and CO_2 emissions has actually been achieved in a number of countries over certain periods of time, but only on the basis of production or territorial emissions. Once the global ecological impact of their economic development is taken into account (i.e. "net decoupling"), only relative decoupling remains.

7 Richard Wilkinson and Michael Marmot can be credited for opening this avenue of research, now widely pursued in governmental and international institutions.
8 Some portions of this section are taken from Laurent 2020.
9 For more on the capabilities approach applied to environmental justice, see Schlosberg (2007).
10 Households in the top 10% in the US have a carbon footprint three times higher than that of households in the bottom 10% (Laurent 2020).
11 For more on this see www.cancer-environnement.fr/
12 UNDRR: www.undrr.org/about-undrr
13 Despite this, the Americas reported the highest economic losses, representing 88% of the total cost from 93 disasters. China, US, and India were the hardest hit countries in terms of occurrence with 25, 20, and 15 events respectively.
14 Blaikie et al. (2004) made the case for the need to set apart "natural events" and "natural disasters."
15 On those two points, see Pastor et al. (2006).
16 The authors are able to determine that in Italy, three earthquakes prior to 1915 claimed over 120,000 lives, and since then there have been fewer than 10,000 deaths. In California, since the 1906 San Francisco earthquake that killed 3,000 people, there have been only 350 deaths. In Japan, since the 1923 earthquake that left nearly 120,000 dead, there have been only 30,000 deaths despite a very high number of earthquakes (190 earthquakes of magnitude greater than 6).
17 African-Americans are 52% more likely than whites to live in exposed neighbourhoods, Asians 32%, and Hispanics 21%, see Jesdale, Morello-Frosch and Cushing (2013).
18 "Individuals may start with very different endowments at birth because of events that happened to them during a critical period: The nine months that they were in utero. In turn, endowments at birth have been shown to be predictive of adult outcomes and of the outcomes of the next generation." Currie adds, "Mechanisms underlying the perpetuation of lower socioeconomic status: Poor and minority children are more likely to be in poor health at birth, partly because their mothers are less able to provide a healthy fetal environment. Poor health at birth is associated with poorer adult outcomes, which in turn provide less than optimal conditions for the children of the poor" (p. 6). Toxicologists refer to the first 1,000 days in the life of a child (gestation and the first two years of life) as the period when he or she is critically exposed or protected from environmental nuisances and pollution that can impact her or his life for decades.

Bibliography

Agyeman, Julian, Robert D. Bullard, and Bob Evans. 2002. "Exploring the Nexus: Bringing Together Sustainability, Environmental Justice and Equity". *Space and Polity* 6 (1): 77–90. doi:10.1080/13562570220137907.

Alvaredo, Facundo, Lucas Chancel, Thomas Piketty, Emmanuel Saez, and Gabriel Zucman, eds. 2018. *World Inequality Report 2018*. Cambridge, MA: The Belknap Press of Harvard University Press.

Andersson, Krister, and Arun Agrawal. 2011. "Inequalities, Institutions, and Forest Commons". *Global Environmental Change* 21 (3): 866–875. doi:10.1016/j.gloenvcha.2011.03.004.

Baland, J.-M., and J.-P. Platteau. 1997. "Wealth Inequality and Efficiency in the Commons. Part I: The Unregulated Case". *Oxford Economic Papers* 49 (4): 451–482. doi:10.1093/oxfordjournals.oep.a028620.

Barrett, Christopher B., Alexander J. Travis, and Partha Dasgupta. 2011. "On Biodiversity Conservation and Poverty Traps". *Proceedings of the National Academy of Sciences* 108 (34): 13907–13912. doi:10.1073/pnas.1011521108.

Bechtel, Michael M., and Kenneth F. Scheve. 2013. "Mass Support for Global Climate Agreements Depends on Institutional Design". *Proceedings of the National Academy of Sciences* 110 (34): 13763–13768. doi:10.1073/pnas.1306374110.

Berry, Audrey, and Eloi Laurent. 2019. "Taxe carbone, le retour, à quelles conditions?". *Sciences Po*. https://spire.sciencespo.fr/notice/2441/5j4beego4m8vk98ao7kolj4865.

Blaikie, Piers, Terry Cannon, Ian Davis, and Ben Wisner. 2004. *At Risk: Natural Hazards, People's Vulnerability, and Disasters* (2nd Edition). New York, NY: Routledge.

Boyce, James K. 1994. "Inequality as a Cause of Environmental Degradation". *Ecological Economics* 11 (3): 169–178.

Boyce, James K. 2002. *The Political Economy of the Environment*. Cheltenham, UK: Edward Elgar.

Boyce, James K., Klara Zwickl, and Michael Ash. 2016. "Measuring Environmental Inequality". *Ecological Economics* 124 (avril): 114–123. doi:10.1016/j.ecolecon.2016.01.014.

Chancel, L. 2020. *Unsustainable Inequalities. Social Justice and the Environment.* Cambridge, MA: Belknap Press of Harvard University Press.

Currie, Janet. 2011. "Inequality at Birth: Some Causes and Consequences". *American Economic Review* 101 (3): 1–22. doi:10.1257/aer.101.3.1.

Dasgupta, P., and V. Ramanathan. 2014. "Pursuit of the Common Good". *Science* 345 (6203): 1457–1458. doi:10.1126/science.1259406.

Deguen, Séverine, Claire Petit, Angélique Delbarre, Wahida Kihal, Cindy Padilla, Tarik Benmarhnia, Annabelle Lapostolle, Pierre Chauvin, and Denis Zmirou-Navier. 2015. "Neighbourhood Characteristics and Long-Term Air Pollution Levels Modify the Association Between the Short-Term Nitrogen Dioxide Concentrations and All-Cause Mortality in Paris". *PLoS One* 10 (7): e0131463. doi:10.1371/journal.pone.0131463.

Dobson, A. 1999. *Justice and the Environment: Conceptions of Environmental Sustainability and Theories of Distributive Justice.* Oxford and New York: Oxford University Press.

Epstein, Gerald. 2016. "Financialization". In Louis-Philippe Rochon and Sergio Rossi, eds., *An Introduction to Macroeconomics; A Heterodox Approach to Economic Analysis.* Northampton, MA: Edward Elgar Publishers, pp. 319–335.

European Commission. 2017. "Attitudes of European Citizens Towards the Environment". In *Special Eurobarometer*, no. 468. Brussels: European Commission, October.

European Environmental Agency. 2018. *Air Quality in Europe: 2018 Report.* Publications Office. https://data.europa.eu/doi/10.2800/777411.

European Environmental Agency. 2019. *Unequal Exposure and Unequal Impacts: Social Vulnerability to Air Pollution, Noise and Extreme Temperature in Europe.* Luxembourg: Publications Office.

Farzan, Shohreh F., Susan Korrick, Zhigang Li, Richard Enelow, A. Jay Gandolfi, Juliette Madan, Kari Nadeau, and Margaret R. Karagas. 2013. "In Utero Arsenic Exposure and Infant Infection in a United States Cohort: A Prospective Study". *Environmental Research* 126 (octobre): 24–30. doi:10.1016/j.envres.2013.05.001.

Gough, Ian. 2017. *Heat, Greed and Human Need: Climate Change, Capitalism and Sustainable Wellbeing.* Cheltenham: Edward Elgar Publishing.

Hatzfeld, D., Jackson, J., and Tucker, B. 2009. "Can We Minimize Earthquake Disasters". mimeo.

Jesdale, Bill M., Rachel Morello-Frosch, and Lara Cushing. 2013. "The Racial/Ethnic Distribution of Heat Risk – Related Land Cover in Relation to Residential Segregation". *Environmental Health Perspectives* 121 (7): 811–817. doi:10.1289/ehp.1205919.

Klooster, Daniel. 2000. "Institutional Choice, Community, and Struggle: A Case Study of Forest Co-Management in Mexico". *World Development* 28 (1): 1–20. doi:10.1016/S0305-750X(99)00108-4.

Krugman, Paul. 2002. "For Richer". *The New York Times Magazine*, 20 October.

Laurent, Éloi. 2011a. *Social-écologie.* Paris: Flammarion.

Laurent, Éloi. 2011b. "Issues in Environmental Justice Within the European Union". *Ecological Economics* 70 (11): 1846–1853. doi:10.1016/j.ecolecon.2011.06.025.

Laurent, Éloi. 2015. *Social-Ecology: Exploring the Missing Link in Sustainable Development.* Documents de Travail de l'OFCE 2015–07, Observatoire Francais des Conjonctures Economiques (OFCE). https://ideas.repec.org/p/fce/doctra/1507.html.

Laurent, Éloi. 2018. *Measuring Tomorrow: Accounting for Well-Being, Resilience, and Sustainability in the Twenty-First Century.* Princeton and Oxford: Princeton University Press.

Laurent, Éloi. 2020. *The New Environmental Economics: Sustainability and Justice.* Cambridge and Medford, MA: Polity Press.

Leigh, J.A., ed. 1967. *Correspondence complète de Jean Jacques Rousseau*, R. Spang, trans. Geneva: Institut et Musée Voltaire, vol. 4, pp. 37–50, Rousseau to Voltaire, 18 August 1756.

Lelieveld, Jos, Klaus Klingmüller, Andrea Pozzer, Ulrich Pöschl, Mohammed Fnais, Andreas Daiber, and Thomas Münzel. 2019. "Cardiovascular Disease Burden from Ambient Air Pollution in Europe Reassessed Using Novel Hazard Ratio Functions". *European Heart Journal* 40 (20): 1590–1596. doi:10.1093/eurheartj/ehz135.

Liu, Lee. 2012. "Environmental Poverty, a Decomposed Environmental Kuznets Curve, and Alternatives: Sustainability Lessons from China". *Ecological Economics* 73 (janvier): 86–92. doi:10.1016/j.ecolecon.2011.10.025.

McCarty, Nolan M., Keith T. Poole, and Howard Rosenthal. 2006. *Polarized America: The Dance of Ideology and Unequal Riches*. The Walras-Pareto Lectures. Cambridge, MA: MIT Press.
Milanović, Branko. 2016. *Global Inequality: A New Approach for the Age of Globalization*. Cambridge, MA: The Belknap Press of Harvard University Press.
Mohai, Paul, and Robin Saha. 2015. "Which Came First, People or Pollution? A Review of Theory and Evidence from Longitudinal Environmental Justice Studies". *Environmental Research Letters* 10 (12): 125011. doi:10.1088/1748-9326/10/12/125011.
Motesharrei, Safa, Jorge Rivas, and Eugenia Kalnay. 2014. "Human and Nature Dynamics (HANDY): Modeling Inequality and Use of Resources in the Collapse or Sustainability of Societies". *Ecological Economics* 101 (mai): 90–102. doi:10.1016/j.ecolecon.2014.02.014.
Ostrom, Elinor. 2010. "Beyond Markets and States: Polycentric Governance of Complex Economic Systems". *American Economic Review* 100 (3): 641–672. doi:10.1257/aer.100.3.641.
Otto, Ilona M., Kyoung Mi Kim, Nika Dubrovsky, and Wolfgang Lucht. 2019. "Shift the Focus from the Super-Poor to the Super-Rich". *Nature Climate Change* 9 (2): 82–84. doi:10.1038/s41558-019-0402-3.
Parrique, T., et al. 2019. *Decoupling Debunked: Evidence and Arguments Against Green Growth*. A Report Reviewing the Empirical and Theoretical Literature to Assess the Validity of the Decoupling Hypothesis, European Environmental Bureau. https://degrowth.org/2019/11/19/european-environmental-bureau-eeb-report-decoupling-debunked-evidence-and-arguments-against-green-growth-as-a-sole-strategy-for-sustainability/.
Pascal, M., et al. 2013. "Assessing the Public Health Impacts of Urban Air Pollution in 25 European Cities: Results of the Aphekom Project". *Science of the Total Environment* 449 (avril): 390–400. doi:10.1016/j.scitotenv.2013.01.077.
Pastor, Manuel, et al. 2006. *In the Wake of the Storm: Environment, Disaster and Race After Katrina*. New York: Russell Sage Foundation.
Princen, Thomas. 1997. "The Shading and Distancing of Commerce: When Internalization Is Not Enough". *Ecological Economics* 20 (3): 235–253. doi:10.1016/S0921-8009(96)00085-7.
Prüss-Üstün, Annette, et al. 2016. *Preventing Disease Through Healthy Environments: Towards an Estimate of the Environmental Burden of Disease*. Geneva: World Health Organization.
Schlosberg, David. 2007. *Defining Environmental Justice: Theories, Movements, and Nature*. Oxford and New York: Oxford University Press.
Science for Environment Policy. 2018. "What Are the Health Costs of Environmental Pollution?". *Future Brief* 21.
Serret, Ysé, and Nick Johnstone, éds. 2006. *The Distributional Effects of Environmental Policy*. Cheltenham and Northampton, MA: Edward Elgar, OECD.
Sinha, Avik, Oana Driha, and Daniel Balsalobre-Lorente. 2020. "Tourism and Inequality in Per Capita Water Availability: Is the Linkage Sustainable?" *Environmental Science and Pollution Research* 27 (9): 10129–10134. doi:10.1007/s11356-020-07955-6.
UNEP. 2010. *Mainstreaming the Economics of Nature: A Synthesis of the Approach, Conclusions and Recommendations of TEEB*. The Economics of Ecosystems & Biodiversity. Geneva: UNEP.
United Nations World Commission on Environment and Development, ed. 1987. *Report of the World Commission on Environment and Development: Our Common Future*. Oxford: Oxford University Press.
World Bank. 2009. *Reshaping Economic Geography: World Development Report 2009*. Washington, DC: World Bank.
WWF. 2018. *Living Planet Report – 2018: Aiming Higher*, M. Grooten and R.E.A. Almond, eds. Gland, Switzerland: WWF.

4

A SOCIO-METABOLIC PERSPECTIVE ON (MATERIAL) GROWTH AND INEQUALITY

Anke Schaffartzik and Fridolin Krausmann

Introduction: global growth and inequality

The Great Acceleration describes the global rise in resource use and in wastes and emissions since World War II (Steffen et al. 2015). Global material extraction – that is, agricultural and forestry harvest as well as mining for metals, fossil energy carriers, and non-metallic minerals – was increased from approximately 23 billion tons in 1970 to 92 billion tons in 2017 (UNEP 2019b): Per year, 70 billion tons *more* were extracted in 2017 than in 1970. This growth accelerated over time: It took 24 years to add 23 billion tons to annual extraction between 1970 and 1994 and then approximately the same time span (25 years) to add twice that amount (46 billion tons). This overarching growth trajectory powerfully communicates that society-nature relations are in crisis, with the global economy pushing hard on or even surpassing planetary boundaries (Rockström et al. 2009).

This global growth is the cumulative effect of diverging regional patterns: of expanding extraction, production, and consumption in some parts of the world as well as of stagnating or even decreasing material flows in others. At the regional and national level, extremely high per capita extraction and/or consumption in some places is juxtaposed with extremely low per capita values elsewhere (Schaffartzik et al. 2014; Pothen and Schymura 2015). In many of the wealthy, mature industrialized economies, material extraction has been stagnating (Wiedenhofer et al. 2013) with material imports, especially of fossil energy carriers, playing an increasingly important role for production and consumption (Schaffartzik and Pichler 2017). The wealthier an economy, the more material resources it tends to require (Duro, Schaffartzik, and Krausmann 2018) and the more greenhouse gas emissions are linked to its consumption (Chakravarty et al. 2009). Materials extracted elsewhere, especially in the middle-income economies, are needed in the production of the imported goods (and also services) (Wiedmann et al. 2015). Those economies directly and indirectly supplying not only material resources but also, for example, energy and human labor to the global economy tend to have lower levels of consumption. The extraction and processing of resources for export appears to aggravate rather than alleviate international inequalities (Schaffartzik, Duro, and Krausmann 2019).

Where extraction expands, land is claimed not only for mining, agriculture, or forestry but also for the energy and transport infrastructures and the processing plants through which resources are channeled. Extractive expansion commonly excludes other claims – for subsistence,

DOI: 10.4324/9780367814533-5

for other extractive uses, for conservation – and is heavily contested (Temper, Del Bene, and Martinez-Alier 2015; also see Martinez Allier, "Environmental conflicts" and Alfredo Rosete, "Southeast Asia, agriculture, and contested forms of land rights: lessons from the Philippines," this volume). In the face of this contestation, extractive expansions are enforced through political, legal, and economic mechanisms, through cultural appeals, and through outright violence (Bene, Scheidel, and Temper 2018; White et al. 2012; Navas, Mingorria, and Aguilar-González 2018). Growth-led capitalist expansion relies on and reinforces existing socio-ecological inequalities – in the interrelated terms of access to resources, exposure to environmental burdens, decision-making power and economic influence, income and wage-dependency, and so on.

In this chapter, we investigate the international material inequalities that are simultaneously the outcome of and prerequisite for the globally unsustainable patterns of growth described as the Great Acceleration. To do so, we require a theoretical approach that allows us to consider society not only in terms of social relations, perhaps as related to growth-sustaining inequalities in capital and income (Piketty 2014; Milanovic 2012), but as dependent upon material and energy inputs, transformations, and the thermodynamically unavoidable outputs of wastes and emissions for their biophysical reproduction. As such, societies – in an analogy to living organisms – have a metabolism (Fischer-Kowalski and Haberl 2015) that is interrelated with their internal social organization and contingent upon relations with other socio-economic systems. We present these concepts in the section "Concepts: A Socio-Metabolic Perspective." By examining the quality and extent of social metabolism as well as its changes over time, we derive patterns in metabolic inequality as one aspect of international socio-ecological inequality. The section "Material Inequality" offers an overview of some of these empirical results. We provide a brief discussion of the role of inequality for sustainability transformations in "The Inertia of Growth." From considering inequality within the growth debate, we derive a need to assess the socio-ecological benefits and impacts of any further resource use expansion, requiring a balance between the two.

Concepts: a socio-metabolic perspective

One of the principles of strong sustainability is that money and financial capital are no substitute for the functioning of ecosystems and biogeochemical cycles and for the (other) resources required for human and societal reproduction (Neumayer 2013). While money may buy resources, it cannot be used in their stead to construct shelter, to provide nutritional energy, or as a source of the comforts of indoor lighting and heating or cooling. Social ecology (of the Vienna variety, see Fischer-Kowalski and Weisz (2016)) understands society as socio-economically (or socio-culturally) and biophysically reproduced. Within a society, it is not possible for a socio-economic program (such as growth-led capitalist expansion) to be pursued without consequences for society's material and energy stocks and flows and the related exchange with the natural environment. Accordingly, such programs always have not only intended but also unintended environmental outcomes. Society not only is the outcome of social processes but also exerts tangible influence on its environment, effecting material and energetic changes therein. In turn, society is also directly affected by changes in the environment. This is so because society consists of human beings as living, bodily organisms who use buildings and infrastructures, machines and tools, a variety of (durable) consumer goods, and even animals to "make use" of their environment. In amassing and sustaining society in this biophysical sense, continuous material and energy inputs are required, must be distributed and transformed, and are eventually discharged as wastes and emissions. These material and energy flows can be understood as the manifestation of a society's metabolism (Fischer-Kowalski and Haberl 2015).

This metabolism can be charted with the tool of material and energy flow accounting, essentially a bookkeeping method for the inputs and outputs of a given socio-economic system (Krausmann et al. 2017; Fischer-Kowalski et al. 2011). A socio-economic system can be defined by two types of boundaries: those between the system and its (natural) environment and those with other socio-economic systems. Accordingly, inputs into the system can stem from the environment (through extraction) or other economies (through imports). These inputs are processed, transformed, and distributed within the socio-economic system, with a share of these flows integrated into societal stocks of buildings, infrastructures, and machinery (such as limestone processed into cement and used in construction) while others are almost directly transformed into wastes and emissions (such as coal burned to generate energy and producing ash and gas). Next to wastes and emissions as outputs to the environment, outputs can also go to other socio-economic systems as export commodities. Material and energy flow accounting (Fischer-Kowalski et al. 2011) investigates these flows and yields a variety of indicators which can be compared across systems in order to generate an understanding of material inequalities. These include:

1 domestic extraction – those materials extracted through agriculture, forestry, and mining and then further processed or used – as a measure of the intervention into the environment on a given territory,
2 domestic material consumption – domestic extraction plus imports minus exports, accounted for with their actual mass upon crossing the boundary into or out of the system in question – as a measure of the material consumed and (eventually) transformed into wastes and emissions on a given territory, and
3 the material footprint – the material extraction, no matter where in the world it occurs, required to meet final demand for goods and services in a socio-economic system (Wiedmann et al. 2015) – as a measure of not only the direct but also the indirect appropriation or provisioning of material resources.

These three main indicators provide complementary perspectives on the material resource requirements of societal reproduction. They can theoretically be calculated for systems as small as a household and as large as the entire world. In our investigation of international material inequality and its role in growth-led development, we apply the material flow accounting framework at the national level. We are thereby able to capture some of the widely diverging trends underlying the globally observed Great Acceleration, at the expense of not being able to consider subnational inequalities that match and even exceed international inequalities (e.g., Wiedenhofer et al. 2016; also see Tony Reames, "Energy inequality"). In examining international inequalities through the lens of social metabolism, we contextualize – wherever possible – findings on current patterns of extraction, consumption, and appropriation or provisioning with the historical development of the resource use patterns of the system in question. Current material flows required for societal reproduction are also the result of material (and wider resource) use patterns of the past. Stagnation of per capita material extraction and consumption or even saturation with some of the central stock-building materials in those economies that industrialized early by international comparison (Fishman, Schandl, and Tanikawa 2016; Bleischwitz et al. 2018), for example, must be understood in the context of the decades (Mayer, Haas, and Wiedenhofer 2017) or even centuries of historical material accumulation. A society's current material stocks represent past and shape present and future resource use. While extensive built-up stocks may mean that less stock-building materials (e.g., stone and steel) are required than is the case in a country just building up its industries and infrastructures, more materials and energy may be needed for the foreseeable future in order to maintain and operate these stocks (Krausmann et al. 2017).

In this sense, the Great Acceleration is also based on the building up of material stocks (infrastructures, buildings, and increasingly also consumer products) after World War II and the material and energy then required to use and maintain these stocks. This role of stocks as a destination or a point of transmission for material and energy flows also means that they spatially distribute flows to where stocks are either built up or used and maintained.

We consider the international patterns of material flows in terms of metabolic profiles, that is, the average composition and per capita dimensions of the indicators considered. Differences in metabolic profiles do not only occur across space but also across time for the same system and correspond to (far-reaching) changes in the societal organization of the biophysical reproduction that they represent. Specific metabolic profiles are considered typical of so-called metabolic regimes (Krausmann, Weisz, and Eisenmenger 2016) with their typified organization around dominant energy sources (uncontrolled and controlled use of solar energy for hunters and gatherers and agrarian societies and fossil energy for industrial societies). Shifts between these regimes can be thought of as socio-ecological transitions, as far-reaching fundamental changes not only in the resource base of a society but also in the corresponding social organization (Fischer-Kowalski 2011). Within the global economy, a transition in one place may necessitate or limit socio-metabolic change elsewhere, potentially aggravating or reducing inequalities. Transformative future change will have to consist in a redistribution as well as reduction of global resource use (Scheidel and Schaffartzik 2019).

Material inequality

Material inequality in the global economy

The Great Acceleration of global material extraction described in the Introduction is the aggregate result of different underlying trajectories. The high-income countries expanded their extraction of material resources from approximately 12 billion tons per year in 1970 to 19 billion tons in the years before the 2007/08 economic and financial crisis, which was associated with strong decline in extraction. In 2010, this group of countries accounted for 25% of global extraction (Figure 4.1). The dissolution of the Soviet Union was associated with a drop in the extraction in the (upper) middle-income countries, but since this coincides with the acceleration of extraction in China, the global growth trajectory continues uninterrupted in this period. It is in particular growth in China's extraction that also "compensates" the slump in the high-income countries following the 2007/08 crisis. In 2010, almost half of the global population lived in low- or lower-middle-income countries (including India), accounting for only 21% of global extraction. This potential for extractive expansion (Schaffartzik and Pichler 2017) has important implications for upholding growth-led development, to which we will return.

The Great Acceleration (especially in its second phase of even faster acceleration) of resource extraction with its far-reaching environmental and social impacts is clearly visible in Figure 4.1, as is the fact that it is an acceleration at the global level involving different subglobal trends. The effect that this global extractive expansion has on international material inequality depends, of course, not only on how much is extracted where but also on who has access to these resources. A very simple distinction in access lies in whether extraction is for domestic consumption or for consumption elsewhere, that is, used in the production of exported goods and services.

In the following, we contrast material extraction with the material footprint, that is, with the amount of material extraction required for domestic final consumption in the investigated

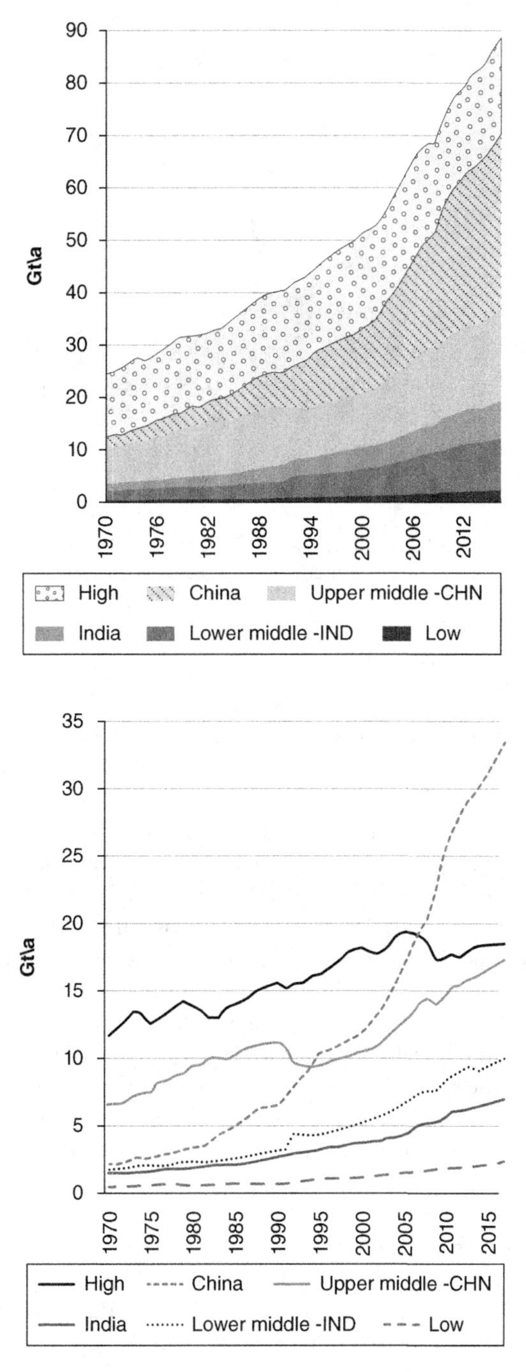

Figure 4.1 Material extraction 1970–2017 by World Bank (2016) 2010 income-based groupings and China and India. In 2017, approximately 10% of the world population lived in the low-income countries, 15% each in the upper-middle-income countries without China and in the high-income countries, and 20% each in the lower-middle-income countries without India, in India, and in China.

Source: Data from UNEP (2019b).

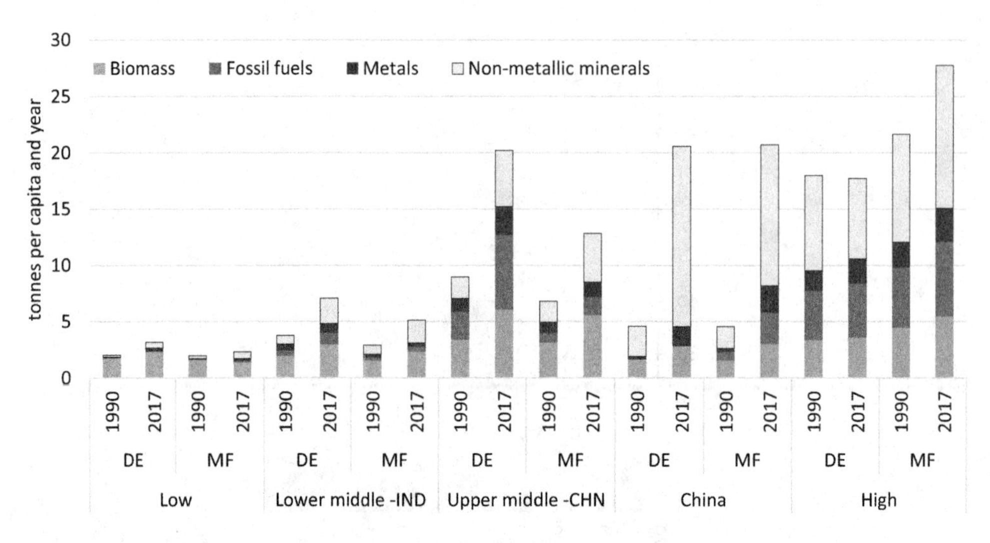

Figure 4.2 Domestic extraction (DE) and the material footprint (MF) of 2010 income-based country groupings and China in 1990 and 2017 in tons per capita and year

Source: Data from UNEP (2019b).

countries and country groupings. We also add some detail to our investigation of which materials are extracted by turning to the four main material resource categories of biomass (extracted through agriculture and forestry, mainly), fossil energy carriers (coal, oil, and gas), metals (extracted as metal ores and traded as concentrates or metal products), and non-metallic minerals (most of which is accounted for by materials used in bulk for construction). This view of global material flows is highly indicative of an increasing polarization as a type of inequality (Figure 4.2). The low- and lower-middle-income countries, including India, were home to half of the global population in 2017 and appropriated (as reflected by the material footprint) between 2 and 7 tons of material per capita. Biomass constituted the majority of these material flows. The growth that had occurred since 1990 was slight. In the wealthier countries with upper-middle- or high-income growth, already higher levels of extraction became much more pronounced and occurred especially for non-metallic minerals and to a lesser extent for fossil energy carriers. During the quarter century traced here, China moved out of a metabolic profile similar to that of the economies with lower income into one that, in terms of magnitude although not quite of composition, is similar to the high-income country grouping. Throughout the entire period under investigation here, China's material footprint was lower than its domestic extraction – the economy was a net provider of material resources to the global economy. By 2017, however, these two indicators had become very similar. The high-income economies, on the other hand, have remained net appropriators of material resources from the global economy; their material footprint is consistently higher than their domestic extraction.

Based on these generalized metabolic profiles, we can identify patterns of growth in the high-income countries' final demand (growth of the material footprint, Figure 4.2) as requiring extractive expansion for export elsewhere, especially in the middle-income countries (Figure 4.1). This extractive expansion relies on claims to land realized in the face of controversy, dispossessing the local population from their livelihood resources, oftentimes violently (Martinez Allier, "Environmental conflicts").

International highly unequal material resource flows underlie the Great Acceleration as the overwhelmingly disastrous global growth trajectory. These inequalities simultaneously result from and support power relations: The observed material flows are inextricably linked to colonialism and modern-day trade relations and agreements (Shiva and Mies 2014; Infante-Amate and Krausmann 2019; Hornborg 2003).

Metabolic regimes and transitions

Although the need to reduce material resource consumption (Akenji et al. 2016; UNEP 2016) and the generation of wastes and emissions (e.g., of greenhouse gas emissions, see IPCC 2018) is increasingly acknowledged, the transition currently occurring at the global level is one of increased material extraction, of intensifying claims to land for resource extraction, and of growing international trade flows. This transition in material flow patterns is part of a larger socio-metabolic transition that also encompasses the energy system. At the global level, an ongoing transition to a fossil energy system can be observed, with the development and use of renewable energy sources currently still confined to pockets of the fossil energy system (Schaffartzik and Fischer-Kowalski 2018).

In material terms, the stagnation or even decline of (per capita) material and energy flows in those economies that industrialized early (Wiedenhofer et al. 2013; Bleischwitz et al. 2018; Fishman, Schandl, and Tanikawa 2016) appears to be based on both enormous amounts of extraction in the past and present-day appropriation of resources extracted elsewhere. The imperial mode of living is collateralized by past and present claims to resources (Brand and Wissen 2012). Even though it relies on unlimited growth in a physically finite world and can clearly not be universally implemented, Western industrialization remains a blueprint for "development," tied to the acknowledged need for economic and – by extension – material growth. As this growth-led trajectory continues globally, it implies continuing along a path of persistent inequality. Some countries or regions use significantly less of global resources and sinks than others do: They extract far less (in the case of the low-income countries) and consume less and also emit less into the global environment. In contrast to the high-income countries, they provide their material resources to the global economy. Even leaving aside the expected growth of the global population, generalizing the current material footprint of the high-income countries (of approximately 28 tons per capita and year; Figure 4.2) to the entire global population would require an annual material resource extraction of 197 billion tons, more than 100 billion tons in addition to what is currently already extracted each year. On each and every square kilometer of the earth's land surface, over 1,300 tons of material resources would have to be extracted every year. And all these materials would eventually be transformed to wastes and emissions (see the "Concepts" section). The shift to the "industrialized" metabolic profile would have an impact not only because of its extent but also because of its composition. Whereas the economies with low and lower incomes currently mainly extract (and appropriate) biomass, the industrialized countries rely on large amounts of non-renewable minerals, including construction minerals, a broad range of metals, and fossil energy carriers, with their corresponding legacies (Krausmann et al. 2017) and environmental impacts (IPCC 2018). Large amounts of the currently extracted stock-building materials continue to go into the infrastructures that prescribe further use of fossil fuels: roads, airports, and power plants, for example (Krausmann, Wiedenhofer, and Haberl 2020). The use of established and currently planned fossil fuel–based structures alone would, by 2030, result in annual emissions twice as high as the amount that would allow for a realistic chance of limiting global heating to 1.5°C (UNEP 2019a).

Although the development narrative reiterated by governments and government-based institutions all around the world continues to be one of the poorer or "less developed" countries catching up to the wealthier ones, this is – from a materials and mass perspective – not possible. Growth requires and exacerbates inequality, in material terms but also along other socio-metabolic and socio-economic dimensions. As a global trajectory, it currently pairs extreme poverty with extreme wealth, dispossession with appropriation (Martinez-Alier 2003), and peripheral with core economies in the sense of dependency and world systems theory (Frank 1966). In this sense, "catching-up development" – based on the assumption that the world's wealthiest countries set the example that should and can be followed by the world's poorer countries – must be understood to be a myth rather than a plan (Mies 1993). In order not to further limit the access to resources and thus the option space, especially for the poor and dispossessed segments of the global population, massive dematerialization and decarbonization of the world's wealthiest economies are a necessity (Scheidel and Schaffartzik 2019; Rodríguez-Labajos et al. 2019).

Inequality in control over material flows

So far, we have argued strongly from a socio-metabolic perspective for the necessity of material degrowth in order to alleviate (or at least not further exacerbate) existing inequalities. We now turn to the role of the predominance of the economic system in shaping resource use and the implications for the role of power relations in both upholding this system and shaping resource use patterns. Next to material inequalities, asymmetrical power relations rely on far-reaching inequalities in the (international) division of labor based on wealth, gender, race, and a rural-urban as well as a North-South divide. In our current economic system, power – amongst other implications – is also the power to externalize costs, providing an important incentive for upholding existing power relations (Shiva and Mies 2014). Resources such as clean water and air and soil fertility as well as (reproductive) labor can be used (up) without financial compensation – often, this is possible because those who should be compensated have insufficient access to decision-making power (an access that is often reclaimed through innovative, powerful, and generally dangerous interventions in ecological distribution conflicts, see Martinez Allier, "Environmental conflicts"). The international patterns of extraction and appropriation, which we discussed in the section "Material Inequality in the Global Economy," and the externalization of costs are important cornerstones of international trade as ecologically unequal exchange (Hornborg 1998). Ecologically unequal exchange analytically considers the net transfer of resources from the Global South to the Global North, from the poorer to the wealthier countries, and the asymmetrical monetary flows that accompany these transfers. Seen from an integrated socio-metabolic and political-economic perspective, ecologically unequal exchange mediates between different but functionally dependent metabolic profiles and along asymmetrical power relations. Current patterns in international trade mean that access to resources and sinks improves for some countries while it deteriorates for others, linking expanding to constricting option spaces. Power asymmetries between economies play a role in determining which costs can be externalized, where they are internalized, and who ultimately has to bear these costs. This, in turn, has a decisive impact on the profit margins on exports and hence on the asymmetrical monetary flows that are understood to contribute to ecologically unequal exchange.

We established in the "Concepts" section that the current Western industrial metabolic profile must be understood as the result of historical processes, including the material accumulation of the past. We also noted that – as long as societies' material stocks are to be further maintained and used – a dependency on resource inputs is locked in. Rather than giving leverage to the places where resources are extracted, the stock and flow relationships have hitherto implied that whoever controls the stocks also controls the flows. On the one hand, this has to do with the extraction, processing, transport, and even use of much of the material extracted globally being dependent on (large-scale) machinery and high amounts of technical energy. Without machines and energy inputs, metals ores could not be used to the current extent, for example. Most fossil energy carriers, including the expanding unconventional ones, are – in the absence of machines and appliances – of no use to humans. From harvest or mining to processing, transport, storage, retail, and manipulation during consumption and use, materials increasingly flow through stocks of buildings, infrastructures, and machinery. Accordingly, current patterns of ownership of (or comparable forms of control over) stocks are decisive in the distribution of resource use and of the services thereby generated. Where it occurs, extractive expansion typically coincides not only with changed material flows (for example the shift from subsistence biomass to market minerals) but also with changed control over these resource flows. Decision-making power shifts to economically and politically (more) powerful actors.

Decision-making power with regard to resource flows can be equated with responsibility for those decisions and accordingly also for resource flows. This is a fundamentally different interpretation from the one commonly offered by neoliberal politics and based on neoclassical economic principles. Based on the understanding that demand drives supply, responsibility for environmentally and socially destructive patterns of production and consumption is often assigned to the consumers: If they cease demanding unsustainable products and services, these will no longer be produced. This interpretation has been referred to as consumer scapegoatism (Akenji 2014), because it blames consumers for something that is not within their power to change.

The inertia of growth

The globally unsustainable and internationally unequal patterns of material resource extraction and use are shaped at the intersection of social metabolism and political economy. Past and current global growth in material flows – as part of the Great Acceleration – have depended on net transfers of resources, often along a trajectory of power. Since World War II and especially since the dissolution of the Soviet Union and the end of the Cold War, the volume of net transfers has increased dramatically with much of the extractive expansion occurring for export. While world systems theory and ecologically unequal exchange tend to generically conceptualize these asymmetrical flows as stemming from the Global South and destined for the Global North, the socio-metabolic perspective reveals that those countries with the lowest per capita incomes (commonly thought to represent the Global South) are far less integrated in the global economy than are the wealthier upper-middle-income countries (this is also true for agricultural trade, Krausmann and Langthaler 2019). This is not to say that exports might not play an important role for the individual low-income economies but that they – from the perspective of the net-importing high-income countries – are not (yet) very relevant as suppliers. For many resources, this crucially limits the leverage of those countries and regions currently providing materials to the global economy: Possibilities for extractive expansion may appear plentiful, especially in the Global South. As we have demonstrated, this perspective is based on the notion that currently established (non-)uses of land and resources can be displaced as needed. The maintenance and even the possible future change of global resource use patterns support current asymmetrical power relations. In this sense, the (global) political economy with its inequalities in wealth and decision-making power is also structured according to material flows and material stocks and the control over flows they imply. Western industrialization has been (and is) based on massive accumulation of largely inflexible stocks (i.e., stocks that do not lend themselves easily to alternative, more sustainable, and socially useful modes of use) and the concentration of decision-making power with regard to resource flows. To a large extent, the generation and distribution of societal benefits from resource use is shaped within the confines of stocks and flows.

From a socio-metabolic perspective, potential lock-ins into patterns of high and growing global resource use with substantial subglobal (and also subnational) inequalities is forged not only by institutions and actors and their vested interests but also by past and present patterns of resource use. From the emergence of colonialism until today, the ability to control transport infrastructures and the ability to control resource flows have been closely interlinked. The accumulation of material resources – regardless of where they were extracted – in certain places then gives rise to the need to similarly channel further resource flows. Material stocks and flows have a tremendous impact on the potentials of and barriers to sustainability transformations.

Conclusion: transforming growth

The material growth that stands at the starting point of our deliberations is not a process of change of the global economy and the inequalities it harbors but a system of their preservation. It is not the growth which – according to the neoclassical maxim – like a rising tide would lift all boats but which distributes resources in such a way that unlimited appropriation for the imperial mode of living becomes possible within the physical confines of planet earth. Global growth is not a symptom of underlying contraction and convergence but of increasing polarization as a form of inequality (Duro, Schaffartzik, and Krausmann 2018). In this sense, global patterns of material resource consumption are disconnected from human needs' satisfaction. They do not entail the distribution of the available resources where they are needed most. This implies that future societal transformations must and can address not only the curbing of material and energy inputs and outputs and hence of degradation and pollution of the environment but also the organization and reconfiguration of society at large, as centered around human needs.

With the aim of enabling sustainability transformations, material growth may be necessary to satisfy human needs and improve human well-being, but even then, these benefits must be gauged against the incurred resource flows and legacy effects, opposing those forms of growth in which no acceptable balance is struck (airports, coal mining, thermal power plants) and favoring decentralized, flexible options with high returns for human well-being.

Bibliography

Akenji, Lewis. 2014. "Consumer Scapegoatism and Limits to Green Consumerism." *Journal of Cleaner Production, Special Volume: Sustainable Production, Consumption and Livelihoods: Global and Regional Research Perspectives* 63 (January): 13–23. https://doi.org/10.1016/j.jclepro.2013.05.022.

Akenji, Lewis, Magnus Bengtsson, Raimund Bleischwitz, Arnold Tukker, and Heinz Schandl. 2016. "Ossified Materialism: Introduction to the Special Volume on Absolute Reductions in Materials Throughput and Emissions." *Journal of Cleaner Production* (April). https://doi.org/10.1016/j.jclepro.2016.03.071.

Bene, Daniela del, Arnim Scheidel, and Leah Temper. 2018. "More Dams, More Violence? A Global Analysis on Resistances and Repression Around Conflictive Dams Through Co-Produced Knowledge." *Sustainability Science* 13 (3): 617–633. https://doi.org/10.1007/s11625-018-0558-1.

Bleischwitz, Raimund, Victor Nechifor, Matthew Winning, Beijia Huang, and Yong Geng. 2018. "Extrapolation or Saturation – Revisiting Growth Patterns, Development Stages and Decoupling." *Global Environmental Change* 48 (January): 86–96. https://doi.org/10.1016/j.gloenvcha.2017.11.008.

Brand, Ulrich, and Markus Wissen. 2012. "Global Environmental Politics and the Imperial Mode of Living: Articulations of State – Capital Relations in the Multiple Crisis." *Globalizations* 9 (4): 547–560. https://doi.org/10.1080/14747731.2012.699928.

Chakravarty, Shoibal, Ananth Chikkatur, Heleen de Coninck, Stephen Pacala, Robert Socolow, and Massimo Tavoni. 2009. "Sharing Global CO2 Emission Reductions Among One Billion High Emitters." *Proceedings of the National Academy of Sciences* 106 (29): 11884–11888. https://doi.org/10.1073/pnas.0905232106.

Duro, Juan Antonio, Anke Schaffartzik, and Fridolin Krausmann. 2018. "Metabolic Inequality and Its Impact on Efficient Contraction and Convergence of International Material Resource Use." *Ecological Economics* 145 (March): 430–440. https://doi.org/10.1016/j.ecolecon.2017.11.029.

Fischer-Kowalski, Marina. 2011. "Analyzing Sustainability Transitions as a Shift between Socio-Metabolic Regimes." *Environmental Innovation and Societal Transitions* 1 (1): 152–159. https://doi.org/10.1016/j.eist.2011.04.004.

Fischer-Kowalski, Marina, and Helmut Haberl. 2015. "Social Metabolism: A Metric for Biophysical Growth and Degrowth." In *Handbook of Ecological Economics*, edited by Joan Martinez-Alier and Roldan Muradian, 100–138. Edward Elgar Publishing. www.elgaronline.com/view/9781783471409.00009.xml.

Fischer-Kowalski, Marina, F. Krausmann, S. Giljum, S. Lutter, A. Mayer, S. Bringezu, Y. Moriguchi, H. Schütz, H. Schandl, and H. Weisz. 2011. "Methodology and Indicators of Economy-Wide Material Flow Accounting." *Journal of Industrial Ecology* 15 (6): 855–876. https://doi.org/10.1111/j.1530-9290.2011.00366.x.

Fischer-Kowalski, Marina, and Helga Weisz. 2016. "The Archipelago of Social Ecology and the Island of the Vienna School." In *Social Ecology: Society-Nature Relations Across Time and Space*, edited by Helmut Haberl, Marina Fischer-Kowalski, Fridolin Krausmann, and Verena Winiwarter, vol. 5, 3–28. Cham: Springer International Publishing. https://doi.org/10.1007/978-3-319-33326-7_1.

Fishman, Tomer, Heinz Schandl, and Hiroki Tanikawa. 2016. "Stochastic Analysis and Forecasts of the Patterns of Speed, Acceleration, and Levels of Material Stock Accumulation in Society." *Environmental Science & Technology* 50 (7): 3729–3737. https://doi.org/10.1021/acs.est.5b05790.

Frank, Andre Gunder. 1966. "The Development of Underdevelopment." *Monthly Review* 18 (4): 17–31.

Hornborg, Alf. 1998. "Towards an Ecological Theory of Unequal Exchange: Articulating World System Theory and Ecological Economics." *Ecological Economics* 25 (1): 127–136. https://doi.org/10.1016/S0921-8009(97)00100-6.

Hornborg, Alf. 2003. "The Unequal Exchange of Time and Space: Toward a Non-Normative Ecological Theory of Exploitation1." *Journal of Ecological Anthropology; Athens* 7 (1): 4–10.

Infante-Amate, Juan, and Fridolin Krausmann. 2019. "Trade, Ecologically Unequal Exchange and Colonial Legacy: The Case of France and Its Former Colonies (1962–2015)." *Ecological Economics* 156 (February): 98–109. https://doi.org/10.1016/j.ecolecon.2018.09.013.

IPCC. 2018. *Global Warming of 1.5 °C, an IPCC Special Report on the Impacts of Global Warming of 1.5 °C above Pre-Industrial Levels and Related Global Greenhouse Gas Emission Pathways, in the Context of Strengthening the Global Response to the Threat of Climate Change, Sustainable Development, and Efforts to Eradicate Poverty*. Special Report SR15. Geneva, Switzerland: Intergovernmental Panel on Climate Change. www.ipcc.ch/report/sr15/.

Krausmann, Fridolin, and Ernst Langthaler. 2019. "Food Regimes and Their Trade Links: A Socio-Ecological Perspective." *Ecological Economics* 160: 87–95.

Krausmann, Fridolin, Heinz Schandl, Nina Eisenmenger, Stefan Giljum, and Tim Jackson. 2017a. "Material Flow Accounting: Measuring Global Material Use for Sustainable Development." *Annual Review of Environment and Resources* 42: 647–675.

Krausmann, Fridolin, Helga Weisz, and Nina Eisenmenger. 2016. "Transitions in Sociometabolic Regimes Throughout Human History." In *Social Ecology: Society-Nature Relations Across Time and Space*, edited by Helmut Haberl, Marina Fischer-Kowalski, Fridolin Krausmann, and Verena Winiwarter, 63–92. Human-Environment Interactions. Cham: Springer International Publishing. https://doi.org/10.1007/978-3-319-33326-7_3.

Krausmann, Fridolin, Dominik Wiedenhofer, and Helmut Haberl. 2020. "Growing Stocks of Buildings, Infrastructures and Machinery as Key Challenge for Compliance with Climate Targets." *Global Environmental Change* 61: 102034.

Krausmann, Fridolin, Dominik Wiedenhofer, Christian Lauk, Willi Haas, Hiroki Tanikawa, Tomer Fishman, Alessio Miatto, Heinz Schandl, and Helmut Haberl. 2017b. "Global Socioeconomic Material Stocks Rise 23-Fold Over the 20th Century and Require Half of Annual Resource Use." *Proceedings of the National Academy of Sciences* 114 (8): 1880–1885. https://doi.org/10.1073/pnas.1613773114.

Martinez-Alier, Joan. 2003. *The Environmentalism of the Poor: A Study of Ecological Conflicts and Valuation*. Edward Elgar Publishing.

Mayer, Andreas, Willi Haas, and Dominik Wiedenhofer. 2017. "How Countries' Resource Use History Matters for Human Well-Being – An Investigation of Global Patterns in Cumulative Material Flows from 1950 to 2010." *Ecological Economics* 134 (April): 1–10. https://doi.org/10.1016/j.ecolecon.2016.11.017.

Mies, Maria. 1993. "The Myth of Catching-Up Development." In *Ecofeminism*, edited by Maria Mies and Vandana Shiva. London: Zed Books.

Milanovic, Branko. 2012. *Global Income Inequality by the Numbers: In History and Now an Overview*. Policy Research Working Papers. The World Bank. http://elibrary.worldbank.org/doi/abs/10.1596/1813-9450-6259.

Navas, Grettel, Sara Mingorria, and Bernardo Aguilar-González. 2018. "Violence in Environmental Conflicts: The Need for a Multidimensional Approach." *Sustainability Science* 13 (3): 649–660. https://doi.org/10.1007/s11625-018-0551-8.

Neumayer, Eric. 2013. *Weak Versus Strong Sustainability: Exploring the Limits of Two Opposing Paradigms*, Fourth Edition. Cheltenham and Northampton: Edward Elgar Publishing.

Piketty, Thomas. 2014. *Capital in the Twenty-First Century*. Cambridge: Harvard University Press.

Pothen, Frank, and Michael Schymura. 2015. "Bigger Cakes with Fewer Ingredients? A Comparison of Material Use of the World Economy." *Ecological Economics* 109 (January): 109–121. https://doi.org/10.1016/j.ecolecon.2014.10.009.

Rockström, Johan, Will Steffen, Kevin Noone, Åsa Persson, F. Stuart Chapin, Eric F. Lambin, Timothy M. Lenton, et al. 2009. "A Safe Operating Space for Humanity." *Nature* 461 (7263): 472–475. https://doi.org/10.1038/461472a.

Rodríguez-Labajos, Beatriz, Ivonne Yánez, Patrick Bond, Lucie Greyl, Serah Munguti, Godwin Uyi Ojo, and Winfridus Overbeek. 2019. "Not So Natural an Alliance? Degrowth and Environmental Justice Movements in the Global South." *Ecological Economics* 157 (March): 175–184. https://doi.org/10.1016/j.ecolecon.2018.11.007.

Schaffartzik, Anke, Juan Antonio Duro, and Fridolin Krausmann. 2019. "Global Appropriation of Resources Causes High International Material Inequality – Growth Is Not the Solution." *Ecological Economics* 163 (September): 9–19. https://doi.org/10.1016/j.ecolecon.2019.05.008.

Schaffartzik, Anke, and Marina Fischer-Kowalski. 2018. "Latecomers to the Fossil Energy Transition, Frontrunners for Change? The Relevance of the Energy 'Underdogs' for Sustainability Transformations." *Sustainability* 10 (8): 2650. https://doi.org/10.3390/su10082650.

Schaffartzik, Anke, Andreas Mayer, Simone Gingrich, Nina Eisenmenger, Christian Loy, and Fridolin Krausmann. 2014. "The Global Metabolic Transition: Regional Patterns and Trends of Global Material Flows, 1950–2010." *Global Environmental Change* 26 (May): 87–97. https://doi.org/10.1016/j.gloenvcha.2014.03.013.

Schaffartzik, Anke, and Melanie Pichler. 2017. "Extractive Economies in Material and Political Terms: Broadening the Analytical Scope." *Sustainability* 9 (7): 1047. https://doi.org/10.3390/su9071047.

Scheidel, Arnim, and Anke Schaffartzik. 2019. "A Socio-Metabolic Perspective on Environmental Justice and Degrowth Movements." *Ecological Economics* (March). https://doi.org/10.1016/j.ecolecon.2019.02.023.

Shiva, Vandana, and Maria Mies. 2014. *Ecofeminism*. London: Zed Books.

Steffen, Will, Wendy Broadgate, Lisa Deutsch, Owen Gaffney, and Cornelia Ludwig. 2015. "The Trajectory of the Anthropocene: The Great Acceleration." *The Anthropocene Review* 2 (1): 81–98. https://doi.org/10.1177/2053019614564785.

Temper, Leah, Daniela Del Bene, and Joan Martinez-Alier. 2015. "Mapping the Frontiers and Front Lines of Global Environmental Justice: The EJ Atlas." *Journal of Political Ecology* 22 (1): 255–278.

UNEP. 2016. *Global Material Flows and Resource Productivity, Authors: Heinz Schandl, Marina Fischer-Kowalski, Jim West, Stefan Giljum, Monika Dittrich, Nina Eisenmenger, Arne Geschke, Mirko Lieber, Hanspeter Wieland, Anke Schaffartzik, Fridolin Krausmann, Sylvia Gierlinger, Karin Hosking, Manfred Lenzen, Hiroki Tanikawa, Alessio Miatto, Tomer Fishman*. Nairobi: United Nations Environment Programme, International Resource Panel. www.resourcepanel.org/reports/global-material-flows-and-resource-productivity.

UNEP. 2019a. *Emissions Gap Report 2019*. Nairobi: United Nations Environment Programme.

UNEP. 2019b. *Natural Resources: Resource Efficiency Indicators*. International Resource Panel of the United Nations Environmental Programme (UNEP). http://environmentlive.unep.org/downloader.

White, Ben, Saturnino M. Borras, Ruth Hall, Ian Scoones, and Wendy Wolford. 2012. "The New Enclosures: Critical Perspectives on Corporate Land Deals." *The Journal of Peasant Studies* 39 (3–4): 619–647. https://doi.org/10.1080/03066150.2012.691879.

Wiedenhofer, Dominik, Dabo Guan, Zhu Liu, Jing Meng, Ning Zhang, and Yi-Ming Wei. 2016. "Unequal Household Carbon Footprints in China." *Nature Climate Change*. https://doi.org/10.1038/nclimate3165.

Wiedenhofer, Dominik, Elena Rovenskaya, Willi Haas, Fridolin Krausmann, Irene Pallua, and Marina Fischer-Kowalski. 2013. "Is There a 1970s Syndrome? Analyzing Structural Breaks in the Metabolism of Industrial Economies." *Energy Procedia* 40: 182–191. https://doi.org/10.1016/j.egypro.2013.08.022.

Wiedmann, Thomas O., Heinz Schandl, Manfred Lenzen, Daniel Moran, Sangwon Suh, James West, and Keiichiro Kanemoto. 2015. "The Material Footprint of Nations." *Proceedings of the National Academy of Sciences* 112 (20): 6271–6276. https://doi.org/10.1073/pnas.1220362110.

World Bank. 2016. *World Bank GNI Per Capita Operational Guidelines & Analytical Classifications*. https://datahelpdesk.worldbank.org/knowledgebase/articles/906519-world-bank-country-and-lending-groups.

5

THE HISTORY OF ENVIRONMENTAL AND ENERGY ECONOMICS THROUGH THE LENS OF POLITICAL ECONOMY

Antoine Missemer

Introduction

Adopting a political economy perspective when exploring the history of economic thought leads to opening the eyes to new insights. The challenge is no longer simply to read past ideas, paradigms, and theories from the point of view of their intellectual roots and of our understanding of economic processes. The political context, the distributional effects of the proposals, and their underlying social questions become important. On energy and environmental matters, which involve systemic issues (i.e. all sectors of the economy are impacted) and which cover ethical and political concerns (polluter-pays-principle, intergenerational equity, etc.), political economy perspectives seem particularly relevant. Basically, three different ways of intersecting the history of economic thought and political economy can be considered.

First, it can mean focusing on the authors and corpuses that historically claimed to be part of 'political economy' in the old sense of the term. In 1615, Antoine de Montchrétien was one of the firsts to use the expression 'political economy' ('*économie politique*'). At the time, the word 'economy' was generic, used for all sorts of good management or organization of things. The addition of the adjective 'political' served to show the national dimension, subject to the sovereign's wishes, of the production and distribution of wealth (Schabas and De Marchi 2003). Economists were *political* economists until the end of the 19th century, when a new generation of scholars, later called the marginalists, pushed for renaming their discipline 'economics', in order, according to them, to leave aside values and ideologies to focus on allegedly purely scientific questions. The relevance and realism of such a project are questionable, obviously. But it means that a first way of articulating the history of economic thought with political economy is to focus on early economic ideas, from the physiocrats in the 18th century to the late classical economists in the 19th century. In that period, energy and environmental issues were not addressed in the same way as today. People talked about their surroundings but not *the* environment, and energy sources were examined separately (wood, coal, waterpower) without a global view of *energy issues*. This does not prevent the historian of ideas to explore the articulation between early political economy and the natural world: Linnaeus's œconomy of nature, Ricardo's rent theory, John Stuart Mill's conception of nature, and W. Stanley Jevons's

 DOI: 10.4324/9780367814533-6

coal question are all examples that have attracted attention (Christensen 2004; Schabas 2005; DesRoches 2015; Wolloch 2016; Missemer 2017). Generally speaking, one may also notice that political economy emerged around the same moment as early environmental concerns, in the context of Western proto-industrialization (Albritton Jonsson 2013; Warde 2018; Kelly 2019). Examining this co-occurrence can be insightful.

A second intersection between the history of economic thought and political economy, ignoring the chronological split of marginalism, leads to deal with all corpuses, ancient and modern, that paid attention to political, institutional, and social concerns in their analysis of economic processes. In environmental and energy affairs, this covers corpuses that were interested in the unequal access to natural resources, in the distributional consequences of energy policies, in the detrimental effects of the polluter-pays-principle, in the social embeddedness of environmental economics, and so on. The historian of ideas is thus led to explore, for instance, socio-energetics in the early 20th century, John Kenneth Galbraith's incursion into environmental issues in the 1960s–1970s, and the roots of modern social ecological economics (Spash 1995; Douai and Plumecocq 2017; Franco 2018; Chirat 2020). The idea here is to examine only a part of past economic thought, that which dealt with questions dear to political economy as it is defined today.

The third and last option is to consider more traditional episodes in the history of economic thought and to reanalyze them through the lens of political economy. In terms of environmental ideas and theories, this means coming back to classic texts in the mainstream history of natural resource economics or externalities to show what a political economy sight can bring to our understanding of the discipline. Examples here are numerous, from the enigma of the disappearance of land from the production function, to the institutional networks that influenced the research conducted at Resources for the Future (RFF) in the 1950s–1960s, including the motives that led Harold Hotelling to establish his famous theoretical 'rule' in 1931 (Czech 2009; Banzhaf 2019; Franco, Gaspard, and Mueller 2019; Gaspard and Missemer 2019; Ferreira da Cunha and Missemer 2020). That way of intersecting the history of environmental and energy economics with political economy may be the most stimulating one, because it forces us to look at corpuses in a different way, to highlight the contributions of a political economy perspective to our understanding of the history of economics.

This chapter adopts this latter option, exploring two specific examples in the history of environmental and energy economics, to show how political economy can enrich our examination of past paradigms, ideas, theories, concepts, and models. The first section discusses the emergence of the concept of natural capital in the 1900s, which can be classically analyzed in connection with the history of capital theory. A political economy perspective adds new information about the context in which the concept appeared and about its implications in terms of the relationship between human beings and the rest of nature. The next section returns to the first experiences of measuring the energy-growth coupling, in the late 1920s, this episode being scrutinized through the history of production theory and through the methodological challenges of inter-index correlations. A political economy view not only sheds light on the context of the emergence of the project carried out at the time, but also leads to a redefinition of the very nature of the project, from an apparently purely energetic to a social and distributional one.

Natural capital in the 1900s

The concept of natural capital is now one of the most used and commented on in sustainability sciences, including environmental and ecological economics. Natural resources and biophysical processes are defined as a self-regenerated capital producing useful services to human beings for the satisfaction of their needs (DesRoches 2015). The concept of natural capital finds itself at the frontier of ecology and economics, commonly employed in academia but also in public expertise and policy making (Fenech et al. 2003; Helm 2019). It is also a structuring concept for the distinction between weak and strong sustainability: the possibility of substituting natural capital with other types of capital (human, manufactured) is at the foundation of competing views of sustainable growth and of intergenerational justice (Neumayer 1999; Barbier 2019). Because it leads to considering the natural world from the point of view of capitalism and human production processes, the concept of natural capital has also been strongly criticized for participating in the noxious commodification of nature (Sullivan 2017; Smessaert, Missemer, and Levrel 2020). Basically, it is a metaphor, mobilizing the economic lexicon and economic theories to deal with natural resources and ecosystems.

The origins of the concept of natural capital are usually identified in the 1970s–1980s (Akerman 2003; Bell 2005; Nadal 2016), when Ernst F. Schumacher (1973) apparently coined it and when David Pearce (1988) pushed for its adoption by ecological economists. However, deeper historical investigation allows us to find old conceptions of nature very close to that conveyed by the concept of natural capital. Linnaeus's view of the œconomy of nature, in the 18th century, has been related to a capitalistic conception of the natural world (DesRoches 2018). It seems there is even in Adam Smith's *Wealth of Nations* (1776) something reminiscent of a vision of nature in the form of a self-regenerated capital (Wolloch 2020). More directly, the expression 'natural capital' actually already existed in the 19th century, albeit with different meanings from those of today (Missemer 2018). Natural capital was used to gather all agents of production created by God or nature (i.e. natural resources and labor force), or it was employed to denounce the concentration of land property in a few hands. The expression 'natural capital' in our modern sense was born later, though well before the 1970s.

It appeared in the writings of Alvin S. Johnson, who was a young economist in the 1900s and who would become famous a few years later as one of the founding fathers of the New School for Social Research in New York. Johnson wrote a couple of textbooks in the 1900s. In his 1909 *Introduction to Economics*, he clearly posited the distinction between artificial and natural capital:

> A generation ago practically all economists restricted the term "capital" to productive wealth that has been produced by industry, such as machines, stocks of materials, etc. Productive wealth, the origin of which cannot be traced to man's industry, was usually classified under the heading "natural agents," or simply under "land" . . . In everyday language men speak of investing capital in land, as of investing capital in buildings or machinery. This usage will be followed in this book; wherever it is necessary to distinguish between the two classes of productive wealth, we shall call the one artificial capital, the other natural capital.
>
> (Johnson 1909, 197)

Johnson elaborated on the role of natural capital in production processes and on its accumulation (1909, 214). His conceptual innovation was noticed by reviewers, who however did not perceive the potential scope of the new concept (Carver 1910). Uses of the expression 'natural capital' were more and more numerous in the 1910s and 1920s, and then decreased in the 1930s, until Schumacher relaunched the concept, probably without knowing about its first life.

Describing and explaining the emergence of the concept of natural capital in the early 20th century is within the reach of the historian of economic thought who can focus on the theoretical context in which Johnson wrote his book. Capital theory was then at a momentum. Since the mid-19th century, capital has always been a subject of controversy among economists, over its definition, its forms (funds, machinery, stocks, etc.), and its role in production. The debates between Austrian and American economists, from the 1880s onwards, were still vivid when Johnson started his academic career. In particular, John Bates Clark (1899) played a significant role in both the renewed capital theory and in Johnson's curriculum – Clark was his supervisor at Columbia University (Samuels 2008). What mattered in Clark's conception of capital was the distinction between the funds of capital and capital goods. These two categories were, according to him, a way of understanding the circulation of capital and its concrete forms in production processes. The common feature of all forms of capital was, in Clark's framework, their capacity to produce things. A sum of money could be considered as capital if it represented productive capacities or was intended for investment. And the concrete assets of a company (machines, buildings, etc.) would become capital goods as soon as they participated in the production process. Johnson inherited from this comprehensive vision of capital, centered on the idea of productivity of the financial and physical assets under consideration.

The other important theoretician who influenced Johnson's general conception of capital was Irving Fisher. This can be noticed throughout Johnson's writings, with explicit references. It is confirmed by the tribute that Johnson (1952) made to Fisher in his autobiography. In the first capital theory controversy, Fisher (1906) had a specific position. He argued that capital had to be defined in a very flexible way, based on the concept of wealth. The important distinction, for Fisher, was not between financial assets and capital goods but between income and capital, that is, flows and stocks of wealth. This perspective implied that any artificial or natural riches could be considered as capital. Likewise, 'resources' and 'assets' were seen as synonymous, opening the door to the application of capital theory to *natural* resources – Fisher described how discounting could be applied to wine, forests, and mines. Fisher did not use the word 'natural capital', but he had a very extensive vision of capital, enabling his readers and followers to include a wide variety of goods and riches into the category 'capital'. Johnson was influenced by this extensive view, seeing it as an opportunity to add new epithets to the word 'capital'.

In other words, Johnson coined the concept of natural capital under the double influence of Clark's emphasis on the productive nature of capital and of Fisher's extensive definition of wealth. All the theoretical conditions were in place to allow Johnson to take the final step leading to the establishment of the concept of natural capital.

The historian of economic thought can find the results of this inquiry sufficient. Johnson's innovation, which appeared firstly as a historical coincidence, finds its explanation in the theoretical context of the time and in Johnson's own intellectual path. Lessons can also be learned for contemporary issues, for a better understanding of the place of the concept of natural capital as a mainstream or challenging concept in economic thought (Missemer 2018). More generally, drawing from capital theory to deal with environmental and ecological issues offers both opportunities and drawbacks (Victor 1991). Scrutinizing the early history of the concept of natural capital provides an interesting hindsight. Should we stop here? We could, but nourishing the history of economic thought with concerns that are dear to political economy can offer a complementary perspective that we would be wrong to ignore.

In the 1900s, when Johnson coined the concept of natural capital, American environmentalism was still in its infancy. It appeared in the mid-19th century through the writings of Ralph W. Emerson, Henry D. Thoreau, and George P. Marsh, who had warned their contemporaries about the destructive powers of the natural world by human activities. As well established now, this early environmentalism gave birth to two competing views in the late 19th century: John Muir's preservationism and Gifford Pinchot's conservationism (Hays 1959; Fox 1981).

Muir, whose ideas participated to the creation of the first national parks in the United States, considered that parts of the natural environment had to be fully protected, with no human intervention at all. Pinchot, who had been trained at the *École Nationale des Eaux et Forêts* in Nancy (France), was a forester more inclined to exploit natural resources, provided it was done in a reasonable, rational way. Those two versions of environmentalism obtained a large political audience after the election of Theodore Roosevelt to the US presidency. Roosevelt's terms, known as the Progressive Era, were based on the will to promote public good and to contain the private, economic interests of trusts. Preservationism and conservationism were two (environmental) ways of incorporating common good into the public agenda.

Until the mid-1900s, Roosevelt oscillated between Muir and Pinchot, consulting both to imagine his policy for natural resources and environmental protection. The famous Hetch-Hetchy controversy, in 1905, disturbed that balance. An engineering project was set up to secure the water supply in the San Francisco Bay area. The construction of a dam in the Hetch-Hetchy valley was envisaged. Preservationists were opposed to the project because the Hetch-Hetchy valley was located inside Yosemite National Park and because it was considered as a particularly remarkable natural space. Conservationists promoted the project, seeing it as an opportunity to improve the living conditions of the people of San Francisco. Roosevelt decided in favor of the dam, and appointed Pinchot to the head of the US Forest Service. Conservationism triumphed over preservationism and took the lead in the design of American environmental policies.

Throughout the1900s–1910s, conservationism became the national doctrine to deal with environmental issues. Pinchot clearly defined his movement as a utilitarian movement, aiming to "the greatest good to the greatest number for the longest time" (1910, 48). This utilitarian, instrumental view of nature spread over all policies and was adopted by scientists, experts, and policymakers from various horizons. In 1908–1909, the National Conservation Commission and the report that came out from it were the climax of the conservation movement, involving representatives of all parts of the American society (National Conservation Commission 1909; Van Hise 1910).

It is in that context that Johnson started his academic career and coined the concept of natural capital. In his textbook, he only mentioned conservation briefly, when detailing the need to encourage "the development of industries that make no drain upon the natural resources of a country" and to retard "the development of industries that destroy such resources" (1909, 367). He met Pinchot once, but this meeting had no direct influence on his conception of nature. However, it is significant that Johnson's general political and intellectual context was marked by conservationism. A utilitarian, instrumental, capitalistic vision of the natural world was in the zeitgeist. This certainly played a role in Johnson's conceptualization.

Other capitalistic expressions dealing with the natural world appeared in the 1900s–1910s, in the conservationist context. Economist Ralph H. Hess (1918, 131) made a clear parallel between the conservation of natural resources and the accumulation of capital. Pinchot himself (1921, 163) talked of the "forest capital" of the United States. And ornithologists involved in the examination of the role of birds in agriculture frequently mentioned the *services of birds* in the same fashion as we talk today about *ecosystem services* in relation to natural capital (Kronenberg 2014).

At the turn of the 1910s, conservationism became more and more invested by people not really familiar with natural resources and the protection of the environment. Conservation soon was used as a generic term to deal with public good and equality. From environmentalism, conservationism became a political doctrine applied to many social subjects, in line with Roosevelt's Progressive Era (Pinchot 1937). Through the lens of political economy, this shows how much Johnson's conceptual proposal was embedded in a utilitarian trend – conservationism – that went beyond environmental issues, becoming a general ideology. The historian of economic thought is then able to understand with better accuracy the reasons why the concept of natural capital appeared at that moment, and what it conveyed in terms of instrumental views of nature, beyond the mere application of capital theory to natural resources.

Decoupling in the 1920s

Natural capital's invention helps us see the contribution of a political economy perspective to the understanding of the theoretical and political context of ideas. But this is only one part of the concerns of political economists, who are also and mainly interested in the distributional effects of economic processes and in the social and political issues behind the scene of the economic life. Looking at our second example – decoupling – in the traditional history of environmental economics shall allow us to illustrate this latter dimension.

Decoupling refers to the relationship between energy production or consumption and economic output, as measured by GDP. In the light of historical data, if GDP grows at the same rate as energy production or consumption, we speak of an energy-growth *coupling*. If we manage to partially disconnect the two trends, with energy production or consumption growing less rapidly than GDP, we speak of *relative decoupling*. If, finally, we succeed in observing GDP growth with stable or even decreasing energy production or consumption, then we speak of *absolute decoupling*. Decoupling can also be understood in relation to greenhouse gas (GHG) emissions, in the context of climate change mitigation. Historically, the very close coupling between energy and output after the Second World War relaxed in the 1970s, with relative decoupling occurring in some Western economies. However, the feasibility of an absolute decoupling between energy, GHG emissions, and economic growth for the whole world remains controversial (Fischer-Kowalski et al. 2011; Ward et al. 2016; Mardani et al. 2019).

Roots of these modern debates can be found in the 1920s, at the Brookings Institution (Missemer and Nadaud 2020). The geologist and statistician Frederick G. Tryon, joined by a few colleagues, investigated the relationship between energy consumption and economic output, with precise measurements and theoretical concerns regarding the role of energy in production processes. Tryon's seminal article, published in 1927, has sometimes been mentioned in the literature but rarely been examined in detail (e.g. Berndt 1978; Cleveland et al. 1984). Tryon clearly argued that energy as a whole, not only coal, petroleum, or waterpower, had to be scrutinized to see the role it played in economic development. In the opening of his paper, he wrote:

> Anything as important in industrial life as power deserves more attention than it has yet received from economists. . . . The great advance in material standards of life in the last century was made possible by an enormous increase in the consumption of energy, and the prospect of repeating the achievement in the next century turns perhaps more than on anything else on making energy cheaper and more abundant. A theory of production that will really explain how wealth is produced must analyze the contribution of this element of energy. These considerations have prompted the Institute of Economics [of the Brookings Institution] to undertake a reconnaissance in the field of power as a factor of production.
>
> (Tryon 1927, 271)

Tryon's project, therefore, was explicit, and it embraced both empirical and theoretical dimensions, from the measurement of the impact of power in everyday life to the amendment of the theory of production in economics. As the mention of the Institute of Economics makes clear, Tryon's project was related to a broader program set up at the Brookings Institution. There was a tradition there to explore energy issues, from coal to waterpower (Hamilton and Wright 1925; Moulton, Morgan, and Lee 1929). And Tryon's specific undertaking was one among others mentioned in the reports of the organization (Moulton 1929, 17–19). Through many publications, Tryon explored various aspects of energy issues, so including what we now call decoupling (e.g. Tryon 1929; Tryon and Rogers 1930; Tryon and Eckel 1932; Tryon and Schoenfeld 1933).

As with the emergence of the concept of natural capital, it is possible to draw on a classic history of environmental and energy economics to study Tryon's project and proposals. In the 1920s, the theoretical context was marked by many innovations in the field of production theory. Production functions were not stabilized yet, but the tendency was to ignore natural resources and sources of power in the mathematical representations of production. Charles Cobb and Paul Douglas mentioned the need for including "the third factor of natural resources in [their] equations" (1928, 165), but they did not complete the task, and their basic formula would soon become the reference point in production theory without any mention of natural resources or power. It is thus easy to understand Tryon's theoretical ambition, even if it finally did not succeed – energy has so far not become a factor of production in the standard theory.

Tryon's methodological proposals to achieve a satisfying measurement of the coupling between energy consumption and economic output are also worth examining. To show the correlation between the two variables, Tryon needed two series of index, one for energy and the other for production. Regarding the latter, he used Carl Snyder's "index of the volume of trade" (1925). Snyder worked at the Federal Reserve Bank of New York and tried throughout the 1920s to build a synthetic index of production for the United States. He started with extrapolations from a few big cities to arrive at increasingly precise estimates for the whole country. On the energy side, Tryon worked with the young economist and statistician Carroll R. Daugherty from the University of Pennsylvania. He met him in 1924–1925 when Daugherty was still a student, inviting him to work on a yardstick to gather different sources of energy in a single index. Daugherty published his results in 1928. As a 1927 letter from Tryon to Harold Moulton shows, Tryon got access to Daugherty's data before publication. Daugherty focused on "prime movers" (1928, 13), that is, primary energy, and he only took into account movement, not heat, in his calculations. That obliged Tryon to make corrections to Daugherty's index before examining the correlation with Snyder's production index. He included heat in the data, using BTU as the common measurement. And he tried to convert available capacities into effectively used machines. He was finally able to compare his two series of data and to draw two curves on a single graph to highlight the correlation between the variables (Tryon 1927, 277). Tryon concluded that a strong coupling between energy consumption and economic output did exist from the late 19th century to the mid-1920s, observing a slight relative decoupling after 1917.

Interestingly, one may find in Tryon the same methodological difficulties as today to conclude on the exact magnitude of coupling or decoupling and to infer any causal relationship between the variables – a step that would have been essential to him to translate his empirical findings into theoretical proposals regarding the theory of production (Ayres 2001; Cahen-Fourot and Durand 2016). Likewise, Tryon (1927, 281) struggled to explain the lags that he observed between his two series of data, which suggests that modern disputes over the econometric treatments of decoupling are rather due to the very object being studied (the energy-growth correlation) than to the (now sophisticated) tools at our disposal – uncertainties have not disappeared for a century.

Telling the early history of decoupling in energy economics without convening political economy apparently already provides insightful results, both historiographically speaking and for contemporary research (Missemer and Nadaud 2020). However, we would be wrong to stop here, once again, because political economy offers a perspective that is likely to shed light on the context of Tryon's project and that also helps us to better define what was the real interest of the Brookings Institution when it funded such a research. In other words, it can lead the historian of thought to shift his or her eyes from his or her initial subject of investigation.

The Brookings Institution was created in December 1927 as the merging of three organizations: the Institute of Economics, the Institute for Government Research, and the Brookings Graduate School of Economics and Government. All these entities participated in the development of institutionalism in the United States (Rutherford 2011). In fact, Tryon was not initially a member of the Brookings Institution. He worked at the US Bureau of Mines and was appointed to the Brookings as a temporary member, from the mid-1920s to 1933–1934. There, he met important representatives of the institutionalist movement: Harold Moulton but also Walton H. Hamilton, Isador Lubin, and Edwin G. Nourse. Institutionalism, in the legacy of Thorstein Veblen, John Rogers Commons, and Wesley C. Mitchell, had the ambition to associate empirical realism with a will to conduct economic and social reforms. We can trace its origins in German historicism and, in a sense, in the old political economy before it was supplanted by marginalism (Hédoin 2013). What mattered in institutionalist studies was to obtain policy-relevant results. The Brookings Institution clearly emphasized this goal in the presentation of its *raison d'être*:

> To play its part in connection with the formulation of sound national policies, the Institution will maintain a number of research institutes, covering . . . the whole range of the social sciences. . . . Such cooperative research should tend to bring a new unity to the humanistic sciences, and promote a greater realism in economic, social, and political thought. The investigations of the Institution will be concerned with enduring problems of theoretical significance as well as with questions of more immediate public import.
>
> (Brookings Institution 1928, 3–4)

These contextual elements lead us to conclude that Tryon's decoupling project was embedded in institutionalism, probably hiding the ambition to exert some influence on public policies. This concerned energy policies in the 1920s–1930s. Already in 1924, Tryon indeed argued that exploring energy issues conducted to question the role of federal authorities in the management of power. In particular, he defended the implementation of national tools for monitoring energy policy, which did not exist at the time because of a segmented view of energy sectors (coal, oil, etc. separately). Tryon's authorship in the report *America's Capacity to Produce*, providing an institutionalist response to the Great Depression, can also be considered as a signal of his involvement in policy affairs (Nourse et al. 1934).

More fundamentally, adopting a political economy perspective on Tryon's decoupling project helps us realize the ultimate motives that conducted a small group of scholars, at the Brookings Institution, to work on the relationship between energy and output. When searching for the social effects of economic processes related to energy, the question of the substitution of the labor force by mechanical power soon appears. Daugherty already had this subject in mind while he worked on his index of energy. And in fact, in his seminal article, Tryon explicitly argued that this was the ultimate question raised by the role of energy in production. What mattered was to estimate "the total consumption of power in all forms, and the aggregate degree of replacement of human labor by power machines" (1927, 273). The Brookings Institution had an interest in energy issues to design sound energy policies. But it certainly also had this interest because of the social question that was behind the scene, that is, the risk to see unemployment increase because of the massive use of mechanical power.

The Great Depression made social issues related to energy even more visible. Tryon (1936, 437) denounced the overspecialization of mining areas, conducting to the appearance of a "stranded population" when the wells and companies shut down. With Margaret H. Schoenfeld, he renewed his call for a federal energy policy to better distribute production capacities and thus better fight against territorial inequalities in terms of access to energy (Tryon and Schoenfeld 1933).

Overall, adding a political economy perspective to the historical examination of Tryon's project therefore leads us to see a series of social issues that attracted the attention of the protagonists and that we would miss by limiting ourselves to a more classical exegesis, centered on the history of production theory and of decoupling in energy economics. Even more interestingly, the true social question that motivated Tryon and the Brookings Institution – the substitution of labor by mechanical power – leads us to shift our eyes from the history of energy to the history of unemployment and of the fears associated with technical progress, from the Luddites to today's digital revolution. In other words, it displaces the historical corpus in which to study and locate the Brookings program. This is a major change, which shows the immense role that political economy can play in the historian's work.

Conclusion

Like any historian's work, the work of the historian of economic thought is not confined to a frozen narration, once and for all, of past events that would exist fully independently of the way we look at them. Historical investigation means selection and interpretation. The usual reading that can be done of past economic ideas insists on the meanders of economic analysis, which evolved with theoretical advances (and setbacks), in multiple historical contexts.

Looking at the history of environmental and energy economics through the lens of political economy allows us to move our eyes to new insights, emphasizing the political, institutional, and social motives to and consequences of economic processes. As mentioned in the introduction, this lens can be used in different ways, and this chapter has covered only one use, via only two examples among many others. The full benefits of political economy for the historian of economic thought would therefore merit further examination.

Nevertheless, it is already clear that political economy allows us, first, to have a better understanding of the context in which ideas were produced, and second, to redefine certain objects of research, which at first glance seemed to belong to the history of subfield A (e.g. the history of decoupling), but which in fact also belonged to the history of subfield B (e.g. the history of the social consequences of technical progress). Convening political economy is therefore not anecdotal but can profoundly modify the historian's points of attention.

There is a call today for environmental and ecological economics to mobilize more political economy. This book contributes to this. The history of environmental and energy economics would also benefit from being part of this approach, in order to enrich our understanding of past paradigms, ideas, theories, and models that still structure our contemporary economic conceptions of nature.

Bibliography

Akerman, Maria. 2003. 'What Does "Natural Capital" Do? The Role of Metaphor in Economic Understanding of the Environment'. *Environmental Values* 12: 431–448.

Albritton Jonsson, Fredrik. 2013. *Enlightenment's Frontier: The Scottish Highlands and the Origins of Environmentalism*. New Haven and London: Yale University Press.

Ayres, Robert U. 2001. 'The Minimum Complexity of Endogenous Growth Models: The Role of Physical Resource Flows'. *Energy* 26 (9): 817–838.

Banzhaf, H. Spencer. 2019. 'The Environmental Turn in Natural Resource Economics: John Krutilla and "Conservation Reconsidered"'. *Journal of the History of Economic Thought* 41 (1): 27–46.

Barbier, Edward B. 2019. 'The Concept of Natural Capital'. *Oxford Review of Economic Policy* 35 (1): 14–36.

Bell, Derek R. 2005. 'Environmental Learning, Metaphors and Natural Capital'. *Environmental Education Research* 11 (1): 53–69.

Berndt, Ernst R. 1978. 'Aggregate Energy, Efficiency, and Productivity Measurement'. *Annual Review of Energy* 3: 225–273.

Brookings Institution. 1928. *The Brookings Institution, Devoted to Public Service Through Research and Training in the Humanistic Sciences*. Washington, DC: Brookings Institution.

Cahen-Fourot, Louison, and Cédric Durand. 2016. 'La Transformation de La Relation Sociale à l'énergie Du Fordisme Au Capitalisme Néolibéral'. *Revue de La Régulation* 20. https://doi.org/10.4000/regulation.12015.

Carver, Thomas Nixon. 1910. 'Review of Introduction to Economics, by Alvin S. Johnson'. *The Economic Bulletin* 3 (1): 21–23.

Chirat, Alexandre. 2020. 'L'Économie Intégrale de John Kenneth Galbraith (1933–1983): Une Analyse Institutionnaliste Historique Américaine Des Mutations de La Société Industrielle'. PhD Thesis, University of Lyon, Lyon.

Christensen, Paul P. 2004. 'Economic Thought, History of Energy in'. In *Encyclopedia of Energy*, edited by Cutler J. Cleveland, vol. 2, 117–130. Amsterdam: Elsevier. https://doi.org/10.1016/B0-12-176480-X/00033-4.

Clark, John Bates. 1899. *The Distribution of Wealth: A Theory of Wages, Interest and Profits*. New York: Palgrave Macmillan.

Cleveland, Cutler J., Robert Costanza, Charles A. S. Hall, and Robert Kaufmann. 1984. 'Energy and the U.S. Economy: A Biophysical Perspective'. *Science* 225 (August): 890–897.

Cobb, Charles, and Paul Douglas. 1928. 'A Theory of Production'. *American Economic Review* 18 (1): 139–165.

Czech, Brian. 2009. 'The Neoclassical Production Function as a Relic of Anti-George Politics: Implications for Ecological Economics'. *Ecological Economics* 68 (8–9): 2193–2197. https://doi.org/10.1016/j.ecolecon.2009.04.009.

Daugherty, Carroll Roop. 1928. 'The Development of Horsepower Equipment in the United States'. In *Power Capacity and Production in the United States*, edited by Carroll Roop Daugherty, Albert Howard Horton, and Royal William Davenport, 5–112. Washington, DC: Government Printing Office.

DesRoches, C. Tyler. 2015. 'The World as a Garden: A Philosophical Analysis of Natural Capital in Economics'. PhD Thesis, University of British Columbia, Vancouver.

DesRoches, C. Tyler. 2018. 'On the Historical Roots of Natural Capital in the Writings of Carl Linnaeus'. *Research in the History of Economic Thought and Methodology* 36 (C): 103–117.

Douai, Ali, and Gaël Plumecocq. 2017. *L'Économie Écologique*. Paris: Repères La Découverte.

Fenech, Adam, Jay Foster, Kirk Hamilton, and Roger Hansell. 2003. 'Natural Capital in Ecology and Economics: An Overview'. *Environmental Monitoring and Assessment* 86: 3–17. https://doi.org/10.1023/A:1024046400185.

Ferreira da Cunha, Roberto, and Antoine Missemer. 2020. 'The Hotelling Rule in Non-Renewable Resource Economics: A Reassessment'. *Canadian Journal of Economics* 53 (2): 800–820. https://doi.org/10.1111/caje.12444.

Fischer-Kowalski, Marina, Mark Swilling, Ernst Ulrich von Weizsäcker, Yong Ren, Yuichi Moriguchi, Wendy Crane, Fridolin Krausmann, et al. 2011. *Decoupling Natural Resource Use and Environmental Impacts from Economic Growth*. New York: United Nations Working Group on Decoupling.

Fisher, Irving. 1906. *The Nature of Capital and Income*. London and New York: Palgrave Macmillan.

Fox, Stephen R. 1981. *The American Conservation Movement: John Muir and His Legacy*. Madison: University of Wisconsin Press.

Franco, Marco P. V. 2018. 'Searching for a Scientific Paradigm in Ecological Economics: The History of Ecological Economic Thought, 1880s – 1930s'. *Ecological Economics* 153: 195–203.
Franco, Marco P. V., Marion Gaspard, and Thomas M. Mueller. 2019. 'Time Discounting in Harold Hotelling's Approach to Natural Resource Economics: The Unsolved Ethical Question'. *Ecological Economics* 163: 52–60.
Gaspard, Marion, and Antoine Missemer. 2019. 'An Inquiry into the Ramsey-Hotelling Connection'. *European Journal of the History of Economic Thought* 26 (2): 352–379.
Hamilton, Walton Hale, and Helen Russell Wright. 1925. *The Case of Bituminous Coal*. New York: Palgrave Macmillan.
Hays, Samuel P. 1959. *Conservation and the Gospel of Efficiency*. Pittsburgh: University of Pittsburgh Press.
Hédoin, Cyril. 2013. *L'Institutionnalisme Historique et La Relation Entre Théorie et Histoire En Économie*. Paris: Classiques Garnier.
Helm, Dieter. 2019. 'Natural Capital: Assets, Systems, and Policies'. *Oxford Review of Economic Policy* 35 (1): 1–13. https://doi.org/10.1093/oxrep/gry027.
Hess, Ralph H. 1918. 'Conservation and Economic Evolution'. In *The Foundations of National Prosperity: Studies in the Conservation of Permanent National Resources*, edited by Richard T. Ely, Ralph H. Hess, Charles K. Leith, and Thomas Nixon Carver, 93–184. New York: Palgrave Macmillan.
Johnson, Alvin S. 1909. *Introduction to Economics*. Boston, MA: D.C. Heath & Co.
Johnson, Alvin S. 1952. *Pioneer's Progress: An Autobiography*. New York: The Viking Press.
Kelly, Duncan. 2019. *Politics and the Anthropocene*. Cambridge: Polity Press.
Kronenberg, Jakub. 2014. 'What Can the Current Debate on Ecosystem Services Learn from the Past? Lessons from Economic Ornithology'. *Geoforum* 55: 164–177. https://doi.org/10.1016/j.geoforum.2014.06.011.
Mardani, Abbas, Dalia Streimikiene, Fausto Cavallaro, Nanthakumar Loganathan, and Masoumeh Khoshnoudi. 2019. 'Carbon Dioxide (CO2) Emissions and Economic Growth: A Systematic Review of Two Decades of Research from 1995 to 2017'. *Science of the Total Environment* 649: 31–49. https://doi.org/10.1016/j.scitotenv.2018.08.229.
Missemer, Antoine. 2017. *Les Économistes et La Fin Des Énergies Fossiles (1865–1931)*. Paris: Classiques Garnier.
Missemer, Antoine. 2018. 'Natural Capital as an Economic Concept, History and Contemporary Issues'. *Ecological Economics* 143: 90–96.
Missemer, Antoine, and Franck Nadaud. 2020. 'Energy as a Factor of Production: Historical Roots in the American Institutionalist Context'. *Energy Economics* 86: 104706. https://doi.org/10.1016/j.eneco.2020.104706.
Moulton, Harold G. 1929. *The Brookings Institution, Report of the President for the Year Ending June Thirtieth 1929*. Washington, DC: Brookings Institution.
Moulton, Harold G., Charles S. Morgan, and Adah L. Lee. 1929. *The St. Lawrence Navigation and Power Project*. Washington, DC: Brookings Institution.
Nadal, Alejandro. 2016. 'The Natural Capital Metaphor and Economic Theory'. *Real-World Economics Review* 74: 64–84.
National Conservation Commission. 1909. *Report of the National Conservation Commission*, vol. I. Washington, DC: Government Printing Office.
Neumayer, Eric. 1999. *Weak Versus Strong Sustainability: Exploring the Limits of Two Opposing Paradigms*. Cheltenham and Northampton: Edward Elgar Publishing.
Nourse, Edwin G., Frederick G. Tryon, Horace B. Drury, Maurice Leven, Harold G. Moulton, and Cleona Lewis. 1934. *America's Capacity to Produce*. Washington, DC: Brookings Institution.
Pearce, David W. 1988. 'Economics, Equity and Sustainable Development'. *Futures* 20 (6): 598–605.
Pinchot, Gifford. 1910. *The Fight for Conservation*. New York: Doubleday, Page & Co.
Pinchot, Gifford. 1921. 'The Economic Significance of Forestry'. *The North American Review* 213 (783): 157–167.
Pinchot, Gifford. 1937. 'How Conservation Began in the United States'. *Agricultural History* 11 (4): 255–265.
Rutherford, Malcom. 2011. *The Institutionalist Movement in American Economics, 1918–1947*. New York: Cambridge University Press.
Samuels, Warren J. 2008. *Johnson, Alvin Saunders (1874–1971)*. www.dictionaryofeconomics.com/article?id=pde2008_J000018.

Schabas, Margaret. 2005. *The Natural Origins of Economics*. Chicago and London: University of Chicago Press.

Schabas, Margaret, and Neil De Marchi. 2003. 'Introduction to Oeconomies in the Age of Newton'. *History of Political Economy* 35 (S1): 1–13. https://doi.org/10.1215/00182702-35-suppl_1-1.

Schumacher, Ernst Friedrich. 1973. *Small Is Beautiful: Economics as if People Mattered*. London: Blond & Briggs.

Smessaert, Jacob, Antoine Missemer, and Harold Levrel. 2020. 'The Commodification of Nature, a Review in Social Sciences'. *Ecological Economics* 172: 106624.

Smith, Adam. 1776. *An Inquiry into the Nature and Causes of the Wealth of Nations*, vol. 2. London: W. Strahan & T. Cadell.

Snyder, Carl. 1925. 'The Revised Index of the Volume of Trade'. *Journal of the American Statistical Association* 20 (151): 397–404.

Spash, Clive L. 1995. 'The Political Economy of Nature'. *Review of Political Economy* 7 (3): 279–293. https://doi.org/10.1080/09538259500000042.

Sullivan, Sian. 2017. 'On "Natural Capital", "Fairy Tales" and Ideology'. *Development and Change* 48 (2): 397–423. https://doi.org/10.1111/dech.12293.

Tryon, Frederick G. 1924. 'The Underlying Facts of the Coal Situation in the United States'. *Proceedings of the Academy of Political Science in the City of New York* 10 (4): 149–172.

Tryon, Frederick G. 1927. 'An Index of Consumption of Fuels and Water Power'. *Journal of the American Statistical Association* XXII (159): 271–282.

Tryon, Frederick G. 1929. *The Trend of Coal Demand*. Columbus: Ohio State University Press.

Tryon, Frederick G. 1936. 'The Chances in Mining'. In *Migration and Economic Opportunity. The Report of the Study of Population Redistribution*, edited by Carter Goodrich, Bushrod W. Allin, C. Warren Thornthwaite, Hermann K. Brunck, Frederick G. Tryon, Daniel B. Creamer, Rupert B. Vance, and Marion Hayes, 421–439. Philadelphia: University of Pennsylvania Press.

Tryon, Frederick G., and Edwin C. Eckel, eds. 1932. *Mineral Economics: Lectures Under the Auspices of the Brookings Institution*. New York and London: McGraw-Hill.

Tryon, Frederick G., and H. O. Rogers. 1930. 'Statistical Studies of Progress in Fuel Efficiency'. In *Transactions Second World Power Conference (Berlin)*, edited by F. zur Nedden and C. T. Kromer, 343–365. Berlin: Vdi-Verlag Gmbh.

Tryon, Frederick G., and Margaret H. Schoenfeld. 1933. 'Utilization of Natural Wealth: Mineral and Power Resources'. In *Recent Social Trends in the United States: Report of the President's Research Committee on Social Trends*, edited by Wesley C. Mitchell, 59–90. New York and London: McGraw-Hill.

Van Hise, Charles Richard. 1910. *The Conservation of Natural Resources in the United States*. New York: Palgrave Macmillan.

Victor, Peter A. 1991. 'Indicators of Sustainable Development: Some Lessons from Capital Theory'. *Ecological Economics* 4: 191–213.

Ward, James D., Paul C. Sutton, Adrian D. Werner, Robert Costanza, Steve H. Mohr, and Craig T. Simmons. 2016. 'Is Decoupling GDP Growth from Environmental Impact Possible?' *Plos One* 11 (10).

Warde, Paul. 2018. *The Invention of Sustainability. Nature and Destiny, c. 1500–1870*. Cambridge: Cambridge University Press.

Wolloch, Nathaniel. 2016. *Nature in the History of Economic Thought: How Natural Resources Became an Economic Concept*. London and New York: Routledge.

Wolloch, Nathaniel. 2020. 'Adam Smith and the Concept of Natural Capital'. *Ecosystem Services* 43: 101097. https://doi.org/10.1016/j.ecoser.2020.101097.

Global and regional political economy of the environment

6

GLOBAL ENVIRONMENTAL AND CLIMATE JUSTICE MOVEMENTS

David N. Pellow

Introduction: the facts of global environmental inequalities

The widespread existence of disproportionate environmental threats in socially marginalized communities around the U.S. has now been documented by scholars as a global reality. That is, the intersection of environmental risks and social inequalities – known variously as environmental injustice, environmental inequality, or environmental racism – has been observed at multiple geographic scales in communities of color, Indigenous communities, immigrant communities, low-income/working-class communities, and global South communities around the world (Agyeman and Ogneva-Himmelberger 2009; Grineski and Collins 2017; Mohai et al. 2009). Examples of the kinds of environmental inequalities observed include communities located near landfills and municipal and industrial hazardous waste dumps (Bullard 2000; Bullard and Wright 2012; Hoover 2018; Taylor 2009) and neighborhoods and schools heavily impacted by toxic agricultural pesticide drift (Harrison 2011), coal-fired power plants, prisons and jails, and major transportation routes (Bullard et al. 2004; Opsal and Malin 2019; Pellow 2018). While much of the documentation of environmental injustices centers on urban areas, in fact many rural communities are heavily affected by being in close proximity to mining and other industrial-scale extraction zones for coal, natural gas, oil, gold, and other substances, which often results in ecological disorganization associated with massive siphoning of natural materials from the earth, toxic chemicals used in these processes, and mining waste dumping (Ashwood 2018; Bell 2016). Environmental injustices in workplaces around the world are commonplace as well, with employees in hazardous industries often forced to choose between their well-being and a paycheck in sectors such as energy, waste management, and electronics (Akese and Little 2018; Ash and Boyce 2018; Malin 2015). These environmental inequalities harm both ecosystems and human health in the form of higher rates of morbidity and mortality, in general, and of increases in cancer, asthma, and respiratory disorders in particular (Brulle and Pellow 2006; Pastor et al. 2007).

In this chapter, I present some of the major findings and trends in the literature on environmental justice (EJ) studies, from local to global geographic scales. This includes a consideration of scholarship on food justice and how that issue is linked to environmental and climate change concerns. I also consider the urgency of engaging racial capitalism and settler colonialism as drivers of environmental injustice that social movements must confront if significant change is

DOI: 10.4324/9780367814533-7

to be achieved. There are a number of exciting new directions in the field of environmental justice studies that I examine here as well. Next, I detail what I believe are the key topics that multidisciplinary environmental studies scholars must grapple with over the massive challenge of global climate change, especially as it produces highly uneven and unjust consequences across social geographies. I then explore the critical environmental justice framework as a way of extending the conceptual and political depth and breadth of EJ studies. I apply the CEJ framework to the "wicked problem" of global anthropogenic climate change, with a particular focus on the global climate justice movement. The goal is to demonstrate how the field of EJ studies is grappling with what are the most significant socioecological threats of our era. In the last section of the paper, I offer a plea for reframing the challenge of climate change through a lens that amplifies the justice-related nature of this global problem: genocide.

Environmental injustice and food systems

A number of scholars have framed the lack of access to healthy, culturally appropriate, and nutritious foods – as well as democratic participation in and control over our broader food systems – as an environmental justice challenge (Gottlieb and Joshi 2010). A lack of food justice and food sovereignty haunts the same communities facing toxic threats long associated with environmental justice concerns (Mihesuah and Hoover 2019). Food justice scholarship has documented that "access to healthy, culturally appropriate, sustainably grown food is mediated by inequalities of race and class" and that, in response, the food justice movement mobilizes "to dismantle the classist and racist structural inequalities that are manifest in the consumption, production, and distribution of food" (Mares and Alkon 2011: 75). In other words, food justice movements respond to the widespread lack of decent food and economic opportunities in marginalized communities by promoting autonomy, security, and justice in food systems (Schlosberg and Collins 2014). Alkon and Agyeman (2011) rightly point out that there has long been an unfortunate disconnect between environmental justice and food justice research, and their work is credited with encouraging other scholars to bring those literatures together in meaningful ways. For example, food, climate, and energy inequalities are linked since climate disruptions have resulted in dramatic changes to agricultural productivity and (in)stability for farmers, farmworkers, and consumers around the world, particularly in global South and immigrant communities. Moreover, the ways in which food is produced, distributed, and accessed are always connected to energy systems, so when the latter are based on fossil fuels versus renewable sources, there are clear and layered implications for social and environmental justice.

Environmental injustices, racial capitalism, and settler colonialism

Recently, a number of scholars have suggested that a focus on the distributional consequences associated with environmental injustice is important but fails to sufficiently address the deeper, underlying driving forces behind this phenomenon. Accordingly, two interrelated areas of focus have received much greater attention by these scholars: racial capitalism and settler colonialism (Heynen 2016). Racial capitalism (Robinson 2000) is a concept that begins with the proposition that one cannot separate race and racism from capitalism's origins, evolution, and current dynamics – that racism has fueled capitalism from its beginning and has worked to give it strength, structure, and resilience over the centuries. Thus, racism is a structuring logic of capitalism because that economic system requires and thrives off the generation of various categories of social difference (such as race) to enable and maintain the inequalities that constitute its foundation (Pulido et al. 2016).

Closely related to racial capitalism, settler colonialism is reflected in the historical and ongoing structures of social and ecological domination associated with the invasion of Indigenous lands and territories. Indigenous studies scholar Kyle Powys Whyte defines settler colonialism as

> complex social processes in which at least one society seeks to move permanently onto the terrestrial, aquatic, and aerial places lived in by one or more other societies who already derive economic vitality, cultural flourishing, and political self-determination from the relationships they have established with the plants, animals, physical entities, and ecosystems of those places.
>
> (Whyte 2017: 157)

Or, put more simply, it is the occupation or control over land, water, aerial space, and people by an external power. And since this is a process and structure that involves control over both people and nonhuman natures, Whyte argues, "settler colonialism *is* an environmental injustice. . . [because] the U.S. settlement process aims directly at undermining the ecological conditions required for indigenous peoples to exercise their cultures, economies, and political self-determination" (Whyte 2017: 165, emphasis added). Environmental sociologist J. M. Bacon describes these processes as "colonial ecological violence" because settler colonialism "disrupts Indigenous eco-social relations . . . which results in particular risks and harms experienced by Native peoples and communities." Colonial ecological violence involves both "the ferocious and spectacular assaults on Native people through environmental damage" (Bacon 2018: 6) associated with land theft, the extermination of fauna and flora, as well as the more structurally produced forms of violence that occur as settlers seek to erase Indigenous ecologies and replace them with new ecological practices such as small family farms, commercial agriculture and forestry practices, mining, and cities. Settler colonialism is a framework that undergirds all environmental justice struggles in the U.S., whether directly involving Indigenous peoples or not, because EJ conflicts always involve land and resources entangled with histories of conquest (Hoover 2017; Voyles 2015). Or as Audra Simpson and Andrea Smith put it, "a logic of settler colonialism [i.e. elimination] structures the world for everyone, not just for native peoples" (Simpson and Smith 2014: 13).

A result of these structures is the impoverishment of both Indigenous lands (and water) and peoples, which is a particular form of environmental injustice because it reflects social hierarchies and conquests of ecosystems between nations rather than ethnic groups and represents not just racist policies and practices but violations of treaties and Indigenous sovereignty as well (Gilio-Whitaker 2019). Thus EJ scholars call for centering more radical, transformative theorizing and thinking into our analyses of the causes of and solutions to environmental injustice (Sze 2020), and racial capitalism and settler colonialism figure prominently in those formulations. As Gilio-Whitaker argues, "EJ for Indigenous peoples . . . must be capable of a political scale beyond the homogenizing, assimilationist, capitalist State. It must conform to a model that can frame issues in terms of their colonial condition and can affirm decolonization" (Gilio-Whitaker 2019: 25).

Finally, as Kari Norgaard and James Fenelon argue, the supreme irony of ignoring the histories and contemporary practices of settler colonialism is, "Indigenous peoples hold real alternatives in the form of technologies, epistemologies, social structures, moral codes and ecologies themselves that are critically needed to respond to ecological crises today" (Norgaard and Fenelon forthcoming).

New directions

While the majority of work in the field emphasizes the role of race/racism and social class inequalities in relation to environmental hazards, a growing number of scholars are exploring how gender, sexuality, and (dis)ability figure into the ways people construct and experience environmental risk and privilege (Alaimo 2010; Ray 2013; Voyles 2015). Sarah Ray extends the reach of EJ studies by arguing that, in addition to being the disproportionate targets of environmental harm, many marginalized communities are, in the dominant cultural imaginary, cast as the "ecological other" – as threats to nature and national, racial, and corporeal purity because they are different and are "unfit" to address our ecological challenges. The ecological other, then, is the antithesis of the ecological citizen – the green consumer who behaves responsibly in order to tackle our ecological crises. Therefore, if environmentalism is invested in discourses and visions of environmental purity, it is also deeply invested in ideas of cultural and bodily purity – the able body – through means and practices rooted in sexism and racism. Some scholars find that there are strong correlations between gender equality and ecologically protective practices. For example, Christina Ergas and Richard York (2012) find that carbon emissions are lower in nations where women have higher political status, thus efforts to improve gender equality and gender justice will likely be more effective if they work synergistically with campaigns to address ecological harm. Ergas and York also find that nations with greater military spending also have higher carbon emissions than other nations, supporting the longstanding work of ecofeminist scholars who have argued for an important linkage between masculinist policy making and ecological harm (Gaard 2017).

As much as all of the aforementioned forms of environmental injustice and their driving forces have been clearly established in the literature, none seems more urgent and "wicked" than anthropogenic climate change or global climate disruption. I explore this topic in greater depth next.

The facts of climate change and disruption, and climate injustice

It is now known unequivocally that significant warming of the atmosphere is occurring, coinciding with increasing levels of atmospheric carbon dioxide. And while there is a scientific consensus that anthropogenic climate change is severely affecting ecosystems, habitats, oceans, weather patterns, and human communities around the planet, it is also the case that these phenomena are unfolding in highly uneven ways.

For instance, while contributing the least of anyone to the causes of climate disruption, people of color, women, Indigenous communities, and global South nations often bear the brunt of climate disruption in terms of ecological, economic, and health burdens – thereby giving rise to the concept of *climate injustice* (Roberts and Parks 2007; Ciplet et al. 2015). These communities are among the first to experience the effects of climate disruption, which can include greater vulnerability to "natural" disasters like extreme weather events and droughts, agricultural decline, rising sea levels, increases in respiratory illness and infectious disease, heat-related morbidity and mortality, and large spikes in energy costs. The effects of climate injustice have been evident for years (see also Kalkstein 1992). Flooding from severe storms, rising sea levels, and melting glaciers affect millions in Asia and Latin America, while sub-Saharan Africa is experiencing sustained droughts. Consider that nearly 75 percent the world's annual CO_2 emissions come from the global North, where only 15 percent of the global population resides. If historic responsibility for climate change is taken into account, global North nations have consumed more than three times their share of the atmosphere (in terms of the amount of emissions that we can safely put into the atmosphere) while the poorest 10 percent of the world's population has contributed less than 1 percent of carbon emissions. Thus, the struggle for racial, gender, and economic justice is inseparable from any effort to combat climate disruption.

Climate justice is a vision aimed at dissolving and alleviating the unequal burdens created by climate change. The topic of climate justice is a major point of tension in both U.S. and international policy efforts to address climate disruption because it would require wealthy nations that have contributed the most to the problem to take on greater responsibilities for solutions. For many observers, the path is clear: for humanity's survival, for justice, and for sustainability, they maintain that we must reduce our emissions and consumption here at home in the global North. Sociologist Corrie Grosse goes further and defines climate justice as follows: a fossil-free world with a healthy climate and just society, one with an economic system other than capitalism and where decisions about development are democratic and grounded in understanding the interdependency of social justice and a healthy planet (Grosse 2017).

In recent years, there has been a significant growth in the field of climate justice studies (Bhavnani et al. 2019; Ciplet et al. 2015; Dunlap and Brulle 2015; Okereke 2010; Schlosberg 2012; Schlosberg and Collins 2014), which reports similar patterns of inequity found in earlier environmental justice scholarship. With respect to the drivers of this crisis, some scholars argue that the contemporary "wicked problem" of global climate disruption was initiated by the Industrial Revolution – that series of European and Euro-American conquests of Indigenous peoples and lands and the enslavement and forced labor of vast swaths of people across the global South, which also ushered in the Anthropocene (Heynen 2016; Whyte 2017). Accordingly, efforts to confront the roots of this crisis will have to go beyond what is presently being considered in national and global policy circles.

Millions of activists and organizations around the world have formed grassroots community groups and national and transnational social movement networks to resist and reverse the scourge of climate disruption and climate injustice. They regularly organize petitions, protest marches, sit-ins, and other forms of civil disobedience and participate in the annual United Nations conferences on climate change. The climate justice movement is a particularly large formation within the broader global environmental justice movement.

In the next section, I have two goals: (1) explore the critical environmental justice framework as a means of advancing EJ studies in a globalizing world and (2) apply that framework to offer a structured narrative of the climate justice movement's evolution and limitations.

The critical environmental justice framework

Critical EJ studies (CEJ) is a perspective put forth by a number of scholars intended to address some limitations and tensions within EJ studies (Adamson 2011; Pellow 2018; Pellow and Brulle 2005). These include four major concerns. The first concern is the need to examine *multiple* forms of inequality and their intersections (Malin and Ryder 2018). EJ scholars have a tendency to limit our analyses to race or class, while a small but growing group of scholars has explored the role of gender, sexuality, and disability in EJ studies (Alaimo 2010; Buckingham and Kulcur 2010; Gaard 2017; Ray 2013). Moreover, the category *species* remains, at best, at the margins of the field. So CEJ insists on a focus across a broad range of categories of difference and is inclusive of more-than-human beings and things. This framework also seeks to challenge the forms of inequality and institutional violence that correspond with these social categories, which would include heteropatriarchy, all forms of racism (not just white supremacy), ableism, speciesism/dominionism, transphobia, and colonialism, so this project is firmly aimed at confronting dominance in all forms.

The second concern of CEJ is that the EJ studies literature could be improved through greater attention to *multiscalar* approaches. *Scale* is important because it allows us to understand how environmental injustices are facilitated by decision makers who behave as if places where hazards are produced "out of sight and out of mind" are somehow irrelevant to the health of people and ecosystems at the original sites of decision-making power and sites of consumption (Voyles 2015). Attention to scale also assists us in observing how grassroots *responses* to environmental injustices draw on spatial frameworks, networks, and knowledge to make the connections between hazards in one place and harm in another. CEJ studies thus advocates multiscalar methodological and theoretical approaches to studying EJ issues in order to better comprehend the complex spatial and *temporal* causes, consequences, and possible resolutions of EJ struggles.

The third concern of CEJ is the degree to which state power is taken as a given or as unproblematic. EJ scholarship generally looks to the state and its legal systems to deliver justice. But as numerous studies have demonstrated, the track record of state-based regulation and enforcement of environmental and civil rights legislation in communities of color has not been promising. In its more than a quarter century history of processing environmental discrimination complaints, the USEPA's Office of Civil Rights has reviewed some 300 complaints filed by communities living in the shadows of polluting industry and has never once made a formal finding of a civil rights violation. In a groundbreaking study, Jill Lindsey Harrison (2019) found that, within the very government agencies responsible for promoting, overseeing, and applying environmental justice principles and policies in vulnerable communities, many staff members thwart their colleagues' efforts to implement EJ policies, thus reinforcing institutional and society-wide racism. Yet, a number of scholars contend that the EJ movement continues to seek justice through a system that was arguably never intended to provide justice for marginalized peoples and nonhuman natures (Benford 2005; Pulido et al. 2016). The concern here is that such an approach runs the risk of leaving intact the very power structures that produced environmental injustice in the first place. My review of the EJ studies literature concludes that, while scholars are generally critical of the abuse of power by dominant institutions with respect to the perpetration of environmental injustices, they generally do not question the existence of those institutions. In other words, the approach is reformist rather than abolitionist, and this is concerning for me and many other scholars who view certain institutions (for example, the military) as an existential threat to social and environmental justice (Aikau 2015; Heynen 2016). CEJ seeks to push our analyses and actions beyond the state, beyond capital, and beyond the human via a broad anti-authoritarian perspective.

Finally, the fourth concern of CEJ is the largely unexamined question of the *expendability* of human and nonhuman populations facing threats from states, industries, and other political economic forces. Traditional EJ studies suggests that various marginalized human populations are treated – if not viewed – as inferior and less valuable to society than others. In the book *Black and Brown Solidarity*, John Márquez introduces the concept "racial expendability" to argue that black and brown bodies are, in the eyes of the state and its constituent legal system, generally viewed as criminal, deficient, threatening, and deserving of violent discipline and even obliteration (Márquez 2014). Laura Pulido (2017) describes this as the process whereby producing social difference becomes central to creating value, wherein some groups are defined as disposable. CEJ studies makes this theme explicit by arguing that these populations are marked for erasure and early death and that that ideological and institutional othering is linked to the more-than-human world as well. CEJ counters that perspective by viewing these threatened bodies, populations, and spaces as *indispensable* to building socially and environmentally just and resilient futures for us all. In other words, we should all be seen as members of a global community who are capable of participating in efforts to make a more livable and equitable planet.

Taken together, these four concerns or pillars of the CEJ framework are intended to advance our thinking and action for environmental justice in productive and impactful directions. Next, I consider how this framework might be applied to the climate justice movement.

Applying critical environmental justice to climate justice movements

Pillar one: multiple categories of difference

The climate justice movement has done an exemplary job of extending consideration to a much broader range of social categories than is traditionally associated with EJ research and politics. In October of 2002, during the Eighth COP – the Conference of the Parties of the United Nations Framework Convention on Climate Change – in New Delhi, India, thousands of climate justice activists took to the streets to march for their cause: "The protestors affirmed that 'climate change is a human rights issue' affecting 'our livelihoods, our health, our children and our natural resources'" (Roberts and Parks 2009: 386; see also Khastagir 2002). In this single statement, the climate justice movement extended its reach beyond the traditional EJ categories of race and class, to include age or youth (e.g. children), the human (i.e. "human rights"), and the nonhuman (e.g. "natural resources"). Regarding the importance of naming the categories of nonhuman species, Patrick Bond and Michael Dorsey describe climate justice as "the new movement that best fuses a variety of progressive political-economic and political-ecological currents to combat the most serious threat humanity *and most other species* face in the 21st century" (Bond and Dorsey 2010: 286, emphasis added). As Schlosberg and Collins (2014) note, after the momentous events of Hurricane Katrina, environmental justice and climate justice activists began to reflect more deeply on "the ecological damage done to surrounding ecosystems that have led to greater vulnerabilities for *both human communities and the nonhuman environment*" (emphasis added, see also Ross and Zepeda 2011).

In 2002, an international coalition of non-governmental organizations (NGOs) drafted and released the Bali Principles of Climate Justice, which seek to redefine climate change from a human rights and environmental justice perspective. While covering an ambitious range of topics, these principles make clear that, for many people, the climate issue is a matter of life and death, and that perhaps the gravest injustice associated with this phenomenon is that those who suffer the greatest harm are the least responsible for contributing to the problem. The principles consider the causes of climate change and offer a far-reaching vision for solutions.

For example, the Bali Principles suggest that any move forward in global climate policy must be inclusive of all peoples, especially persons from those communities most affected by climate disruption. Principles #12 and #20 state the following: "Climate Justice affirms the right of all people, including the poor, women, rural and indigenous peoples, to have access to affordable and sustainable energy" and

> Climate Justice recognizes the right to self-determination of Indigenous Peoples, and their right to control their lands, including sub-surface land, territories and resources and the right to the protection against any action or conduct that may result in the destruction or degradation of their territories and cultural way of life.

The Preamble of the Bali Principles contains a statement that offers a truly expansive understanding of how broad climate injustice reaches across a diverse swath of humanity. It states,

"Whereas the impacts of climate change are disproportionately felt by small island states, women, youth, coastal peoples, local communities, indigenous peoples, fisherfolk, poor people and the elderly."

Pillar two: multiscalar analyses

The call for addressing "ecological debt" or "climate debt" is a powerful illustration of how the climate justice movement has articulated and mobilized around multiscalar analyses, both geographic and temporal. Climate debt is the dividend or reparations owed by wealthy global North nations to lower-income global South nations for centuries of "ecologically unequal exchange" wherein energy and a spectrum of ecological materials were and continue to be siphoned from South to North, contributing to poverty, economic and political instability, and environmental harm in its wake in the South and unparalleled riches in the North. Trade between global North and South is generally highly imbalanced, with the South exporting large quantities of goods to the North whose prices do not account for the social and ecological costs involved in their production (Jorgenson 2009; Rice 2007). Ecological materials involved in these unequal relationships include, for example, oil, natural gas, trees and agricultural products, minerals, marine life, and genetic matter. Ecologically unequal exchange also produced the vast majority of global carbon emissions, thus contributing massively to anthropogenic climate change. In other words, the financial wealth and privileges we find in (over)"developed" nations is often the direct result of materials that have been extracted from "developing" nations through means that many scholars and activists would define as unjust and unfair because they are so unfavorable to the South (Martinez-Alier 2003). This framing of the challenge of climate disruption has taken hold within climate justice movements and has resulted in a generative way to propose solutions because it is an analysis that centers inequality at its core (Roberts and Parks 2009). For example, Principle #7 of the Bali Principles of Climate Justice states, "Climate Justice calls for the recognition of a principle of ecological debt that industrialized governments and transnational corporations owe the rest of the world as a result of their appropriation of the planet's capacity to absorb greenhouse gases." Additionally, Principles #8 and #9 call for accountability for those actors responsible for creating ecological debt (primarily corporations from the global North) and justice for those who have suffered from climate impacts (primarily peoples of the global South).

From a CEJ perspective, ecological debt and ecologically unequal exchange are examples of multiscalar thinking because they demonstrate how spatially disparate global North and global South communities, economies, and environments are linked to each other through historical and contemporary exploitative relationships, as well as how those relationships have contributed to global climate change. Additionally, the climate justice movement itself is an inspiring embodiment of multiscalar activism and organizing, connecting activists in local communities on the frontlines and fencelines of fossil fuel extraction to national, transnational, and global social movement networks like the Environmental Justice and Climate Change Initiative, the Rising Tide Coalition for Climate Justice, Via Campesina, the Climate Justice Alliance, and the Climate Justice Now! Network. These groups share information, tactics, strategies, and documents and collaborate to keep fossil fuels in the ground in communities around the globe while also advocating for policy changes at all governmental scales.

Pillar three: justice beyond the state?

This is perhaps where the climate justice movement is in greatest need of improvement and evolution. Principle #4 of the Bali Principles of Climate Justice states, "Climate Justice affirms that *governments* are responsible for addressing climate change in a manner that is both democratically accountable to their people and in accordance with the principle of common but differentiated responsibilities" (emphasis added). Consider Swedish youth climate activist Greta Thunberg's statements during COP-25 in Madrid, Spain, in December 2019 that the climate movement seeks to urge "the people with the power" to make the changes necessary to address the climate crisis. She was applauded by her fellow activists for repeatedly invoking this logic. In other words, the millions of activists around the globe who see themselves as leaders for the cause of climate justice are apparently only powerful to the extent that they can persuade elected officials to make key policy changes. This is an extraordinary abdication of leverage and reflects the climate justice movement's (as well as the broader EJ movement's) general and tragically impoverished approach to social change (Pulido et al. 2016). This centering of governments or states as the responsible party for addressing climate change is in keeping with long traditions of social movement organizing, which are premised on a theory of change that views elected officials and the legal apparatus they direct as the most impactful levers of reform. That makes a great deal of sense in contexts where all that is needed is *reform*. Unfortunately, in the case of global anthropogenic climate change, something much more transformative is required to confront and reverse this crisis.

To its credit, there has been a well-developed anti-capitalist discourse and framework among climate justice activists for many years. For example, leading climate justice activist-scholars Patrick Bond and Michael Dorsey write that a number of leaders in this movement agree that achieving climate justice "will require society moving from a fossil-fuel-dependent capitalism to eco-socialism" (Bond and Dorsey 2010: 292). The Preamble of the Bali Principles of Climate Justice contains the following statement: "Whereas market-based mechanisms and technological 'fixes' currently being promoted by transnational corporations are false solutions and are exacerbating the problem." At COP-8 in New Delhi, activists declared, "We reject the market-based principles that guide the current negotiations to solve the climate crisis: Our World is Not for Sale!" (Khastagir 2002). In April 2010, some 20,000 activists from around the world convened in Cochabamba, Bolivia, for the World People's Conference on Climate Change and the Rights of Mother Earth. They declared that capitalism was the chief problem and held up Indigenous cultural and economic practices as a solution. The message was that the dominant model of social and economic organization is based on infinite growth and has produced climate disruption because it is rooted in "the submission and destruction of human beings and nature" (World People's Conference 2010).

However, as laudable and necessary as these anti-capitalist framings are for addressing the root causes of climate disruption, without a similarly abolitionist orientation toward the state, these efforts will be severely limited if not counterproductive. I should also note that the strong anti-capitalist orientation among many climate justice activists and movement organizations stands in total contrast to the generally pro-capitalist posture of mainstream environmental and climate policy groups, who tend to share a consensus that emissions trading and carbon markets are an effective means of tackling the scourge of climate disruption.

Some activist groups have in fact practiced a range of tactics, some of which go beyond the traditional reliance on the state. For example, groups like Environmental Rights Action have successfully used nonviolence to advance the goals of climate justice in Nigeria. More controversial is the Movement for the Emancipation of the Niger Delta (MEND), an insurgency that used armed struggle and guerrilla warfare to pressure oil companies into leaving the region, including taking oil workers hostage. By many accounts, they have been successful. Through this combination of approaches, the nation's oil production was reduced significantly and Shell Oil was forced to leave the Ogoniland area in 2008, a region that it had thoroughly polluted and whose Indigenous peoples it had repeatedly harmed through its operations (Bond and Dorsey 2010). A year later, the families of activist Ken Saro-Wiwa and his eight colleagues who were executed by the Nigerian government for their activism against Shell Oil were awarded a sum of $15.5 million through an Alien Tort Claims Act. This is not to suggest that climate justice activists take up arms; rather, I want to underscore that a broader spectrum of tactics and strategies could produce greater efficacy.

Returning to COP-8 in 2002, there was a seed of radical change planted in the streets of New Delhi when protesters declared that they would "advocate for *and practice* sustainable development" (Khastagir 2002, emphasis added). A greater emphasis on practice would likely go a long way to reducing the climate justice movement's dependence on governments for making the changes needed to tackle climate change.

Pillar four: from expendability to indispensability

The ethic of indispensability flows through the entire body of the Bali Principles of Climate Justice. For example, the very first principle reads, "Affirming the sacredness of Mother Earth, ecological unity and the interdependence of all species, Climate Justice insists that communities have the right to be free from climate change, its related impacts and other forms of ecological destruction."

But perhaps there is no better example of a commitment to indispensability in the climate justice movement than Indigenous resistance movements. There are numerous concepts and frameworks from within Indigenous, non-Western communities and cultures that offer alternatives to the dominant Western capitalist approach to economic development, and they have played significant roles in influencing the language and vision of climate justice movements. For example, the Quechua concept of *sumak kawsay* (also closely associated with *buen vivir*) roughly translates to "good living," and can be defined as "the sum of actions conducive to a certain way of life that remains mostly resistant to the long colonial occupation and the effects of its aftermath" (Acosta 2013: 2). *Sumak kawsay* emerged from non-capitalist, Indigenous communities and is a framework for breaking away from hegemonic Western values and practices associated with anthropocentrism, neoliberalism, and extractivism. In fact, *sumak kawsay* is not only a response to the dominant logic of capitalist development; it also rejects the very premise of development itself – that societies must follow a linear path of progress from pre-modern to modern, from underdeveloped to developed, and so on. Instead, *sumak kawsay* rests on the idea of a set of relationships among humans and between humans and more-than-human natures that is rooted in respect rather than domination – in other words, indispensability. We witnessed the articulation of similar ideas during the Indigenous-led #NoDAPL movement on the Standing Rock Sioux reservation in the Dakotas. For example, the phrase *mni wiconi* or "water is life" became a rallying cry for Indigenous activists and their allies who sought to prevent this fossil fuel pipeline project (the Dakota Access Pipeline) from moving forward because, in addition to promoting practices that contribute to climate disruption, it also threatened Indigenous sovereignty and the health of massive bodies of water. In that context, *mni wiconi* signaled not only an opposition to the DAPL project but also an opposition to the broader development model that it reflects. These actions and discourses powerfully reflect the idea of indispensability – a counter to the logic of expendability that undergirds practices and policies that produce environmental and climate injustices.

Thus, the critical environmental justice framework is a useful way of thinking through key dimensions of the struggle for climate and environmental justice around the world and for suggesting creative forms of action that might address this challenge.

Climate change as a wicked problem of genocidal proportions

In this final section, I want to raise what may be a controversial point, but one that I believe is important, given the urgency of our social and ecological predicament in an epoch of global climate disruption.

Environmental studies scholars and advocates for environmental sustainability have often struggled to explain why the level of public concern (and action) regarding critical ecological challenges is not as high as one might expect (or hope), given the severity of our global environmental crises. As an environmental sociologist, I have studied the importance of framing these issues and how language matters a great deal with respect to the ability to capture people's imaginations and motivate them to act. Toward that end, I propose that we consider reframing the grand challenge of anthropogenic climate change and disruption not only as an ecological crisis, but also as a social crisis that has genocidal implications. How might one substantiate such a claim? Consider the United Nations Convention on the Prevention and Punishment of the Crime of Genocide. That convention draws on multiple measures of evidence to define genocide, including actions or policies that cause serious bodily or mental harm to members of a group and killing members of a group. This is my focus here.

The scholarly research on climate change reveals quite clearly that climate disruptions affect communities, nations, and regions of the globe very differently, with Indigenous peoples and global South nations hit the hardest – hence the term "climate injustice." Consider the fact that African societies are less responsible for climate disruption due to their lower per capita energy consumption and greenhouse gas generation. In fact, the European Union, United States, Canada, Australia, and Russia are responsible for nearly 70 percent of all global carbon emissions, whereas sub-Saharan Africa is responsible for only 2 percent. And yet the impacts associated with climate change in Africa include reduced agricultural productivity as a result of rising temperatures, rising food insecurity, hunger, political conflict and war, and refugee crises. Similarly, climate change exacerbates the consequences of colonization that has affected Indigenous peoples in the United States for centuries. Specifically, erratic weather patterns and changing temperatures have disrupted Indigenous peoples' ability to maintain traditional ways of life, including subsistence fishing and hunting. The fact that Indigenous peoples in the U.S. have the highest rates of poverty in that nation and skyrocketing rates of mental and physical illness means that their vulnerability to climate-induced disasters and extreme weather events is heightened. These trends are amplified by actions taken by the oil and gas industries – among the primary institutional drivers of anthropogenic climate change – that routinely target Indigenous communities for extractive development projects in violation of Native sovereignty in the U.S. and around the world. All of these practices contribute to *serious bodily and mental harm* to people in these communities and to *loss of life* and livelihoods. And since the science of anthropogenic climate change is clear, the culpability and accountability for these effects is likewise gaining clarity.

Reframing climate change as an environmental and social crisis with genocidal implications has the potential to dramatically and positively impact the tone and urgency of the public discourse and policy making on this matter. While the vast majority of the literature suggests that substantial and meaningful progress toward environmental justice can be achieved within existing social, political, and economic structures, as this chapter demonstrates, more recent scholarship is pushing beyond those limits. Julie Sze's concept of "restorative environmental justice," for example, "is explicitly decolonial and integrative, including humans as animals and imagining humans and nonhuman nature in nonextractive modes" (Sze 2020: 79). Indigenous studies scholar Nick Estes (2019) demonstrates that the struggle for climate justice is inseparable from the struggle for Indigenous sovereignty, which will require a fundamental transformation of the economy, government, and so much more in the U.S. In other words, if we are to credibly address the genocidal and ecocidal realities of climate change and disruption, we absolutely must be creative, daring, and innovative in ways that move us outside of the traditional modes of thinking and acting that go deeper to the root causes of our climate crisis.

Conclusions

This chapter presents an overview of some of the most important ideas and developments in the environmental justice studies literature, including a consideration of the struggles for food justice and against racial capitalism and settler colonialism. Environmental justice studies has evolved by leaps and bounds in recent years, borrowing from and shaping literatures on food studies, race and ethnic studies, Indigenous studies, and climate change, among others. These are just some of the many new directions in which the field is moving, and I focus much of my attention on the theme of global anthropogenic climate change as one of the most urgent and consequential environmental justice issues of our time. Climate change is a justice issue for a myriad of reasons, but primarily because (1) it has historically been largely fueled by racial capitalism and settler colonialism; (2) it has deeply uneven and harmful impacts on marginalized communities; and (3) those marginalized communities have contributed the least to the problem of climate change because they produce extremely low levels of carbon emissions. I explore the challenge of global climate change and injustice through the framework of critical environmental justice, which facilitates a more complex and potentially transformative way of understanding, documenting, and confronting the climate crisis through (1) an expansion of the social categories of difference associated with impacted populations (both human and more than human); (2) a multiscalar analysis of the problem that captures some of its key geographic and temporal dimensions; (3) an engagement with post-statist analyses that refuse to restrict our grasp of the drivers or embrace of solutions to state-centric models; and (4) a normative call for the recognition of a multiracial, multinational, and multispecies solidarity as crucial for building ecologically sustainable and socially equitable futures. I use this framework for thinking through the promise and potential of the global climate justice movement, thus offering a way of linking theory and practice to address what many scholars and activists believe is the defining public policy issue of our time. Finally, I articulated what I believe could be a motivating frame for understanding the impacts and urgency of global climate change: genocide. While the frames of environmental racism and climate injustice have clearly and powerfully shaped the consciousness, motivations, and actions of climate scientists, scholars, and climate justice activists around the world, given the fact that there is so little time to implement the structural changes necessary to confront the climate crisis, the use of a genocide framing is not only accurate with respect to the impacts of this phenomenon on human populations but also offers a much-needed amplified degree of gravity and weight to the public policy discourse around this dilemma.

References

Acosta, Alberto. 2013. "Ecuador: Building a Good Life – Sumak Kawsay." *La Linea del Fuego*, January 24. Available at: https://upsidedownworld.org/archives/ecuador/ecuador-building-a-good-life-sumak-kawsay/.

Adamson, Joni. 2011. "Medicine Food: Critical Environmental Justice Studies, Native North American Literature, and the Movement for Food Sovereignty." *Environmental Justice* 4(4): 213–219.

Agyeman, Julian and Yelena Ogneva-Himmelberger (Eds.). 2009. *Environmental Justice and Sustainability in the Former Soviet Union*. Cambridge: The MIT Press.

Aikau, Hokulani. 2015. "Following the Alaloa Kipapa of Our Ancestors: A Trans Indigenous Futurity Without the State (United States or Otherwise)." *American Quarterly* 653–661.

Akese, Grace and Peter Little. 2018. "Electronic Waste and the Environmental Justice Challenge in Agbogbloshie." *Environmental Justice* 11(2): 77–83.

Alaimo, Stacy. 2010. *Bodily Natures: Science, Environment, and the Material Self.* Bloomington: Indiana University Press.

Alkon, Alison Hope and Julian Agyeman. 2011. *Cultivating Food Justice: Race, Class, and Sustainability.* Cambridge: The MIT Press.

Ash, Michael and James K. Boyce. 2018. "Racial Disparities in Pollution Exposure and Employment at US Industrial Facilities." *Proceedings of the National Academy of Sciences* 115(42): 10636–10641.

Ashwood, Loka. 2018. *For-Profit Democracy: Why the Government Is Losing the Trust of Rural America.* New Haven: Yale University Press.

Bacon, J. M. 2018. "Settler Colonialism as Eco-Social Structure and the Production of Colonial Ecological Violence." *Environmental Sociology* 1–11.

Bell, Shannon Elizabeth. 2016. *King Coal: The Challenges to Micromobilization in Central Appalachia.* Cambridge: The MIT Press.

Benford, Robert. 2005. "The Half-Life of the Environmental Justice Frame: Innovation, Diffusion, and Stagnation." In David N. Pellow and Robert J. Brulle (Eds.), *Power, Justice, and the Environment: A Critical Appraisal of the Environmental Justice Movement.* Cambridge: MIT Press, 37–53.

Bhavnani, Kum-Kum, John Foran, Priya A. Kurian and Debashish Munshi (Eds.). 2019. *Climate Futures: Re-Imagining Global Climate Justice.* London: Zed Books.

Bond, Patrick and Michael K. Dorsey. 2010. "Anatomies of Environmental Knowledge and Resistance: Diverse Climate Justice Movements and Waning Neo-Ecoliberalism." *Journal of Australian Political Economy* 66: 286–316.

Brulle, Robert J. and David N. Pellow. 2006. "Environmental Justice: Human Health and Environmental Inequalities." *Annual Review of Public Health* 27: 103–124, April.

Buckingham, S. and R. Kulcur. 2010. "Gendered Ggeographies of Environmental Justice." In R. Holifield, M. Porter and G. Walker (Eds.), *Spaces of Environmental Justice.* New York: Wiley-Blackwell, Chapter 3.

Bullard, Robert. 2000. *Dumping in Dixie: Race, Class, and Environmental Quality*, third edition. Boulder, CO: Westview Press.

Bullard, Robert, Glenn Johnson, and Angel Torres (Eds.). 2004. *Highway Robbery: Transportation Racism and Routes to Equity.* Boston: South End Press.

Bullard, Robert and Beverly Wright. 2012. *The Wrong Complexion for Protection: How the Government Response to Disaster Endangers African American Communities.* New York: New York University Press.

Ciplet, David, J. Timmons Roberts and Mizan R. Khan. 2015. *Power in a Warming World: The New Global Politics of Climate Change and the Remaking of Environmental Inequality.* Cambridge: The MIT Press.

Dunlap, Riley and Robert J. Brulle (Eds.). 2015. *Climate Change and Society: Sociological Perspectives.* Oxford: Oxford University Press.

Ergas, Christina and Richard York. 2012. "Women's Status and Carbon Dioxide Emissions: A Quantitative Cross-National Analysis." *Social Science Research* 41: 965–976.

Estes, Nick. 2019. *Our History Is the Future: Standing Rock Versus the Dakota Access Pipeline, and the Long Tradition of Indigenous Resistance.* New York: Verso.

Gaard, Greta. 2017. *Critical Ecofeminism.* Lanham, MD: Lexington Books.

Gilio-Whitaker, Dina. 2019. *As Long as Grass Grows: The Indigenous Fight for Environmental Justice, from Colonization to Standing Rock.* Boston: Beacon Press.

Gottlieb, Robert and Anupama Joshi. 2010. *Food Justice.* Cambridge: The MIT Press.

Grineski, Sara E. and Timothy W. Collins. 2017. "Environmental Justice Across Borders: Lessons from the U.S.-Mexico Borderlands." In Ryan Holifield, Jayajit Chakraborty and Gordon Walker (Eds.), *The Routledge Handbook of Environmental Justice.* London: Routledge, 528–542.

Grosse, Corrie. 2017. *Working Across Lines: Resisting Extreme Energy Extraction in Idaho and California.* PhD Thesis. Department of Sociology. University of California, Santa Barbara.

Harrison, Jill Lindsey. 2011. *Pesticide Drift and the Pursuit of Environmental Justice.* Cambridge: The MIT Press.

Harrison, Jill Lindsey. 2019. *From the Inside Out: The Fight for Environmental Justice Within Government Agencies.* Cambridge: The MIT Press.

Heynen, Nik. 2016. "Urban Political Ecology II: The Abolitionist Century." *Progress in Human Geography* 40(6): 839–845.

Hoover, Elizabeth. 2017. *The River Is in Us: Fighting Toxics in a Mohawk Community*. Minneapolis: University of Minnesota Press.
Hoover, Elizabeth. 2018. "Environmental Reproductive Justice: Intersections in an American Indian Community Impacted by Environmental Contamination." *Environmental Sociology* 4(1): 8–21.
Jorgenson, Andrew K. 2009. "The Sociology of Unequal Exchange in Ecological Context: A Panel Study of Lower-Income Countries, 1975–2000." *Sociological Forum* 24(1): 22–46.
Kalkstein, L. S. 1992. "Impacts of Global Warming on Human Health: Heat Stress-related Mortality." In *Global Climate Change: Implications, Challenges and Mitigating Measures*, edited by S. K. Majumdar, L. S. Kalkstein, B. Yarnal, E. W. Miller and L. M. Rosenfield. Philadelphia: Pennsylvania Academy of Science.
Khastagir, N. 2002. "The Human Face of Climate Change: Thousands Gather in India to Demand Climate Justice." CorporateWatch Website, 4 November. Available at: www.corpwatch.org/campaigns/PCD.jsp?articleid=4728.
Malin, Stephanie. 2015. *The Price of Nuclear Power: Uranium Communities and Environmental Justice*. New Brunswick, NJ: Rutgers University Press.
Malin, Stephanie and Stacia Ryder. 2018. "Developing Deeply Intersectional Environmental Justice Scholarship." *Environmental Sociology* 4(1): 1–7.
Mares, Teresa Marie and Alison Hope Alkon. 2011. "Mapping the Food Movement: Addressing Inequality and Neoliberalism." *Environment and Society: Advances in Research* 2: 68–86.
Márquez, John. 2014. *Black-Brown Solidarity: Racial Politics in the New Gulf South*. Austin, TX: University of Texas Press.
Martinez-Alier, Joan. 2003. *The Environmentalism of the Poor: A Study of Ecological Conflicts and Valuation*. Cheltenham: Edward Elgar.
Mihesuah, Devon and Elizabeth Hoover. 2019. *Indigenous Food Sovereignty in the United States: Restoring Cultural Knowledge, Protecting Environments, and Regaining Health*, vol. 18. Norman: University of Oklahoma Press.
Mohai, Paul, David N. Pellow, and J. Timmons Roberts. 2009. "Environmental Justice." *Annual Review of Environment and Resources* 34: 405–430.
Norgaard, Kari and James Fenelon. forthcoming. "Toward an Indigenous Environmental Sociology." Chapter in Beth Caniglia, Andrew Jorgenson, Stephanie Malin, Lori Peek and David Pellow (Eds.), *International Handbook of Environmental Sociology*. Springer.
Okereke, Chukwumerije. 2010. "Climate Justice and the International Regime." *WIREs Climate Change* 1: 462–474.
Opsal, Tara and Stephanie A. Malin. 2019. "Prisons as LULUs: Understanding the Parallels Between Prison Proliferation and Environmental Injustices." *Sociological Inquiry*. doi:10.1111/soin.12290.
Pastor, Manuel, James Sadd and Rachel Morello-Frosch. 2007. *Still Toxic After All These Years: Air Quality and Environmental Justice in the San Francisco Bay Area*. Santa Cruz: Center for Justice, Tolerance, and Community: University of California, February.
Pellow, David N. 2018. *What Is Critical Environmental Justice?* London: Polity Press.
Pellow, David N. and Robert J. Brulle (Eds.). (2005). *Power, Justice, and the Environment: A Critical Appraisal of the Environmental Justice Movement*. Cambridge: The MIT Press.
Pulido, Laura. 2017. "Geographies of Race and Ethnicity II: Environmental Racism, Racial Capitalism and State-Sanctioned Violence." *Progress in Human Geography* 41(4): 524–533.
Pulido, Laura, E. Kohl and N. Cotton. 2016. "State Regulation and Environmental Justice: The Need for Strategy Reassessment." *Capitalism Nature Socialism* 27(2): 12–31.
Ray, Sarah Jaquette. 2013. *The Ecological Other: Environmental Exclusion in American Culture*. Tucson: University of Arizona Press.
Rice, J. 2007. "Ecological Unequal Exchange: International Trade and Uneven Utilization of Environmental Space in the World System." *Social Forces* 85(3): 1369–1392.
Roberts, J. Timmons and Bradley Parks. 2007. *A Climate of Injustice: Global Inequality, North-South Politics, and Climate Policy*. Cambridge: The MIT Press.
Roberts, J. Timmons and Bradley Parks. 2009. "Ecologically Unequal Exchange, Ecological Debt, and Climate Justice." *International Journal of Comparative Sociology* 50(3–4): 385–409.
Robinson, Cedric. 2000. *Black Marxism: The Making of the Black Radical Tradition*. Chapel Hill, NC: University of North Carolina Press.
Ross, J. Ashleigh and Lydia Zepeda. 2011. "Wetland Restoration, Environmental Justice and Food Security in the Lower 9th Ward." *Environmental Justice* 4: 101–108.

Schlosberg, David. 2012. "Climate Justice and Capabilities: A Framework for Adaptation Policy." *Ethics & International Affairs* 26(4): 445–461.

Schlosberg, David and Lisette B. Collins. 2014. "From Environmental to Climate Justice: Climate Change and the Discourse of Environmental Justice." *WIREs Climate Change*. doi:10.1002/wcc.275.

Simpson, Audra and Andrea Smith. 2014. *Theorizing Native Studies*. Durham, NC: Duke University Press.

Sze, Julie. 2020. *Environmental Justice in a Moment of Danger*. Berkeley: University of California Press.

Taylor, Dorceta. 2009. *The Environment and the People in American Cities*. Durham, NC: Duke University Press.

Voyles, Traci. 2015. *Wastelanding: Legacies of Uranium Mining in Navajo Country*. Minneapolis: University of Minnesota Press.

Whyte, Kyle Powys. 2017. "The Dakota Access Pipeline, Environmental Injustice, and U.S. Colonialism." *Red Ink* 19(1): 154–169.

World People's Conference on Climate Change and the Rights of Mother Earth. 2010. "People's Agreement of Cochabamba." Available at: http://pwccc.wordpress.com/2010/04/24/peoples-agreement/.

7

GLOBAL INEQUALITIES AND CLIMATE CHANGE

Céline Guivarch and Nicolas Taconet

Recent decades have seen economic convergence between countries, driven in particular by the rapid development of India and China, although GDP growth rates remain low in some African countries (Firebaugh 2015; Milanovic 2016). In contrast, income inequalities within countries have increased over the same period (Alvaredo et al. 2018). For example, in the United States, the incomes of the poorest 10% have stagnated since the 1980s and those of the richest 1% have grown by an average of 2% per year (Piketty, Saez, and Zucman 2018). Considering both inter-country and intra-country inequalities, income growth since 1990 has been very unevenly distributed among the different income deciles worldwide, as shown by the so-called elephant curve (Milanovic 2016; Alvaredo et al. 2018). At both ends of the distribution, the poorest have benefited little from this growth, while the richest 1% have experienced strong income growth. In between, the increase in the incomes of a large part of the population in emerging economies contrasts with the decline of the middle class in developed countries.

At the same time, global greenhouse gas emissions have increased, and there is already an average global warming of 1.1°C compared to the pre-industrial era, with significant consequences for income inequality. Indeed, climate and inequality are closely linked for several reasons. The climatic and environmental conditions enjoyed by countries partly explain differences in their economic performance (Mellinger, Sachs, and Gallup 2000). Moreover, at both country and individual levels, it is generally the less wealthy who are most vulnerable to the impacts of climate change. The various effects of climate change (heat waves, droughts, sea level rise, etc.) disproportionately affect the less wealthy. They could slow down the expected convergence between countries and make it more difficult to reduce inequality within countries.

In addition, economic inequalities are reflected in the differences in the contribution to greenhouse gas emissions on a global scale. Developed countries, and the richest individuals, by their level of consumption, contributed disproportionately to the increase in temperature. This is a double penalty: those who are most likely to suffer the consequences of climate change contribute the least to the problem (Roberts 2001; Althor, Watson, and Fuller 2016) (IPCC Special Report 1.5, chapter 3), and conversely the most responsible countries are also the least vulnerable (Figure 7.1).

DOI: 10.4324/9780367814533-8

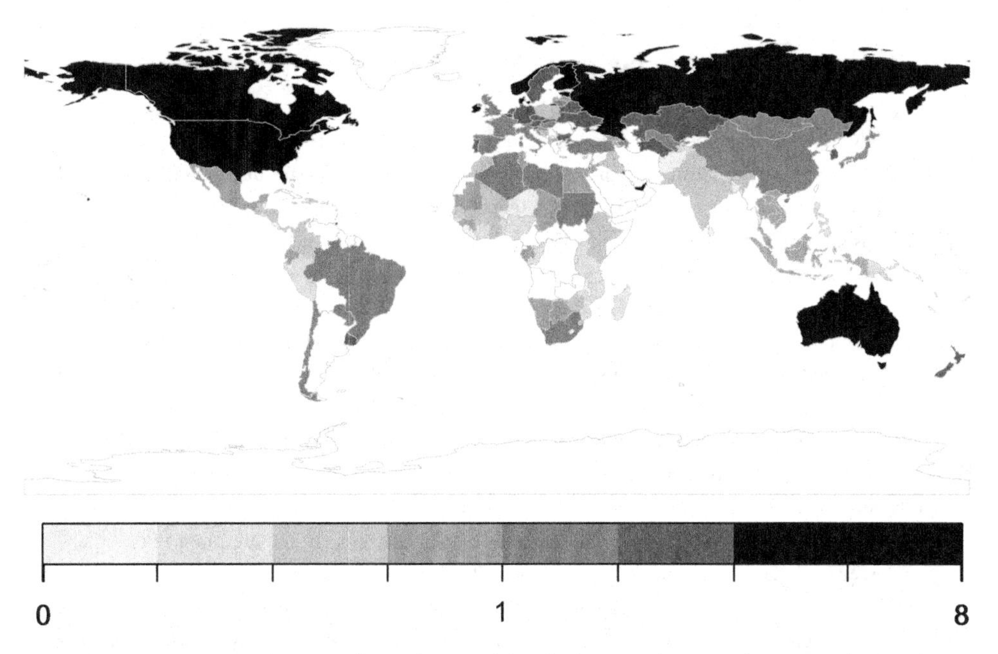

Figure 7.1 Emergence-index ratio (logarithmic scale), which quantifies contribution to climate change over future impacts. A value above 1 means that a country is relatively more responsible for climate change than impacted by it. See Figures 7.2 and 7.3 for the two underlying components of the ratio.

Source: Data from Frame et al. (2019).

Finally, the design and implementation of climate policies to reduce greenhouse gas emissions and adapt to a changing climate also pose questions in terms of inequalities between and within countries. Between countries, it raises the issue of equity in the distribution of mitigation and adaptation actions and their financing. Within countries, climate policies can affect inequalities when their costs weigh more heavily on the most modest or when certain social categories are excluded from their benefits. For example, mitigation policies may increase energy or food prices, with the risk that the poorest would face a decline in their standard of living and that poor countries would slow their development (IPCC Special Report 1.5, chapter 5). On the other hand, climate policies may also reduce inequalities, depending on the policy design. It is thus a matter of understanding under which conditions climate policy can be reconciled with the achievement of development objectives, the reduction of poverty and inequality.

In this chapter, we summarize recent literature on the links between climate change and inequality to show how issues related to climate change impacts and mitigation affect inequalities, both between countries and between individuals. First, we analyze the inequalities in exposure and vulnerability to the impacts of climate change. Then, we look at the inequalities in the contribution to greenhouse gas emissions between countries and between individuals. Finally, we show how inequalities in the face of climate change can shed light on the equity of the distribution of actions to combat climate change.

Box 7.1: Defining inequalities

The study of inequalities focuses on how certain benefits are distributed within a society (distributive justice) and on the fairness of processes by which these benefits are distributed (procedural justice). In the economic sense of the term, inequality is often understood as the extent to which income is unequally distributed among individuals in a population or between countries. It can be measured using indicators such as the Gini index, which measures the gap between the observed distribution of income and an ideal egalitarian distribution where every individual would receive the same income. It is also possible to analyze the situation of a given proportion of the poorest households and compare it with the situation of the richest. However, income provides a limited view of economic inequalities: wealth, both land and financial assets, is often more concentrated than income and is therefore an important source of inequality between individuals. Wealth inequalities have generally increased in recent decades, and the share of wealth held by the richest 1% has risen from 28% in 1980 to 33% in 2017 (Alvaredo et al. 2018).

Moreover, inequalities are not limited to purely economic aspects and are often multidimensional (see IPCC, Fifth Assessment Report, Group 2, chapter 13). Other types of social inequalities can strongly influence people's living conditions and opportunities. Moreover, inequalities are not limited to purely economic aspects and are often multidimensional (see IPCC, Fifth Assessment Report, Group 2, chapter 13). Other types of social inequalities can strongly influence people's living conditions and opportunities (Crow, Zlatunich, and Fulfrost 2009; Sen 1997), such as access to health, education, and participation in decision-making, as well as racial or gender inequalities, which can exclude social groups from access to jobs or social services. Finally, inequalities can be of an environmental nature, through differentiated access to certain natural resources, services provided by nature, or exposure to pollution externalities.

Box 7.2: Typology of inequalities linked to climate change

We can distinguish different types of inequalities related to the environment (Laurent 2011):

- *Exposure and access inequalities* deal with the unequal distribution of environmental quality for different individual and social groups, whether negatively (exposure to environmental nuisance or hazard) or positively (access to environmental amenities). In the case of climate change, individuals and countries are, and will be unequally affected by the consequences of climate change. (See first section of this chapter.)
- *Impact inequalities* reflect the differential contribution to environmental degradation, for instance in greenhouse gas emissions which are responsible for climate change. (See second section.)
- *Policy effect inequalities* occur when environmental policies are implemented. Mitigation or adaptation actions can amplify inequalities, for example because their costs may weigh more on the poorest households or because certain categories may be excluded from their benefits. (See third section.)
- *Policymaking inequalities* may exist because of unequal involvement and empowerment of individuals and groups in decisions regarding their environment.

Poor countries and poor households are the most vulnerable to the impacts of climate change

Inequalities exist outside of any consideration related to climate change. Yet, like many factors such as institutions, education, labour markets, or social structures, climate plays a role in people's living conditions, since it affects some sources of income (especially from agriculture), can lead to the destruction of homes or physical capital, and has an impact on well-being and health. Not all individuals are affected in the same way by climate change: the physical impacts will be different from one region to another. In addition, economic impacts depend on the socioeconomic vulnerability of individuals and countries. In general, poor countries and poor individuals are the most *vulnerable* to the impacts of climate change: they are more exposed, are more sensitive, and have a lesser ability to adapt (Figure 7.2). Climate change is already exacerbating inequalities and may exacerbate them further.

The physical impacts are already greater in poor countries, and they will be even more so in the future (IPCC, Special Report 1.5, chapter 3). Because of their location, poor countries are more exposed to the various effects of climate change: water stress, drought intensity, heat waves, loss of agricultural yields, or degradation of natural habits. Some authors estimate, using indicators that take into account these effects of climate change, that 90% of exposure to climate risks falls on Africa and Southeast Asia (Byers et al. 2018), and the poorest individuals within these regions are the most at risk.

For the agricultural sector, studies show that the impacts of climate change are negative overall, particularly in the low latitude regions in which developing countries are concentrated

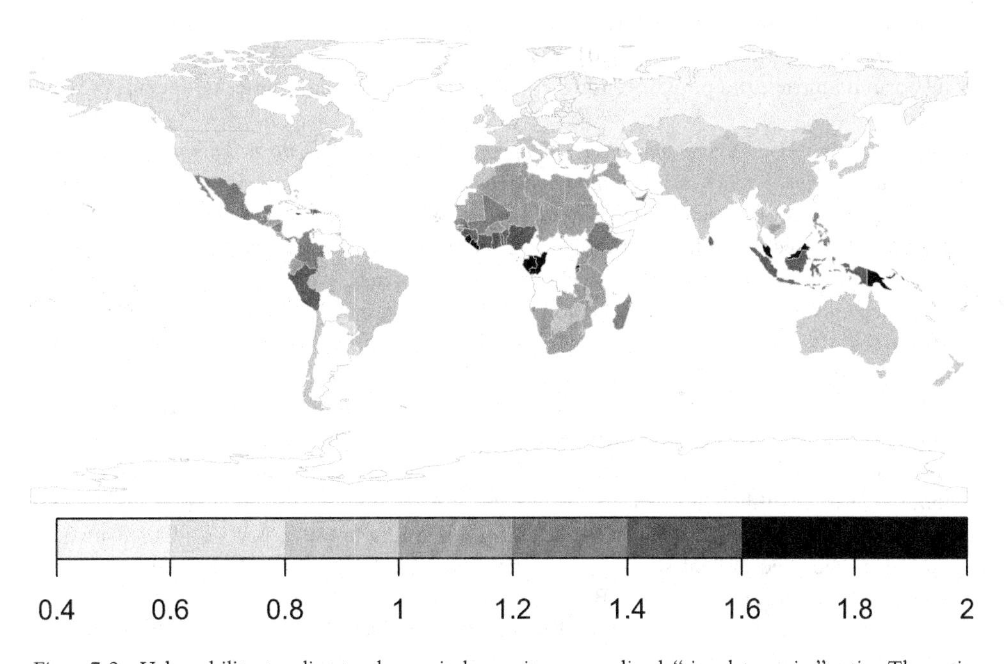

Figure 7.2 Vulnerability to climate change index, using normalized "signal-to-noise" ratio. The ratio indicates how much temperature will increase compared to observed historical variability and thus the sensitivity to climate change.

Source: Data from Frame et al. (2019).

(Rosenzweig et al. 2014). The differentiated effect between countries has already been observed: although climate change has reduced agricultural yields in most regions (Lobell, Schlenker, and Costa-Roberts 2011), some developed countries, notably in Europe, have benefited from this warming, for example Scotland (Gregory and Marshall 2012) and other Northern European countries (Supit et al. 2010).

Various indicators also illustrate this unequal distribution of physical impacts. The daily temperature extremes expected as a result of climate change are located in less developed areas (Harrington et al. 2016). While there is uncertainty at the global level about the evolution of water resources due to climate change, the regions in which water stress is expected to increase are disadvantaged areas, particularly in North Africa (Gosling and Arnell 2016). Ecosystems are also disproportionately affected in poor areas. Tropical ecosystems are usually adapted to narrow ecological conditions, whereas those in temperate zones can adapt to greater climate variations, which they experience during the year. Tropical ecosystems are therefore threatened by smaller temperature variations. For this reason, limiting the global temperature increase to 1.5°C rather than 2°C would benefit the poorest countries (King and Harrington 2018).

Within countries, poor communities or households are also located in areas with higher climate risk, for which land is often more affordable or because they offer opportunities in terms of access to employment, education, or health. People in these communities may be forced to live in flood-prone or risky delta areas ('World's 15 Countries with the Most People Exposed to River Floods' 2015; Brouwer et al. 2007). In cities, informal settlements are frequently located in areas subject to climatic hazards, for example in Dhaka (Braun and Aßheuer 2011), or on slopes likely to experience mudslides, as in South America (Painter 2007). In particular, the poorest are disproportionately located in areas at risk of urban flooding or drought, and the number of people exposed to such risk could increase by about 10% in 2030 in the absence of emission reductions (Jongman et al. 2015). The same is true for exposure to extremes of heat: as in warm countries, the poorest tend to concentrate in areas with higher temperatures (Park et al. 2018).

Moreover, the *same physical impacts do not result in the same damage, due to differences in sensitivity and adaptive capacities between countries and between individuals*. The greater *sensitivity* of poor countries to the impacts of climate change is due in part to the importance of the agricultural, forestry, and fishing sectors in the economy. A significant proportion of the population is directly dependent on activities that may be affected by climate change, particularly the poorest whose survival depends on natural capital at hand rather than on physical or human capital (Huq et al. 2010) and who benefit from many services provided by nature, which may be threatened by climate change.

The poorest are also highly vulnerable to extreme events such as natural disasters, which are likely to increase with climate change. They live in lower-quality homes and are therefore more sensitive to climatic hazards. Cumulative repair costs can represent a larger share of their income than for wealthier households, as was the case following the Mumbai floods in 2005 (Patankar 2015). Although the number of natural disasters between low- and high-income countries has been equivalent since the 1970s, the number of deaths is 10 times higher in the poorest countries (Strömberg 2007). Beyond income, institutions also play an important role in protecting people from natural disasters (Kahn 2005). The difference in vulnerability between rich and poor countries is decreasing but still remains considerable: for the period 2007–2016, the mortality rate due to natural disasters is about four times higher in poor countries (Formetta and Feyen 2019).

Box 7.3: Hurricane Harvey

The case of Hurricane Harvey, which hit Texas in 2017, shows that developed countries are also vulnerable to extreme weather events. The hurricane and its torrential rains killed about 100 people and caused damage estimated at about $100 billion. The poorest suffered most of the damage, as low-income households were concentrated in flood-prone areas (Reeves 2017). It was also more difficult for them to relocate (Boustan et al. 2017). Most did not have insurance, which can push them into poverty in an enduring way.

According to the IPCC, hurricane intensity is likely to increase with climate change. In particular, the annual probability of Texas experiencing rainfall comparable to Hurricane Harvey will increase to 18% by the end of the 21st century in the most pessimistic greenhouse gas emission scenario, compared to only 1% for the period 1980–2000 (Emanuel 2017).

Finally, the poorest households are at risk of suffering from the various health effects of climate change, via heat waves (Ahmadalipour, Moradkhani, and Kumar 2019) or the spread of diseases (malaria, dengue). Heat waves affect different social groups unevenly. During the 2003 heat wave in Europe, beyond the demographic factor (90% of deaths were people above 65 years old), mortality was higher for the lowest social categories (Borrell et al. 2006). This heat wave could be an average summer at the end of the century in high emission scenarios.

The poorest also face *indirect impacts*, such as higher food prices resulting from lower agricultural yields or extreme weather events (Hallegatte and Rozenberg 2017). They are particularly sensitive to changes in these prices, since they spend a large share of their income on food. Rising prices could threaten food security in some regions, particularly in sub-Saharan Africa or South Asia, which would increase poverty in these regions (Hertel 2015). Income can also be affected when labour productivity declines due to high temperatures, particularly for outdoor work (Deryugina and Hsiang 2014; Heal and Park 2016).

For all these impacts, the poorest have lower adaptive capacities, and climate change exacerbates pre-existing difficulties. Most of the time, they people do not benefit from insurance mechanisms or access to basic health services that can mitigate price or income shocks. In the case of damage caused by a natural disaster, such as a storm or flood, they must draw on their own assets. With fewer assets, it is more difficult for them to cope with risk. Their assets are also less diversified: for poor urban households, housing constitutes the bulk of their assets (Moser 2007) and is at risk in case of extreme events. For poor rural households, their capital lies mostly in herds, and they may be lost during a drought (Nkedianye et al. 2011). In the event of climatic hazards, the poorest are also more affected by diseases such as malaria and waterborne diseases (Hallegatte et al. 2015). An environmental shock results in long-term effects, increasing their changes of falling into poverty traps (Carter et al. 2007). Thus, climate change acts as a risk amplifier for the poorest.

These inequalities in vulnerabilities are linked to other socioeconomic dynamics, at both the social group level and the country level. Vulnerability is multidimensional and can be accentuated by different forms of discrimination against certain groups, based on gender, race, or class. In many developing countries, women are responsible for collecting water and firewood, making them vulnerable to the effect of global warming (Egeru, Kateregga, and Majaliwa 2014;

IPCC, Fifth Assessment Report, Working Group II, chapter 13). Beyond the dimension of income, race, family structure, and level of education play a role in how individuals are affected by natural disasters, as was the case during Hurricane Katrina (Elliott and Pais 2006; Logan 2006; Masozera, Bailey, and Kerchner 2007; Myers, Slack, and Singelmann 2008). This situation is reinforced by the fact that disadvantaged groups have less decision-making power and thus may benefit less from public resources.

Climate change is therefore likely to exacerbate existing inequalities. Greater impact from climate change for the poorest can already be measured at all scales. Climate change has increased inequalities between countries, and one study suggests that the ratio between the last and first deciles would be 25% lower if there had been no climate change (Diffenbaugh and Burke 2019). The impact of climate change disproportionately affects the most disadvantaged within countries between different regions and within cities. Without action to limit climate change, its impacts will continue to amplify inequalities – between and within countries – and could undermine development and poverty eradication (King and Harrington 2018; Bathiany et al. 2018; Hallegatte and Rozenberg 2017). A World Bank report estimates that an additional 100 million people could fall into poverty in 2030 because of climate change (Hallegatte et al. 2015). Managing global warming is therefore a prerequisite for sustainable improvement in living conditions.

Rich countries and individuals contribute disproportionately to climate change

While the poorest countries and individuals are the most vulnerable to the impacts of climate change, it is the richest who are responsible for the majority of greenhouse gas emissions, whose accumulation in the atmosphere causes climate change.

While some emerging countries have begun to overtake developed countries in terms of current total emissions – China is by now the largest emitter of carbon dioxide (Quéré et al. 2018) – there remains a disparity between developed and developing countries in terms of emissions per capita and total historical emissions, and thus contributions to observed global warming. Territorial greenhouse gas emissions remain today mainly linked to the level of wealth and development of countries: relative to population, emissions in the United States reach nearly 20 tCO_2-eq/person/year, those in the European Union and China are close to 8 tCO_2-eq/person/year, those in India just over 2 tCO_2-eq/person/year, and those in Senegal or Burkina Faso, for example, are between 1 and 2 tCO_2-eq/person/year (Ritchie and Roser 2017).

If emissions from the production of goods are reallocated to countries where the goods are consumed, the gap between developed and developing countries widens further (Peters et al. 2011; Karstensen, Peters, and Andrew 2013; Caro et al. 2014). Developed countries are indeed net importers of emissions "incorporated" into trade, and emerging and developing countries are exporters.

Finally, if we try to attribute to countries the historical responsibility for the additional radiative forcing or global warming observed today (Figure 7.3), the contribution of developed countries is greater than it is based solely on current emissions, because, having been the first to initiate the industrial revolution, they have caused the accumulation of greenhouse gases in the atmosphere for longer. Depending on the choice of year from which to start accounting for emissions, inclusion or exclusion of emissions from land-use change (including deforestation) and gases other than CO_2, countries relative contributions change significantly (Höhne et al. 2011; Den Elzen et al. 2013; Matthews et al. 2014; Matthews 2016). Nevertheless, it appears that the historical responsibility for the observed warming is mainly borne by industrialized

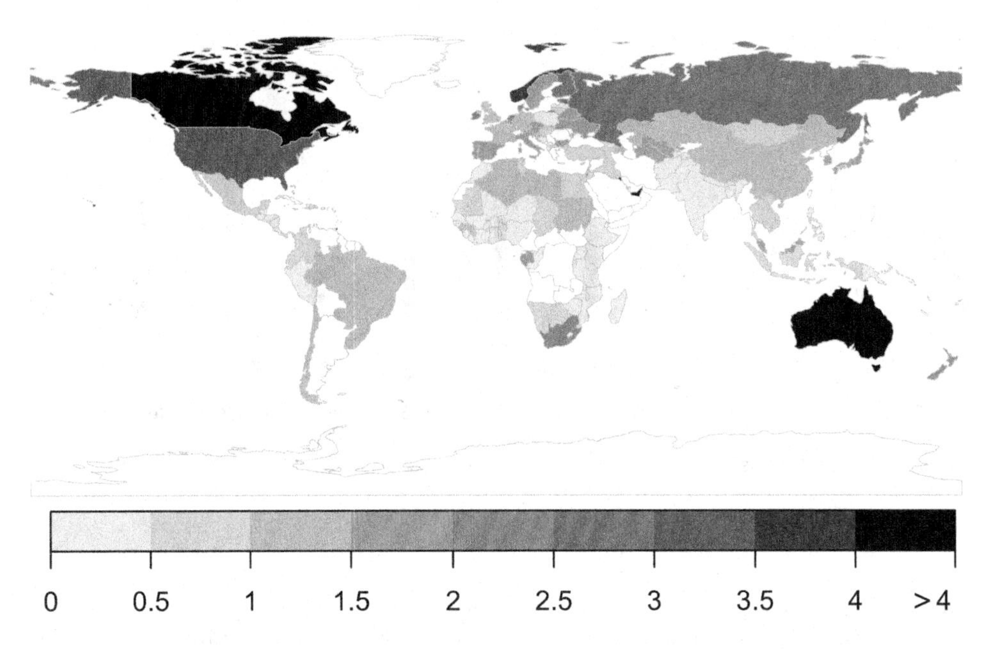

Figure 7.3 Normalized cumulated per capita emissions for the 130 countries for which population is larger than 1 million. This index quantifies countries' historical responsibility for global warming.

Source: Data from (Frame et al. (2019).[1]

countries (which account for more than 55% of cumulative emissions since 1850), but also by countries with high levels of deforestation. The share of historical responsibility attributable to emerging and developing countries is gradually increasing, particularly those of China and India, and could exceed that of developed countries by 2030 (Ward and Mahowald 2014).

Within countries, there are also large disparities in the carbon footprint of households. If an individual's level of wealth is not the only determinant of his/her emissions (the other determinants being his/her urban/rural location, age, etc.), it remains the first. This has been shown in particular for European (Ivanova et al. 2017; Sommer and Kratena 2017), American, (Jorgenson, Schor, and Huang 2017) and Chinese (Wiedenhofer et al. 2017; Chen et al. 2019) households. In France, households from the highest decile emit almost three times more than do households from the lowest decile (Figure 7.4). The analysis of the Palma "carbon" index –the ratio of emissions of the 10% of the most emitting individuals to those of the 40% least emitting – shows that this ratio is higher in developing countries than in developed countries. Globally, the Palma carbon index is higher than within any country, reflecting a very marked inequality when considering individual emissions beyond territorial boundaries. The rapid development of China and other emerging countries has reduced emissions inequalities between countries in recent decades, but this movement has been accompanied by an increase in emissions inequalities within countries. Thus, today, on a global scale, the 10% of the most emitting households are responsible for about 40% of greenhouse gas emissions, while the 40% that emit the least represent less than 8% of emissions (Piketty and Chancel 2015).

Moreover, not all emissions can be equated from an ethical point of view. Among the emissions, it is indeed necessary to distinguish those linked to basic needs from those that constitute a "luxury" (Shue 1993, 2019). For example, can we consider that a ton of CO_2 emitted to travel

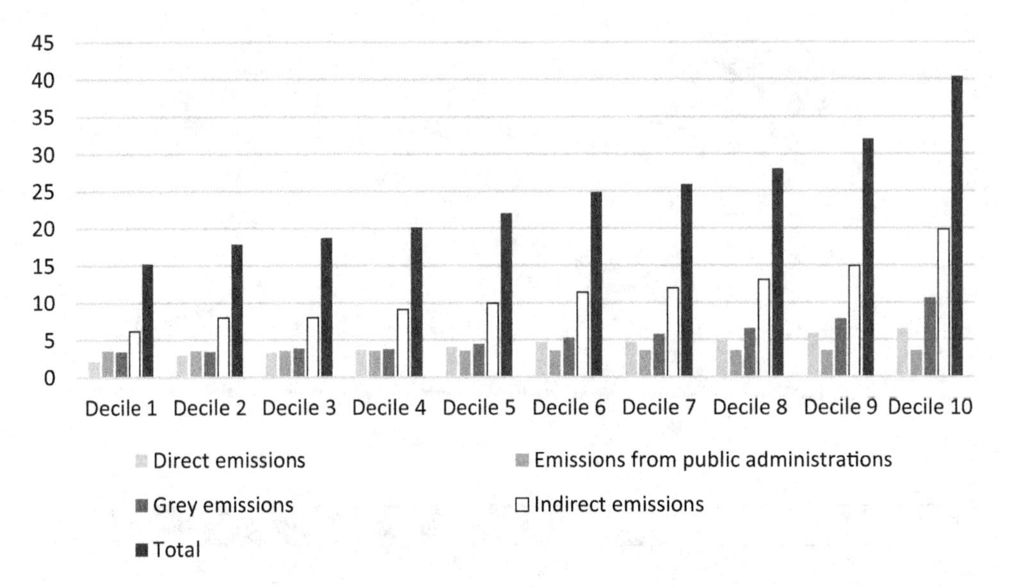

Figure 7.4 Carbon footprint by income decile. The footprint is composed of direct emissions (emitted at the moment of consumption), indirect emissions (emitted during production of the goods or services), grey emissions (occurring upstream from the value chain), and emissions from public administrations. Footprints are calculated at the household level to account for composition effect. When the analysis is done at the individual level, the increase in footprint with income is slightly reduced.

Source: Ademe et al., 2019, La fiscalité carbone aux frontières.

to a distant holiday destination and a ton of CO_2 emitted to produce a staple food are to be considered on the same level? It relates to the principle of equity, which has been present in the texts of international climate negotiations since the 1992 United Nations Framework Convention on Climate Change and is reflected in the Paris Agreement, which stresses that "action and response to climate change and the effects of climate change are intrinsically linked to equitable access to sustainable development and poverty eradication".

Based on the capacity and basic needs approach, some authors (Rao and Baer 2012; Rao and Min 2018b; O'Neill et al. 2018) have interpreted this principle of equitable access to sustainable development by defining a set of universal, irreducible, and essential material conditions for achieving basic human well-being, as well as associated indicators and quantitative thresholds. They define a "decent living standard" (Rao and Baer 2012; Rao and Min 2018b), or a "safe and just" development space (O'Neill et al. 2018), through indicators measuring the satisfaction of basic human needs (adequate nutrition, housing, access to health care, education, etc.). They then quantify the energy needs and emissions associated with these indicators. There is a consensus that the eradication of extreme poverty or universal access to energy can be achieved without representing significant greenhouse gas emissions (Tait and Winkler 2012; Pachauri 2014; Chakravarty and Tavoni 2013; Rao, Riahi, and Grubler 2014; Pachauri et al. 2013). However, studies give divergent results on the direction of the effect of a reduction in inequality on emissions, leading to an increase or decrease in emissions (Hubacek, Baiocchi, Feng, and Patwardhan 2017; Grunewald et al. 2017; Rao and Min 2018a). However, the absolute effect remains moderate: (Rao and Min 2018b) limit to 8% the maximum plausible increase in emissions that would accompany the reduction of the global Gini coefficient from its current level of 0.55 to a level of 0.3.

Finally, several studies conclude that reaching higher income levels, beyond exiting extreme poverty, and achieving more qualitative social objectives are associated with higher emissions (Hubacek, Baiocchi, Feng, Castillo et al. 2017; Scherer et al. 2018; O'Neill et al. 2018). This requires policies that can take into account both mitigation and inequality reduction objectives, including focusing on the carbon intensity of lifestyles (Scherer et al. 2018), paying attention to sufficiency and equity (O'Neill et al. 2018), and targeting people at the other end of the social scale – the super-rich (Otto et al. 2019).

Distributional effects and equity in actions to respond to climate change

Given the strong ties between climate change and inequality which have been mentioned, it is essential to articulate policies to reduce greenhouse gas emissions with their effects on current and future inequalities. Taking into account the distributional effects of mitigation, in terms of both the distribution effects of benefits due to avoided climate change impacts and the distribution of mitigation costs, can help clarify the level of ambition of climate policies and the fairness of mitigation actions and their financing between different countries. Mitigation and adaptation policies can indeed have regressive or progressive effects, either increasing or decreasing inequalities and poverty, depending on how they are designed and implemented.

The disproportionate impacts of future climate damages warrant more ambitious mitigation policies. Reducing emissions today limits future risks for the most vulnerable to experience extreme events or impacts on their health. The reduction of future inequalities can thus be seen as a "co-benefit" of mitigation. This benefit can be measured using an economic analysis tool called the social cost of carbon, which corresponds to the discounted value of the avoided damages and to the value given to mitigation actions. This value is used in particular to carry out cost-benefit analysis of public policies and public investment projects or to design a carbon tax. Determining this value raises philosophical and ethical questions about how risk is taken into account and how inequalities are valued (Fleurbaey et al. 2019), but the fact that the impacts fall more heavily on the lowest income groups gives them more weight. This can increase the value of attenuation by a factor between 2 and 10 (Dennig et al. 2015; Anthoff and Emmerling 2018) (see Figure 7.5). The magnitude of this effect may be limited when the costs of mitigation disproportionately affect the most vulnerable (Budolfson et al. 2017). However, even when costs are shared regressively between countries, mitigation can still reduce inequalities in the long term in many socioeconomic scenarios (Taconet, Méjean, and Guivarch 2020).

Defining the fair distribution of mitigation actions, and their financing between countries, is difficult, because of both the difficulty of taking into account the different levels of interactions between inequality and climate and the given different world views on what is fair (Pottier et al. 2017). In climate negotiations, countries have sought during the various COPs to define the equitable distribution of emission reductions between countries and international financing obligations, respecting both countries' historical responsibility and their different capacities. This has notably led to the adoption of the principle of "Common but Differentiated Responsibility" first in the United Nations Framework Convention on Climate Change (UNFCCC) and then in the Kyoto Protocol. But many questions arise to make this concept operational. Should we compensate countries that will be more affected by climate change (Cian et al. 2016)? How do we take into account the need for development while limiting the temperature increase to 2°C

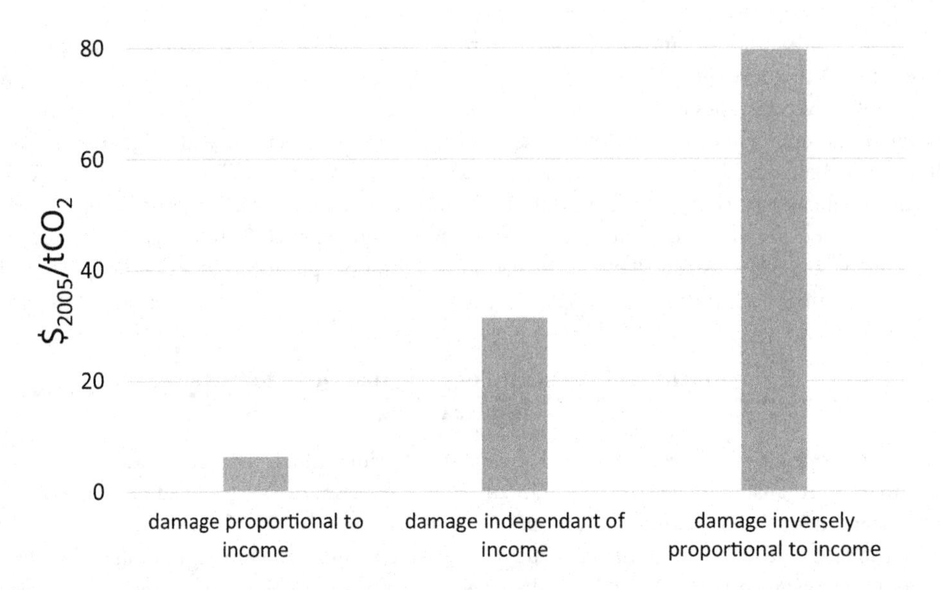

Figure 7.5 Social cost of carbon in 2005, depending on the assumption made on the relationship between climate change damage and income

Source: Results from Dennig et al. (2015).

(Winkler, Letete, and Marquard 2013)? How do we assign responsibility for emissions between production and consumption? Should priority be given to the poorest, and how can exemptions be created for emissions to meet the basic needs of the poorest (Rao 2014; Chakravarty et al. 2009)? Should inequalities due to carbon externality be treated with those outside the climate issue (Gosseries 2005)?

The recognition of the historical responsibility of developed countries led the Kyoto Protocol to impose emission reductions only on so-called Annex 1 countries and to propose North-South financing mechanisms, such as the Clean Development Mechanism and technology transfer. The Kyoto Protocol was to be a first step towards a universal emission reduction agreement, which was to enter into force after 2012. The top-down approach of emission reductions burden-sharing was abandoned after the Conference of the Parties in Copenhagen in 2009, due to the impossibility of agreeing on a fair share for all. Under the Paris Agreement, it is up to each country to define its contribution to emission reductions through *nationally determined contributions* (NDCs). If NDCs were exactly achieved, they would contribute to a reduction in per capita emissions inequalities between countries by 2030, with a reduction for the main OECD countries and an increase for emerging and developing countries (Benveniste et al. 2018) (Figure 7.6). Nevertheless, the resulting emissions in 2030 would be too high to be compatible with the Paris Agreement objective of containing the increase in global average temperature well below +2°C compared to pre-industrial levels. Compared to a more ambitious short-term emission reduction scenario, NDCs are unfavourable in terms of intergenerational equity, but also in terms of future intra-generational equity because future generations would have to bear the cost of very rapid emission reductions after 2030 and/or greater impacts of climate change – these impact hitting primarily the poorest (Liu, Fujimori, and Masui 2016). In view of the revision of the NDCs, which should lead to increased ambition, several studies (Robiou du Pont et al. 2017; Kartha et al. 2018; Berg et al. 2019) have assessed the current NDCs against the

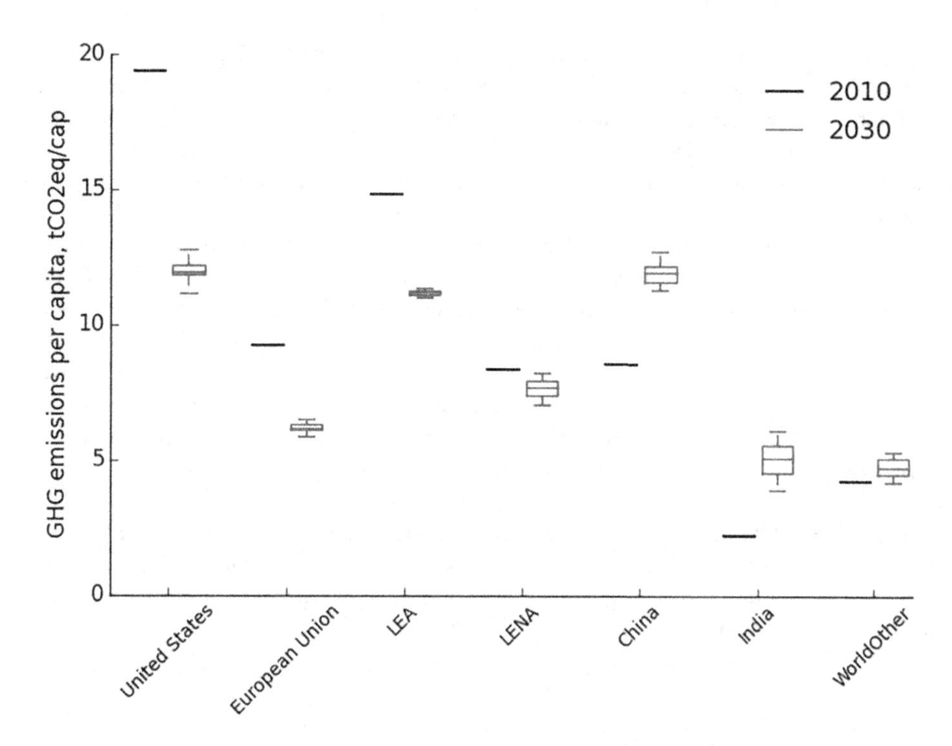

Figure 7.6 Evolution of per capita greenhouse gas emissions (in tCO_2-eq per capita), between 2010 (line) and 2030 based on exact realization of NDCs (boxplot) for different countries or groups of countries. The uncertainty range for 2030 are presented with 5th percentile, 1st quartile, median, 3rd quartile, and 95th percentile. LEA stands for *large emitters* which have NDCs with *absolute* reduction compared to a reference year (Australia, Brazil, Canada, Japan, Kazakhstan, Russia, and Ukraine). LENA is for *large emitters* which have NDCs with *no absolute* targets (Egypt, Indonesia, Iran, South Korea, Malaysia, Mexico, South Africa, Taiwan, Thailand, Turkey, and United Arab Emirates).

Source: Supplementary material from Benveniste et al. (2018).

main proposed mitigation burden-sharing criteria (convergence of emissions per capital, equality of cumulative emissions per capita, capacity to pay, etc.). The emissions that would be allocated to a given country vary widely across criteria, and some criteria lead to negative emissions budgets for developed countries (see for instance www.ccalc.ethz.ch or paris-equity-check.org).

The question of equity and fairness of the ambition for the NDCs will keep on playing a role in international negotiations, and the long-term objective set by the Paris Agreement requires each country to move towards carbon neutrality, at a pace that depends on its specific capacities. Equity is now more about financing (Holz, Kartha, and Athanasiou 2018).

Finally, actions to reduce greenhouse gas emissions and adapt to a changing climate must not overlook their own impact on inequality and on poverty. Climate policies induce costs and benefits for different individuals within a country. These policies can be regressive, that is, the cost expressed as a share of income is greater for the poorest (Bento 2013). Indeed, these policies raise the prices of emissions-intensive goods, which account for a larger share of the poorest people's spending. The shift to cleaner technologies, which are sometimes more capital intensive, also affects income. These effects depend on both the type of policy instrument and how it is implemented.

For example, emissions taxation induces important distributive effects. These effects are more significant in some sectors, such as transport, and in developed countries than in developing countries, where energy consumption by low-income households is low (Dorband et al. 2019; Ohlendorf et al. 2018). The impact of a tax also depends on the effects on labour and capital income (Goulder et al. 2019), on how consumers react to price changes, and to income changes over the course of their lifetimes (Ohlendorf et al. 2018). In France, a carbon tax on the transport and housing sectors was introduced in 2014 and its level is to increase each year, with a risk for car-dependent households or living in poorly insulated housing. The effect of a tax at 30 euros per CO_2t (its 2017 level) thus increases the number of people in fuel poverty by about 6% (Berry 2019). However, the introduction of a tax is accompanied by additional tax revenues, the use of which determines its fairness (see Figure 7.7). The increase in fuel poverty induced by the carbon tax can be offset by redistributing part of the revenues to households: it is sufficient to use 15% of the revenue to cancel out the effect on fuel poverty. Although the lowest 10% of households can on average benefit from redistribution, there is still a large proportion of households whose situation is deteriorating due to great heterogeneity within deciles (Douenne 2020). Likewise, the effect of emissions permits depend on the allocation rules – with free allocation favouring owners of polluting companies (Dinan and Rogers 2002; Parry 2004). Finally, tax reforms to remove fossil fuel subsidies can be beneficial if they are replaced by direct transfers (Durand-Lasserve et al. 2015; Vogt-Schilb et al. 2019).

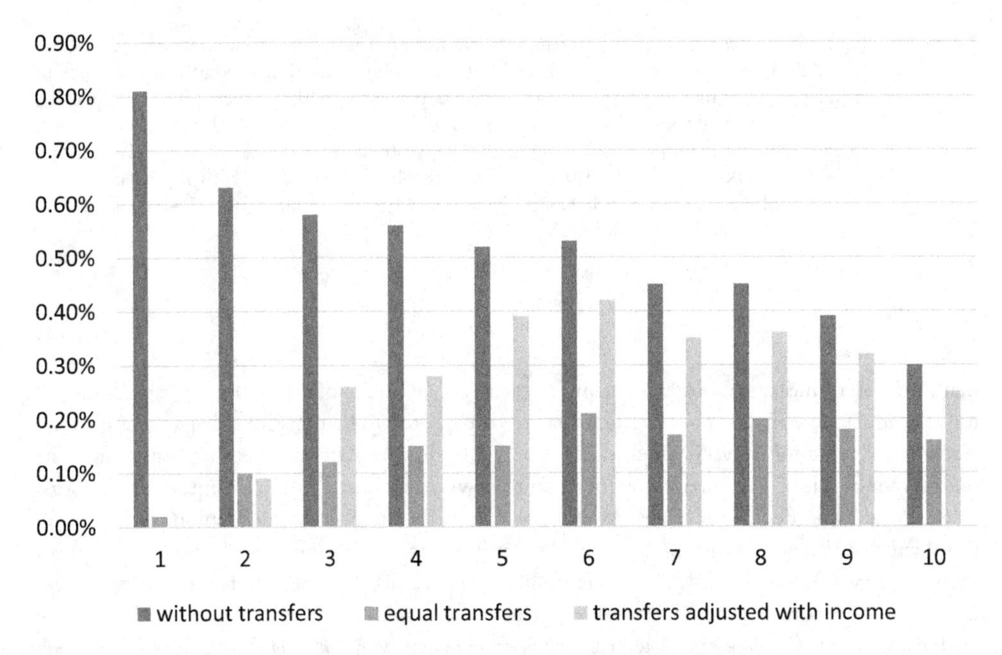

Figure 7.7 The case without transfers is the average share of income from French household, by decile, spent on carbon tax. The other cases correspond to the share after redistribution of part of the revenues to compensate the regressivity generated by the tax (regressivity is based on Suits index): either on an equal per household basis (in such a case, 59% of revenues need to be redistributed), or with transfers inversely proportional to income (in such a case, only 33% of revenues need to be distributed).

Source: Berry (2019).

Other public policies aimed at reducing emissions can have a negative effect on the poorest. Energy efficiency standards for vehicles, while saving emissions, also increase the cost of purchasing vehicles (Levinson 2016). To reach the same emission reduction, standards can be more regressive than taxes (Fullerton 2017). Similarly, energy efficiency standards in the construction sector in California have had a negative effect on the poorest and have resulted in a reduction in the surface area of their homes (Bruegge, Deryugina, and Myers 2019). The distributional effect of subsidies for renewable energies vary according to their design, in particular on the way prices are set in the electricity market, and the ability of producers to pass on costs to consumers (Reguant 2018). Finally, tax credits for the installation of solar panels or the purchase of electric vehicles can benefit the richest. In the United States, 60% of the different "green" tax credits between 2006 and 2013 went to the richest 20% (Borenstein and Davis 2016).

Some mitigation policies affect the poorest through effects on food prices. For instance, the development of biofuel can have a detrimental effect on food security (Hasegawa et al. 2018; Fujimori et al. 2019). The use of land for biofuel production raises food prices and can have negative impacts, particularly in low-income regions such as sub-Saharan Africa and South Asia. It could also lead to deforestation and dispossess communities of their land.

Conversely, some mitigation policies have co-benefits for the most vulnerable. Indeed, the combustion of fossil fuels releases local pollutants such as fine particles or nitrogen oxides that cause cardiorespiratory diseases (Smith et al. 2013). The socially disadvantaged communities are the most exposed to these global health risks (Hajat, Hsia, and O'Neill 2015). They could therefore benefit from the reduction in internal combustion engine vehicles or restrictions on coal use. Similarly, the use of more efficient furnaces reduces greenhouse gas emissions while improving air quality and thus the health of users (Rao et al. 2013).

Adaptation policies face analogous challenges and can have important effects on low-income households. Some adaptation actions can reduce the vulnerability of the poorest to climate hazards, such as conversion to more resilient crops. The development of financial services for the most vulnerable, from which they are often excluded, improves their ability to cope with unforeseen events, particularly climatic ones. Indexation of cash transfers to food prices could also help households during food prices spikes (Hallegatte et al. 2015). However, adaptation spending sometimes focuses more on protecting physical capital than on people at risk (Jorgenson et al. 2016). As such, some public decision-making tools such as cost-benefit analyses, which only take into account future benefits and not how they are distributed, may favour projections with the highest monetary benefits to the detriment of those that provide better protection for the most vulnerable. Taking into account welfare effects, not just absolute monetary benefits, would better ensure that projects that protect the poorest are financed.

Conclusion

Climate change acts as an inequality amplifier by disproportionally affecting the most disadvantaged at all scales, who are both more exposed and more vulnerable to the impacts of climate change. Taking into account these inequalities in impact gives more value to actions to mitigate greenhouse gas emissions and should lead to more ambitious mitigation policies.

To the extent that emission levels differ between countries and individuals and that the costs of reducing emissions and the benefits of avoided impacts are unequally distributed among individuals and between countries, equity issues within each generation are essential to define fair low-carbon pathways that respect the needs of present generations and the interests of future generations (Klinsky et al. 2017; Klinsky and Winkler 2018).

Emission reduction policies can also have impacts for the poorest. At the international level, the aim is to reduce emissions without impeding access to development, particularly in the least developed countries, and to support poverty eradication. Within a country, reducing emissions raises the question of a just transition. Depending on the type of public policies that are put in place, the most modest can be disproportionately affected or on the contrary can benefit from the policies.

Studies on the subject show that climate and equality do not necessarily oppose one another and that there are ways to articulate climate policies and social justice. This requires first recognizing potential conflicts between social justice and climate policies, and second setting up supporting and compensation mechanisms.

Note

1 Countries for which data is missing are left blank. Note that we removed values for the Central African Republic due to an apparent error in the data.

Bibliography

Ademe, Paul Malliet, Ruben Haalebos, and Emeric Nicolas. 2019. 'La fiscalité carbone aux frontières: ses impacts redistributifs sur le revenu des ménages français'. https://ideas.repec.org/p/spo/wpmain/infohdl2441-4m3adkbdj69ti8jp2r95u0a2ko.html.

Ahmadalipour, Ali, Hamid Moradkhani, and Mukesh Kumar. 2019. 'Mortality Risk from Heat Stress Expected to Hit Poorest Nations the Hardest'. *Climatic Change* 152 (3): 569–579. https://doi.org/10.1007/s10584-018-2348-2.

Althor, Glenn, James E. M. Watson, and Richard A. Fuller. 2016. 'Global Mismatch Between Greenhouse Gas Emissions and the Burden of Climate Change'. *Scientific Reports* 6 (February): 20281. https://doi.org/10.1038/srep20281.

Alvaredo, Facundo, Lucas Chancel, Thomas Piketty, Emmanuel Saez, and Gabriel Zucman. 2018. 'World Inequality Report 2018'. http://wir2018.wid.world/.

Anthoff, David, and Johannes Emmerling. 2018. 'Inequality and the Social Cost of Carbon'. *Journal of the Association of Environmental and Resource Economists* 6 (2): 243–273. https://doi.org/10.1086/701900.

Bathiany, Sebastian, Vasilis Dakos, Marten Scheffer, and Timothy M. Lenton. 2018. 'Climate Models Predict Increasing Temperature Variability in Poor Countries'. *Science Advances* 4 (5): eaar5809. https://doi.org/10.1126/sciadv.aar5809.

Bento, Antonio M. 2013. 'Equity Impacts of Environmental Policy'. *Annual Review of Resource Economics* 5 (1): 181–196. https://doi.org/10.1146/annurev-resource-091912-151925.

Benveniste, Hélène, Olivier Boucher, Céline Guivarch, Hervé Le Treut, and Patrick Criqui. 2018. 'Impacts of Nationally Determined Contributions on 2030 Global Greenhouse Gas Emissions: Uncertainty Analysis and Distribution of Emissions'. *Environmental Research Letters* 13 (1): 014022. https://doi.org/10.1088/1748-9326/aaa0b9.

Berg, Nicole J. van den, Heleen L. van Soest, Andries F. Hof, Michel G. J. den Elzen, Detlef P. van Vuuren, Wenying Chen, Laurent Drouet, et al. 2019. 'Implications of Various Effort-Sharing Approaches for National Carbon Budgets and Emission Pathways'. *Climatic Change* (February). https://doi.org/10.1007/s10584-019-02368-y.

Berry, Audrey. 2019. 'The Distributional Effects of a Carbon Tax and Its Impact on Fuel Poverty: A Microsimulation Study in the French Context'. *Energy Policy* 124 (January): 81–94. https://doi.org/10.1016/j.enpol.2018.09.021.

Borenstein, Severin, and Lucas W. Davis. 2016. 'The Distributional Effects of US Clean Energy Tax Credits'. *Tax Policy and the Economy* 30 (1): 191–234. https://doi.org/10.1086/685597.

Borrell, Carme, Marc Marí-Dell'Olmo, Maica Rodríguez-Sanz, Patrícia Garcia-Olalla, Joan A. Caylà, Joan Benach, and Carles Muntaner. 2006. 'Socioeconomic Position and Excess Mortality During the Heat Wave of 2003 in Barcelona'. *European Journal of Epidemiology* 21 (9): 633–640. https://doi.org/10.1007/s10654-006-9047-4.

Boustan, Leah Platt, Matthew E. Kahn, Paul W. Rhode, and Maria Lucia Yanguas. 2017. 'The Effect of Natural Disasters on Economic Activity in Us Counties: A Century of Data'. National Bureau of Economic Research. https://www.nber.org/papers/w23410.

Braun, Boris, and Tibor Aßheuer. 2011. 'Floods in Megacity Environments: Vulnerability and Coping Strategies of Slum Dwellers in Dhaka/Bangladesh'. *Natural Hazards* 58 (2): 771–787.

Brouwer, Roy, Sonia Akter, Luke Brander, and Enamul Haque. 2007. 'Socioeconomic Vulnerability and Adaptation to Environmental Risk: A Case Study of Climate Change and Flooding in Bangladesh'. *Risk Analysis* 27 (2): 313–326. https://doi.org/10.1111/j.1539-6924.2007.00884.x.

Bruegge, Chris, Tatyana Deryugina, and Erica Myers. 2019. 'The Distributional Effects of Building Energy Codes'. *Journal of the Association of Environmental and Resource Economists* 6 (S1): S95–S127.

Budolfson, Mark, Francis Dennig, Marc Fleurbaey, Asher Siebert, and Robert H. Socolow. 2017. 'The Comparative Importance for Optimal Climate Policy of Discounting, Inequalities and Catastrophes'. *Climatic Change* 145 (3–4): 481–494. https://doi.org/10.1007/s10584-017-2094-x.

Byers, Edward, Matthew Gidden, David Leclère, Juraj Balkovic, Peter Burek, Kristie Ebi, Peter Greve, et al. 2018. 'Global Exposure and Vulnerability to Multi-Sector Development and Climate Change Hotspots'. *Environmental Research Letters* 13 (5): 055012. https://doi.org/10.1088/1748-9326/aabf45.

Caro, Dario, Anna LoPresti, Steven J. Davis, Simone Bastianoni, and Ken Caldeira. 2014. 'CH4 and N2O Emissions Embodied in International Trade of Meat'. *Environmental Research Letters* 9 (11): 114005.

Carter, Michael R., Peter D. Little, Tewodaj Mogues, and Workneh Negatu. 2007. 'Poverty Traps and Natural Disasters in Ethiopia and Honduras'. *World Development* 35 (5): 835–856. https://doi.org/10.1016/j.worlddev.2006.09.010.

Chakravarty, Shoibal, Ananth Chikkatur, Heleen de Coninck, Stephen Pacala, Robert Socolow, and Massimo Tavoni. 2009. 'Sharing Global CO2 Emission Reductions Among One Billion High Emitters'. *Proceedings of the National Academy of Sciences* 106 (29): 11884–11888. https://doi.org/10.1073/pnas.0905232106.

Chakravarty, Shoibal, and Massimo Tavoni. 2013. 'Energy Poverty Alleviation and Climate Change Mitigation: Is There a Trade Off?' *Energy Economics* 40: S67–S73.

Chen, Jiandong, Chong Xu, Lianbiao Cui, Shuo Huang, and Malin Song. 2019. 'Driving Factors of CO2 Emissions and Inequality Characteristics in China: A Combined Decomposition Approach'. *Energy Economics* 78: 589–597.

Cian, E. De, A. F. Hof, G. Marangoni, M. Tavoni, and D. P. van Vuuren. 2016. 'Alleviating Inequality in Climate Policy Costs: An Integrated Perspective on Mitigation, Damage and Adaptation'. *Environmental Research Letters* 11 (7): 074015. https://doi.org/10.1088/1748-9326/11/7/074015.

Crow, Ben, Nichole Zlatunich, and Brian Fulfrost. 2009. 'Mapping Global Inequalities: Beyond Income Inequality to Multi-Dimensional Inequalities'. *Journal of International Development* 21 (8): 1051–1065. https://doi.org/10.1002/jid.1646.

Den Elzen, Michel G. J., Jos G. J. Olivier, Niklas Höhne, and Greet Janssens-Maenhout. 2013. 'Countries' Contributions to Climate Change: Effect of Accounting for All Greenhouse Gases, Recent Trends, Basic Needs and Technological Progress'. *Climatic Change* 121 (2): 397–412.

Dennig, Francis, Mark B. Budolfson, Marc Fleurbaey, Asher Siebert, and Robert H. Socolow. 2015. 'Inequality, Climate Impacts on the Future Poor, and Carbon Prices'. *Proceedings of the National Academy of Sciences* 112 (52): 15827–15832. https://doi.org/10.1073/pnas.1513967112.

Deryugina, Tatyana, and Solomon M. Hsiang. 2014. 'Does the Environment Still Matter? Daily Temperature and Income in the United States'. National Bureau of Economic Research. https://www.nber.org/papers/w20750.

Diffenbaugh, Noah S., and Marshall Burke. 2019. 'Global Warming Has Increased Global Economic Inequality'. *Proceedings of the National Academy of Sciences* 116 (20): 9808–9813.

Dinan, Terry M., and Diane Lim Rogers. 2002. 'Distributional Effects of Carbon Allowance Trading: How Government Decisions Determine Winners and Losers'. *National Tax Journal* 55 (2): 199–221.

Dorband, Ira Irina, Michael Jakob, Matthias Kalkuhl, and Jan Christoph Steckel. 2019. 'Poverty and Distributional Effects of Carbon Pricing in Low- and Middle-Income Countries – A Global Comparative Analysis'. *World Development* 115 (March): 246–257. https://doi.org/10.1016/j.worlddev.2018.11.015.

Douenne, Thomas. 2020. 'The Vertical and Horizontal Distributive Effects of Energy Taxes: A Case Study of a French Policy'. *The Energy Journal* 41 (3). https://doi.org/10.5547/01956574.41.3.tdou.

Durand-Lasserve, Olivier, Lorenza Campagnolo, Jean Chateau, and Rob Dellink. 2015. *Modelling of Distributional Impacts of Energy Subsidy Reforms: An Illustration with Indonesia.* SSRN Scholarly Paper ID 2633472. Rochester, NY: Social Science Research Network. https://papers.ssrn.com/abstract=2633472.

Egeru, Anthony, Eseza Kateregga, and Gilber Jackson Mwanjalolo Majaliwa. 2014. 'Coping with Firewood Scarcity in Soroti District of Eastern Uganda'. *Open Journal of Forestry* 4 (1): 70.

Elliott, James R., and Jeremy Pais. 2006. 'Race, Class, and Hurricane Katrina: Social Differences in Human Responses to Disaster'. *Social Science Research* 35 (2): 295–321.

Emanuel, Kerry. 2017. 'Assessing the Present and Future Probability of Hurricane Harvey's Rainfall'. *Proceedings of the National Academy of Sciences* 114 (48): 12681–12684. https://doi.org/10.1073/pnas.1716222114.

Firebaugh, Glenn. 2015. 'Global Income Inequality'. In *Emerging Trends in the Social and Behavioral Sciences*. Hoboken, NJ: John Wiley & Sons, Inc. https://doi.org/10.1002/9781118900772.etrds0149.

Fleurbaey, Marc, Maddalena Ferranna, Mark Budolfson, Francis Dennig, Kian Mintz-Woo, Robert Socolow, Dean Spears, and Stéphane Zuber. 2019. 'The Social Cost of Carbon: Valuing Inequality, Risk, and Population for Climate Policy'. *The Monist* 102 (1): 84–109. https://doi.org/10.1093/monist/ony023.

Formetta, Giuseppe, and Luc Feyen. 2019. 'Empirical Evidence of Declining Global Vulnerability to Climate-Related Hazards'. *Global Environmental Change* 57 (July): 101920. https://doi.org/10.1016/j.gloenvcha.2019.05.004.

Frame, David J., Luke J. Harrington, Jan S. Fuglestvedt, Richard J. Millar, Manoj M. Joshi, and Simon Caney. 2019. 'Emissions and Emergence: A New Index Comparing Relative Contributions to Climate Change with Relative Climatic Consequences'. *Environmental Research Letters* 14 (8): 084009. https://doi.org/10.1088/1748-9326/ab27fc.

Fujimori, Shinichiro, Tomoko Hasegawa, Volker Krey, Keywan Riahi, Christoph Bertram, Benjamin Leon Bodirsky, Valentina Bosetti, Jessica Callen, Jacques Després, and Jonathan Doelman. 2019. 'A Multi-Model Assessment of Food Security Implications of Climate Change Mitigation'. *Nature Sustainability* 2 (5): 386.

Fullerton, Don. 2017. *Distributional Effects of Environmental and Energy Policy*. London: Routledge.

Gosling, Simon N., and Nigel W. Arnell. 2016. 'A Global Assessment of the Impact of Climate Change on Water Scarcity'. *Climatic Change* 134 (3): 371–385. https://doi.org/10.1007/s10584-013-0853-x.

Gosseries, Axel. 2005. 'Cosmopolitan Luck Egalitarianism and the Greenhouse Effect'. *Canadian Journal of Philosophy* 35 (sup1): 279–309. https://doi.org/10.1080/00455091.2005.10716857.

Goulder, Lawrence H., Marc A. C. Hafstead, GyuRim Kim, and Xianling Long. 2019. 'Impacts of a Carbon Tax Across US Household Income Groups: What Are the Equity-Efficiency Trade-Offs?' *Journal of Public Economics* 175 (July): 44–64. https://doi.org/10.1016/j.jpubeco.2019.04.002.

Gregory, Peter J., and Bruce Marshall. 2012. 'Attribution of Climate Change: A Methodology to Estimate the Potential Contribution to Increases in Potato Yield in Scotland Since 1960'. *Global Change Biology* 18 (4): 1372–1388. https://doi.org/10.1111/j.1365-2486.2011.02601.x.

Grunewald, Nicole, Stephan Klasen, Inmaculada Martínez-Zarzoso, and Chris Muris. 2017. 'The Trade-Off Between Income Inequality and Carbon Dioxide Emissions'. *Ecological Economics* 142: 249–256.

Hajat, Anjum, Charlene Hsia, and Marie S. O'Neill. 2015. 'Socioeconomic Disparities and Air Pollution Exposure: A Global Review'. *Current Environmental Health Reports* 2 (4): 440–450. https://doi.org/10.1007/s40572-015-0069-5.

Hallegatte, Stephane, Mook Bangalore, Marianne Fay, Tamaro Kane, and Laura Bonzanigo. 2015. *Shock Waves: Managing the Impacts of Climate Change on Poverty*. Washington, DC: World Bank Publications.

Hallegatte, Stephane, and Julie Rozenberg. 2017. 'Climate Change Through a Poverty Lens'. *Nature Climate Change* 7 (4): 250–256.

Harrington, Luke J., David J. Frame, Erich M. Fischer, Ed Hawkins, Manoj Joshi, and Chris D. Jones. 2016. 'Poorest Countries Experience Earlier Anthropogenic Emergence of Daily Temperature Extremes'. *Environmental Research Letters* 11 (5): 055007. https://doi.org/10.1088/1748-9326/11/5/055007.

Hasegawa, Tomoko, Shinichiro Fujimori, Petr Havlík, Hugo Valin, Benjamin Leon Bodirsky, Jonathan C. Doelman, Thomas Fellmann, et al. 2018. 'Risk of Increased Food Insecurity Under Stringent Global Climate Change Mitigation Policy'. *Nature Climate Change* 8 (8): 699–703. https://doi.org/10.1038/s41558-018-0230-x.

Heal, Geoffrey, and Jisung Park. 2016. 'Reflections – Temperature Stress and the Direct Impact of Climate Change: A Review of an Emerging Literature'. *Review of Environmental Economics and Policy* 10 (2): 347–362. https://doi.org/10.1093/reep/rew007.

Hertel, Thomas W. 2015. 'Food Security under Climate Change'. *Nature Climate Change* 6 (1): 10.

Höhne, Niklas, Helcio Blum, Jan Fuglestvedt, Ragnhild Bieltvedt Skeie, Atsushi Kurosawa, Guoquan Hu, Jason Lowe, Laila Gohar, Ben Matthews, and Ana Claudia Nioac De Salles. 2011. 'Contributions of

Individual Countries' Emissions to Climate Change and Their Uncertainty'. *Climatic Change* 106 (3): 359–391.

Holz, Christian, Sivan Kartha, and Tom Athanasiou. 2018. 'Fairly Sharing 1.5: National Fair Shares of a 1.5 C-Compliant Global Mitigation Effort'. *International Environmental Agreements: Politics, Law and Economics* 18 (1): 117–134.

Hubacek, Klaus, Giovanni Baiocchi, Kuishuang Feng, and Anand Patwardhan. 2017. 'Poverty Eradication in a Carbon Constrained World'. *Nature Communications* 8 (1): 912.

Hubacek, Klaus, Giovanni Baiocchi, Kuishuang Feng, Raúl Muñoz Castillo, Laixiang Sun, and Jinjun Xue. 2017. 'Global Carbon Inequality'. *Energy, Ecology and Environment* 2 (6): 361–369.

Huq, Mainul, Malik Fida Khan, Kiran Pandey, Manjur Murshed Zahid Ahmed, Zahirul Huq Khan, Susmita Dasgupta, and Nandan Mukherjee. 2010. *Vulnerability of Bangladesh to Cyclones in a Changing Climate: Potential Damages and Adaptation Cost*. Washington, DC: World Bank Publications.

Ivanova, Diana, Gibran Vita, Kjartan Steen-Olsen, Konstantin Stadler, Patricia C. Melo, Richard Wood, and Edgar G. Hertwich. 2017. 'Mapping the Carbon Footprint of EU Regions'. *Environmental Research Letters* 12 (5): 054013.

Jaggard, K. W., A. Qi, and M. A. Semenov. 2007. 'The Impact of Climate Change on Sugarbeet Yield in the UK: 1976–2004'. *The Journal of Agricultural Science* 145 (4): 367–375. https://doi.org/10.1017/S0021859607006922.

Jongman, Brenden, Hessel C. Winsemius, Jeroen C. J. H. Aerts, Erin Coughlan de Perez, Maarten K. van Aalst, Wolfgang Kron, and Philip J. Ward. 2015. 'Declining Vulnerability to River Floods and the Global Benefits of Adaptation'. *Proceedings of the National Academy of Sciences* 112 (18): E2271–E2280. https://doi.org/10.1073/pnas.1414439112.

Jorgenson, Andrew, Juliet Schor, and Xiaorui Huang. 2017. 'Income Inequality and Carbon Emissions in the United States: A State-Level Analysis, 1997–2012'. *Ecological Economics* 134 (April): 40–48. https://doi.org/10.1016/j.ecolecon.2016.12.016.

Kahn, Matthew E. 2005. 'The Death Toll from Natural Disasters: The Role of Income, Geography, and Institutions'. *Review of Economics and Statistics* 87 (2): 271–284.

Karstensen, Jonas, Glen P. Peters, and Robbie M. Andrew. 2013. 'Attribution of CO2 Emissions from Brazilian Deforestation to Consumers between 1990 and 2010'. *Environmental Research Letters* 8 (2): 024005.

Kartha, Sivan, Tom Athanasiou, Simon Caney, Elizabeth Cripps, Kate Dooley, Navroz K. Dubash, Teng Fei, et al. 2018. 'Cascading Biases Against Poorer Countries'. *Nature Climate Change* 8 (5): 348–349. https://doi.org/10.1038/s41558-018-0152-7.

King, Andrew D., and Luke J. Harrington. 2018. 'The Inequality of Climate Change from 1.5 to 2°C of Global Warming'. *Geophysical Research Letters* 45 (10): 5030–5033. https://doi.org/10.1029/2018GL078430.

Klinsky, Sonja, Timmons Roberts, Saleemul Huq, Chukwumerije Okereke, Peter Newell, Peter Dauvergne, Karen O'Brien, et al. 2017. 'Why Equity Is Fundamental in Climate Change Policy Research'. *Global Environmental Change* 44 (May): 170–173. https://doi.org/10.1016/j.gloenvcha.2016.08.002.

Klinsky, Sonja, and Harald Winkler. 2018. 'Building Equity in: Strategies for Integrating Equity into Modelling for a 1.5 C World'. *Philosophical Transactions of the Royal Society A: Mathematical, Physical and Engineering Sciences* 376 (2119): 20160461.

Laurent, Eloi. 2011. 'Issues in Environmental Justice Within the European Union'. *Ecological Economics* 70 (11): 1846–1853.

Levinson, Arik. 2016. 'Are Energy Efficiency Standards Less Regressive Than Energy Taxes?' Working Paper. https://www.nber.org/papers/w22956.

Liu, Jing-Yu, Shinichiro Fujimori, and Toshihiko Masui. 2016. 'Temporal and Spatial Distribution of Global Mitigation Cost: INDCs and Equity'. *Environmental Research Letters* 11 (11): 114004.

Lobell, David B., Wolfram Schlenker, and Justin Costa-Roberts. 2011. 'Climate Trends and Global Crop Production Since 1980'. *Science* 333 (6042): 616–620. https://doi.org/10.1126/science.1204531.

Logan, John R. 2006. *The Impact of Katrina: Race and Class in Storm-Damaged Neighborhoods*. Frankfurt am Main: Campus Verl.

Masozera, Michel, Melissa Bailey, and Charles Kerchner. 2007. 'Distribution of Impacts of Natural Disasters Across Income Groups: A Case Study of New Orleans'. *Ecological Economics* 63 (2–3): 299–306.

Matthews, H. Damon. 2016. 'Quantifying Historical Carbon and Climate Debts Among Nations'. *Nature Climate Change* 6 (1): 60–64. https://doi.org/10.1038/nclimate2774.

Matthews, H. Damon, Tanya L. Graham, Serge Keverian, Cassandra Lamontagne, Donny Seto, and Trevor J. Smith. 2014. 'National Contributions to Observed Global Warming'. *Environmental Research Letters* 9 (1): 014010.

Mellinger, Andrew D., Jeffrey D. Sachs, and John Luke Gallup. 2000. 'Climate, Coastal Proximity, and Development'. *The Oxford Handbook of Economic Geography* 169: 194.

Milanovic, Branko. 2016. *Global Inequality: A New Approach for the Age of Globalization*. Cambridge: Harvard University Press.

Moser, Caroline. 2007. 'Asset Accumulation Policy and Poverty Reduction'. *Reducing Global Poverty: The Case for Asset Accumulation* 83–103.

Myers, Candice A., Tim Slack, and Joachim Singelmann. 2008. 'Social Vulnerability and Migration in the Wake of Disaster: The Case of Hurricanes Katrina and Rita'. *Population and Environment* 29 (6): 271–291.

Nkedianye, David, Jan de Leeuw, Joseph O. Ogutu, Mohammed Y. Said, Terra L. Saidimu, Shem C. Kifugo, Dickson S. Kaelo, and Robin S. Reid. 2011. 'Mobility and Livestock Mortality in Communally Used Pastoral Areas: The Impact of the 2005–2006 Drought on Livestock Mortality in Maasailand'. *Pastoralism: Research, Policy and Practice* 1 (1): 17.

Noack, Frederik, Sven Wunder, Arild Angelsen, and Jan Börner. 2015. *Responses to Weather and Climate: A Cross-Section Analysis of Rural Incomes*. Policy Research Working Papers. Washington, DC: The World Bank. https://doi.org/10.1596/1813-9450-7478.

Ohlendorf, Nils, Michael Jakob, Jan Christoph Minx, Carsten Schröder, and Jan Christoph Steckel. 2018. *Distributional Impacts of Climate Mitigation Policies – A Meta-Analysis*. SSRN Scholarly Paper ID 3299337. Rochester, NY: Social Science Research Network. https://papers.ssrn.com/abstract=3299337.

O'Neill, Daniel W., Andrew L. Fanning, William F. Lamb, and Julia K. Steinberger. 2018. 'A Good Life for All Within Planetary Boundaries'. *Nature Sustainability* 1 (2): 88.

Otto, Ilona M., Kyoung Mi Kim, Nika Dubrovsky, and Wolfgang Lucht. 2019. 'Shift the Focus from the Super-Poor to the Super-Rich'. *Nature Climate Change* 9 (2): 82.

Pachauri, Shonali. 2014. 'Household Electricity Access a Trivial Contributor to CO 2 Emissions Growth in India'. *Nature Climate Change* 4 (12): 1073.

Pachauri, Shonali, Bas J. van Ruijven, Yu Nagai, Keywan Riahi, Detlef P. van Vuuren, Abeeku Brew-Hammond, and Nebojsa Nakicenovic. 2013. 'Pathways to Achieve Universal Household Access to Modern Energy by 2030'. *Environmental Research Letters* 8 (2): 024015.

Painter, James. 2007. 'Deglaciation in the Andean Region'. *Human Development Report 2008*. hdr.undp.org/en/content/deglaciation-andean-region.

Park, Jisung, Mook Bangalore, Stephane Hallegatte, and Evan Sandhoefner. 2018. 'Households and Heat Stress: Estimating the Distributional Consequences of Climate Change'. *Environment and Development Economics* 23 (3): 349–368. https://doi.org/10.1017/S1355770X1800013X.

Parry, Ian W. H. 2004. 'Are Emissions Permits Regressive?' *Journal of Environmental Economics and Management* 47 (2): 364–387. https://doi.org/10.1016/j.jeem.2003.07.001.

Patankar, Archana. 2015. *The Exposure, Vulnerability, and Ability to Respond of Poor Households to Recurrent Floods in Mumbai*. Washington, DC: The World Bank.

Peters, Glen P., Jan C. Minx, Christopher L. Weber, and Ottmar Edenhofer. 2011. 'Growth in Emission Transfers via International Trade from 1990 to 2008'. *Proceedings of the National Academy of Sciences* 108 (21): 8903–8908.

Piketty, Thomas, and L. Chancel. 2015. *Carbon and Inequality: From Kyoto to Paris. Trends in the Global Inegality of Carbon Emissions (1998–2013) & Prospects for an Equitable Adaptation Fund*. Paris: School of Economics.

Piketty, Thomas, Emmanuel Saez, and Gabriel Zucman. 2018. 'Distributional National Accounts: Methods and Estimates for the United States'. *The Quarterly Journal of Economics* 133 (2): 553–609. https://doi.org/10.1093/qje/qjx043.

Pottier, Antonin, Aurélie Méjean, Olivier Godard, and Jean-Charles Hourcade. 2017. 'A Survey of Global Climate Justice: From Negotiation Stances to Moral Stakes and Back'. *International Review of Environmental and Resource Economics* 11 (1): 1–53. https://doi.org/10.1561/101.00000090.

Quéré, Corinne, Robbie Andrew, Pierre Friedlingstein, Stephen Sitch, Judith Hauck, Julia Pongratz, Penelope Pickers, Jan Ivar Korsbakken, Glen Peters, and Josep Canadell. 2018. 'Global Carbon Budget 2018'. *Earth System Science Data* 10 (4): 2141–2194.

Rao, Narasimha D. 2014. 'International and Intranational Equity in Sharing Climate Change Mitigation Burdens'. *International Environmental Agreements: Politics, Law and Economics* 14 (2): 129–146. https://doi.org/10.1007/s10784-013-9212-7.

Rao, Narasimha D., and Paul Baer. 2012. '"Decent Living" Emissions: A Conceptual Framework'. *Sustainability* 4 (4): 656–681.
Rao, Narasimha D., and Jihoon Min. 2018a. 'Less Global Inequality Can Improve Climate Outcomes'. *Wiley Interdisciplinary Reviews: Climate Change* 9 (2): e513.
Rao, Narasimha D., and Jihoon Min. 2018b. 'Decent Living Standards: Material Prerequisites for Human Wellbeing'. *Social Indicators Research* 138 (1): 225–244. https://doi.org/10.1007/s11205-017-1650-0.
Rao, Narasimha D., Keywan Riahi, and Arnulf Grubler. 2014. 'Climate Impacts of Poverty Eradication'. *Nature Climate Change* 4 (9): 749.
Rao, Shilpa, Shonali Pachauri, Frank Dentener, Patrick Kinney, Zbigniew Klimont, Keywan Riahi, and Wolfgang Schoepp. 2013. 'Better Air for Better Health: Forging Synergies in Policies for Energy Access, Climate Change and Air Pollution'. *Global Environmental Change* 23 (5): 1122–1130.
Reeves, Eleanor Krause, and V. Richard. 2017. 'Hurricanes Hit the Poor the Hardest'. *Brookings* (blog), 18 September. www.brookings.edu/blog/social-mobility-memos/2017/09/18/hurricanes-hit-the-poor-the-hardest/.
Reguant, Mar. 2018. 'The Efficiency and Sectoral Distributional Impacts of Large-Scale Renewable Energy Policies'. *Journal of the Association of Environmental and Resource Economists* 6 (S1): S129–S168. https://doi.org/10.1086/701190.
Ritchie, Hannah, and Max Roser. 2017. 'CO2 and Greenhouse Gas Emissions'. *Our World in Data*, May. https://ourworldindata.org/co2-and-other-greenhouse-gas-emissions.
Roberts, J. Timmons. 2001. 'Global Inequality and Climate Change'. *Society & Natural Resources* 14 (6): 501–509. https://doi.org/10.1080/08941920118490.
Robiou du Pont, Yann, M. Louise Jeffery, Johannes Gütschow, Joeri Rogelj, Peter Christoff, and Malte Meinshausen. 2017. 'Equitable Mitigation to Achieve the Paris Agreement Goals'. *Nature Climate Change* 7 (1): 38–43. https://doi.org/10.1038/nclimate3186.
Rosenzweig, Cynthia, Joshua Elliott, Delphine Deryng, Alex C. Ruane, Christoph Müller, Almut Arneth, Kenneth J. Boote, et al. 2014. 'Assessing Agricultural Risks of Climate Change in the 21st Century in a Global Gridded Crop Model Intercomparison'. *Proceedings of the National Academy of Sciences* 111 (9): 3268–3273. https://doi.org/10.1073/pnas.1222463110.
Scherer, Laura, Paul Behrens, Arjan de Koning, Reinout Heijungs, Benjamin Sprecher, and Arnold Tukker. 2018. 'Trade-Offs Between Social and Environmental Sustainable Development Goals'. *Environmental Science & Policy* 90: 65–72.
Sen, Amartya K. 1997. 'From Income Inequality to Economic Inequality'. *Southern Economic Journal* 64 (2): 384–401. https://doi.org/10.2307/1060857.
Shue, Henry. 1993. 'Subsistence Emissions and Luxury Emissions'. *Law & Policy* 15 (1): 39–60.
Shue, Henry. 2019. 'Subsistence Protection and Mitigation Ambition: Necessities, Economic and Climatic'. *The British Journal of Politics and International Relations* 21 (2): 251–262.
Smith, Kirk R., Howard Frumkin, Kalpana Balakrishnan, Colin D. Butler, Zoë A. Chafe, Ian Fairlie, Patrick Kinney, et al. 2013. 'Energy and Human Health'. *Annual Review of Public Health* 34 (1): 159–188. https://doi.org/10.1146/annurev-publhealth-031912-114404.
Sommer, Mark, and Kurt Kratena. 2017. 'The Carbon Footprint of European Households and Income Distribution'. *Ecological Economics* 136: 62–72.
Strömberg, David. 2007. 'Natural Disasters, Economic Development, and Humanitarian Aid'. *Journal of Economic Perspectives* 21 (3): 199–222.
Supit, I., C. A. van Diepen, A. J. W. de Wit, P. Kabat, B. Baruth, and F. Ludwig. 2010. 'Recent Changes in the Climatic Yield Potential of Various Crops in Europe'. *Agricultural Systems* 103 (9): 683–694. https://doi.org/10.1016/j.agsy.2010.08.009.
Taconet, Nicolas, Aurélie Méjean, and Céline Guivarch. 2020. 'Influence of Climate Change Impacts and Mitigation Costs on Inequality between Countries'. *Climatic Change*, February. https://doi.org/10.1007/s10584-019-02637-w.
Tait, Louise, and Harald Winkler. 2012. 'Estimating Greenhouse Gas Emissions Associated with Achieving Universal Access to Electricity for All Households in South Africa'. *Journal of Energy in Southern Africa* 23 (4): 8–17.
Vogt-Schilb, Adrien, Brian Walsh, Kuishuang Feng, Laura Di Capua, Yu Liu, Daniela Zuluaga, Marcos Robles, and Klaus Hubaceck. 2019. 'Cash Transfers for Pro-Poor Carbon Taxes in Latin America and the Caribbean'. *Nature Sustainability* 2 (10): 941–948. https://doi.org/10.1038/s41893-019-0385-0.
Ward, D. S., and N. M. Mahowald. 2014. 'Contributions of Developed and Developing Countries to Global Climate Forcing and Surface Temperature Change'. *Environmental Research Letters* 9 (7): 074008. https://doi.org/10.1088/1748-9326/9/7/074008.

Wiedenhofer, Dominik, Dabo Guan, Zhu Liu, Jing Meng, Ning Zhang, and Yi-Ming Wei. 2017. 'Unequal Household Carbon Footprints in China'. *Nature Climate Change* 7 (1): 75–80. https://doi.org/10.1038/nclimate3165.

Winkler, Harald, Thapelo Letete, and Andrew Marquard. 2013. 'Equitable Access to Sustainable Development: Operationalizing Key Criteria'. *Climate Policy* 13 (4): 411–432. https://doi.org/10.1080/14693062.2013.777610.

'World's 15 Countries with the Most People Exposed to River Floods'. 2015. World Resources Institute, 5 March. www.wri.org/blog/2015/03/world-s-15-countries-most-people-exposed-river-floods.

Chapters from IPCC reports

IPCC, Fifth Assessment Report, Working Group II, Chapter 13.

Olsson, L., M. Opondo, P. Tschakert, A. Agrawal, S.H. Eriksen, S. Ma, L.N. Perch, and S.A. Zakieldeen, 2014: Livelihoods and poverty. In: *Climate Change 2014: Impacts, Adaptation, and Vulnerability. Part A: Global and Sectoral Aspects. Contribution of Working Group II to the Fifth Assessment Report of the Intergovernmental Panel on Climate Change* [Field, C.B., V.R. Barros, D.J. Dokken, K.J. Mach, M.D. Mastrandrea, T.E. Bilir, M. Chatterjee, K.L. Ebi, Y.O. Estrada, R.C. Genova, B. Girma, E.S. Kissel, A.N. Levy, S. MacCracken, P.R. Mastrandrea, and L.L. White (eds.)]. Cambridge University Press, Cambridge, United Kingdom and New York, NY, USA, pp. 793–832

IPCC, Special Report 1.5, Chapter 3.

Hoegh-Guldberg, O., D. Jacob, M. Taylor, M. Bindi, S. Brown, I. Camilloni, A. Diedhiou, R. Djalante, K.L. Ebi, F. Engelbrecht, J. Guiot, Y. Hijioka, S. Mehrotra, A. Payne, S.I. Seneviratne, A. Thomas, R. Warren, and G. Zhou, 2018: Impacts of 1.5°C Global Warming on Natural and Human Systems. In: *Global Warming of 1.5°C. An IPCC Special Report on the impacts of global warming of 1.5°C above pre-industrial levels and related global greenhouse gas emission pathways, in the context of strengthening the global response to the threat of climate change, sustainable development, and efforts to eradicate poverty* [Masson-Delmotte, V., P. Zhai, H.-O. Pörtner, D. Roberts, J. Skea, P.R. Shukla, A. Pirani, W. Moufouma-Okia, C. Péan, R. Pidcock, S. Connors, J.B.R. Matthews, Y. Chen, X. Zhou, M.I. Gomis, E. Lonnoy, T. Maycock, M. Tignor, and T. Waterfield (eds.)]. In Press.

IPCC, Special Report 1.5, Chapter 5.

Roy, J., P. Tschakert, H. Waisman, S. Abdul Halim, P. Antwi-Agyei, P. Dasgupta, B. Hayward, M. Kanninen, D. Liverman, C. Okereke, P.F. Pinho, K. Riahi, and A.G. Suarez Rodriguez, 2018: Sustainable Development, Poverty Eradication and Reducing Inequalities. In: *Global Warming of 1.5°C. An IPCC Special Report on the impacts of global warming of 1.5°C above pre-industrial levels and related global greenhouse gas emission pathways, in the context of strengthening the global response to the threat of climate change, sustainable development, and efforts to eradicate poverty* [Masson-Delmotte, V., P. Zhai, H.-O. Pörtner, D. Roberts, J. Skea, P.R. Shukla, A. Pirani, W. Moufouma-Okia, C. Péan, R. Pidcock, S. Connors, J.B.R. Matthews, Y. Chen, X. Zhou, M.I. Gomis, E. Lonnoy, T. Maycock, M. Tignor, and T. Waterfield (eds.)]. In Press.

8

NATURAL DISASTERS, POVERTY AND INEQUALITY

New metrics for fairer policies

Stéphane Hallegatte and Brian Walsh

Introduction

Worldwide, natural disasters pose a growing threat to economic and political stability. According to Munich Re, economic losses to natural disasters averaged US $187 billion per year from 2009 to 2018, a 30 percent increase over the inflation-adjusted 30-year average (Munich Re 2019). This increase is driven largely by economic growth and urbanization, poverty and inequality, and climate change, each of which presents unprecedented challenges in the decades ahead.

In a world of massive inequalities within and across countries, the increase in aggregate economic losses cannot inform us on the real impact of these disasters. In the conventional practice of disaster risk management, the severity of disasters is measured by their direct damages, or the replacement cost of assets damaged or destroyed by a shock. Other dimensions – such as the impact on health, education, or quality of life – are not usually incorporated into disaster loss estimates or in cost-benefit analysis of possible risk reduction interventions.

One implication of the use of economic or asset losses as a measure of disaster impacts is that disaster risk management (DRM) strategies tend to favor the wealthy, central business districts, and other clusters of valuable assets. Interventions targeting poor people, who have few assets to start with, cannot generate large gains in terms of avoided asset losses and are therefore discouraged by this metric. And while this prioritization makes sense from a purely monetary perspective, it disincentivizes attractive investments in the poorest areas, even when small interventions could significantly reduce the stunting of children (Dercon and Porter 2014), disease transmission (Yonson, Noy, and Gaillard 2018; Erman et al. 2018), absenteeism from work and school, lost wages, and many other types of disaster impacts on well-being (Hallegatte and Rozenberg 2017).

At a macroeconomic level, asset losses also obscure the relationship between vulnerability and development. Economic growth increases the value of assets and thus tends to increase disaster losses, when they are measured in asset losses (Kahn 2005; Schumacher and Strobl 2011; Hallegatte et al. 2016). But development and higher incomes also make people more resilient: the long-term impacts of disasters on communities' well-being and prospects depend not only on direct impacts (asset losses), but also on the accessibility of financial tools (e.g., social transfers, formal and informal post-disaster support, savings, insurance, and access to credit). Households that lack access to these tools will struggle to cope with shocks and could

DOI: 10.4324/9780367814533-9

fall into chronic poverty as a result (Carter and Barrett 2006). In short, complete reliance on asset losses obscures the role of poverty reduction as a tool to reduce disaster impacts (Hallegatte et al. 2016) and impedes the development of DRM strategies that can be integrated into larger development agendas.

Disasters have complex and diverse consequences that can be measured (and, increasingly, anticipated) in terms of recovery times, economic (income and consumption) losses, poverty incidence, or welfare and well-being losses, among other metrics. Each of these metrics provides a different perspective on disaster costs. In contrast to direct damages, many of these impacts of natural disasters accrue disproportionately to poor households. This is because income shocks can force the poor to make difficult decisions between food, housing, education and healthcare, and reconstruction. As a result of these trade-offs, poor households take longer to recover from disasters and are more likely to face long-term consequences.

The next section will trace the evidence for the impacts of disasters on poverty, accounting for both human and economic costs, which are well documented in case studies. In the third section, we will flip the perspective and consider the ways in which poverty exacerbates the effects of natural disasters. The fourth section introduces the concepts of well-being losses and socioeconomic resilience and quantified metrics to measure them. It then uses these concepts to explain how these metrics can lead to more effective and efficient DRM strategies. Finally, the fifth section summarizes the policy implications of the socioeconomic resilience framework.

Traditional economic assessments do not capture the full impact of disasters on poor people

Income and economic consumption are distributed very unequally in the world. In 2017, and using purchasing power parity (PPP) exchange rates, the GDP of sub-Saharan Africa was around $6 trillion, that is, 4 percent of the world total of $141 trillion. In other terms, the economy of sub-Saharan Africa is of the same size as the five richest cities in the world (Tokyo, New York City, Los Angeles, Seoul, and London). It means that even tragic disasters in sub-Saharan Africa are unlikely to have economic losses that compare with recent events in high-income countries (such as Hurricanes Katrina, Sandy, and Harvey in the US). It does not mean, however, that disasters in sub-Saharan Africa are less important or less impactful on people's well-being.

The same issue is valid within countries. In Guatemala, the income of people in the bottom 20 percent of the population represents only 4 percent of the national income. It means that these people are five times poorer than the average. Even a massive loss of income or assets for this group cannot have a large impact on national GDP, which again does not mean that this loss is not important for the well-being and long-term prospects of a significant fraction of the population.

These considerations suggest the need to have a closer look at how disasters affect poverty and poor people, to make sure those impacts – which are unlikely to be well measured by GDP or income impacts – are given due consideration in risk assessments and in the design of risk management policies. This section examines this question.

Disasters have visible impacts on local poverty

Poverty increases in the direct aftermath of a disaster are widely documented. This section provides a short review of case studies that document this effect, for various hazard categories, regions, and timescales.

In Bolivia, the incidence of poverty climbed by 12 percent in Trinidad City after the 2006 floods, a fivefold increase compared with the national average (Perez-De-Rada and Paz 2008). Examining the ex post impacts of Hurricane Mitch, which struck Nicaragua in 1998, Jakobsen (2012) found that poorer households faced a larger *absolute* decline in productive assets immediately after Mitch. Furthermore, among those households affected by Mitch, the share of asset-poor households (those who own less than a given asset-poverty line) increased from 75 percent in 1998 to 80 percent in 2001.

Among households hit by Tropical Storm Agatha in 2010 in Guatemala, consumption per capita fell by 5.5 percent, increasing poverty by 14 percent (Baez et al. 2016). Whereas previous studies typically focused on the impacts of Agatha in rural areas, Baez et al. (2016) document the sharp impacts of Agatha in urban areas of Guatemala, where poverty increased by 18 percent, mainly because of higher food prices. Meanwhile, Ishizawa and Miranda (2019) find that an increase of one standard deviation in the intensity of a hurricane in Central America increases moderate and extreme poverty levels by 1.5 percentage points. Finally, a recent meta-analysis of 38 such studies found that incomes are consistently reduced by natural disasters (Karim and Noy 2014).

Beyond the immediate impact after a disaster, evidence suggests that natural disasters increase poverty over the medium and long term. Glave, Fort, and Rosemberg (2008) studied exposure to disasters and poverty from 2003 to 2008 at the provincial level in Peru. They found that one extra disaster per year increased poverty rates by 16–23 percent. At the municipal level in Mexico, Rodriguez-Oreggia and his colleagues (2013) found that floods and droughts increased poverty levels between 1.5 and 3.7 percent between 2000 and 2005. And in Ecuador, Calero, Maldonado, and Molina (2008) found that from 1970 to 2007 exposure to drought increased the incidence of poverty by 2 percent on average.

In Asia, Akter and Mallick (2013) surveyed households in coastal communities affected by Cyclone Aila in 2010 in the southwest of Bangladesh. Unemployment skyrocketed, from 11 percent in 2009 to 60 percent in 2010, and the poverty headcount rate increased from 41 percent before the storm to 63 percent afterward. In a recent analysis of the 2011 floods in Bangkok, Thailand, Noy, Nguyen, and Patel (2014) report a large decrease in the agricultural and total income of poor households, compared with those with greater wealth. And even households that were not directly affected by the floods experienced a significant decrease in income – a spillover effect of the flood. In their study in the Philippines, Safir, Piza, and Skoufias (2013) found that low precipitation (below one standard deviation) decreases consumption by 4 percent, and all of the decrease occurs in food consumption, suggesting potential health impacts through undernutrition.

Disasters can have permanent impacts on human capital and well-being through education and health, with poor children as the main victims

Disasters force poor households to make choices that can have detrimental long-term effects. Recurrent events, such as urban floods in informal settlements, have impacts on the health of adults and children and have large cumulative impacts on poor people, even if each event is relatively small (Erman et al. 2018). Such events lead in particular to missed days at school for children and missed days at work for adults because traveling to the workplace is impossible or because adults (mostly women) stay home to take care of sick children.

Impacts on education are prevalent. In Africa, enrollment rates have declined 20 percent in regions affected by drought (Jensen 2000). Similar post-disaster impacts on health and education have been found in Asia, Latin America, and elsewhere (Baez, de la Fuente, and Santos 2010; Maccini and Yang 2009). In Mexico, once children have been taken out of school, even just for a temporary shock such as a flood, they are 30 percent less likely to proceed with their education, compared with children who remain in school (De Janvry et al. 2006). The impacts of the 1970 Ancash earthquake in Peru on educational attainment can be detected even for the children of mothers affected at birth, demonstrating that the effects of large disasters can extend even to the next generation (Caruso and Miller 2015).

Evidence also suggests that disasters have acute impacts on health, either directly or indirectly, through lower post-disaster consumption. After the 2004 floods in Bangladesh, more than 17,000 cases of diarrhea were registered (Chibani-Chennoufi et al. 2004), and the 1998 cholera epidemic in West Bengal, India, was attributed to the earlier floods (Sur et al. 2000). In Pakistan, the incidence of infectious disease and diarrhea increased as a result of the impact of the 2010 floods on the quality of the water. Ongoing efforts to eradicate polio were also interrupted, further setting back this goal (Warraich, Zaidi, and Patel 2011).

In sub-Saharan Africa, asset-poor households respond to weather shocks by reducing the quality of the nutrition provided to their children (Alderman, Hoddinott, and Kinsey 2006; Dercon and Porter 2014; Hoddinott 2006; Yamano, Alderman, and Christiaensen 2005), and they are less likely to take sick children for medical consultations (Jensen 2000). These behaviors have short- and long-term impacts, particularly for children younger than 2. Six months after a drought, children in households reducing nutrition were 0.9 centimeters shorter than other children (Yamano, Alderman, and Christiaensen 2005), and the stature of children in these households was permanently lowered by 2–3 centimeters (Alderman, Hoddinott, and Kinsey 2006; Dercon and Porter 2014).

In Central America, major disasters have also reduced investments in human capital. After Hurricane Mitch hit Nicaragua in 1998, the probability of child undernourishment in regions affected by the hurricane increased by 8.7 percent, and child labor force participation increased by 5.6 percent (Baez and Santos 2007). In Guatemala, Storm Stan increased the probability of child labor by 7.3 percent in departments hit by the storm (Bustelo 2011). Natural disasters also increase the multidimensional poverty index through a deterioration of "education conditions" and "child and youth conditions," as demonstrated by Sanchez and Calderon (2014) for Colombia from 1976 to 2005.

From case studies to a global estimate: disasters contribute to global poverty

This collection of case studies suggests that disasters have a significant impact on poor people and contribute to poverty. But by how much? It's difficult to extrapolate from case studies to global estimates, due to the heterogeneity in hazard distribution and vulnerability.

Although it remains impossible to quantify the full effect of natural disasters on the number of impoverished, it is possible to assess the short-term impacts of income losses (see Hallegatte et al. 2016). To do so, a counterfactual scenario was built of what people's income in developing countries would be in the absence of natural disasters. This scenario uses surveys of 1.4 million households, which are representative of 1.2 billion households and 4.4 billion people in 89 countries. The analysis concludes that if all disasters could be prevented next year, 26 million fewer people would be in extreme poverty – that is, living on less than $1.90 a day. Although this estimate is subject to large uncertainties and cannot capture all impacts, including those on health, education, and savings, it still shows how severely natural hazards affect poverty.

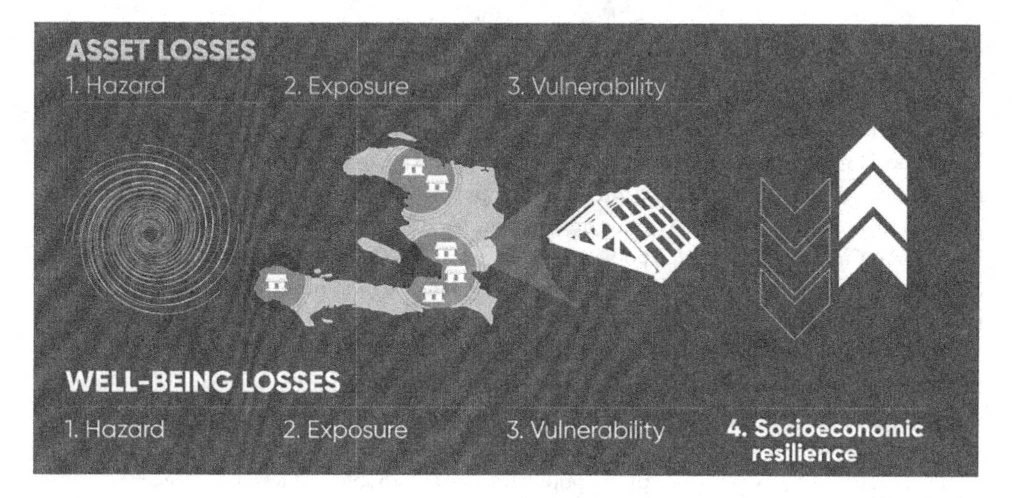

Figure 8.1 A comprehensive framework to understand the impact of natural hazards on well-being

Source: Hallegatte et al. (2016).

Poor people are disproportionately affected by natural disasters

One reason why disaster impacts on poverty are significant, probably more than impacts on GDP, is that disasters affect poor people more. Natural disasters hit poor people particularly hard for multiple reasons. Some of these reasons are linked to people's exposure to natural hazards (the probability to be affected by a hazard); others are linked to their vulnerability (the impact on people's assets and livelihoods when they are affected); and finally some are linked to people's socioeconomic resilience (their ability to cope with and recover from the shock). This section explores these three dimensions (Figure 8.1).

Exposure: poor people are often (but not always) more likely to be affected by natural hazards

In many places, poor people are more likely to be affected by a natural hazard than the rest of the population. In particular, poor people are often exposed to frequent, low-intensity events, such as the recurrent floods that affect many cities with insufficient drainage infrastructure. These events do not attract media interest and are poorly documented, but they can have significant cumulative impacts, especially through their effects on health. In Vietnam's Mekong Delta, 38 percent of the region's poor but only 29 percent of the region's nonpoor live in frequently flooded areas (Lam-Dao et al. 2011).

This pattern also exists for major disasters. After Cyclone Aila hit Bangladesh in 2009, a post-disaster survey of 12 villages on the southwest coast found that 25 percent of poor households in these villages were exposed to the cyclone, whereas only 14 percent of nonpoor households were (Akter and Mallick 2013). However, this pattern is not universal. After the 2011 floods in Kenya, almost everyone in the Bunyala District – poor and nonpoor – was affected (Opondo 2013). And in at least two documented cases, poor people were less exposed: after Hurricane Mitch struck Honduras in 1998, more than 50 percent of nonpoor households but only 22 percent of poor households were affected (Carter et al. 2007), and a similar pattern was observed after the 2011 floods in Thailand (Noy and Patel 2014).

Hallegatte et al. (2016) perform a global analysis of poverty and exposure to disasters and conclude that the relationship between poverty and disaster exposure depends on the type of hazard, local geography, and institutions. In most countries (representing about 60 percent of the population of the analyzed countries), poor people are more exposed to floods than the population average. This bias is only present among urban households, suggesting that it is land scarcity in cities that forces poor people to settle in dangerous areas. In parallel, around 85 percent of the analyzed population live in countries in which poor people are overexposed to drought. Finally, poor people are more exposed to extreme high temperature in 37 out of 52 countries (representing 56 percent of the population). Many of these countries are already hot. Cooler countries exhibit a smaller bias, and in some cool countries a negative bias because in these cool countries the nonpoor tend to settle in areas with higher temperatures, which are climatically more desirable.

These results suggest a sorting of the population into desirable and less desirable areas within a country, with wealthier households typically living in desirable areas and poorer households in less desirable ones.

For floods, another important issue is the availability of protective infrastructure such as dikes and drainage systems. FLOPROS (flood protection standards), a global open and collaborative database, has illustrated the lack of infrastructure to protect poor people (Scussolini et al. 2016). People in low-income countries – especially those with GDP per capita of less than $5,000 in purchasing power parity exchange rates – are significantly less protected than those in richer countries. This difference in protection alone can explain a factor 100 difference in flood risks between poor and rich countries (even before population vulnerability is considered). There are differences within countries as well, even if we cannot quantify them at this stage. Too often, investments – including those in disaster risk reduction – are directed toward the relatively wealthier areas at the expense of poorer neighborhoods. This effect can amplify the exposure gap between poor and nonpoor households and generate pockets of high risk.

Vulnerability: poor people lose more (in relative terms) when they are affected by a natural shock

People's vulnerability – that is, how much they lose when they are hit – is a critical determinant of the impacts of natural disasters. When poor people are affected, the share of their wealth lost is two to three times that of the nonpoor, largely because of the nature and vulnerability of their assets and livelihoods (see Figure 8.2).

Why is it that poor people lose relatively more? First, poor people tend to have less diversified portfolios: they hold a larger percentage of their assets in material form and save "in kind." The first savings of poor urban dwellers often take the form of investments in their home, which may be vulnerable to natural hazards such as floods or landslides (Moser and Felton 2007). Many rural poor use livestock as savings, despite their vulnerability to drought (Nkedianye et al. 2011). The nonpoor, who have higher financial access, are able to spatially diversify and save in financial institutions, and their savings are thus better protected from natural hazards.

In addition to the portfolio composition effect, the quality of assets owned by poor people is lower. An example is housing stock: households living in slums or informal settlements constructed of wood, bamboo, and mud and occupying steep slopes will sustain greater damage from a natural disaster than households whose homes are made of stone or brick. In coastal communities

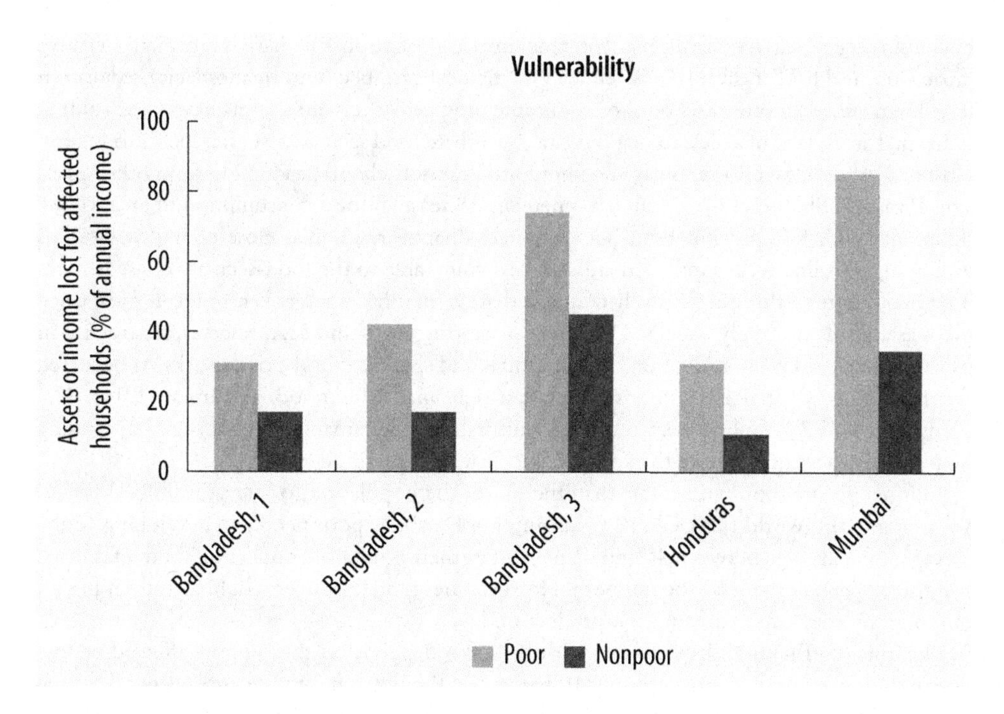

Figure 8.2 Percent of assets or income lost due to a disaster, for poor and nonpoor households: Bangladesh, Honduras, and Mumbai, India

Source: Hallegatte et al. (2016).

Other Sources: Del Ninno (2001) for the 1998 floods in Bangladesh (Bangladesh 1); Brouwer et al. (2007) for floods in southeast Bangladesh (Bangladesh 2); Rabbani, Rahman, and Faulkner (2013) for flooding due to cyclone Sidr in 2007 and Aila in 2009 (Bangladesh 3); Carter et al. (2007) for hurricane Mitch in Honduras; and Patankar and Patwardhan (2016) for the 2005 great flood in Mumbai.

Note: Each study has a different definition of "poor" and "nonpoor" in its sample. Vulnerability depends on the type of hazard and context in which it occurs; even within the same country (Bangladesh), vulnerability measures vary based on location and severity of flooding. The first three studies use percent of income loss as a metric, while the Honduras case uses asset loss and the Mumbai case uses asset, income, and repair loss. For Honduras, the graph reflects asset losses relative to total assets.

in southwest Bangladesh following Cyclone Aila, 76 percent of households in kacha houses (traditional homes built of mud and bamboo) reported structural damage – far above the 47 percent for those in pucca houses (built of concrete and wood) (Akter and Mallick 2013). A global analysis, based on the Global Building Inventory database from PAGER (Jaiswal et al. 2011), shows that, on average globally, the poorest 20 percent in terms of consumption are 1.8 times more likely than the average person to live in dwellings in the "fragile" category (Hallegatte et al. 2016).

Socioeconomic resilience: poor people are less able to cope with and recover from disasters

The very fact that they are poor makes poor people less able to cope with income losses. A 50 percent drop in income has very different consequences for two households living on $1,000 and $30,000 a year. In particular, poorer households cannot cut back on luxury consumption or delay consumption the way wealthier households can, and in many countries they are close to the subsistence level, which means that reducing consumption can have immediate negative impacts on health (if food intake is reduced or medical care becomes unaffordable), education (if children are taken out of school), or economic prospects (if essential assets have to be sold).

In rural areas, lack of access to markets can exacerbate food security issues: if local production is lost to a drought or a flood, isolated communities cannot rely on production from other areas. Safir, Piza, and Skoufias (2013) found a 4 percent decrease in food consumption in areas of the Philippines with low precipitation, but this effect disappears in areas close to highways. This finding suggests that well-connected areas are less vulnerable to the food-security consequences of natural disasters. But even in well-connected areas, natural disasters can result in food price spikes as a result of supply shocks. Disasters can destroy crops and seed reserves, destroying in turn productive assets in agricultural communities and sparking food price shocks, as occurred after the unprecedented 2010 floods in Pakistan (Cheema, Mehmood, and Imran 2016). The floods destroyed 2.1 million hectares of agricultural land, decimating production and sending the price of wheat up to more than 50 percent above the preflood level.

Poor people are more vulnerable than the rest of the population to increases in food prices. According to the World Bank Global Consumption Database, poor people in developing countries spend on average between 40 and 60 percent of their household budget on food – far more than the 25 percent spent by the nonpoor. However, net food producers could gain from higher food prices if they can maintain their production levels.

The impact of natural disasters on well-being also depends on the support affected people receive. In low-income countries, only 19 percent of the bottom quintile are covered by social safety net systems (Ivaschenko et al. 2018). After they are hit by a shock, poor people receive less post-disaster support than do nonpoor people. For example, in response to the floods and landslides in Nepal in 2011, only 6 percent of the very poor sought government support, compared with almost 90 percent of the well-off (Gentle et al. 2014).

Even when poor households receive support, the amounts received are often too small to enable better coping strategies. In Bangladesh, following the 1998 Great Flood, 66 percent of households in the bottom quintile received transfers, compared with 33 percent in the top quintile, and 53 percent of the flood-exposed households received transfers, compared with 34 percent of nonflood-exposed households (Del Ninno 2001). Although the targeting was relatively good, the transfer amounts were small: only 4 percent of the total household monthly expenditure for poor households and 2 percent for all households. Household borrowing highlights this limit: poor households affected by the flood borrowed six to eight times more than the level of government transfers.

Post-disaster support often fails to provide the poorest with enough resources because of their lack of voice and influence. As different categories of the population compete for help after a disaster, those with better connections are likely to get more, and more timely, support. When poor people are excluded from governance and have no say in the decision-making process, support is less likely to be timely or adequate. In case studies on Thailand, it was found that the majority of government support after a flood benefited the well-off, with 500 baht per capita (about $14) going to the richest quartile, compared with 200 baht per capita for the poorest quartile (Noy, Nguyen, and Patel 2014).

The need for a better measurement of disaster impacts – modeling disasters at the household level

In summary, poor people are disproportionately affected by disasters because they are often more likely to be affected by a shock, they lose more when they are affected, they have lower capacity to cope with their losses, and they receive less external support for recovery. These biases make it extremely problematic that our main metric of disaster severity is asset losses, since any impact on the poorest people is unlikely to be visible with this metric.

This section proposes an approach to measure natural disasters in a way that gives more visibility to the impact of poor people and therefore to better capture the real welfare- or well-being-related impacts of disasters. It is based on a series of papers, including some global analyses (Hallegatte et al. 2016) and country studies, in the Philippines, Sri Lanka, and Fiji (Walsh and Hallegatte 2019, 2020; Government of Fiji 2017).

The methodology starts from the traditional metric (asset losses), but estimated at the household level, and then moves on to income losses, then consumption losses, and finally welfare losses. Comparing these different metrics at the global or national scale offers new evidence that using asset losses only leads to an underestimation of the welfare impact of disaster and to policies that can be not only unfair, but also less efficient (in welfare terms).

A traditional metric: asset losses

Typically, the metric of asset losses is the one used to measure disaster severity. For instance, the amount of asset losses is what makes headlines in newspapers after a disaster. The main reinsurers publish every year an assessment of the total asset losses during the year, as cited in the introduction to this chapter. While these assessments sometimes include agricultural production losses (e.g., the value of the crops lost to a flood or a drought) and business interruptions (i.e., the inability of firms to produce in the immediate aftermath of a disaster), these additional components are similarly focusing on the pre-disaster value of what has been lost or damaged. Similarly, risk assessments are generally limited to average annual asset losses in the area of interest.

To provide a fair assessment of the well-being impact of disaster, however, providing the aggregate asset losses is not enough. One needs to consider who is affected and how aggregate losses are distributed among households. This is what is done in Hallegatte et al. (2016) (considering two categories of households per country in a simple model) and Walsh and Hallegatte (2020) (considering 40,000 households in the Philippines).

Using household-level data on exposure (Where are people living? Are they exposed to floods or earthquakes?) and vulnerability (In what type of dwelling do people live? How much asset do they have?), these studies estimate the distribution of asset losses and usually find that poor people tend to lose a larger fraction of their assets than do richer individuals. However, the absolute value of the losses is larger for richer people: in a subset of 117 countries, the bottom 20 percent in terms of income (one of many possible definitions of the poorest segment of the population) experiences "only" 11 percent of average annual asset losses. It means that poor people experience asset losses that are half of the country average.

Moving from aggregated asset losses to household-level asset losses already provides a much more granular view of disaster impacts as well as a better starting point to assess disaster impacts and design risk management interventions.

Income losses

Since asset losses are only a partial measure of the impact of disasters, it is possible to extend the analysis to explore how asset losses translate into income losses at the household level. In this process, the analysis moves from a *stock* analysis to a *flow* analysis, and the result becomes time dependent through the recovery and reconstruction period.

Over the short term, total income losses are likely to decrease in proportion to total asset losses, and total income losses can be estimated as the product of the total asset losses and the *average* productivity of capital (Hallegatte and Vogt-Schilb 2019).[1] A pastoralist losing one-third of his or her herd is likely to lose one-third of the income derived from it.

Since housing and public infrastructure represent a significant part of disaster damages, it is also critical to account for the loss in housing and infrastructure services, even when these are not exchanged on a market. For instance, a household who owns a dwelling will experience the equivalent of an income loss if their dwelling is destroyed and stops generating housing services (something that's often missed in economic statistics in low- and middle-income countries). Also, the services provided by roads and bridges is usually not traded on a market (with the exception of toll roads and bridges), but the loss of services when they are damaged can affect well-being in a significant manner (see Hallegatte et al. 2019 for an estimate of the economic and health implications of infrastructure disruptions).

At the individual level, focusing on one household, it is important to account for the fact that people are affected by the loss of assets they do not own but use to generate their income. This includes public assets, such as road and the power grid (and environment and natural capital) and some assets that are owned by other households, such as factories. In the methodology proposed here, the solution to ensure that all relevant asset losses are considered is to estimate the value of the assets household use to generate their income (including the value of housing and infrastructure services) based on their income and an estimate of the average productivity of capital in the considered country (see Walsh and Hallegatte 2020 for details). As a result, the loss of a factory will affect not only the owner(s) of the factory, but also all the workers who depend on this asset for their income.

Moving from household-level asset losses to household-level income level makes it possible to look not only at the immediate consequences of a disaster (damage and destruction), but also at the full recovery and reconstruction process. It redefines a disaster from an event during which an event causes damages (from a few minutes for an earthquake to months or more for droughts) to a much longer event that encompasses years or decades of recovery and reconstruction. This broader definition means that the severity of the disaster depends not only on the extent of the damages, but also on the duration of the recovery and reconstruction period, which in turn depends on the ability of the affected communities to respond and rebuild. While a more efficient reconstruction leaves asset losses unchanged, it can significantly reduce income losses to disasters (Hallegatte, Rentschler, and Walsh 2018).

Consumption losses

To better understand well-being losses, income losses can then be translated into consumption losses, accounting for the response to the disaster. Two dimensions are important.

First, households often experience a drop in income after a disaster (there are exceptions such as people working in the construction sector), but they also have to use part of their income to replace the asset they have lost. For instance, they need to replace or fix their roof after a storm; to replace their appliance after a flood; or to replace livestock after a drought. It means the consumption losses can often be larger than income losses, a major difference with pure income shocks (e.g., due to fluctuation of demand) (World Bank 2013).

The pace at which households replace their lost asset depends on their characteristics, but it can take years or more before poor households can restore their asset stock to the pre-shock level (Dercon and Porter 2014). This reconstruction spending explains why household expenditures are often found to increase after a disaster (Erman et al. 2018; Noy, Nguyen, and Patel

2014). This increase in spending does not mean that their well-being increases compared with a no-disaster counterfactual. Instead, they correspond to forced spending (or "defensive expenditures"). In the assessment of consumption losses, we therefore remove reconstruction spending from the income stream.

Second, households have instruments to smooth their consumption when they experience a shock, such as the use of savings, formal and informal insurance (Kunreuther 1996; Skoufias 2003), remittances (Le De, Gaillard, and Friesen 2013), ad hoc post-disaster transfers, and the scaling up of social protection (Siegel and de la Fuente 2010). These mechanisms can replace some of the lost income after a disaster and reduce the resulting consumption losses. Some of them are transfers across time (like the use of savings), others are risk-sharing mechanisms across people or households that also transfer consumption across time (like formal or informal insurance), and finally others are pure transfers (such as humanitarian aid).

In practice, estimating the dynamics of consumption losses is difficult. In our approach, this is done by assuming that households determine the optimal pace of reconstruction to minimize the long-term welfare losses, through a trade-off between a quick reconstruction (which increases income fast but at the expense of short-term consumption) and a limited drop in near-term non-reconstruction consumption (but a drop that will last longer, since the recovery of the asset stock will take longer). Figure 8.3 illustrates one reconstruction pathway, showing both the drop in income, the larger drop in consumption due to reconstruction needs, and the role of savings and post-disaster support (PDS) in mitigating the consumption shock.

Moving from income losses to consumption losses allows for a better accounting of people's socioeconomic resilience and access to coping mechanisms and instruments, such as savings, insurance, or social protection. Here again, the broader definition of the disaster and its losses

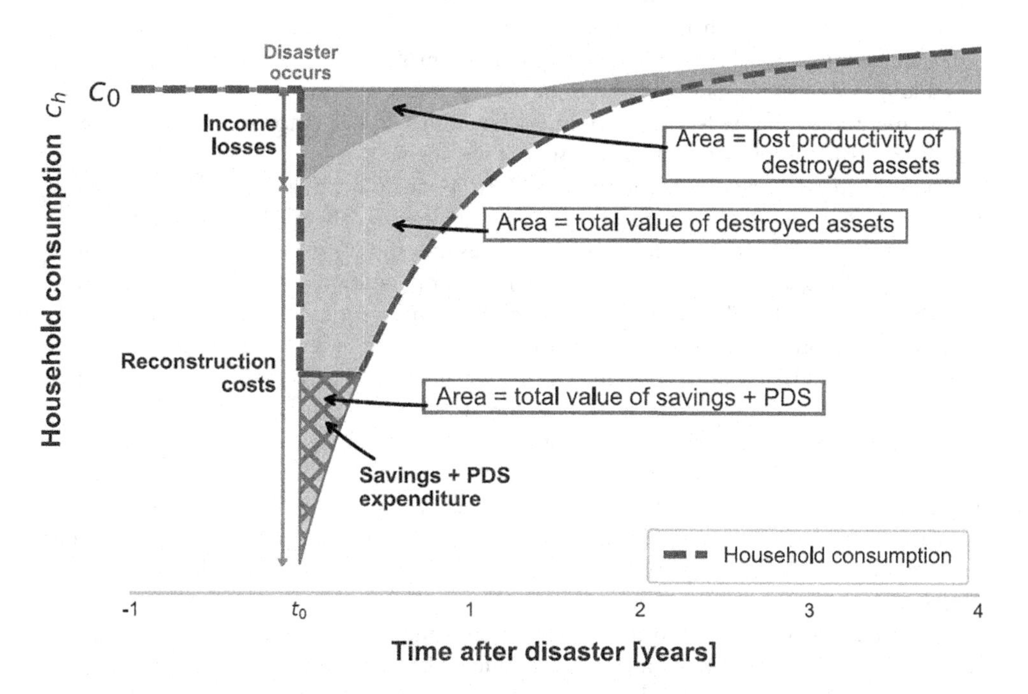

Figure 8.3 Illustrative consumption pathway during the reconstruction period, showing both the drop in income, the larger drop in consumption due to reconstruction needs, and the role of savings and post-disaster support (PDS) in mitigating the consumption shock

highlights not only larger welfare losses, but also new opportunities to reduce them through a new set of interventions (from financial inclusion to social protection).

Welfare (or well-being) losses and other related metrics

As stated earlier, it is well accepted that the impact of the same consumption loss (say $1,000) translates into different well-being consequences for a rich or a poor household, so a different metric is needed to capture these consequences.

One first option to look at well-being impacts is to focus on poverty headcount. This metric has many shortcomings. In particular, it measures only what happens to a small share of the population. For instance, it is independent of what happens to people already in poverty before the disaster: since they are in poverty already, their fate cannot affect the number of people falling in poverty. But it has the advantage of being simple and easy to communicate, even to non-experts.

The calculation of the household-level impact of the disaster on consumption makes it easy to calculate the number of people who will be (temporarily) in poverty due to the shock, as illustrated in Figure 8.4 for a Yolanda-like (100-year) typhoon making landfall in the Eastern Visayas region of the Philippines. In that case, the modeling exercise estimates that about 176,800 people, that is, 4 percent of the region's population, would fall in poverty. Such an estimate provides a measure of the storm severity that is a good complement to the pure monetary losses.

A poverty headcount is useful but does not capture many important factors, hence the need for a more comprehensive metric that would still be able to account for the disproportionate impacts on poor people. This is the objective of the "well-being losses."

At a country level, well-being losses can be measured as *equivalent consumption losses*, defined as the decrease in national aggregate consumption (optimally shared across the population) that would lead to *the same decrease in welfare* as the actual, individual losses from the disaster. (Note that we use welfare and well-being interchangeably in this text; in practice, welfare is the economic term that refers to a traditional measure of well-being.)

While $1 in asset or consumption losses affects a poor individual more than a rich one, well-being losses are defined such that a $1 well-being loss affects the rich and the poor equally. Well-being losses are calculated from consumption losses using a constant relative risk aversion function (CRRA; Wakker 2008). This operation translates into welfare units the value of a household's consumption at each point in its unique recovery, with decreasing returns to represent the fact that increasing consumption by $1 increases the well-being of a poor individual more (compared with a rich person). The difference in the welfare generated by $1 of consumption is a simple proxy for the continuum from survival consumption (the very first units of consumption that have the largest impact on well-being) to luxury consumption (which increases welfare less and less). This continuum is described in practice by the marginal utility of consumption. The elasticity of the marginal utility describes how the marginal utility of consumption decreases as income grows (in other terms, how much less a hedge fund manager in London cares about one dollar than does a Haitian farmer).

In global assessments, the elasticity has often been assumed to be between 1 and 3 (Dasgupta 2012; Heal and Miller 2014). Here, we use a value of 1.5 (consistent with, e.g., Evans 2005). This choice will always be partly arbitrary. It represents objective factors, like the unquestionable fact that the impact of losing one dollar on the quality of life of an individual depends on his or her wealth. But it also depends on values and political choices, such as *whether* societies want to

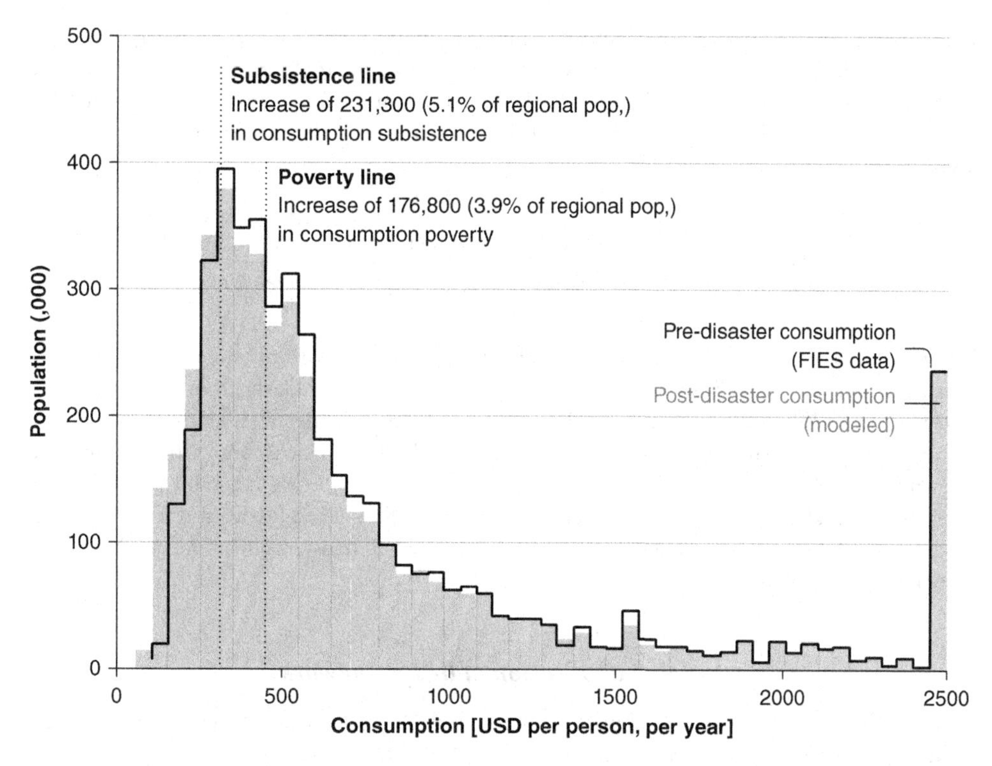

Figure 8.4 Shift in the consumption distribution after a 100-year typhoon landfall in the Eastern Visayas region of the Philippines

Source: Walsh and Hallegatte (2020).

eradicate extreme poverty, provide decent quality of life to all, and ensure that children are given the opportunities they deserve. In a sense, this elasticity represents our "aversion for inequality" or our "preference for an equal society."

Moving from consumption losses to well-being losses, using a traditional welfare function, is a simple and practical way of accounting from the common sense idea that poor people suffer more when experiencing the same monetary loss than do richer people. Most importantly, it offers a way to maximize the welfare benefits from disaster risk management interventions and prevents these interventions from being captured by the richest households who experience the largest asset losses.

Socioeconomic resilience

The ratio of asset and welfare losses is an important indicator: it measures the ability of the affected population to cope with and recover from $1 in asset losses without experiencing large well-being losses and is what we refer to as "socioeconomic resilience."

$$\text{Socioeconomic resilience} = \frac{\text{Asset losses}}{\text{Welfare losses}}$$

If socioeconomic resilience is 50 percent, then well-being losses are twice as large as asset losses. That is, $1 in asset losses from a disaster is equivalent to $2 in consumption losses, perfectly shared across the population. As illustrated in Figure 8.1, socioeconomic resilience can be considered a driver of the risk to well-being, along with the three usual drivers of risk assessment:

$$Risk\ to\ well-being = \frac{Expected\ asset\ losses}{Socioeconomic\ resilience} = \frac{(Hazard) \times (Exposure) \times (Vulnerability)}{Socioeconomic\ resilience}$$

The socioeconomic resilience measure used here captures part of the United Nations definition of resilience: the ability to resist, absorb, accommodate, and recover from the effects of a hazard in a timely and efficient manner. But it does not cover all the areas discussed in research on resilience (see Barrett and Constas 2014; Engle et al. 2014). For example, this framework does not take into account direct human impacts (such as death, injuries, and psychological impacts), cultural and heritage losses (such as destruction of historical assets), social and political destabilization, and environmental degradation (such as when disasters affect industrial facilities and create local pollution).

A global application of this framework

In Hallegatte et al. (2016), we apply this framework in more than 117 countries through a simple model exercise showing that well-being losses from natural disasters (river floods, coastal floods due to storm surge, windstorms, earthquakes, and tsunamis) are larger than asset losses.

According to the *United Nations Global Assessment Report on Disaster Risk Reduction* – the so-called GAR (UNISDR 2015) – total asset losses from natural disasters in these countries average $327 billion a year. But because disaster losses are concentrated on a small share of country populations, imperfectly shared, and affect more poor people (who have limited ability to cope with them), well-being losses are larger than asset losses. Hallegatte and Rozenberg (2017) estimate that well-being losses are equivalent in terms of well-being to a $520billion drop in consumption, uniformly distribution across the population. This is 60 percent larger than what asset losses suggest.

Risk to well-being decreases with country income (Figure 8.5b). This decrease is mostly driven by better protection against floods, higher-quality buildings, and widespread early warning systems in wealthier countries, but resilience also matters. Figure 8.5a also shows that, overall, resilience grows with GDP per capita. The fact that rich countries are more resilient than poor countries is not a surprise. But resilience varies widely across countries of similar wealth because it depends on many other factors, including inequality and safety nets.

Globally, poor people are disproportionately affected by well-being losses: people in the bottom 20 percent experience only 11 percent of total asset losses but 47 percent of well-being losses. Thus, poor people experience asset losses that are only half of what the average person experiences, but well-being losses that are more than twice as large as those experienced by the average person. It suggests that targeting poorer people with disaster risk reduction interventions – such as dikes and drainage systems – would generate lower gains in avoided asset losses but larger gains in well-being.

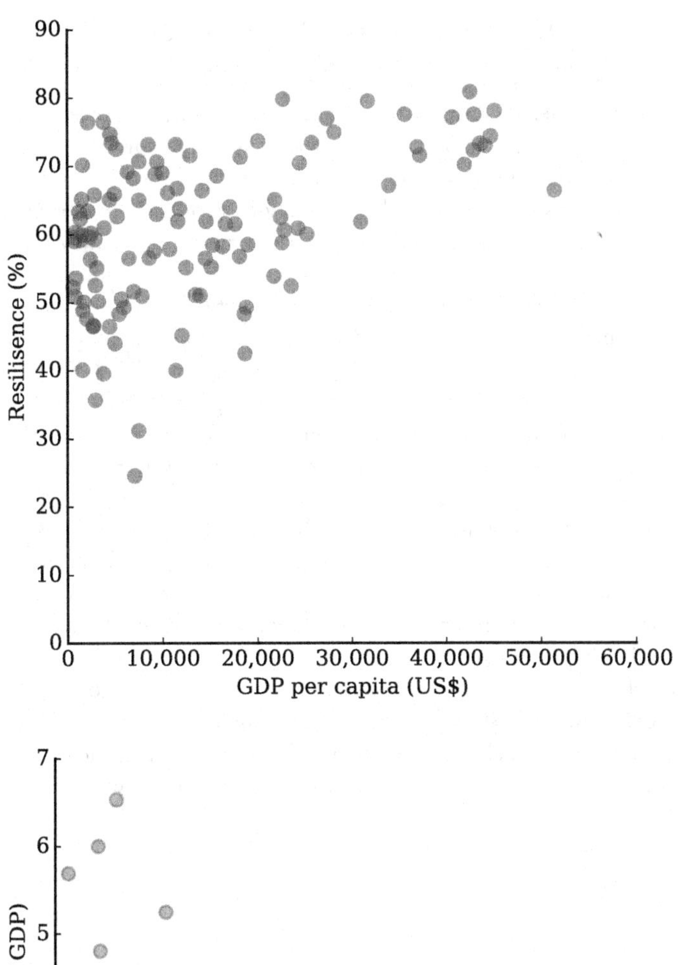

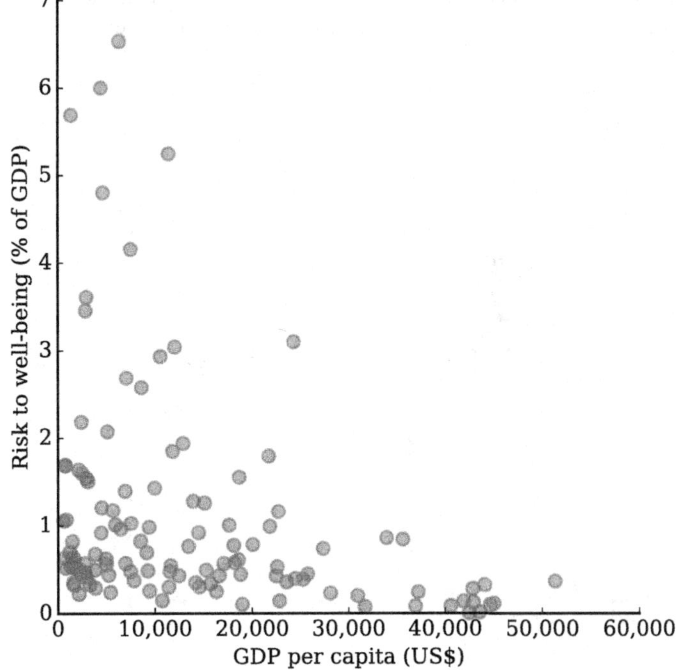

Figure 8.5 Socioeconomic resilience tends to increase with income, while risk to well-being decreases: (a) socioeconomic resilience, (b) risk to well-being

Source: Hallegatte and Rozenberg (2017).

Implication for policies and actions

When deciding where to invest in risk management or resilience, what matters is not only how much benefit a project generates, but also who benefits. To ensure that investments in resilience are distributed fairly across the population, one option is to measure the impacts of disasters and infrastructure disruptions using a metric that accounts for the socioeconomic status of the affected populations. This section explains how using socioeconomic resilience and well-being losses can help decide where to invest, but also which interventions to favor.

A recent analysis in the Philippines employed a multimetric assessment of disaster risks at the regional level using (1) traditional asset losses; (2) poverty-related measures such as poverty headcount; (3) well-being losses for a balanced estimate of the impact on poor and rich households; and (4) socioeconomic resilience, an indicator that measures the ability of the population to cope with and recover from asset losses (Walsh and Hallegatte 2020).

In the Philippines, the most important interventions will take place in the Manila area if asset losses are the main measure of disaster impacts (Figure 8.6). Other regions become priorities if the policy objectives are expressed in terms of poverty incidence and well-being losses. In particular, a risk mitigation policy focusing on preventing impacts on poverty would focus on the Bicol region more than in Calabarzon, because it hosts many more people who are near poor and less resilient and thus more vulnerable to falling into poverty in case of disasters. Mindanao appears as a priority in terms of socioeconomic resilience (because of its socioeconomic context, it is the least resilient region of the country), but not so much in terms of risk, because of its much lower exposure to typhoons. It means that the region will not be affected often by large shocks, but it will struggle to recover and suffer from large well-being losses when it does. For informed policy making, assessments of national risk and identification of critical infrastructure need to account for multiple policy objectives and, therefore, use a set of metrics that goes beyond asset losses.

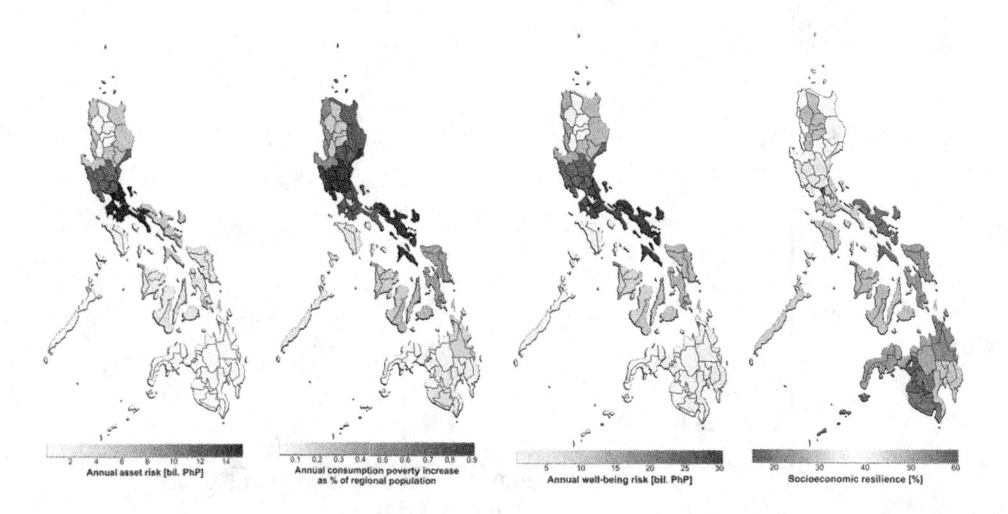

Figure 8.6 Different measures of natural risks in the Philippines highlight different priorities for interventions: (a) annual asset risk, (b) number of people falling into poverty every year, (c) annual well-being risk, (d) socioeconomic resilience

Source: Walsh and Hallegatte (2020).

No matter how much countries try to reduce people's exposure to natural hazards or to make assets more resistant to hazards such as earthquakes and floods, natural risk cannot be reduced to zero. Disasters will continue to inflict damage, and so it is critical to supplement actions on exposure and vulnerability with improvements in the ability of people to cope with the shocks that cannot be avoided despite efforts to reduce exposure or vulnerability.

Of course, one challenge is that these measures do not yield any measurable benefits, if the benefits of disaster risk management are measured in terms of the conventional metric, asset losses. Insurance, social protection, and remittances do not reduce the direct damages that a disaster may cause. However, they can significantly reduce the welfare impacts of such a disaster.

One advantage of using well-being losses as a metric to measure disaster impacts is that it makes it possible to assess and compare measures that reduce asset losses (e.g., building a dike) and measures that increase socioeconomic resilience (e.g., providing insurance to the population at risk). By expanding the range of policies that are considered to reduce disaster impacts, the hope is to create cheaper and more efficient policy packages that are also more equitable.

Returning to the Yolanda-like hurricane event discussed earlier, one can illustrate the benefits of post-disaster support: expected wind damage to household assets in the Eastern Visayas region is valued at US $633 million. Well-being losses from the same event are valued at US $2,176 million. If it is assumed that the government can provide a total of US $187 million in post-disaster support, distributed uniformly among all affected households, then the first quintile would see its well-being losses halved (while the impact on the richest quintile is almost invisible). In total, post-disaster support reduces well-being losses to US $1,265 million, a 42 percent decrease relative to the nominal simulation.

Because post-disaster support does not impact asset losses, such programs cannot be subjected to traditional cost-benefit analyses that focus on avoided asset losses. Indeed, an efficient cash transfer does not directly reduce the exposure of the population to floods or hurricanes, and it does not reduce the physical damages when an earthquake hits. Its disaster-related benefit-cost ratio, if expressed in avoided asset losses divided by the cost, is zero. However, cash transfers do increase the socioeconomic resilience of the region from 29 to 50 percent, because it makes the population better able to cope with and recover from any physical damages from a natural disaster. It means that, without any change in asset losses, a cash transfer can reduce the well-being impact of disasters by close to a factor two. As a result, the benefit-to-cost ratio of this intervention is estimated at 4.9, if its disaster-related benefits are expressed in avoiding well-being losses.

Conclusion

This chapter has outlined a risk assessment based on an expanded framework, which includes in the analysis the ability of affected households to cope with and recover from disaster asset losses and which uses well-being losses as a measure of disaster severity to complement asset losses. This framework adds to the three usual components of a risk assessment – hazard, exposure, and vulnerability – a fourth component, socioeconomic resilience. And to the traditional measure of disaster severity (mostly asset losses), it adds the possibility of looking at income losses, consumption losses, and well-being losses.

Measuring disaster impacts using these additional metrics helps quantify the benefits of interventions that may not reduce asset losses but do reduce well-being consequences by making the population more resilient. These interventions include financial inclusion, social protection, and more generally the provision of post-disaster support to affected households. By expanding our accounting of disaster impacts and quantifying the benefits of resilience-building measures, these new metrics expand the DRM toolbox and help identify more effective opportunities to manage natural risks.

The metric of well-being losses is of particular importance because it captures the impact of disasters in a way that is not biased toward the richer people and regions, like other economic metrics. Comparing various metrics shows how the regions and communities identified as priorities for interventions differ depending on how risk is measured. While a simple cost-benefit analysis based on asset losses would drive risk reduction investments toward the richest regions and areas, a focus on poverty or well-being losses accounts for a broader set of disaster impacts and leads to a set of priorities that are both fairer for poor individuals and better integrated with the broader development agenda.

Note

1 Note that a simple inclusion of capital losses into a traditional growth model would lead to a different calculation in which income losses are the product of the asset losses and the *marginal* productivity of capital. Hallegatte and Vogt-Schilb (2019) provides theoretical and empirical evidence – based on the heterogeneity of capital assets and the network effect within an economy – that using the average productivity is a better approximation.

References

Akter, Sonia, and Bishawjit Mallick. "The poverty – vulnerability – resilience nexus: Evidence from Bangladesh." *Ecological Economics* 96 (2013): 114–124.

Alderman, Harold, John Hoddinott, and Bill Kinsey. "Long term consequences of early childhood malnutrition." *Oxford Economic Papers* 58, no. 3 (2006): 450–474.

Baez, Javier, Alejandro De la Fuente, and Indhira Vanessa Santos. "Do natural disasters affect human capital? An assessment based on existing empirical evidence." (2010). https://papers.ssrn.com/sol3/papers.cfm?abstract_id=1672172.

Baez, Javier E., Leonardo Lucchetti, Maria E. Genoni, and Mateo Salazar. "Gone with the storm." (2016). https://openknowledge.worldbank.org/handle/10986/21396.

Baez, Javier E., and Indhira V. Santos. *Children's vulnerability to weather shocks: A natural disaster as a natural experiment.* New York: Social Science Research Network (2007).

Barrett, Christopher B., and Mark A. Constas. "Toward a theory of resilience for international development applications." *Proceedings of the National Academy of Sciences* 111, no. 40 (2014): 14625–14630.

Brouwer, Roy, Sonia Akter, Luke Brander, and Enamul Haque. "Socioeconomic vulnerability and adaptation to environmental risk: A case study of climate change and flooding in Bangladesh." *Risk Analysis: An International Journal* 27, no. 2 (2007): 313–326.

Bustelo, Monserrat. *Bearing the burden of natural disasters: Child labor and schooling in the aftermath of the tropical storm stan in Guatemala.* Urbana-Champaign: University of Illinois (2011).

Calero, C., R. Maldonado, and A. Molina. "Disaster risk and poverty in Latin America: The case of ecuador." Background paper of the ISDR/RBLAC-UNDP Project on Disaster Risk and Poverty in Latin America (2008).

Carter, Michael R., and Christopher B. Barrett. "The economics of poverty traps and persistent poverty: An asset-based approach." *The Journal of Development Studies* 42, no. 2 (2006): 178–199.

Carter, Michael, Peter Tewodaj Mogues Little, and W. Negatu. "Poverty traps and the long-term consequences of natural disasters in Ethiopia and Honduras." *World Development* 35 (2007).

Caruso, Germán, and Sebastian Miller. "Long run effects and intergenerational transmission of natural disasters: A case study on the 1970 Ancash earthquake." *Journal of Development Economics* 117 (2015): 134–150.

Cheema, Abdur Rehman, Abid Mehmood, and Muhammad Imran. "Learning from the past." *Disaster Prevention and Management* 25 (2016).

Chibani-Chennoufi, Sandra, Josette Sidoti, Anne Bruttin, Marie-Lise Dillmann, Elizabeth Kutter, Firdausi Qadri, Shafiqul Alam Sarker, and Harald Brüssow. "Isolation of Escherichia coli bacteriophages from the stool of pediatric diarrhea patients in Bangladesh." *Journal of Bacteriology* 186, no. 24 (2004): 8287–8294.

Dasgupta, P. *Time and the generations. Climate change and common sense: Essays in honour of Tom Schelling*, eds. R. W. Hahn and A. Ulph. New York: Oxford University Press (2012).

De Janvry, Alain, Frederico Finan, Elisabeth Sadoulet, and Renos Vakis. "Can conditional cash transfer programs serve as safety nets in keeping children at school and from working when exposed to shocks?" *Journal of Development Economics* 79, no. 2 (2006): 349–373.

Del Ninno, Carlo, ed. *The 1998 floods in Bangladesh: Disaster impacts, household coping strategies, and response*, vol. 122. International Food Policy Research Institute (IFPRI) (2001). https://ideas.repec.org/p/fpr/resrep/122.html.

Dercon, Stefan, and Catherine Porter. "Live aid revisited: Long-term impacts of the 1984 Ethiopian famine on children." *Journal of the European Economic Association* 12, no. 4 (2014): 927–948.

Engle, Nathan L., Ariane de Bremond, Elizabeth L. Malone, and Richard H. Moss. "Towards a resilience indicator framework for making climate-change adaptation decisions." *Mitigation and Adaptation Strategies for Global Change* 19, no. 8 (2014): 1295–1312.

Erman, A., et al. *The road to recovery: The role of poverty in the exposure, vulnerability and resilience to floods in Accra*. Washington, DC: The World Bank (2018).

Evans, D. J. "The elasticity of marginal utility of consumption: Estimates for 20 OECD countries." *Fiscal Studies* 26 (2005): 197–224. doi:10.1111/j.1475-5890.2005.00010.x.

Gentle, Popular, Rik Thwaites, Digby Race, and Kim Alexander. "Differential impacts of climate change on communities in the middle hills region of Nepal." *Natural Hazards* 74, no. 2 (2014): 815–836.

Glave, Manuel, Ricardo Fort, and Cristina Rosemberg. "Disaster risk and poverty in Latin America: The Peruvian case study." (2008). https://core.ac.uk/download/pdf/51209083.pdf.

Government of Fiji. *Climate vulnerability assessment: Making Fiji climate resilient*. Washington, DC: World Bank (2017).

Hallegatte, Stephane, Jun Rentschler, and Julie Rozenberg. *Lifelines: The resilient infrastructure opportunity*. Washington, DC: The World Bank (2019).

Hallegatte, Stephane, and Julie Rozenberg. "Climate change through a poverty lens." *Nature Climate Change* 7, no. 4 (2017): 250–256.

Hallegatte, Stéphane, Jun Rentschler, and Brian Walsh. *Building back better: Achieving resilience through stronger, faster, and more inclusive post-disaster reconstruction*. Washington, DC: The World Bank, 2018.

Hallegatte, Stephane, and Adrien Vogt-Schilb. "Are losses from natural disasters more than just asset losses?" In *Advances in spatial and economic modeling of disaster impacts*. Cham: Springer, pp. 15–42 (2019).

Hallegatte, Stephane, Adrien Vogt-Schilb, Mook Bangalore, and Julie Rozenberg. *Unbreakable: building the resilience of the poor in the face of natural disasters*. Washington, DC: The World Bank (2016).

Heal, G. M., and A. Miller. 2014. "Agreeing to disagree on climate policy." *Proceedings of the National Academy of Sciences* 111, no. 10 (March): 3695–3698. doi:10.1073/pnas.1315987111.

Hoddinott, John. "Shocks and their consequences across and within households in rural Zimbabwe." *The Journal of Development Studies* 42, no. 2 (2006): 301–321.

Ishizawa, Oscar A., and Juan Jose Miranda. "Weathering storms: Understanding the impact of natural disasters in Central America." *Environmental and Resource Economics* 73, no. 1 (2019): 181–211.

Ivaschenko, Oleksiy, Claudia P. Rodriguez Alas, Marina Novikova, Carolina Romero, Thomas Vaughan Bowen, and Linghui Zhu. *The state of social safety nets 2018*. Washington, DC: The World Bank Group (2018).

Jaiswal, K. S., David J. Wald, Paul S. Earle, Keith A. Porter, and Mike Hearne. "Earthquake casualty models within the USGS prompt assessment of global earthquakes for response (PAGER) system." In *Human casualties in earthquakes*. Dordrecht: Springer, pp. 83–94 (2011).

Jakobsen, Kristian Thor. "In the eye of the storm – the welfare impacts of a hurricane." *World Development* 40, no. 12 (2012): 2578–2589.

Jensen, Robert. "Agricultural volatility and investments in children." *American Economic Review* 90, no. 2 (2000): 399–404.

Kahn, Matthew E. "The death toll from natural disasters: The role of income, geography, and institutions." *Review of Economics and Statistics* 87, no. 2 (2005): 271–284.

Karim, Azreen, and Ilan Noy. "Poverty and natural disasters: A meta-analysis." (2014). https://ideas.repec.org/p/vuw/vuwecf/3234.html.

Kunreuther, Howard. "Mitigating disaster losses through insurance." *Journal of Risk and Uncertainty* 12, no. 2–3 (1996): 171–187.

Lam-Dao, Nguyen, Viet Pham-Bach, Minh Nguyen-Thanh, Mai-Thy Pham-Thi, and Phung Hoang-Phi. "Change detection of land use and riverbank in Mekong Delta, Vietnam using time series remotely sensed data." *Journal of Resources and Ecology* 2, no. 4 (2011): 370–374.

Le De, L., J. C. Gaillard, and W. Friesen. "Remittances and disaster: A review." *International Journal of Disaster Risk Reduction* 4 (2013): 34–43.

Maccini, Sharon, and Dean Yang. "Under the weather: Health, schooling, and economic consequences of early-life rainfall." *American Economic Review* 99, no. 3 (2009): 1006–1026.

Moser, Caroline, and Andrew Felton. "The construction of an asset index measuring asset accumulation in Ecuador." *Chronic Poverty Research Centre Working Paper* 87 (2007).

Munich Re. "Natural catastrophes factsheet 2019." (27 January 2020). https://www.iii.org/fact-statistic/facts-statistics-global-catastrophes.

Nkedianye, David, Jan de Leeuw, Joseph O. Ogutu, Mohammed Y. Said, Terra L. Saidimu, Shem C. Kifugo, Dickson S. Kaelo, and Robin S. Reid. "Mobility and livestock mortality in communally used pastoral areas: The impact of the 2005–2006 drought on livestock mortality in Maasailand." *Pastoralism: Research, Policy and Practice* 1, no. 1 (2011): 17.

Noy, Ilan, Cuong Nhu Nguyen, and Pooja Patel. "Floods and spillovers: Households after the 2011 great flood in Thailand." *Economic Development and Cultural Change* 69, no. 2 (2014).

Opondo, Denis Opiyo. "Erosive coping after the 2011 floods in Kenya." *International Journal of Global Warming* 5, no. 4 (2013): 452–466.

Patankar, A., and A. Patwardhan. "Estimating the uninsured losses due to extreme weather events and implications for informal sector vulnerability: A case study of Mumbai, India." *Natural Hazards* 80, no. 1 (2016): 285–310. doi:10.1007/s11069-015-1968-3.

Perez-De-Rada, E., and D. Paz. *Análisis de la relación entre amenazas naturales y condiciones de vida: El Caso de Bolivia.* New York: ISDR/RBLAC Research Project on Disaster Risk and Poverty (2008).

Rabbani, G., S. H. Rahman, and L. Faulkner. "Impacts of climatic hazards on the small wetland ecosystems (ponds): Evidence from some selected areas of coastal Bangladesh." *Sustainability* 5 (2013): 1510–1521. doi:10.3390/su5041510.

Rodriguez-Oreggia, Eduardo, Alejandro De La Fuente, Rodolfo De La Torre, and Hector A. Moreno. "Natural disasters, human development and poverty at the municipal level in Mexico." *The Journal of Development Studies* 49, no. 3 (2013): 442–455.

Safir, Abla, Sharon Faye Piza, and Emmanuel Skoufias. *Disquiet on the weather front: The welfare impacts of climatic variability in the rural Philippines.* Washington, DC: The World Bank (2013).

Sanchez, Fabio, and S. Calderon. *Natural disasters and multidimensional poverty in Colombia.* Washington, DC: World Bank Workshop on Climate Change and Poverty in the LAC Region (2014).

Schumacher, Ingmar, and Eric Strobl. "Economic development and losses due to natural disasters: The role of hazard exposure." *Ecological Economics* 72 (2011): 97–105.

Scussolini, Paolo, Jeroen C. J. H. Aerts, Brenden Jongman, Laurens M. Bouwer, Hessel C. Winsemius, Hans de Moel, and Philip J. Ward. "FLOPROS: An evolving global database of flood protection standards." *Natural Hazards & Earth System Sciences* 16, no. 5 (2016).

Siegel, P., and Alejandro De La Fuente. "Mainstreaming natural disaster risk management into social protection policies (and vice versa) in Latin America and the Caribbean." *Social Policy* 6 (2010): 131–159.

Skoufias, Emmanuel. "Economic crises and natural disasters: Coping strategies and policy implications." *World Development* 31, no. 7 (2003): 1087–1102.

Sur, D., P. Dutta, G. B. Nair, and S. K. Bhattacharya. "Severe cholera outbreak following floods in a Northern district of West Bengal." *Indian Journal of Medical Research* 112 (2000): 178.

UNISDR, GAR. *Global assessment report on disaster risk reduction, making development sustainable: The future of disaster risk management.* Geneva: United Nations (2015).

Wakker, Peter P. "Explaining the characteristics of the power (CRRA) utility family." *Health Economics* 17, no. 12 (2008): 1329–1344.

Walsh, Brian, and Stephane Hallegatte. "Measuring natural risks in the Philippines: Socioeconomic resilience and wellbeing losses." *Economics of Disasters and Climate Change* 4 (2020).

Walsh, Brian, and Stephane Hallegatte. *Socioeconomic resilience in Sri Lanka: Natural disaster poverty and wellbeing impact assessment.* Washington, DC: The World Bank (2019).

Warraich, Haider, Anita K. M. Zaidi, and Kavita Patel. "Floods in Pakistan: A public health crisis." *Bulletin of the World Health Organization* 89 (2011): 236–237.

World Bank. *World development report 2014: Risk and opportunity. Managing risk for development.* Washington, DC: World Bank (2013).

Yamano, Takashi, Harold Alderman, and Luc Christiaensen. "Child growth, shocks, and food aid in rural Ethiopia." *American Journal of Agricultural Economics* 87, no. 2 (2005): 273–288.

Yonson, Rio, Ilan Noy, and J. C. Gaillard. "The measurement of disaster risk: An example from tropical cyclones in the Philippines." *Review of Development Economics* 22, no. 2 (2018): 736–765.

9

CONTRACTS AND DISPOSSESSION

Agribusiness venture agreements in the Philippines

Alfredo R. M. Rosete

Introduction

The economic literature has argued that partnerships between smallholders and agribusiness firms can propel rural development. These benefits accrue through two channels. First, agribusiness firms can create linkages between the farm and non-farm sectors. Since high-valued crops are cultivated for export rather than direct consumption, smallholders need to purchase fertilizers and pesticides to meet market standards. Further, in order to bring crops to markets, producers need to form links with transport and logistics firms (Briones 2015; Reardon & Barrett 2000; Wang et al. 2014). Second, agribusiness firms can help relax constraints that prevent smallholders from using their holdings to generate viable livelihoods (Otsuka et al. 2016). For example, agribusiness firms provide inputs such as fertilizers, chemicals, and infrastructure, so that smallholders no longer need to enter disadvantageous credit arrangements with creditors and vendors who exercise market power (see e.g. Ghosh & Ray 2016; Shami 2012).

Despite these possible advantages, scholars (Amanor 2012; Cáceres 2015; Jiwan 2013) and civil society groups (see e.g. Colchester et al. 2011) have reported cases where governments and agribusiness firms take coercive or deceptive actions, allowing them to evict, threaten, or underpay smallholders. Coercive activities, however, do not explain all adverse contracts. Smallholders may voluntarily enter contracts where they lose rights to their land to relieve costs, credit constraints, and risks (Wilson 1986; Singh 2002). In other words, the constraints that agribusiness firms are supposed to alleviate can themselves become avenues for dispossession.

This chapter describes the conditions under which voluntary agribusiness partnerships become avenues of dispossession through the experience of agrarian reform beneficiaries (ARBs) in the Southern Mindanao region of the Philippines. I will show that the institutional environment and the organizations that govern the ARBs set the parameters they consider when bargaining with agribusiness partners. These, in turn, generate incentives for ARBs to choose contracts where they effectively cede rights to their land.

This study is significant for three reasons. First, it uses the experience of ARBs to show how institutional and political conditions affect the parameters of bargaining in the context of agribusiness investment. Second, the study focuses on contracts to argue that it is configurations of rights in a contract, rather than contract types, that lead to effective dispossession.

DOI: 10.4324/9780367814533-10

The theoretical framework conceptualizes property rights as a set of abilities over an asset. Such a framework has often been applied to problems of natural resource use (Cole & Cole 2002, Cole & Grossman 2002). However, to my knowledge, it has not been used in the context of agribusiness contracting. Finally, the study complements current literature on land acquisitions by showing how even voluntary transactions can result in dispossession.[1] Finally, the findings of this chapter point to social and economic vulnerabilities that prevent smallholders from undertaking more environmentally sustainable small-scale, labor-intensive agriculture, and maintain the dominance of industrial agriculture.

The chapter is arranged as follows: the second section will set out the context of agribusiness partnerships, its relationship to land reform in the Philippines, and the theoretical framework for the analysis. The third section describes the fieldwork and methodology. The fourth section discusses the findings. I begin by summarizing each group's histories and how these histories resulted in certain attributes. I then discuss how these attributes enhanced their ability to get more favorable contracts with their agribusiness partners. The fifth section discusses the consequences of these contracts on the ARBs' incomes and control over their holdings. The last section discusses how the findings inform policies for agribusiness partnerships.

Agribusiness partnerships: context and theory

In 1988, after the ouster of Ferdinand Marcos, the Philippine House of Representatives passed the Comprehensive Agrarian Reform Law (CARL). Land reform was a policy promise of the new Aquino regime to civil society organizations following the death of 13 farmers during a protest dubbed the "Mendiola Massacre." While allegedly written with minimal consultation from peasant and farmer organizations (Putzel 1992), its stated objectives are to achieve "a more equitable distribution of land" founded on "the rights of farmers and regular farmworkers to own directly or collectively the lands they till, or, in the case of other farmworkers, receive a just share of the fruits thereof"(Republic of the Philippines 2009, Ch. 1, Sec. 2). To implement CARL, the Department of Agrarian Reform (DAR) created the Comprehensive Agrarian Reform Program (CARP).

Soon after the implementation of CARP in 1988, the DAR observed that smallholders were entering into a variety of contracts with agribusiness investors. Recognizing their potential benefits, the Philippine government sought to encourage these arrangements, attempting to put in place regulations that would ensure ARB welfare under such contracts (DAR 2006b). These arrangements were termed agribusiness venture agreements (AVAs). The official definition of an AVA is an "entrepreneurial collaboration between ARBs and private investors to implement an agribusiness venture on lands distributed under CARP" (DAR 1996, 2006a). Through these agreements, ARBs can raise the productivity of their holdings through accessing upstream markets, capital, and farming technology. Further, ARBs can supposedly seek the state's assistance, since the DAR's local governments keep files of the certificates of land ownership award (CLOA), and the contracts that ARBs and agribusiness entities sign.

A 2010 report released by the Philippines' Inter-Agency Committee on Institutional Arrangements for Land Management and Rural Development[2] states that AVAs cover approximately 1.2 million hectares or 12.41% of total agricultural land and 29.64% of all land distributed under CARP as of 2009[3] (cited by IBON 2013). The large disparity between this figure and the DAR's official list of 50,103 hectares was acknowledged when I spoke to the DAR authorities in the town of Santo Tomas. When I had quoted these figures, one of them quipped, "*Parang dito lang yata, ganyan na kadami*" ("Maybe in this province alone, there's already that many"). They said they were still catching up with listing all AVAs.

The optimism of agrarian reform officials regarding the development potential of AVAs, however, was not widely shared by land reform advocates. In recent years, scholars have compiled cases of AVAs resulting in adverse outcomes for ARBs (see e.g. Borras 2007; Menguita-Feranil 2013; Adam 2013) showing how ARBs in certain areas of the country are compelled to accept contracts that leave them with neither control over land or income from its use. These reports point to an inherent problem with the policy of encouraging agribusiness partnerships for rural development. Behind the AVA policy is the idea that with property rights assigned to the ARBs, partnerships with agribusiness firms can function as a means for them to improve the use of their land, effectively becoming rural entrepreneurs.

This idea stems from the economic literature on property rights. Well-defined property rights are among the bases for contracting. Secure property rights can minimize the risks of losing an asset that is used in production or the risks of losing crops (Auerbach & Azariadis 2015; Besley & Ghatak 2010). When property rights are secure, partners to a contract can ensure that their investments are safe. Secure, well-defined property rights do not mean, however, that an asset holder will have control once the contract is in force. In reality, contracts also reassign certain abilities to each partner that can determine who effectively controls land. This is because property rights are a set of abilities that agents have over an object or asset in question (Hodgson 2015; Arruñada 2017). Control is a matter of what abilities one party holds over land. In agribusiness contracting, the possible abilities that constitute property rights are:

- *Determining User*: The ability to determine who can work on the land, cultivate crops, and harvest. This ability extends to determining who can be employed in a partnership.
- *Determining Use*: Determine what is planted, or being able to determine whether crops will be changed.
- *Determining Methods*: Determine how crops are cultivated, including the types of fertilizer and infrastructure used.
- *Determining Withdrawal*: Determine the duration of any contract involving the land or the ability to withdraw the use of the land in a partnership.

Table 9.1 shows the differences among three broad contract types that are widely used in agribusiness partnerships (Majid Cooke et al. 2011; Majid Cooke 2012; Cramb & Curry 2012) in terms of *remuneration* and *control* (De la Cruz 2012). Remuneration refers to the sources of income that each partner gets from the transaction. Control refers to the abilities that a contract confers to each partner. We can think of lease and growership contracts as two extremes. Under a lease contract, the smallholder forfeits all control over her land and most production decisions. In the growership contract, the investor acts as a buyer of the smallholder's crops, while the smallholder makes most of the decisions on how to cultivate, what methods to use, and whom to employ. However, the investor can use monitoring mechanisms such as checking the state of crops, restricting the types of fertilizers used, and consulting with smallholders to ensure crop delivery. In between these two extremes are what I broadly call joint ventures. The decision rights that are traded in such intermediary contracts differ considerably. Some joint ventures are framed as management contracts where the investor advises and monitors cultivation. Others are arranged so that the investor builds necessary infrastructure and then transfers these to smallholders for a share of the profits. In this case, the division of control is temporal. During the building period, the investor may have all the decision-making power. After that, he may act as a consultant to a growership.

A partnership results in dispossession when those who rightfully own land by law completely lose the ability to benefit from or control their holdings. Who ends up with control

Table 9.1 Different contract types and the allocation of abilities. Arrows indicate intermediary types between contracts. Entries in "Control" cells indicate who holds the rights.

	Remuneration		*Control*			
Structure of Incorporation	*Smallholder*	*Investor*	*User*	*Use*	*Methods*	*Withdraw*
Lease ↑	Wage + Lease	Profits	Joint	Investor	Investor	Investor
Joint Venture ↓	Profit Share	Profit Share	Joint	Joint	Joint	Joint
Growership	Profits from sale to Investor	Export Profits	Smallholder	Joint	Joint	Joint

Source: Rosete (2020).

depends on the conditions of contracting. These operate through two channels illustrated in Table 9.2: the ARBs' *bargaining position* and their *bargaining power*. By bargaining position, I mean what the ARBs know or perceive to be their next best alternative to a given contract. By bargaining power, I mean the formal and informal rules that govern the bargaining process between ARBs and their potential partners. Among the conditions that operate through the bargaining position are their knowledge of alternative contracts, the perceived risks, costs, and benefits associated with cultivating Cavendish bananas, and the prospects of fulfilling their obligations under CARP. Conditions that work through bargaining power are the extent that the DAR adjudicates disputes fairly and the extent to which they have access to legal institutions.

While the rules, laws, and policies governing AVAs should be uniform for all ARBs, the quality of the ARB advocates and organizations are not. The different ARB groups can develop attributes that determine how they understand their bargaining positions and participate in enhancing their bargaining power. Table 9.2 summarizes how these attributes function in the bargaining process. On the first row of Table 9.2 are circumstances generated by the law and governance. These are outside the purview of any ARB groups or individuals. Amortization payments weaken an ARB's bargaining position since it is the cost of land ownership. The adjudication of disputes and contracts determines how fairly policymakers, government officials, and courts weigh the interests of ARBs. The second row enumerates the organizational or group attributes that affect bargaining power and bargaining position.

Some attributes affect only one of the conditions of bargaining. AVA education helps ARBs compare the costs, benefits, and risks of different types of contracts, allowing them to accurately assess their bargaining position. Organizational debt can make ARBs reluctant to take risks and more inclined toward contracts where they bear little or no costs in production. This can be consequential in selecting between a lease and a growership contract, since ARBs bear most of the costs in the latter. Legal counsel enhances bargaining power by helping ARBs understand the rules, regulations, and processes needed for contracting. They may also help ARBs anticipate whether the language of a contract creates circumstances that are detrimental to their interests.

Membership consultation affects both bargaining power and bargaining position. Ostrom (2000) and Wilson et al. (2013) argue that in order to take collective action, groups need to have avenues to inform and consult their members on key decisions. While such democratic mechanisms can be costly (Pozzobon & Zylbersztajn 2013), they help in developing solidarity

Table 9.2 The conditions of bargaining that law and governance and organizations generate

	Bargaining position	*Bargaining power*
Law and Governance	Amortization	Adjudication of: • Disputes • Contracts
Organizational/Group Attributes	AVA education Organizational debt Member consultation External advocates	Independent legal counsel Member consultation External advocates

Source: Author's own schematization.

among members and cohesion among a group's objectives (Forney & Häberli 2017). Effective collective action enhances both a group's bargaining power and its bargaining position. Effective collective action enhances the ARBs' bargaining position by easing the process of pooling resources for inputs and coordinating knowledge about AVAs, agribusiness, and banana cultivation. Second, effective collective action can also ease coordination for activities such as sit-ins and marches, which could make public officials more responsive to the demands and grievances of ARBs (Borras & Franco 2013).

External advocates such as NGOs and peasant organizations can help enhance a group's bargaining power by providing legal counsel and coordinating collective action with larger organizations and causes. Coordinating with such forces could enhance the profile of individual ARB cases and force authorities to be responsive to public opinion. Finally, they can coordinate campaigns with other forces (trade unions, etc.), which can encourage ARBs and make policymakers more responsive to their demands (Alden 2014). External advocates may also help ARBs navigate the financial system and understand alternative contract types, enhancing their bargaining position.

The attributes of the ARB groups determine the strength of their bargaining power or bargaining position. When a group develops attributes that enhances either bargaining power or bargaining position, it can maintain more abilities over its holdings and a greater degree of control.

Methodology

The fieldwork took place in the municipality of Santo Tomas and the city of Tagum in the province of Davao Del Norte and the district of Pantukan in the province of Compostela Valley. Both provinces are in the Davao region of the Philippines. The study uses both purposive and snowball sampling, selecting ARB groups that partnered with agribusiness firms. Seven ARB groups participated in focus group discussions, while 71 ARBs participated in individual interviews.[4] While key informant interviews involved the elected leadership of the different organizations, my focus groups involved both leadership and regular members.[5]

Individual interviews yielded information on individual income, housing, and asset ownership. This information is used as anecdotal evidence to relate some of the adverse effects of agribusiness contracting, but this information has no statistical power. However, the anecdotes are illustrative of the effect of contractual types and their consequences for the ARBs.

Structured focus group discussions gave the histories of each group's formation, contracting, and disputes and relationships with authorities and agribusiness partners. Each group also provided copies of their different contracts that contain the rights and abilities, which are my primary outcomes of concern.

The lands awarded to all the ARBs in the study are currently under a lease or growership contract. The study focuses on lease and growership contracts because of their prevalence among the officially listed AVAs that cultivate Cavendish bananas. Of these 61 AVAs, 29 out of 61 (45.9%) are under a lease contract, while 28 out of 61 (48%) are under growership arrangements.[6]

All the agribusiness partnerships in this study cultivate and export Cavendish bananas. The focus on one crop ensures that crop volatility, seasonal, and cost differences do not affect the differences between the respondents.[7] Moreover, the Philippines is the third largest exporter of Cavendish bananas after India and China. Cavendish bananas are worth about $884 million annually, or 14% of the country's total agricultural exports. The total land area devoted to Cavendish banana cultivation in the country is 82,202 hectares, and the Davao region cultivates 46,681 (56.78%) of these. In the DAR's official tally, AVAs that are devoted to the cultivation of Cavendish bananas account for 8,717 ha. Thus, the listed AVAs account for 18.7% of land area devoted to Cavendish banana cultivation in the region and 10.6% of land area devoted to Cavendish banana cultivation in the Philippines.[8]

Table 9.3 gives a summary of each group's AVA and the size of the land area involved and the length of time that each contract is supposed to last. The shortest length of time for a lease contract is 25 years. The average age of an ARB is about 50 years old (Reyes 2002). Moreover, these contracts are renewable at the discretion of the lessee, not the lessor. Thus, the conditions of these contracts may persist up to the lifetimes of the ARBs' heirs. Should the heirs want to

Table 9.3 A summary of current contracts with land area, length of time, and whether there are disputes surrounding the AVAs. "Y" indicates an ongoing dispute. Note that HIJO and TRAIN are owned by the same family.

Location	*Type*	*Group*	*Total land area (ha)*	*Indiv. plot size (ha)*	*Year of CLOA*	*AVA year*	*Length (years)*	*AVA with former owner*	*AVA dispute*
Santo Tomas	Grower	M1	47.52	1.02	2004	2007	10		
		M2	89.76	1.02	2004	2007	10		
		D	799.56	1.04	2002	2002	30 (renewable)	Y	
	Lessor (Group)	S1		1.04	2002	2002		Y	Y
		S2		1.04	2002	2002		Y	Y
Tagum		H	294.33	0.8	1997	2014	60	Y	
Pantukan	Lessor (Ind)	P	32.5 +	2.5 0.6	1992	2009	25 (renewable)		Y

Source: Rosete (2020).

undertake a different form of farming or crop, they can be blocked by the agribusiness firm. Of note are the last two columns. Under the column "AVA with former owner," a Y means that the agribusiness partners are the former owners of the land before the implementation of CARP. In the last column, the Y indicates a disputed contract where the ARBs have taken some action to nullify their contracts or renegotiate the terms. In the municipality of Santo Tomas, Davao Del Norte, for example, the lessor groups have been in a partnership with the former owners of their land since they received their CLOA. Further, all disputed AVAs are currently under lease contracts.

Findings

In principle, all ARBs face the same institutional environment. Their AVAs are supposedly regulated by the DAR and monitored to ensure that they fulfill the objectives of CARP. Further, ARBs must also pay for amortization unless the former landowners freely release land for redistribution. Given these institutional factors common to all ARBs, group attributes may contribute more to determining the AVA contracts. The columns in Table 9.4 show the attributes that each ARB group had at the time of contracting.

The ARB interviewees from Santo Tomas, Davao Del Norte, were all members of a trade union that was friendly with the agribusiness firm that owned the banana plantation that they would receive under CARP. Though they had very similar starting points, grower groups M1 and M2 diverged from their counterparts in lessor groups D, S1, S2.

Key informant interviews suggest that groups M1 and M2 were marginalized members of the trade union. They were given more unpleasant jobs such as digging canals and forming hills for banana trees.[9] They saw land reform as an opportunity to improve their situation. The leadership of the M groups contacted pro-agrarian reform NGOs who provided them with legal assistance and introduced them to other farmers who had entered agribusiness contracts. Through these contacts, they learned about the different types of AVAs. Eventually, the

Table 9.4 Conditions of each group at the time of negotiating and bargaining over current AVAs. "Y" means that the organization had these.

Location		*Group*	*AVA education*	*Member consultation*	*Ind. legal counsel*	*Org. debt*	*External advocates*
Santo Tomas	Growers	M1	Y	Y	Y	Y	Y
		M2	Y	Y	Y	Y	Y
	Lease (Group)	D					
		S1					
		S2					
Tagum		H	Y	Y	Y	Y	Y
Pantukan	Lease (Ind)	P					

Source: Rosete (2020).

leadership of M1 and M2 persuaded other farm workers to talk to the NGOs, who provided them with the same AVA education. To maintain their coalition of farm workers, they also formed methods of democratic voting, debate, and deliberation that allowed their members to deliberate and build consensus. In 2004, the ARBs of M1 and M2 began a growership agreement with their former employers. They eventually dissolved this agreement in 2007 when they found a new agribusiness buyer. Under this new agreement, the ARBs were able to sell a box of class A bananas for $4.10. Of this, $0.85 would go to the cooperative for maintenance, inputs, day-to-day operations, and remaining amortization payments, which amounted to 297,733.43 pesos per hectare.

Groups D, S1, and S2 initially belonged to a single lessor organization whose leadership also led the management-friendly trade union. According to key informant interviews from all three groups, the leadership was comfortable with the status quo and focused on convincing the farm workers that they should enter a lease contract with their current employers. The leadership framed the choice in the following way: if they follow their co-workers in M1 and M2, there will be little financing for operations, amortization payments would be very high, and they bear risks of bad harvests and pests. A lease contract will mean no amortization since their employers will simply transfer the land to them and continue wage payments and additional lease payments by virtue of their land ownership. The leadership of the management-friendly trade union succeeded. In 2002, the members of groups D, S1, and S2 decided to enter a lease contract with their former employers. The lease paid to the ARBs in this contract would rise with the number of boxes starting at $0.07/box for 4,200 boxes or less and ending in $0.11/box for 4,901 boxes or more. This amounts to a range of 13,679 pesos to 25,064 pesos ($294 to $539.11) annually. In addition, if they are employed by the agribusiness partner, they can receive a daily wage of 320 pesos, which is greater than the province's minimum wage at the time of fieldwork of 302 pesos.

Like groups M1 and M2, group H in Tagum City initially saw land reform as a means to improve their situation by owning their own banana farms. They organized their fellow farmworkers, found advocates and lawyers, and negotiated a growership contract. Unlike the M groups, however, their growership contracts were sold to another agribusiness firm, which imposed several conditions that added costs to the ARBs. First, the new buyers installed new irrigation pipes across the farm, which were then billed to the ARBs without being consulted. Second, the ARBs could only buy fertilizers and pesticides from the new buyers. These two conditions were well within the bounds of the contract since the ARBs gave up the ability to determine the methods of cultivation to allow the buyers to have quality control. Since the buyers were now able to exercise monopoly power over the ARBs, they raised the price of the fertilizer. ARBs found themselves in debt due to the high input prices and payments for the irrigation pipes.

Eventually, a typhoon hit the island of Mindanao in 2012, destroying group H's banana plants. The typhoon pushed the group to taking greater debt and closer to not being able to make their amortization payments. Due to these debts, they entered a lease agreement with the former owners of their lands. Under this lease contract, the ARBs would be paid a lease of 17,000 pesos annually in addition to a wage of 320 pesos daily. Group H, however, would retain the ability to hire workers and manage the day-to-day farm operations.

The ARBs of the P group from Compostela Valley were farm workers of a coconut plantation. They decided to continue as coconut farmers after receiving their CLOA in 1992. Eventually, the single buyer of coconuts in the district told them that the price of kopra had fallen from 20 pesos/kilo to 9 pesos/kilo. The lower prices stuck, and the ARBs found themselves unable to pay for their amortization and their daily needs. It was at this time that an agribusiness firm approached the individual ARBs to lease their lands and convert them into banana plantations. The agribusiness partner was willing to pay a five-year advance and cover their missed amortization payments of 8,946.19 pesos annually.

Making their amortization payments and receiving a lump sum for five years of lease payments was tempting, as they were already in debt. The ARBs of the P group did not see the need for an organization, and thus did not build any capacity for collective action. Further, they did not seek out any legal assistance or assistance from external advocates. Their only knowledge of AVAs came from the lease contracts they were presented with individually. The agribusiness firm was able to convince a majority of the ARBs. In 2009, their agribusiness partner sold the lease to another firm without the consent of the ARBs. This new agribusiness partner kept the terms of the original contract. They paid a lease that would increase by 5% every five years.[10] The agribusiness firm also had the choice to hire the ARBs for at least the minimum daily wage of 302 pesos.

Contracts and consequences

One benefit of land ownership and contracts with agribusiness partners can be seen in income. Consistently, individual interviewees cited this as the most important reason to own land. Table 9.5 compares the outcome between a lessor ARB in Santo Tomas and their grower counterparts.

The advantages of being growers, however, are visible if we compare incomes. Growers maintain the ability to determine some methods of cultivation. Even if they shoulder greater costs, growers can experiment on different practices and management techniques that can raise revenue. In Table 9.5, the bi-weekly income is based on an annual yield of 4,000 boxes.[11] It is clear from the table that the growers get a higher income despite shouldering the costs involved in production.[12] This is because growers are owner-cultivators who fully claim the revenue that their lands generate. Lessors, on the other hand, give a greater portion of the revenue to the lessee. However, the lease contract does not simply yield a larger share of the revenue to the agribusiness firm. It also yields greater control.

The contracts of each AVA are described in Table 9.6. The columns correspond to the abilities that ensure control over land. Y means that a group holds this ability under their current contract. From this table, it is evident that growership contracts allowed the ARBs

Table 9.5 A comparison of incomes between the Santo Tomas beneficiaries

Lessor income (4,000 class A boxes/year)	*Grower income (4,000 class A boxes/year)*
Wages (pesos):	**Revenue**:
Daily: 320	Price/box: $3.25
Bi-weekly (pesos/12 days): 3,840	Bi-weekly average yield: 4,000/26 153.85
Exchange rate in 2015: 45.93 pesos/dollar	Total: $615.40
Bi-weekly (USD): $83.61	**Costs:**
Lease payment (BTF):	Bi-weekly (pesos): 9,646
BTF rate (USD): $0.071/box	Exchange rate in 2015: 45.93 pesos/dollar
BTF total (USD): $284.00	Bi-weekly (USD): $106.77
BTF bi-weekly (USD): $11.00	
Bi-weekly income: $94.61	**Bi-weekly net income: $508.63**

Source: Author's fieldwork.

Table 9.6 The rights that each group is able to maintain in their current contracts. "Y" indicates that these rights are given to the beneficiaries. "C" means that they have these rights conditionally.

	Location	*Rights/group*	*Det. rem.*	*Withdraw*	*Det. methods*	*Det. user*
Growers	Santo Tomas	M1	Y	C	C	Y
		M2	Y	C	C	Y
		D	C			
Lease (Group)		S1				
		S2				
	Tagum	H	C			Y
Lease (Ind)	Pantukan	P				

Source: Rosete (2020).

to maintain most abilities over their holdings. Groups that had growership contracts (M1, M2, and, initially, H) developed significant bargaining power and enhanced their bargaining positions. All three groups learned about growerships through NGO contacts and exposure to other farms. By observing and meeting other growers, they learned how to finance their operations through pooling resources and specialized loan programs for cooperatives.[13] They were also able to reach out to lawyers that helped ARBs understand intricacies of contracting and advocates that pressed the DAR to monitor bargaining process and ensure ARB interests. Finally, each group encouraged membership input into the bargaining process to ensure their consent to the contract. However, comparing the experience of M1 and M2 to that of H, growership contracts can have differences. H gave up their ability to determine the methods of cultivation on their lands and went into debt. This debt, coupled with the destruction of their banana farms by the typhoon, forced the ARBs of H to enter a lease contract.

It is evident that ARBs give up more abilities in a lease contract. The interviewees from all lessor groups suggest that they gave up control over land to alleviate burden of debt and amortization payments. The H group, however, was able to negotiate a lease contract where they maintain the right to negotiate the lease and wage and choose who can work on their lands. These were rights that other lessors in this study did not keep.

For groups D, S1, and S2, the actions of the trade union leaders undermined their bargaining position. The ARBs thought they had no other viable option due to the prospect of debt and the supposed lack of financing for growership operations. The leaders simply acquiesced to their former employer's terms since they wanted to maintain the status quo. According to group interviews with members of D, S1, and S2, membership communication was limited to reporting the terms of their former employers and convincing members of the viability of a lease contract. Further, few, if any, efforts were made to expose the ARBs to other AVAs. This made the ARBs think that the only viable option for their lands was the lease contract offered by their former employers. Moreover, the interviewees did not recall any legal counsel assisting in the negotiations. With the union leadership simply acquiescing to the wishes of their former employers for a lease contract, D, S1, and S2 did not develop any bargaining power. Since the groups did not develop bargaining-enhancing attributes, the ARBs entered contracts where they held no rights over their holdings.

In 2004, the ARBs of S1 and S2 decided to assert their land ownership. Members of these groups picketed the DAR for days. Their actions attracted other ARBs and, in 2010, they occupied their lands and attempted to break the lease contracts. Since they ceded the right to withdraw their holdings from the lease contract, their land occupation was deemed illegal. They were violently dispersed and at least one ARB was killed. Since they ceded the ability to determine the users of their land, they were retrenched by their agribusiness partner and are no longer able to work on their holdings.

Despite having an existing organization, the contracts of D, S1, and S2 were similar to that of P who were not organized at the time of contracting. Apart from their status as ARBs, members of the P group did not take measures to enhance their bargaining power. They were not organized and did not negotiate with the help of any outside advocacy or legal counsel. Facing debt because of low *kopra* prices, they hastily entered a lease contract without finding out about alternative AVA types. As a result, they held no control over their holdings.

Initially there was no conflict. The ARBs and their family members were able to work in the AVA. In 2009, the new agribusiness partner heard rumors that one of the ARBs in the P group had a daughter who was a member of a radical trade union. To prevent any type of labor resistance, the new agribusiness partner retrenched all the ARBs and their family members. They were given the reason that since their community has been "infiltrated" by trade unionists, they will inevitably cause problems. Since the AVA contracts gave the right to determine the user of land and withdrawal from the partnership to the agribusiness firm, they could not transfer the AVA, evict the agribusiness firm, or assert their right to work on their lands.

Dispossession

Dispossession results from the abilities that ARBs give up in their contracts with agribusiness firms. If AVAs are supposed to facilitate the objectives of CARP, then not only should ARBs see an appreciable rise in income, but they should also maintain control of their holdings. Groups S1, S2, and P were deprived of the ability to control or benefit from their land because they gave up the ability to select the users of their land and the ability to withdraw their lands from the partnerships.

There are two main reasons why these groups decided to give up these rights. First, the ARBs wanted to alleviate the burden of debt. The ARBs from group P almost lost their lands due to the low price of kopra. ARBs from S1 and S2 accepted the lease terms, initially, because they thought that the amortization payments would be insurmountable and that they would not be able to access any financing for growing bananas. Second, the ARBs entered lease contracts to avoid further debts that can arise from bearing the costs of inadequate harvests or diseases that may afflict the banana plants.[14] In a lease agreement, the agribusiness partners would have had to shoulder these costs. All three groups, however, did not get to study different contractual forms before they negotiated AVA terms. They were not shown how pooled ARB resources and small loans would have provided financing for their operations and helped improve the chances of a growership's success.

Since their agribusiness partners now hold the right to choose who works in the AVA, they were able to prevent the ARBs from working on their holdings. Interviewees from these groups survive through seasonal work in neighboring grower farms, laundry, tailoring, and construction. Some interviewees also used the initial lease payments to buy motorcycles, which they use for delivery and transport jobs. All the interviewees say they would have preferred steady work in the lands that they own since that would be more stable.

While lease contracts result in ARBs giving up a significant subset of rights over their holdings, they do not necessarily lead to dispossession. The experience of group H demonstrates how bargaining power can compensate for a severely weakened bargaining position. They were able to maintain their ability to determine the users of their land by enlisting the help of NGOs and lawyers. These external advocates regularly visited the DAR office to pressure land reform officials to monitor the contracting process. Further, they helped the ARBs negotiate with the new agribusiness lessors for the ARBs' right to choose who can work on their lands. Due to this, they are able to secure employment and even train family members for work on the farm.

Though a lease contract saved the ARBs of group H from outright dispossession, it was a growership contract that had put them at risk. Giving up the ability to determine the methods of cultivation almost led to them losing their lands. Their agreement to buy inputs solely from their agribusiness partner when they had a growership gave the agribusiness firm the power to set input prices. This drove the ARBs into debt – a situation made worse by the typhoon.

The experiences of the lessor ARBs, then, show that dispossession does not depend on whether the ARBs had a lease or growership contract. Rather, dispossession in agribusiness contracts depends on the abilities that smallholders can retain over their lands. Both growership and lease contracts introduce some possibility of smallholders losing rights to their lands. Groups that develop bargaining-enhancing attributes are better able to avoid dispossession.

Discussion and policy implications

The experience of the ARBs shows that the conditions of contracting can have persistent effects on the degree to which smallholders can use their land to generate viable livelihoods. While agribusiness partnerships have the potential to raise income and productivity, the perceived costs and benefits of a contract can force smallholders to give up rights to their land, which preclude them from its use as a means of livelihood and leverage within a partnership. The perception of the costs and benefits of different contractual forms depends on their organizational history and the institutions governing the contracting process. Improving the opportunities of smallholders in agribusiness partnerships, then, entails interventions from both the state and civil society (Borras & Franco 2013).

One implication of the study is that agribusiness partnerships, by themselves, cannot alleviate the constraints that smallholders face. In the case of the lessor ARBs, these material constraints compelled them to give up a substantial degree of control to their agribusiness partners. In the context of negotiating contracts, these material constraints worsen the bargaining position of smallholders. If agribusiness partnerships are to result in viable livelihoods for smallholders, then policymakers have to raise their bargaining position by relieving some material constraints. In the context of the banana cultivators in my fieldwork, it may be wise to give subsidized loans for agricultural and capital inputs. Further, at least in the context of the Philippines, amortization payments have to be less stringent and, perhaps, reduced. The lessor ARBs in my fieldwork all have ceded rights to their land due to the burden of amortization payments. If AVAs are to generate viable livelihoods, then policymakers in the Philippines should re-examine the pricing of lands relative to the capacity of ARBs instead of the avowed valuation of the former owners. Further, the state should re-examine how it can finance amortization payments so as to alleviate the burden for ARBs.

In other contexts, relieving material constraints may also take the form of productive public goods such as farm-to-market roads and purchasing centers where farmers can sell their crops at market price. Both of these measures can alleviate the costs of accessing markets that may compel farmers to rely on monopsonistic crop buyers, raising smallholders' bargaining positions. Roads also have the advantage of generating benefits for other members of the community by giving them access to centers of commerce, jobs, and schools.

Apart from material interventions, policymakers can also ensure that contracts are written so that smallholders maintain the rights to use their lands and withdraw them from adverse partnerships. The reason why the beneficiaries were dispossessed of their lands is that their agribusiness partners can veto their employment. If contracts can ensure that ARBs, like group H, have some capacity to manage and select workers, then they can maintain their employment and avoid dispossession. Admittedly, this may be difficult in the case of the Pantukan beneficiaries, as none of them have had experience in cultivating bananas. However, if the Pantukan beneficiaries were able to withdraw their lands from the agribusiness partnership, then they may be able to find more viable partners. Further, if withdrawing land from a partnership is a credible threat, agribusiness partners would have a greater incentive to ensure that smallholders receive better remuneration within the arrangement.

In 2016, a peasant leader named Rafael Mariano was appointed to head the DAR. While his tenure was short (June 30, 2016–September 6, 2017), the DAR made efforts to address the ongoing disputes of S1 and S2 with their agribusiness partner. The dispute had already reached the Presidential Agrarian Reform Council and was waiting adjudication for six years. All the contracts, including those with group D, were declared void and are now under renegotiation.[15] Since Mariano was not approved by Congress, he stepped down and negotiations seem to have stalled. This experience illustrates the problematic nature of institutions governing agrarian reform and AVAs. The institutions should be robust to whoever takes a leadership position. If beneficiaries are dependent on the type of leadership that governs the DAR, then they may have to wait long periods to viably use their holdings. If the institutions instead had explicit rules to enhance beneficiary bargaining power and bargaining position, then they can deal with potential agribusiness partners without relying on the volatility that changes in the DAR leadership can create.

So far, I have only pointed out regulatory interventions. In my field study, organizations were instrumental in forming the opinions of beneficiaries regarding different contracts. Civil society can facilitate the formation of farmer and peasant organizations to raise their bargaining power vis-à-vis their agribusiness partners. They do not need to organize with forming agribusiness partnerships in mind. What we can learn from the experience of M1 and M2 is that civil society organizations can communicate opportunities to smallholders and facilitate their process of pooling resources, learning methods of management, and becoming familiar with the law.

Finally, both policymakers and civil society have to make interventions together in order to ensure the viability of agribusiness partnerships. Organized smallholders may opt for an unfavorable contract if their organizations do not provide adequate information. In this case, contractual regulations that allow smallholders to renegotiate and determine the users of their land would help them correct for the informational shortcomings. If material assistance is available, but civil society is absent to disseminate this information, smallholders might not be able to maximize opportunities offered by official channels.

The broad guidelines that I propose also benefit smallholders outside the context of agribusiness partnerships. Material interventions such as subsidized loans, farm-to-market roads, and purchasing centers can ease farmers' access to capital and markets regardless of the crops they choose to cultivate. Smallholder organizations can facilitate the creation of cooperatives that share resources and solve collective action problems, depending on the needs of their members.

Thus, ensuring the viability of agribusiness partnerships also means a more comprehensive rural development strategy involving material and institutional interventions by policymakers as well as empowering smallholders by civic organizations (Lahiff et al. 2007). In the end, the way to ensure that agribusiness partnerships are viable for smallholders may mean ensuring that individual farming is viable, even without agribusiness partners.

This study has some limitations, which present avenues for further research. First, one cannot draw generalizable conclusions due to the scope, sample size, and methodology. Instead, the study can inform future empirical and theoretical research on the possible political-economic conditions under which agribusiness partnerships succeed in improving the lives of smallholders. Second, while the study addresses agribusiness partnerships among ARBs, it does not account for counterfactual cases such as non-ARBs in agribusiness partnerships or ARBs that are not in agribusiness partnerships. Introducing ARB status as an additional dimension of variation would expand the possible institutional and legal factors that influence agribusiness partnerships. Finally, the study focuses on the ARBs' experiences and contains no information from their agribusiness partners apart from narratives from focus groups, key informants from ARB organizations, and DAR officials. However, once again, my research does not address the decision of what contracts to offer, but rather which contracts are accepted and why.

Conclusion

In this chapter, I argued that contracts between smallholders and agribusiness firms can be thought of as allocating abilities over land. Depending on the abilities allocated, a contract can result in denying the smallholder the benefits of owning land. That is, contracts can result in dispossession. Dispossession, in turn, depends on the strength of smallholders' bargaining power and bargaining position. I illustrate this dynamic using fieldwork conducted among agrarian reform beneficiaries in the Philippines. ARB groups that developed effective collective action mechanisms and support systems were able to keep more abilities over their holdings. Policies that seek to protect smallholders in contracting with agribusiness firms should ensure that smallholders are able to control who uses their lands and withdraw them from adverse contracts. Further, measures need to be taken to ensure that they are not contracting with weaker bargaining positions or bargaining power.

Notes

1 Arezki et al. (2013) empirically demonstrate that a significant driver in recent land acquisitions is the perceived weakness of property rights institutions in a country where agribusiness firms choose to locate. A reason behind this finding is that agribusiness firms may find it less costly to locate in countries where they can expropriate current land occupants.
2 A committee chaired by the head of the National Economic Development Authority (NEDA) that includes the DAR and the head of the Department of Agriculture. Its function is to streamline efforts by the national government agencies involved in rural development. Its creation is stated in the office of the president's Administrative Order number 34 (AO no. 34).
3 CARP accomplishments in 2009 were at 4,049,016.71 hectares and the country's total agricultural land is at 9.671 million hectares (IBON 2013).
4 All groups identified as cooperatives except for the group in Pantukan, which are beginning to form an association.
5 I made contact with the different groups with the help of the DAR *Unyon ng Manggagawa sa Agrikultura* (UMA) (Agricultural Workers' Union), and *Kilusang Mayo Uno* (May First Movement)-Southern Mindanao Region (KMU-SMR). Both the DAR and UMA provided lists of AVAs from which I selected the groups to contact. Members of KMU-SMR helped me make contacts with groups in Santo Tomas, Davao Del Norte, and Pantukan, Compostela Valley. The groups in Santo Tomas and in

Pantukan gave me the names and addresses of their members, and I went door to door for individual interviews. The ARB group in Tagum city allowed me to interview their members after work hours in their office.

6 The figures here are based on a list of AVAs that I obtained from the DAR through an official request.

7 There are a number of costs in cultivating Cavendish bananas. First, they are chemical- and fertilizer-intensive. Fertilizers can cost up to 990 pesos per 50 kg sack. Under typical growing methods, a hectare of banana uses up to three sacks of fertilizer every 20 days. The farmer must also treat the trees by applying chemicals to guard against fungi and other pests that may prevent the crop from meeting export standards. A typical grower pays for chemicals delivered through aerial spraying on a bi-weekly basis. The farmer must also buy twine for propping up the banana plants and plastic bags to wrap new bunches and protect them against insects. Finally, the farmer must also supply or pay for labor. Typically, the total bi-weekly cost of cultivating bananas is 9,646 pesos ($210) per hectare. These costs were computed using pay receipts from grower ARBs and enumerating prices from agricultural stores around Santo Tomas.

8 The figures quoted on total land area and export revenue were taken from estimates of the Philippine Statistics Authority https://psa.gov.ph/fruits-crops-bulletin.

9 Canals and hills allow excess fertilizer and pesticides to wash away from the plants and need to be re-dug and re-formed. Interviewees report trouble breathing and eye pain when working in the canals.

10 The lease amount was 22,000 as of fieldwork but increases 5% every five years. Net of the amortization payments amounts to 13,058.81. The duration of the contract is 25 years, renewable by the agribusiness firm.

11 I chose this yield because it is at the lower end of the expected yield specified in the lease contracts governing groups D, S1, and S2 shown in Table 9.5.

12 I calculated costs based on the typical yearly cycle of banana growing as narrated to me by managers of two of the grower groups whose members were interviewed. These include fertilizer, twine, aerial spray, and labor. I then obtained sample pay slips with redacted names from growers and crosschecked these with three nearby agricultural supply stores in Santo Tomas.

13 Interviewees did not provide specifics on these but simply stated that they recall having been able to participate in these with information from their NGO contacts.

14 Cavendish bananas are known to have little resistance to diseases caused by fungi like *Sigatoka*. Interviewees from groups H, M1, and M2 state that when a single plant contracts one of these fungal diseases, they would have to undergo a process of burning and chopping the trees to avoid the spread of the fungi. It can mean the loss of thousands of pesos for several months.

15 The latest information from the DAR's news desk can be found in the following link: https://www.dar.gov.ph/articles/news/100798.

References

Adam, J. (2013). Land reform, dispossession and new elites: A case study on coconut plantations in Davao Oriental, Philippines. *Asia Pacific Viewpoint*, *54*(2), 232–245.

Alden, W. L. (2014). The law and land grabbing: Friend or foe? *The Law and Development Review*, 7(2), 1–36.

Amanor, K. S. (2012). Global resource grabs, agribusiness concentration and the smallholder: Two West African case studies. *The Journal of Peasant Studies*, *39*(3–4), 731–749.

Arezki, R., Deininger, K., & Selod, H. (2013). What drives the global "land rush"? *The World Bank Economic Review*, *29*(2), 207–233.

Arruñada, B. (2017). Property as sequential exchange: The forgotten limits of private contract. *Journal of Institutional Economics*, *13*(4), 753–783.

Auerbach, J. U., & Azariadis, C. (2015). Property rights, governance, and economic development. *Review of Development Economics*, *19*(2), 210–220.

Besley, T., & Ghatak, M. (2010). Property rights and economic development. In *Handbook of development economics* (Vol. 5, pp. 4525–4595). Elsevier.

Borras, S. M. (2007). *Pro-poor land reform: A critique*. University of Ottawa Press.

Borras, S. M., & Franco, J. C. (2013). Global land grabbing and political reactions "from below". *Third World Quarterly*, *34*(9), 1723–1747.

Briones, R. M. (2015). Small farmers in high-value chains: Binding or relaxing constraints to inclusive growth? *World Development, 72*, 43–52.
Cáceres, D. M. (2015). Accumulation by dispossession and socio-environmental conflicts caused by the expansion of agribusiness in Argentina. *Journal of Agrarian Change, 15*(1), 116–147.
Colchester, M., Chao, S., Dallinger, J., Toh, S. M., Chan, K., Saptaningrum, I., Ramirez, M. A., & Pulhin, J. (2011). *Agribusiness, large-scale land acquisitions, and human rights in Southeast Asia.* General Publication, Forest People's Programme.
Cole, D. H., & Cole, D. (2002). *Pollution and property: Comparing ownership institutions for environmental protection.* Cambridge University Press.
Cole, D. H., & Grossman, P. Z. (2002). The meaning of property rights: Law versus economics?. *Land Economics, 78*(3), 317–330.
Cramb, R., & Curry, G. N. (2012). Oil palm and rural livelihoods in the Asia – Pacific region: An overview. *Asia Pacific Viewpoint, 53*(3), 223–239.
De la Cruz, R. J. G. (2012). Land title to the tiller: Why it's not enough and how it's sometimes worse. *ISS Working Paper Series/General Series, 534*(534), 1–46.
Department of Agrarian Reform (DAR). (1996). *Administrative order 02-99.* Rules and Regulations Governing Joint Economic Enterprises in Agrarian Reform Areas.
Department of Agrarian Reform (DAR). (2006a). *Administrative order 09-06.* Revised Rules and Regulations Governing Agribusiness Venture Arrangements (AVAs) in Agrarian Reform Areas.
Department of Agrarian Reform (DAR). (2006b). *Case studies on agribusiness venture agreements (AVAs) between investors and agrarian reform beneficiaries (ARBs).* Department of Agrarian Reform Policy and Strategic Research Service Economic and Socio-Cultural Research Division.
Forney, J., & Häberli, I. (2017). Co-operative values beyond hybridity: The case of farmers' organisations in the Swiss dairy sector. *Journal of Rural Studies, 53*, 236–246.
Ghosh, P., & Ray, D. (2016). Information and enforcement in informal credit markets. *Economica, 83*(329), 59–90.
Hodgson, G. M. (2015). Much of the "economics of property rights" devalues property and legal rights. *Journal of Institutional Economics, 11*(4), 683–709.
IBON. (2013). Haciendas remain. *Facts and Figures, 35*(17).
Jiwan, N. (2013). The political ecology of the Indonesian palm oil industry. *The Palm Oil Controversy in Southeast Asia: A Transnational Perspective*, 48–75.
Lahiff, E., Borras, S. M., & Kay, C. (2007). Market-led agrarian reform: Policies, performance and prospects. *Third World Quarterly, 28*(8), 1417–1436.
Majid Cooke, F. (2012). In the name of poverty alleviation: Experiments with oil palm smallholders and customary land in Sabah, Malaysia. *Asia Pacific Viewpoint, 53*(3), 240–253.
Majid Cooke, F., Toh, S., & Vaz, J. (2011). *Community-investor business models: Lessons from the oil palm sector in East Malaysia.* IIED/IFAD/FAO/.
Menguita-Feranil, M. L. (2013). Contradictions of palm oil promotion in the Philippines. *The Palm Oil Controversy in Southeast Asia: A Transnational Perspective*, 97–119.
Ostrom, E. (2000). Collective action and the evolution of social norms. *Journal of Economic Perspectives, 14*(3), 137–158.
Otsuka, K., Nakano, Y., & Takahashi, K. (2016). Contract farming in developed and developing countries. *Annual Review of Resource Economics, 8*, 353–376.
Pozzobon, D. M., & Zylbersztajn, D. (2013). Democratic costs in member-controlled organizations. *Agribusiness, 29*(1), 112–132.
Putzel, J. (1992). *A captive land: The politics of agrarian reform in the Philippines.* Catholic Institute for International Relations.
Reardon, T., & Barrett, C. B. (2000). Agroindustrialization, globalization, and international development: An overview of issues, patterns, and determinants. *Agricultural Economics, 23*(3), 195–205.
Republic of the Philippines. (2009). *An act strengthening the comprehensive agrarian reform program (CARP).* Republic Act No. 9700. Republic of the Philippnes. House of Representatives.
Reyes, C. M. (2002). Impact of agrarian reform on poverty. *Philippine Journal of Development, 29*(2), 63.
Rosete, A. R. (2020). Property, access, exclusion: Agribusiness venture agreements in the Philippines. *Journal of Rural Studies, 79*, 65–73.
Shami, M. (2012). The impact of connectivity on market interlinkages: Evidence from rural Punjab. *World Development, 40*(5), 999–1012.

Singh, S. (2002). Multi-national corporations and agricultural development: A study of contract farming in the Indian Punjab. *Journal of International Development, 14*(2), 181–194.
Wang, H. H., Wang, Y., & Delgado, M. S. (2014). The transition to modern agriculture: Contract farming in developing economies. *American Journal of Agricultural Economics, 96*(5), 1257–1271.
Wilson, D. S., Ostrom, E., & Cox, M. E. (2013). Generalizing the core design principles for the efficacy of groups. *Journal of Economic Behavior & Organization, 90*, S21–S32.
Wilson, J. (1986). The political economy of contract farming. *Review of Radical Political Economics, 18*(4), 47–70.

10

NATURAL RESOURCES, CLIMATE CHANGE AND INEQUALITY IN AFRICA

James C. Murombedzi

Introduction

The use and governance of natural resources are among the most central issues for the daily lives of the majority of Africans. Patterns of rural resource use are fundamental to rural and national economies, as well as to local and global concerns about sustainability. Resource degradation through unsustainable use patterns as well as climate change have resulted in a global environmental and developmental emergency whose impacts include increasing concentration of wealth, growing poverty and inequality.

Institutional histories and political interests fundamentally shape rights over natural resources, and rights are central to the ways in which those resources are used. In this regard, the desire of European powers to capture and exploit African resources played a key role in the transformative process of colonialism (Ocheni and Nwankwo 2012). The core characteristic of the colonial project was the alienation of natural resources and the imposition of new forms of centralized political authority over access to land and resources that had previously been controlled by more localized institutions. The expropriatory processes associated with the imposition of new forms of resource ownership and control occasioned the creation or in some cases the exacerbation of inequalities, which has endured into the present day. Significantly, colonialism was about controlling natural resources, and many of the inequalities on the continent today are reflected in unequal access to the continent's natural resources (ibid.).

Today extreme global economic inequality is largely a result of corporate-led globalization (Joyce 2010; Piketty 2014). The COVID-19 pandemic not only exposes but also further exacerbates the inequality (Stiglitz 2020).[1] In 2006, the first study to tally household wealth worldwide (McGillivray 2006) concluded that the richest 1% of world population owned 39.9%of the world's household wealth, which is greater than the wealth of the world's poorest 95%. Since then, a slew of studies of inequality and maldistribution, focusing *inter alia* on income, employment, assets, access and power inequalities,[2] have all demonstrated that global wealth is increasingly concentrated in the hands of a small wealthy elite. Oxfam (2020) in particular makes a case against the increasing concentration of wealth in a few men as a result of a flawed sexist economic system.[3] The causes of inequality are many but are mostly located in the historical process of production and distribution. Contemporary developments, including environmental degradation and climate change, exacerbate inequalities between high- and low-income

DOI: 10.4324/9780367814533-11

countries, as well as in-country inequalities. In Africa, while precolonial forms of inequality have influenced the continent's encounter with colonialism and been reproduced and sometimes exacerbated in the colonial period, colonialism itself has created new and more extreme forms of inequality that define the continent's social condition today. Colonial dispossession, particularly pronounced in settler colonial societies (Moyo and Yeros 2013), has led to enclosures, created poverty and exacerbated inequalities.

The colonial state in Africa was established to control labour, capital and resources for external European purposes. This set of political objectives resulted in the concentration of central bureaucratic and executive power. The state's powers of coercion were used to limit independent forms of social organization. Governance was not democratic, representative or accountable. States claimed wide powers over natural resources, particularly land, which was generally placed under discretionary bureaucratic control with customary rights subordinated to claims explicitly recognized by the colonial administration. Even under the British indirect rule, this meant concentrating fused executive, legislative and judicial powers in externally recognized local authorities who bolstered what Mamdani (1996) has called "the fist of colonial power." Shivji (1998: 48) notes, "There is a deep structural link between the use and control of resources and the organization and exercise of power. Control over resources is the ultimate source of power."

The colonial instrumentation of Africa in the global economy resulted in a predatory exploitation of its raw materials. This process shaped the continent into a series of enclave economies structured around the control of land and mineral resources, and their supply to the colonial centres. These enclave economies have other fundamental characteristics. Firstly, they are based on the exploitation of a single resource and its exportation in raw unprocessed form, or in partially processed form, for value addition in another economy. Thus for instance many African economies are exporters of a single dominant mineral ore, such as iron ore or copper, or agricultural commodity, such as cotton or palm oil. A second key characteristic is that the enclave is the creation of a small formal labour force in the enclave and the consignment of reproduction of labour outside of the formal economy. This is particularly pronounced in the former settler colonial societies which were structured on an ideology of white supremacy (South Africa, Rhodesia, South West Africa, Portuguese East Africa and Angola), creating an enclave formal economy employing one-fifth of the labour force.[4] The informal sector is characterized by 'customary land tenure' systems where land ownership is effectively vested in the state and thus land use is subject to a high level of state intervention. The state can control commodity and labour markets through economic and extra-economic means of coercion. The formal sector, on the other hand, is characterized by private land tenure systems (freehold or leasehold) which give rights holders a higher degree of control and flexibility regarding land use (Kanyenze and Kondo 2011).

The formal/informal dichotomy in turn informed the huge disparities between the unskilled and semi-skilled native labour force and the skilled non-native labour force. Control over land and mineral resources became the basis on which equity and equality of opportunity – access to nutrition, education and job opportunities – was constructed and highly skewed in favour of the formal sector, resulting in a highly underdeveloped informal sector characterized by low levels of remuneration, low nutrition, high unemployment and high poverty.

The next section explores how contemporary forms of natural resources control have continued almost unchanged from colonial times, entrenching access by new elites and creating new forms of exclusion and inequality. The section seeks to demonstrate how erratic growth and structural adjustment policies of the 1970s and 1980s eroded many of the social policy gains of the early independence period in Africa. The resultant market-driven development approaches informed a logic of hyper-exploitation of natural resources which in turn exacerbated inequalities. This is followed by a section which explores the impacts of climate change on inequality. This section recognizes that although climate impacts are new and accelerating, they amplify existing vulnerabilities and inequalities and contribute to further degradation of natural resources on which the majority of the poor are dependent. Many of these vulnerabilities are historically located, although the new threats also create new forms of exclusion and inequalities. The final section concludes with some suggestions for policy reforms which could inform progress towards greater equity and sustainability.

Contemporary natural resources control and inequality

Most African economies are predominantly agrarian: the bulk of the populations are rural and depend on agriculture for their livelihood. These economies are based on the super-exploitation of peasant labour in order to subsidize labour reproduction for the formal sector. In turn the exploitation of peasant labour becomes increasingly feminized as male labour is absorbed into the formal sector to the exclusion of female labour. This gendered exploitation is reflected in the extreme inequality between men and women, urban and rural, and black and white in the colonial and post-colonial African economies (Kanyenze and Kondo 2011).

Natural resources governance issues such as land and resource tenure continue to underpin evolving relations between states and citizens in the post-colonial era. Post-independence African governments tended to reinforce centralized authority over natural resources as states sought to consolidate the political authority needed to drive modernization processes and to control patronage resources. Post-colonial Africa has continued on a neoliberal development trajectory which not only creates unsustainable use of natural resources, but also exacerbates the historical inequalities of colonial exploitation (Schneider 2003; Gatwiri et al. 2019).

Natural resources, with their historic grounding in the public domain and their high economic values, are central to the patronage interests that allow governing elites to maintain powers and privileges. This political logic shapes natural resource governance patterns across the continent. For example, agricultural policy in agrarian nations has evolved according to political interests bent towards controlling producers' access to markets and inputs in order to extract rent. Similarly, forestry policy and management institutions across the continent are crafted according to central patronage interests in controlling and extracting rents from both formal and informal patterns of trade and utilization of products such as timber and charcoal (Oyono 2004).

While the continent is engaged in various projects to restructure economies away from the historical limitations into more inclusive and equitable trajectories, such transformational initiatives rarely seek to transcend the neoliberal hegemony.[5] Various attempts have been made to promote equitable access to land and natural resources through redistributive processes such as land reform (Cousins and Scoones 2010; Hall and Kepe 2017; Moyo and Yeros 2013), economic empowerment programmes such as in South Africa and so on. However, these policy reforms are occurring in contexts characterized by the enclosure and privatization of public lands (see e.g. Harvey 2003). In Africa, this phase of globalization is creating new enclosures and dispossessions and thus exacerbating the natural resource–based inequalities of the colonial era (Oakland Institute 2011).

Natural resource degradation contributes to growing inequality. Conservative estimates show that industrial gas emissions have increased by almost 50%; more than 300 million ha of forest have been cleared; and many communities in developing countries have lost rights and access to lands and forests to large multinational corporations acting in collaboration with national governments. Although poverty has been reduced in a few industrializing countries, nearly 20% of the world's population remains in absolute poverty (Watts and Ford 2012), and more continue to be impoverished through land and resource expropriations. The commodification and privatization of the environment has accelerated through increased 'green grabs,' carbon sequestration schemes such as REDD+, water privatization, and the creation of new protected areas on lands expropriated from the poor and marginalized, as well as the suppression of indigenous forms of production and consumption.

One of the key attributes of hyper-globalization is accelerated commodification of nature. As Polanyi (1944) prognosticated, the growth of capital depends in part on its ability to create new (and fictitious) commodities. While the emergence of the market economy was marked the replacement of traditional relations with market relations and the commodification of land and labour, the commercialization of the relationship with nature has been detrimental to both nature and humans (Klein 2008; Polanyi 2001). In addition to the inequalities created by natural resource commodification and exploitation, climate change has emerged as a new driver of inequality, driven by neoliberal commodification and the endless pursuit of growth. We turn now to a discussion of climate change and inequality in Africa.

Climate change and inequality

The analytical framework for understanding the relationship between climate change and inequality remains relatively underdeveloped (Lichenko and Silva 2014). Inequality is multidimensional, while the social impacts of climate change are complex, multifaceted and context specific. However, as noted by the Intergovernmental Panel on Climate Change in its Fifth Assessment Report (AR5), "despite the recognition of these complex interactions between climate change and inequality, the literature shows no single conceptual framework that captures them concurrently" (IPCC 2014: 803). As the physical impacts of climate change have become more glaring, focus has also shifted from discourses on the reality of anthropogenic interference with the climate system to understanding the social impacts of climate change and proposing appropriate policy response to aid adaptation to climate impacts.

Consequently, since the turn of the century, greater attention has been paid to the social impacts of climate change, particularly on poverty and vulnerability. Within the global climate governance framework, a World Bank report (2002) presented at the 8th conference of the UNFCCC highlighted the challenges presented by climate change for the achievement of the MDGs. The Stern report (2007) noted that climate change was expected to increase poverty owing to its effects on agriculture, flooding, malnutrition, water resources and health. These have been followed by numerous other studies and reports exploring the social impacts of climate change.[6] Working Group II (WGII) of the IPCC has contributed significantly to the focus on the human dimensions of climate impacts by providing extensive compilation of the evidence on the dynamic interaction between climate change, livelihoods and poverty.

The AR5 of the IPCC concludes that climate change exacerbates inequalities and notes that socially and geographically disadvantaged people – including people facing discrimination based on gender, age, race, class, caste, indigeneity and disability – are particularly affected negatively by climate hazards (IPCC 2014). As noted earlier, exacerbation of inequality can happen through disproportionate erosion of physical, human and social assets. However, as demonstrated by Nazrul Islam and Winkel (2017), climate change creates a vicious cycle whereby initial inequality makes disadvantaged groups suffer disproportionately from the adverse effects of climate change, resulting in greater subsequent inequality through three channels. First, inequality increases the exposure of the disadvantaged social groups to climate hazards. Because of the exposure level, inequality increases the disadvantaged groups' susceptibility to damages caused by climate hazards. Third, inequality decreases these groups' relative ability to adapt and recover from climate change damages.

Climate change has also become the defining context of natural resources use and control in recent times. An inescapable irony of climate change is that those economies whose development has led to climate change are the least vulnerable to its impacts, while countries that emit the least are the most vulnerable to climate change. This exacerbates inequalities between the nations of the developed North and those of the global South. "Climate change is inextricably linked to economic inequality: it is a crisis that is driven by the greenhouse gas emissions of the 'haves' that hits the 'have-nots' the hardest" (Oxfam 2015). GDP growth can be directly linked to carbon emissions, and those economies with the highest rates of GDP are also those that historically have the highest rates of GHG emissions (WMO 2020).

Many countries, particularly the vulnerable, are already experiencing the negative impacts of climate change. Vulnerability to climate change in Africa has its origins in the colonial encounters of the continent and the West. Centuries of European colonialism, including the creation of new nations in service of Western imperial interests, placed the new nations in a vulnerable position. Mass human rights violations, forced labour and then violent exits by colonial powers left many countries in poverty, with weak institutions and primed for internal conflicts. Climate change and colonial history are a toxic combination. The IPCC (2007, 2014, 2019) establish that climate disasters have a greater impact on developing countries with weak infrastructure than on wealthier countries.

Vulnerability to climate change is globally generalized and locally specific. While everyone is vulnerable to the impacts of climate change, developing countries are much more vulnerable because of the structural and historical factors which restrict their abilities to absorb the costs of climate-related events such as droughts, floods and heat waves, as well as to adapt their economies to operate efficiently and sustainably in a changing climate. They also have less ability to take advantage of the opportunities of responding to climate change – such as investment in clean renewable energy and climate-proofing infrastructure and the adoption of smart agriculture options – without external assistance. It is estimated that COVID-19 will cost the world economy up to 5% of GDP. Climate impacts in Africa are already costing most of the continent's economies between 3% and 5% of GDP annually (UNECA/ACPC undated), with some incurring losses of up to 10% of GDP.

Many poor countries are already affected by frequent floods, storms and droughts. Even if the Paris Agreement and other measures succeed in keeping temperature rises below 1.5 degrees, these impacts will continue and even worsen for several decades to come. Recent estimates suggest that climate change will reduce GDP per capita in Africa between 66% and 90% by 2100 (Hsiang 2016). Under a high-warming scenario, West Africa and Eastern Africa would experience a reduction in GDP per capita by about 15% by 2050 (UNECA/ACPC undated).

It is broadly accepted that climate change exacerbates existing inequalities. In the United Nations Framework Convention on Climate Change (UNFCCC) context of the global climate change negotiations, developing countries have historically premised their demands on the 'ecological debt' owed them by developed countries. This ecological debt emanates from the high carbon footprint of the industrialized nations, particularly from resource extraction and fossil fuel burning for energy. Indeed, the ideas of ecological debt and 'ecologically unequal exchange' have led to the development of the climate justice movement (Parks and Roberts 2010), represented in Africa by the Pan African Climate Justice Alliance (PACJA).

There is an observed linear relationship between GDP and emissions. High-income nations are responsible for over 90% of greenhouse gas emissions, while those countries with low cumulative emissions are among the poorest (WMO forthcoming). Africa contributes less than 4% of cumulative GHG concentrations in the atmosphere and is disproportionately vulnerable to climate impacts (Sy 2018). The global North is responsible for about 50% of all materials consumption, and the bulk of these are extracted from the global South. This extraction is made possible by patterns of control and ownership established during the colonial period and maintained in the post-colonial period. The excess development in the North causes de-development in the South. It follows from this that equality between low- and high-income countries cannot be achieved if the global North does not reduce its ecological impact.

Thus, in addition to its impacts on inequalities within countries, climate change increases inequalities between countries. Greater across-country inequality may indeed increase the exposure of the disadvantaged countries to climate hazards and decrease their capacity to build resilience. We can also speculate that the global climate response itself could exacerbate inequalities between rich and poor countries. In terms of the Paris Agreement, all nations must voluntarily commit to limiting greenhouse gas emissions and adapting to the impacts of climate change. For poor countries, these commitments are typically conditional on the availability of external financing. Financing itself is rarely unconditional. Already, African economies are spending up to 9% of their annual GDP (UNECA/ACPC undated) to finance climate adaptations and responses to the impacts of extreme weather and climate events.

Climate actions undertaken in the context of the global climate governance agreement could also exacerbate Africa's vulnerability through a second pathway. As Lord Stern observed, "the effort to control climate change impacts virtually every element of a country's economy so countries have traditionally been nervous about what they're going to be asked to do" (BBC 2020). This is particularly true of high-income countries whose economic, social and environmental activities remain carbon intensive. The pursuit of economic growth drives ever-increasing exploitation of natural resources and GHG emissions. This makes it impossible to decarbonize economies at a rate fast enough to meet the Paris Agreement threshold of limiting global warming to 1.5 degrees Celsius by reducing carbon emissions by 45% by 2030, and reaching net zero emissions by 2050. To achieve this target, industrialized nations must abandon growth as a political and economic objective.

African countries, on the other hand, with low carbon intensity and greater vulnerability to climate impacts, have been most committed to a working climate governance regime. As such the continent has by and large made vastly ambitious commitments to mitigate emissions. These commitments are in some instances clearly prejudicial to their own development. Meanwhile the big emitters are enlarging their carbon footprint.[7] Thus, for example, in the 2020 Petersberg Climate Dialogue,[8] an annual global meeting of environment and climate ministers, leaders undertook to design their COVID-19 pandemic responses in a way that will drive a transition to more sustainable, zero-carbon societies rather than propping up the polluting practices of the past. Already, however, indications are that there is a gap between these optimistic statements and the unfolding realities on the ground. Governments in some of the leading polluting nations are including bailouts for brown energy and excluding green industries from stimulus packages. The US government has mobilized a massive $2 trillion to support industries and workers affected by the pandemic. This compromises their climate commitments and thus potentially exacerbate inequalities with less powerful nations.

Climate change not only exacerbates inequalities between nations, it also amplifies the vulnerabilities of those communities whose livelihoods are directly dependent on access to natural resources. Most of Africa's rural households' livelihoods are based on access to agricultural land and associated natural resources such as forests and water. Access to these resources is already inequitable because of historical factors. Climate change not only increases the related challenges, but also creates new forms of contestation and alienation within the communities for natural resources. Thus pastoral communities and sedentary communities compete for water and grazing lands, and sedentary communities' access to vital forest products is increasingly contested. As rainfall patterns vary, competition for access to land is also increasing. Climate change will affect the productivity of the land, change access to natural resources and create new conditions for further primitive accumulation, as already evidenced in the expansion in land acquisitions for biofuel production on lands belonging to vulnerable communities.

Conclusion: towards an equitable world order?

The use and control of natural resources has historically generated inequality in Africa. Climate change and the responses to it have exacerbated these inequalities. The dominance of the market, representing corporate interests over social and environmental interests, is clearly socially, economically and environmentally unsustainable. Instead of the current production system that emphasizes market mechanisms to allocate the costs and benefits of nature, what is required is a social structure of accumulation that places economic justice over profit and, more practically, institutes an inclusive, sustainable model for growth (Tabb 2012).

Environmental policies in Africa must focus on better environmental protection and social outcomes in order to reduce inequality. The majority of the rural poor, to varying extents, are directly dependent on the environment and natural resources for their livelihoods. Resource extraction negatively impacts them, as does the degradation and loss of biodiversity emanating from climate change and variability. In addition, environmental policies should also ensure tenure security for the poor in order to incentivize the emergence and development of environmental management institutions that are aligned to the livelihood objectives of the poor.

With reference to climate change, it has been proven that it is possible to decouple emissions from GDP by transitioning to renewables (UNECA 2020). However, this alone is not sufficient to ensure the achievement of the 45% reduction of emissions by 2030 recommended by the IPCC (2018). There are many processes around the world today seeking alternatives to the destructive logic of the hegemonic models of production and consumption. These are evident in a wide range of rich experiences in alternative technology, renewable energy and new regulatory regimes that exist in different parts of the world. It is therefore imperative that high-income nations should abandon growth as a political and economic objective, scale down energy use and meet their obligations to finance climate actions in low-income countries in order to limit emissions and support development in low-income countries. Managed reduction of resource use through efficiency, investment in the circular economy, reduction of planned obsolescence and similar measures will contribute towards shifting the focus away from growth. Development in the global South would need to avoid the pitfalls of fossil fuel–intensive and inequitable development of the North. Progressive development trajectories should shift from growth alone, a focus emanating from the structural adjustment programmes of the 1980s and 1990s (Heidhues and Obare 2011), and focus on other imperatives such as human and environmental wellbeing. As the Intergovernmental Panel on Climate Change proposes, "equitable socioeconomic development in Africa may strengthen its resilience to various external shocks, including climate change" (IPCC 2014: 1121).

Notes

1 Stiglitz (2020: 20) notes, "COVID-19 has exposed and exacerbated inequalities between countries just as it has within countries. The least developed economies have poorer health conditions, health systems that are less prepared to deal with the pandemic, and people living in conditions that make them more

vulnerable to contagion, and they simply do not have the resources that advanced economies have to respond to the economic aftermath." https://www.imf.org/external/pubs/ft/fandd/2020/09/pdf/COVID19-and-global-inequality-joseph-stiglitz.pdf; page 20.

2 See e.g. Institute on Taxation and Economic Policy (January 2015) *Who Pays? A Distributional Analysis of the Tax Systems in All Fifty States*; OXFAM (2015) *Wealth: Having It All and Wanting More*, http://policy-practice.oxfam.org.uk/publications/wealth-having-it-all-and-wanting-more-338125; Thomas Piketty (2014) *Capital in the Twenty-First Century*; Stiglitz, J. (2012) *The Price of Inequality: How Today's Divided Society Endangers Our Future*; Alvaredo, F. et al. (2018) *World Inequality Report 2018*.

3 Oxfam (2020) "Economic inequality is out of control. In 2019, the world's billionaires, only 2,153 people, had more wealth than 4.6 billion people. This great divide is based on a flawed and sexist economic system that values the wealth of the privileged few, mostly men, more than the billions of hours of the most essential work."

4 Kanyenze and Kondo (2011) explore the contradiction of Zimbabwe, where a rich, diverse resource base co exists with endemic poverty. They conclude that one reason lies in the colonial economy, which was predicated on an ideology of white supremacy, creating an enclave formal economy employing only one-fifth of the labour force, with the rest forced into informality.

5 Murombedzi, J.C. (2016), *Inequality and Natural Resources in Africa*. World Social Science Report 2016. UNESCO.

6 For a synopsis of studies see e.g. S. Nazrul Islam and J. Winkel (2017). *Climate Change and Social Inequality*. www.un.org/esa/desa/papers/2017/wp152_2017.pdf.

7 In March 2020 China approved five new coal-fired power plants with a total of 7,960 MW (as opposed to 6,310 MW coal-fired power stations approved in the country in all of 2019). In Canada, the government has extended direct tax relief to the Alberta tar sands industry as well as for the renovation of oil wells in Saskatchewan and British Columbia as part of its bailout plan to industries. Australia has also put in place provisions to waive oil and gas exploration fees and approved the expansion of the Acland coal mine. In the UK, the Bank of England has undertaken to buy debt from oil companies as part of its coronavirus stimulus programme. Similarly, the EU has agreed that member states' COVID-19 pandemic response must be aligned with the Union's Green Deal, and the European Central Bank has issued 870 billion euro through its Pandemic Emergency Purchase Programme to buy back bonds to stabilize the Euro. Some of the bonds purchased in the first three weeks of the programme include oil majors Shell, ENI and Total. These patterns of COVID-19 stimulus packages benefitting the fossil fuel industries are also evident in the BRICS countries. Quite clearly, the COVID-19 pandemic is imposing Faustian choices on countries resulting in the more powerful countries embarking on stimulus packages which promote the interests of powerful fossil fuel concerns.

8 Video conference on 29/4/2020.

Bibliography

Acedo, A. (2015) *Climate Change and Capitalism: Challenges of the COP21 Paris and Climate Movements*, 27 August. www.counterpunch.org/2015/08/27/climate-change-and-capitalism-challenges-of-the-cop21-paris-and-climate-movements/.

Alvaredo, F., T. Piketty, E. Saez & G. Zucman (eds.). (2018) *World Inequality Report 2018*. Cambridge: Harvard University Press.

Asefa, Sisay (ed.). (2010) *Globalization and International Development: Critical Issues of the 21st Century*. Kalamazoo, MI: W.E. Upjohn Institute for Employment Research. http://dx.doi.org/10.17848/9781441678829.

Borras, S. M., P. McMichael & I. Scoones (2010) The Politics of Biofuels, Land and Agrarian Change: Editors' Introduction. *The Journal of Peasant Studies* 37 (4), 575–592. doi:10.1080/03066150.2010.512448.

British Broadcasting Corporation. (2020) Has the World Started to Take Climate Change Fight Seriously? *BBC*, 30 September. www.bbc.com/news/science-environment-54347878.

Cousins, B. & I. Scoones (2010) Contested Paradigms of 'Viability' in Redistributive Land Reform: Perspectives from Southern Africa. *The Journal of Peasant Studies* 37 (1), 31–66.

Deininger, Klaus, D. Byerlee, J. Lindsay, A. Norton, H. Selod & M. Stickler (2011) *Rising Global Interest in Farmland: Can it Yield Sustainable and Equitable Benefits?* Agriculture and Rural Development. © World Bank. https://openknowledge.worldbank.org/handle/10986/2263 License: CC BY 3.0 IGO.

Gatwiri, K., J. Amboko & D. Okolla (2019) The Implications of Neoliberalism on African Economies, Health Outcomes and Wellbeing: A Conceptual Argument. *Social Theory & Health* 18 (1), 86–101. Published online 26 June 2019. doi:10.1057/s41285-019-00111-2PMCID: PMC7223727 PMID: 32435159.

Hall, R. & T. Kepe (2017) Elite Capture and State Neglect: New Evidence on South Africa's Land Reform. *Review of African Political Economy* 44, 151.

Harvey, D. (2003) *The New Imperialism*. Oxford: Oxford University Press.

Heidhues, F. & G. Obare (2011) Lessons from Structural Adjustment Programmes and Their Effects in Africa. *Quarterly Journal of International Agriculture* 50 (1), 55–64.

Hsiang, S. (2015) *Economic Risks of Climate Change: An American Prospectus*. New York: Columbia University Press.

Hsiang, S. (2016) Climate Econometrics. *Annual Review of Resource Economics* 8 (1), 43–75.

IPCC. (2007) *Climate Change 2007: Synthesis Report. Contribution of Working Groups I, II and III to the Fourth Assessment Report of the Intergovernmental Panel on Climate Change*. Geneva: IPCC.

IPCC. (2014) *Climate Change 2014: Synthesis Report. Contribution of Working Groups I, II and III to the Fifth Assessment Report of the Intergovernmental Panel on Climate Change*. Geneva: IPCC.

IPCC. (2018) Summary for Policymakers. In: *Global Warming of 1.5°C: An IPCC Special Report on the Impacts of Global Warming of 1.5°C Above Pre-Industrial Levels and Related Global Greenhouse Gas Emission Pathways, in the Context of Strengthening the Global Response to the Threat of Climate Change, Sustainable Development, and Efforts to Eradicate Poverty*. Geneva: IPCC.

IPCC. (2019) Summary for Policymakers. In: *Climate Change and Land: An IPCC special Report on Climate Change, Desertification, Land Degradation, Sustainable Land Management, Food Security, and Greenhouse Gas Fluxes in Terrestrial Ecosystems*. Geneva: IPCC.

Joyce, J. P. (2010) Globalization and Inequality Among Nations. In: A. Sisay (ed.), *Globalization and Internationald Development: Critical Issues for the 21st Century*. Kalamazoo, MI: W.E. Upjohn Institute.

Kanyenze, G. & T. Kondo (eds.). (2011) *Beyond the Enclave. Towards a Pro-Poor and Inclusive Development Strategy for Zimbabwe*. Harare: Weaver Press.

Klein, N. (2008) *The Shock Doctrine*. London: Penguin Books.

Klein, N. (2014) *This Changes Everything: Capitalism vs. the Climate*. New York: Simon and Schuster.

Lander, E. (2012) *The Green Economy: The Wolf in Sheep's Clothing*. www.amandlapublishers.co.za/special-features/the-green-economy/, last visited 27 August 2012.

Lichenko, R. & J. A. Silva (2014) Climate Change and Poverty: Vulnerability, Impacts and Alleviation Strategies. *WIREs Climate Change* 5, 539–556.

Maloney, J. S. *Land Grabs and Food Sovereignty: A Lecture for 21st Century Marxism: International Viewpoint. News and Analysis from the Fourth International*. www.internationalviewpoint.org.

Mamdani, M. (1996) *Citizen and Subject: Contemporary Africa and the Legacy of Late Colonialism*. Princeton: Princeton University Press.

McGillivray, M. (2006) *Inequality, Poverty and Well-Being*. Basingstoke: Palgrave Macmillan. www.palgrave.com/page/detail/inequality-poverty-and-wellbeing-mark-mcgillivray/?isb=9781403987525.

Moyo, S. & P. Yeros (2013) The Resurgence of Rural Movements Under Neo-Liberalism. In: S. Moyo & P. Yeros (eds.), *Reclaiming the Land: The Resurgence of Rural Movements in Africa, Asia and Latin America*. London: Zed Books.

Murombedzi, J. & J. Ribot (2012) *Occupy Nature: Representation as the Basis of Emancipatory Environmentalism*.

Nazrul Islam, S. & John Winkel (2017) *Climate Change and Social Inequality*. Working Paper No. 152S T/ESA /2017/DWP/152October 2017, UN DESA.

Oakland Institute. (2011) *Understanding Land Investment Deals in Africa*. Country Report: Ethiopia. http://media.oaklandinstitute.org/publications.

Ocheni, S. & B. C. Nwankwo (2012) Analysis of Colonialism and Its Impact in Africa. *Cross-Cultural Communication* 8 (3), 46–54.

Oxfam. (2015) *Extreme Carbon Inequality: Why the Paris Climate Deal Must Put the Poorest, Lowest Emitting and Most Vulnerable People First*. Oxfam Media Briefing, 2 December. www.oxfam.de/system/files/oxfam-extreme-carbon-inequality-20151202-engl.pdf.

Oxfam. (2020) *Time to Care: Unpaid and Underpaid Care Work and the Global Inequality Crisis*. Oxford: Oxfam.

Oyono, P. R. (2004) Assessing Accountability in Cameroon's Local Forest Management. Are Representatives Responsive? *African Journal of Political Science* 9 (1), 126–136.

Parks, B. C. & J. T. Roberts (2010) Climate Change, Social Theory and Justice. *Theory Culture Society* 27, 134.

Piketty, T. (2014) *Capital in the Twenty-First Century*. Cambridge: Harvard University Press.

Polanyi, K. (2001) *The Great Transformation: The Political and Economic Origins of Our Time, 1944, 1957*. Boston: Beacon Press.

Schneider, G. E. (2003) Neoliberalism and Economic Justice in South Africa: Revisiting the Debate on Economic Apartheid. *Review of Social Economy* 61 (1), 23–50, March.

Shivji, I. (1998) *Not Yet Democracy: Reforming Land Tenure in Tanzania*. Dar es Salaam: IIED/HAKIARDHI/ Faculty of Law, University of Dar es Salaam.

Stern, N. H. (2007) *The Economics of Climate Change: The Stern Review*. Cambridge: Cambridge University Press.

Stiglitz, J. (2012) *The Price of Inequality: How Today's Divided Society Endangers Our Future*. New York: W.W. Norton.

Stiglitz, J. (2020) Conquering the Great Divide. *IMF Finance and Development*. Fall Issue. www.imf.org/ external/pubs/ft/fandd/2020/09/COVID19-and-global-inequality-joseph-stiglitz.htm.

Sy, A. (2018) *Africa: Financing Adaptation and Mitigation in the World's Most Vulnerable Region*. www.brookings.edu/wp-content/uploads/2016/08/global_20160818_cop21_africa.pdf.

Tabb W. K. (2012) *The Restructuring of Capitalism in Our Time*. New York: Columbia University Press.

UNECA. (2020) *SDG7 Initiative for Africa: Investing in Clean Energy – Brochure*. www.uneca.org/publications/ sdg7-initiative-africa-investing-clean-energy-brochure.

UNECA/ACPC. (undated) *Economic Growth, Development and Climate Change: Summary for Policy Makers*. UNECA. www.uneca.org/sites/default/files/uploaded-documents/ACPC/annex_21a_-_climate_ change_impacts_on_africas_economic_growth_-_summary_for_policy_makers.pdf.

United Nations Department of Economic and Social Affairs. (2007) *Sustainable Consumption and Production: Promoting Climate-Friendly Household Consumption Patterns (United Nations Department of Economic and Social Affairs Division for Sustainable Development Policy Integration and Analysis Branch)*. www.greeningtheblue.org/sites/default/files/Sustainable, last visited 29 August 2012.

United Nations Environmental Programme (UNEP). (2011) *Towards a Green Economy: Pathways to Sustainable Development and Poverty Eradication, a Synthesis for Policy Makers*. www.unep.org/greeneconomy.

Watts, J. & L. Ford (2012) *Rio+20 Earth Summit: Campaigners Decry Final Document*. www.guardian.co.uk/ environment/2012/jun/23/rio-20-earth-summit-document, last visited 28 August 2012.

White, B., S. M. Boras, R. Hall, I. Scoones & W. Wolford (2012) The New Enclosures: Critical Perspectives on Corporate Land Deals. *Journal of Peasant Studies* 39 (3–4), 619–647.

World Bank. (2002) *World Development Report 2003: Sustainable Development in a Dynamic World: Transforming Institutions, Growth, and Quality of Life*. Washington, DC: World Bank.

World Bank. (2013) *Turn Down the Heat: Climate Extremes, Regional Impacts, and the Case for Resilience*. A Report for the World Bank by the Potsdam Institute for Climate Impact Research and Climate Analytics. Washington, DC: World Bank.

World Meteorological Organization. (2020) *State of the Climate in Africa 2019*.

World Meteorological Organization. (forthcoming) *State of Climate over Africa Report*. WMO.

World Rainforest Movement. (2012) *New Crossroads, Same Actors: The Green Economy of the Powerful, Voices of Resistance of Women*. www.wrm.org.uy/plantations/working_conditions.html, last visited 28 August 2012.

11

FROM WESTERN PENNSYLVANIA TO THE WORLD

Environmental injustice and the ethane-to-plastics global production network

Diane M. Sicotte

Although environmental injustice and inequality have been the focus of much scholarship, we are only beginning to understand the connections between environmental injustice and the profitability of polluting and unsustainable industries. Within the global economic order, certain places are defined as sacrifice zones (Lerner 2012), and the externalities of polluting and unhealthy industries are imposed upon the marginalized groups of people who inhabit such places as if they too are expendable (Pellow 2018). This is enabled by lax and selective enforcement of environmental regulations, which results in the preservation of profitability for polluting businesses (Pulido 2017).

In this chapter, I examine a new strand of the global production network for plastics originating in Western Pennsylvania, USA. There, ethane, a fracked gas liquid, is produced and transported by pipeline and export terminal to petrochemical complexes in various nations to manufacture plastic resins, which are used to make plastic goods. At each stage, from the extraction of crude gas to the final disposal of discarded single-use plastic items, ecosystems are disrupted and groups of people suffer exposures to toxic materials. Although this particular strand of the global production network for petrochemicals is by no means the largest, it neatly illustrates two features of local-global environmental injustice: the political economic context behind the increasing production and use of unsustainable products, and the way that the profitability of unsustainable practices is supported by negative externalities imposed upon less powerful groups of people.

According to some researchers, the production of cheap energy (first from coal, then from oil) has, throughout history, provided a long-term "fix" for the falling rate of profit that drives capitalism's periodic crises of underproduction (Moore 2011). But others view overaccumulation, sometimes felt as a glut of investment capital and sometimes as commodities gluts, as the driver of crises within capitalism. Overaccumulation is remedied by investments in the built environment or the opening of new markets (Harvey 2016). Elsewhere, I have argued that the production of cheap ethane in the U.S. is poised to solve the "gas glut" crisis by fueling an increase in worldwide plastics production, shoring up the profitability of the industry (Sicotte 2020). Globally interlinked production chains for commodities (Bair 2005) coordinate the

DOI: 10.4324/9780367814533-12

buildout of the infrastructures required to transport and refine the raw material for plastic resin manufacturing and help to develop new markets for finished plastic goods.

In this chapter, I trace the production of environmental injustice throughout the production network: from the production of fracked ethane in Western Pennsylvania, along high-pressure ethane pipelines, to liquid natural gas export terminals that transform ethane for shipping overseas, to petrochemical plants where ethane is transformed into ethylene to create plastic resin, and finally to sites of disposal for discarded plastics. I conclude by examining the place of these unsustainable products in society and offer some thoughts on phasing out their use.

Fracked ethane

The raw material at the start of the Western Pennsylvania production chain for plastics is ethane, a hydrocarbon gas liquid found in crude natural gas. Compared with other feedstocks used to make plastics (such as naphtha), ethane is more efficiently converted into its plastic building block ethylene, which lowers the cost of plastic resin production (NAS 2016). Before 2009, U.S. gas producers found it too expensive to separate out ethane and produced only 700,000 thousand barrels of ethane per day. But the steep drop in gas prices that year motivated widespread investments in high-pressure pipelines, which could transport gas less expensively than freight trains or tanker trucks (Frittelli et al. 2014) but required that the ethane be separated out to decrease the flammability of the gas. Selling the ethane allowed producers to get back some of the money lost due to low gas prices (Moore 2015). Competition from Saudi gas producers, combined with drillers' need to continue overproducing gas in order to fulfill "deliver or pay" contracts with pipeline companies and refiners (Moore 2015), had caused a gas glut that kept natural gas prices low. Accordingly, in 2018, the U.S. produced a record 1,500,000 barrels of ethane per day (USEIA 2019a). This was the context for the petrochemical industry's announcement that it had invested more than $200 billion in building new pipelines and petrochemical plants to support the production of more plastics (American Chemistry Council 2018).

Much of Pennsylvania overlays the Marcellus Shale gas deposit, where crude gas is particularly rich in ethane (USDOE 2018). Unconventional gas drilling (or "fracking") is a method of horizontal drilling that involves the fracturing of shale rock and the high-pressure injection of chemicals, water and sand in order to force the gas to the surface (Hauter 2016). In the last decade, environmental social movements have risen in protest over fracking in Pennsylvania (Staggenborg 2018), the U.S. (Ladd 2018), and the world (Cheon and Urpelainen 2018). In the Southwest region of the U.S., unconventional gas wells tend to be located near the homes of African American and Hispanic low-income people, but in Pennsylvania the population affected by proximity to fracking operations was found to be predominantly non-Hispanic, white and low-income (Zwickl 2019).

The health of people who live near wellpads where fracking is occurring is at risk from air and water pollution with toxic, flammable and radioactive substances, which threaten the health of people and farm animals (Adgate, Goldstein and McKenzie 2014; Andrews and McCarthy 2014; Hauter 2016; Healy, Stephens and Malin 2019; Kinchy, Parks and Jalbert 2016; Perry 2013; Shonkoff, Hays and Finkel 2014). Pregnant women exposed to fracking activities were more likely to give birth to premature babies, and more fracking activities were more strongly predictive of preterm birth (Casey et al. 2016). Fracking may also be exposing people to endocrine-disrupting chemicals, which can cause abnormalities of the reproductive system, infertility, birth defects of the heart and neural tube, and hormone-related cancers (Kassotis et al. 2016).

Besides the impact on physical health, fracking can impoverish farmers by contaminating the farm's water supply, which can kill or sicken livestock and destroy crops (Malin and DeMaster 2016). Contaminated homes near fracking operations can rapidly lose both their use value and their exchange value (Logan and Molotch 2007), leaving the homeowner homeless (Griswold 2018). While some landowners are enjoying royalties from gas leases, the economic fortunes of the low-income people living nearest gas wells remain unimproved (Clough and Bell 2016). Although compensation for workers in unconventional gas drilling is relatively high, jobs are often of short duration (Hauter 2016). Workers in the industry face health and safety hazards from silica, toxic chemicals, diesel exhaust and traffic. Fatalities were highest for those working for contractors or small companies (Korfmacher et al. 2013).

What is more, the gas drilling industry is financially risky: drilling operations are largely financed by debt, and the price of its products is volatile (Lips 2018). The drilling, pipeline, export terminal and petrochemical plant sectors of the industry are tightly coupled (Perrow 2011) and financially interdependent; if losses are suffered by one sector, the profitability of other sectors is threatened. With so much money at stake, democratic debate over how to protect human health and ecosystems generates controversy and delay, increasing costs and threatening profitability. Democratic decision-making about gas drilling has been prevented by lobbying and federal laws, such as the Energy Policy Act of 2005 that exempted the industry from major environmental laws (Warner and Shapiro 2013), and by oppressive state laws. One such law was Act 13, passed in Pennsylvania in 2013 but struck down as unconstitutional by the Commonwealth Supreme Court in 2016. Among other pro-industry stipulations, Act 13 made it illegal for local municipalities to require that fracking wells be dug at a protective distance from homes, schools and hospitals and prohibited physicians from discussing toxic chemicals defined as "trade secrets" with ailing patients from fracking communities (Andrews and McCarthy 2014). In contrast with federal-level deregulation, these sweeping state laws are an example of "re-regulation," in which laws intended to protect the public were rewritten to protect business interests (Block and Somers 2014).

In Pennsylvania, the emphasis on protecting industry was communicated to state agencies and was the likely reason for their hostile or inadequate responses to resident complaints about hazards from fracking (Hauter 2016; Malin 2014), documented in a Pennsylvania Grand Jury report (Commonwealth of Pennsylvania 2020). The inadequate response from the state reflects both the absence of federal-level regulation (Warner and Shapiro 2013) and the disproportionate power of Pennsylvania's coal, oil and gas industries to influence state-level policy on unconventional gas drilling (Rabe and Borick 2013).

As fracking increased, so has the volume of contaminated water produced by fracking wells. The Marcellus region is geologically unsuited for the industry's preferred disposal method: underground injection of wastewater. In the first years after fracking began, gas corporations sent their wastewater to municipal water treatment plants; but fracking wastewater contains salty brine, toxic chemical compounds, heavy metals and naturally occurring radioactive materials. After it was recognized that municipal treatment plants could not effectively treat large volumes of severely contaminated water, Pennsylvania regulations were tightened, and by 2011 a large portion of fracking wastewater from Pennsylvania was sent to Ohio for underground injection (Lutz, Lewis and Doyle 2013).

Deep well injection of fracking wastewater carries risks such as the leaking of wastewater into groundwater. The injection of wastewater destabilizes bedrock, causing earthquakes (Johnston, Werder and Sebastian 2016; Mix and Raynes 2018). In Ohio, a state where earthquakes had never occurred, 400 small earthquakes occurred during 2013 alone (Ehrman 2016). As was true of people near fracking wells in Pennsylvania, most Ohio residents near injection wells were rural, white and low-income (Silva, Warren and Deziel 2018).

In economic terms, risks and damages from fracking and wastewater disposal are negative externalities, in this case costs of production outsourced onto the public instead of being accounted in the cost of doing business. Profit-driven corporations have strong incentives to maximize externalities (Biglan 2009). Because so few harmed by fracking have been compensated, unconventional gas well drillers have insufficient economic incentives to conduct their activities safely (Centner and Eberhart 2016). The vast majority of the people facing physical or financial threats from fracking and related activities had incomes lower than average, illustrating how negative externalities of production tend to be distributed "in the most politically expedient manner" through the "selective victimization of marginalized people and communities" (Faber 2008, pp. 24–25). With enforcement by state and federal governmental agencies lacking, their only recourse is through environmental justice activism, which publicizes environmental damage and health threats from natural gas drilling (Staggenborg 2018).

High-pressure transmission pipelines

The safest and most inexpensive way to transport large quantities of fracked ethane to refineries and export terminals is through huge, high-pressure transmission pipeline networks. One such pipeline is the problem-ridden Mariner East 2 ethane pipeline, which is currently being built in the suburban counties close to Philadelphia, cutting across land owned mostly by white, middle-class families (Phillips 2021). The Federal Energy Regulatory Commission (FERC) handles the permitting of interstate pipelines. Despite being a federal agency, and thus being directed by Executive Order 12898 (enacted in 1994) to consider environmental justice impacts, FERC has failed to do so (Finkel 2018). FERC has been criticized for considering pre-certification contracts as proof that the pipelines are a necessary convenience for the public (U.S. House Subcommittee on Energy 2020). A certification of public convenience allows pipeline developers to use eminent domain to take the land of unwilling homeowners. But the premise for the use of eminent domain was flimsy at best: the Mariner East pipelines were built to transport ethane, which is not used for household heating and cooking; and the pipelines were large-size transmission pipelines, not the smaller distribution pipelines that facilitate public access to gas (Finkel 2018). Since 1999, FERC has approved all but two of 487 proposed pipelines (U.S. House Subcommittee on Energy 2020). Homeowners' fears about pipelines were realized in 2017, when the Mariner East 1 pipeline leaked 20 barrels of ethane and propane into the soil in Berks County and was later shut down by the Public Utility Commission after sinkholes opened up in a residential neighborhood where the pipeline was being built. Since the start of Mariner East 2 pipeline construction in 2017, the Pennsylvania Department of Environmental Protection has issued more than 60 violations and penalties of $12.6 million against pipeline developer Energy Transfer Partners for polluting wetlands and waterways and destroying more than 10 private wells (Cusick 2018).

The latest site of controversy is the Shell Falcon Pipeline currently under construction, which carries ethane 97.5 miles across Pennsylvania, Ohio and West Virginia, much of it through rugged, mountainous terrain where pipelines cannot simply be laid in a ditch, and developers must drill horizontally through rock. This increases the chances of accidental releases of toxic drilling mud and fluids, which has occurred in over 20 incidents spilling over 5,500 gallons – amounts that Ohio and Pennsylvania regulators suspect to be underestimates (Kelly 2020). Best management practices that could avert destructive spills, releases and accidents exist but are not mandated by either the federal government or state governments due to economic incentives to ignore negative externalities, failure to update regulations that could effectively govern unconventional drilling, lack of scientific information, and the difficulty of proving in court that parties have been damaged by extraction and transportation of natural gas and ethane (Centner and Eberhart 2016).

Similar deficiencies of governance are playing out on a global scale: the Falcon Ethane Pipeline is only one in a massive, worldwide buildout of gas and gas liquid pipelines. Since 2009, an average of 25 new pipelines per year have been added to the global system (Nace, Plant and Browning 2019). Fifteen thousand miles of high-pressure gas pipelines are currently under construction in China, as are 11,000 in Russia, with smaller projects under construction in Europe and the Middle East (Await 2020). In the face of falling demand for fossil fuels, the production of ethane and a sharp increase in plastics production could fill deep revenue holes in the petrochemicals industry while increasing air and water pollution and the production of disposable plastic goods that will add to the world's trash disposal problems.

Liquid Natural Gas (LNG) terminals

In the U.S., the Natural Gas Act of 1938 makes the Department of Energy and the Federal Energy Regulatory Commission (FERC) jointly responsible for administering international energy activities. FERC is responsible for issuing permits for liquid natural gas (LNG) export terminals, which liquefy gas and gas liquids such as ethane for shipping. While most LNG export terminals handle natural gas, ethane transported through the Mariner East 1 and 2 pipelines supply the LNG export terminal at Marcus Hook, PA, which liquefies ethane for shipping overseas. LNG terminals are essential for creating international markets for ethane. A second ethane export terminal is planned for Nederland, Texas, primarily to supply ethane to petrochemical plants in China (U.S. EIA 2019b). The globally interlinked ethane commodity chain enables cheap ethane from the U.S. to fuel increased plastics production in China and elsewhere.

FERC has been criticized by environmental justice activists for regulatory bias and vagueness, for failure to consider cumulative impacts or alternatives to the project, and for only allowing public input at later stages when it is less effective (U.S. House Subcommittee on Energy 2020). Its permitting methods were characterized as a source of procedural injustice by communities fighting pipelines and LNG export terminals (Finley-Brook et al. 2018), which have prevented LNG host communities from participating in deliberations about the risks and benefits of the proposed terminals. Although relatively few communities have risen in protest against LNG terminals (McAdam and Boudet 2012), they present risks from leaks, spills, explosions and fires. Although such accidents are relatively uncommon, they can be disastrous: if a flammable vapor cloud caused by an ethane leak explodes, it would cause an uncontrollable fire (Alderman 2005). LNG export terminals tend to be located in low-income communities such as Marcus Hook, PA, where median household income for 2018 was $34,930, much lower than the U.S. median household income of $60,293 (U.S. Census Bureau 2019). Low-income communities lack the resources to withstand and recover from accidents and disasters, leaving them more vulnerable to physical and financial harms (Walker 2012). Poor coastal communities also face harm from spills and accidents during shipping, as has occurred in communities in Louisiana and Texas near petrochemicals shipping operations (Mah 2015).

Plastic resin factories

The newest plastics factory in the Pennsylvania ethane commodity chain is the Shell Pennsylvania Petrochemical Complex, located in Potter Township in Western Pennsylvania. Currently under construction, it has the capacity to refine ethane into 1.6 million tons of polyethylene per year (Shell Co., n.d.). The two most common types of polyethylene are low-density polyethylene (or LDPE), which is used to manufacture plastic wrap and plastic grocery bags, and high-density polyethylene (HDPE), which is used for crates, bottles, boxes, food containers and toys (PlasticsEurope 2016). Industry analysts have estimated that between 26% (World Economic Forum 2016) and 39.9% (PlasticsEurope 2016) of all plastic produced is used for packaging, materials that are non-durable and meant to be discarded.

The Gulf of Mexico region, not Pennsylvania, produces the lion's share of U.S. plastic resins. In 2012, Texas produced 44% of U.S. resin, and Louisiana produced an additional 7%, compared with 3% in Pennsylvania (American Chemistry Council 2013). By 2016, just three of the six new petrochemical plants planned for the Gulf region were in operation, but U.S. plastics production had already increased 15.84% since 2011 (USEIA 2018). In 2016, China produced 29% of the world's plastics, while 18% were manufactured in North America (PlasticsEurope 2019). Since 2017, India has begun importing about a third of U.S. ethane exports annually (USEIA 2020) and may use the ethane to increase plastics production. Worldwide, people living near petrochemical plants face health and safety risks from fires, explosions, and accidental and chronic chemical releases (Allen 2018; da Rocha et al. 2018; Davies 2018; Jephcote and Mah 2019; Lerner 2012; Mah and Wang 2019). They suffer "slow violence" (Nixon 2011) from the cumulative, long-term effects of exposure to carcinogens, such as benzene, hexane and heptane, and respiratory system irritants, such as sulfuric acid, that trigger asthma attacks (Cutchin 2007; Lerner 2012; Singer 2011; Villarosa 2020; Wright 2005). In the U.S., communities located near huge petrochemical plants tend to be poor and predominantly African American or Hispanic (Lerner 2012; Wright 2005). Regulations limit the amount of toxic chemicals that can be emitted by petrochemical plants, but residents of fenceline communities complain that violations occur regularly (Lerner 2012). In all of these situations, environmental injustice occurs through non-enforcement of existing laws, illustrating how environmental injustice is generated by power imbalances between social movements, the state and capital (Faber 1998).

Plastic trash

While the petrochemical industry's wastes and emissions are obviously toxic, very little is known about the health consequences of allowing plastic garbage to break down into toxic components in the environment, where they leach into drinking water and become part of the food web (Smith et al. 2018). Some of the substances in plastics mimic the effects of female hormones in the bodies of humans and animals (Teuten et al. 2009). Exposure to these substances can cause decreased male fertility, birth defects, reproductive cancers and various health problems such as insulin resistance, obesity and neurobehavioral problems in children (North and Halden 2013).

Since 1950, 8,300 million metric tons of plastic have been produced globally. Seventy-nine percent of this total is still in the global environment, slowly decomposing in the ocean, riverside dumps or landfills. Twelve percent was incinerated, and only 9% recycled (Jambeck et al. 2015). The official plastics recycling rate for the U.S. was 8.5% to 9% as of 2017 (U.S. EPA 2019). However, only 50% of this may really be recycled. Investigative reporters estimated that in 2019 50% of all U.S.-generated plastic waste counted as recycled was actually shipped abroad (McCormick et al. 2019). Compared with paper, metal and glass, plastics are the least recycled substances because their varied chemical composition makes them difficult to recycle into products of equivalent quality.

In the U.S., landfills tend to be located in low-income rural communities, often near the homes of racial/ethnic minority people (Bullard 2000; Pellow 2004; Sze 2006; Taylor 2014).

When plastics are burned in "waste-to-energy" incinerators, some energy can be recovered from the waste, but at the cost of toxic air emissions including chlorine, dioxins and polystyrene. The ash residue from incineration is extremely toxic, containing high concentrations of heavy metals such as lead, arsenic and cadmium (Verma et al. 2016). In 2017, plastics made up 13.14% of the U.S. waste stream, 16% of which (1.24 million tons) were incinerated (U.S. EPA 2019). Eighty percent of U.S. trash incinerators are located in places populated by low-income people of color (Tishman Environment and Design Center 2019).

Until recently, much of the plastic waste generated in the U.S. was shipped to China for use as a feedstock for manufacturing plastic goods. But instead of raw material, this "recycled" plastic was actually part of the global waste circuit in which waste from wealthy countries ends up in poorer regions of the world (Pellow 2007). After the Chinese discovered that most of the imported plastic waste was too contaminated to use, it was burned or dumped in rivers there or in neighboring countries with poor waste disposal infrastructures. The inevitable result was the accumulation of massive amounts of plastic waste in the world's oceans (Dauvergne 2018; Geyer, Jambeck and Law 2017). In 2017, China instituted a ban on the importation of foreign plastic wastes (Li and Chen 2019). In June 2019, India, Malaysia, Poland, Taiwan, Thailand and Vietnam forestalled their own role as lead disposer of U.S. plastic waste by following suit and banning foreign plastic wastes (BBC News 2019). Turkey has become the newest destination for U.S. plastic waste (McCormick et al. 2019), and more of it is likely to find its way to incinerators in marginalized communities in the U.S. Domestic displacement of plastic wastes transfers the waste problem from affluent consumers to poor communities of color already vulnerable to illness (Morello-Frosch et al. 2011), while the export of plastic waste transfers the waste problem to poorer communities abroad that are ill-equipped to deal with it (Clapp 2002). The European Union has reacted by banning the sale of the 10 single-use plastic items most frequently found befouling the ocean (Rankin 2019). But the plastics industry is betting investment monies on the lack of such bans in the U.S., combined with growing demand for disposable plastic bottles, forks, bags and the like in China and India. Without both factors, a large increase in the manufacture of disposable plastic products is unlikely to occur.

Conclusion

Examining the strand of the ethane-to-plastics global supply chain originating in Pennsylvania, USA, reveals the versatility of the uses of hydrocarbons, which is one reason they are so deeply entrenched in the world economy. The global petrochemicals industry generates vast sums of money producing plastics, chemicals, drugs and pesticides, and its investments in ethane crackers, export terminals and petrochemical plants worldwide will ensure that environmental injustices generated by plastics production will continue even if fossil fuel use is phased out.

In rural Pennsylvania, environmental "classism" (Krieg 2005) is a more accurate label for the situation than the environmental racism evident in other regions of the U.S. In recent years, booming hydrocarbon production has expanded sacrifice zones into the backyards of predominantly white, middle-class suburban communities (Klein 2015), such as those currently fighting ethane pipelines in the suburbs of Philadelphia. While such relatively privileged people possess more political and economic power, and thus are generally more successful in imposing delays and costs on pipeline projects, they are still in the path of natural gas and gas liquids infrastructures. Together with poorer and more marginalized people, they have seen their health and safety, the value of their homes and their procedural rights sacrificed to foster the production of gas liquids and plastics.

The products of the global ethane-to-plastics commodity chain are truly (and deliberately) expendable, and at the endpoint of this global production network is all the discarded used plastic packaging, bags and other items. One of the most significant threats to the industry's plan to increase the production of disposable plastic items is global awareness of the extensive pollution of the world's oceans with plastic garbage. The heaps of plastic waste currently decaying in the oceans attests to the utter failure of plastics recycling, the industry's answer to the problem. The lack of demand for recycled plastic goods means that the world will continue to be utterly unable to dispose of the huge volumes of plastics sustainably, even at current levels of production. In order to neutralize public resistance to increased production of disposable goods, overseas shipments of plastic waste must be made invisible by counting them as "recycled" when they are really being shipped to poorer nations with very little capacity to effectively recycle the waste. The suffering of poorer communities and communities of color in the U.S. and abroad, who currently bear the brunt of plastic waste disposal, must continue to be invisible so that more and more disposable plastic items can be manufactured for the world to discard.

Reducing the volume of disposable plastics produced is the only sustainable path forward. But it will be vigorously resisted by the petrochemicals industry, which is a well-financed and politically well-connected opponent. Instituting an EU-style ban on the production and use of disposable plastic items in the U.S. would face many serious obstacles and would have little chance of succeeding unless there is drastic change in the political landscape.

And yet, for all of its power, the petrochemicals industry is not invulnerable. The cost of new regulations or falling demand imposed upon one sector will impact the profitability of other sectors. The boom-and-bust cycle of the petrochemicals business and the tendency toward debt financing results in financial precarity. The industry tends to use undemocratic methods of gaining public consent and to generate environmental injustice and ecosystem disruption at many different places in the world. Both spur public demands for change, creating political vulnerability. The surge of investment in plastics production is still very recent. We do not yet know whether public awareness or political will can block the petrochemicals industry from turning cheap ethane into increased amounts of toxic plastic trash. But as we witness the suffering of far too many people, many of us are hoping it will be so.

Bibliography

Adgate, J.L., Goldstein, B.D. and McKenzie, L.M., 2014. Potential public health hazards, exposures and health effects from unconventional natural gas development. *Environmental Science & Technology*, 48(15), pp. 8307–8320.

Alderman, J.A., 2005. Introduction to LNG safety. *Process Safety Progress*, 24(3), pp. 144–151.

Allen, B.L., 2018. Strongly participatory science and knowledge justice in an environmentally contested region. *Science, Technology, & Human Values*, 43(6), pp. 947–971.

American Chemistry Council, 2013. *Plastic Resins in the United States*. Available at: https://www.packaginggraphics.net/plasticResinInformation/Plastics-Report.pdf (Accessed October 8, 2018).

American Chemistry Council, 2018. *U.S. chemical industry investment linked to shale gas reaches $200 billion*. Available at: www.americanchemistry.com/Media/PressReleasesTranscripts/ACC-news-releases/US-Chemical-Industry-Investment-Linked-to-Shale-Gas-Reaches-200-Billion.html (Accessed October 8, 2018).Andrews, E. and McCarthy, J., 2014. Scale, shale, and the state: Political ecologies and legal geographies of shale gas development in Pennsylvania. *Journal of Environmental Studies and Sciences*, 4(1), pp. 7–16.

Await, J., 2020. International update: Pipeline construction & market trends. *Underground Construction*, 74(11). Available at: https://ucononline.com/magazine/2019/november-2019-vol-74-no-11/features/international-update-pipeline-construction-market-trends (Accessed August 4, 2020).

Bair, J., 2005. Global capitalism and commodity chains: Looking back, going forward. *Competition & Change*, 9(2), pp. 153–180.

BBC News, 2019. *Why some countries are shipping back plastic waste*. Available at: www.bbc.com/news/world-48444874 (Accessed January 6, 2020).

Biglan, A., 2009. The role of advocacy organizations in reducing negative externalities. *Journal of Organizational Behavior Management*, 29(3–4), pp. 215–230.

Block, F. and Somers, M.R., 2014. *The power of market fundamentalism*. Harvard University Press.

Bullard, R.B., 2000. *Dumping in Dixie: Race, class, and environmental quality*. Taylor & Francis.

Casey, J.A., Savitz, D.A., Rasmussen, S.G., Ogburn, E.L., Pollak, J., Mercer, D.G. and Schwartz, B.S., 2016. Unconventional natural gas development and birth outcomes in Pennsylvania, USA. *Epidemiology (Cambridge, MA)*, 27(2), p. 163.

Centner, T.J. and Eberhart, N.S., 2016. The use of best management practices to respond to externalities from developing shale gas resources. *Journal of Environmental Planning and Management*, 59(4), pp. 746–768.

Cheon, A. and Urpelainen, J., 2018. *Activism and the fossil fuel industry*. Routledge.

Clapp, J., 2002. The distancing of waste: Overconsumption in a global economy. In Princen, T., Maniates, M. and Conca, K. (eds.), *Confronting consumption*. MIT Press, pp. 155–176.

Clough, E. and Bell, D., 2016. Just fracking: A distributive environmental justice analysis of unconventional gas development in Pennsylvania, USA. *Environmental Research Letters*, 11(2), p. 025001.

Commonwealth of Pennsylvania Forty-Third Statewide Investigating Grand Jury, 2020. *Report 1 of the forty-third statewide investigating Grand Jury*, February 27. Available at: www.attorneygeneral.gov/wp-content/uploads/2020/06/FINAL-fracking-report-w.responses-with-page-number-V2.pdf (Accessed August 7, 2020).

Cusick, M., 2018. State senator files complaint asking PUC to halt Mariner East pipeline construction. *StateImpact Pennsylvania*, April 27. Available at: https://stateimpact.npr.org/pennsylvania/2018/04/27/state-senator-files-complaint-asking-puc-to-halt-mariner-east-pipeline-construction/ (Accessed August 7, 2020).

Cutchin, M.P., 2007. The need for the "new health geography" in epidemiologic studies of environment and health. *Health & Place*, 13(3), pp. 725–742.

da Rocha, D.F., Porto, M.F., Pacheco, T. and Leroy, J.P., 2018. The map of conflicts related to environmental injustice and health in Brazil. *Sustainability Science*, 13(3), pp. 709–719.

Dauvergne, P., 2018. Why is the global governance of plastic failing the oceans? *Global Environmental Change*, 51, pp. 22–31.

Davies, T., 2018. Toxic space and time: Slow violence, necropolitics, and petrochemical pollution. *Annals of the American Association of Geographers*, 108(6), pp. 1537–1553.

Ehrman, M.U., 2016. Earthquakes in the oilpatch: The regulatory and legal issues arising out of oil and gas operation induced seismicity. *The Georgia State University Law Review*, 33, p. 609.

Faber, D.J. (ed.), 1998. *The struggle for ecological democracy: Environmental justice movements in the United States*. Guilford Press.

Faber, D.J., 2008. *Capitalizing on environmental injustice: The polluter-industrial complex in the age of globalization*. Rowman & Littlefield Publishers.

Finkel, M.L., 2018. *Pipeline politics: Assessing the benefits and harms of energy policy*. ABC-CLIO.

Finley-Brook, M., Williams, T.L., Caron-Sheppard, J.A. and Jaromin, M.K., 2018. Critical energy justice in US natural gas infrastructuring. *Energy Research & Social Science*, 41, pp. 176–190.

Frittelli, J., Andrews, A., Parfomak, P.W., Pirog, R., Ramseur, J.L. and Ratner, M., 2014. *US rail transportation of crude oil: Background and issues for Congress* (Vol. 4). Congressional Research Service.

Geyer, R., Jambeck, J.R. and Law, K.L., 2017. Production, use, and fate of all plastics ever made. *Science Advances*, 3(7), p. e1700782.

Griswold, E., 2018. *Amity and prosperity: One family and the fracturing of America*. Farrar Straus & Giroux Incorporated.

Harvey, D., 2016. *The ways of the world*. Profile Books.

Hauter, W., 2016. *Frackopoly: The battle for the future of energy and the environment*. The New Press.

Healy, N., Stephens, J.C. and Malin, S.A., 2019. Embodied energy injustices: Unveiling and politicizing the transboundary harms of fossil fuel extractivism and fossil fuel supply chains. *Energy Research & Social Science*, 48, pp. 219–234.

Jambeck, J.R., Geyer, R., Wilcox, C., Siegler, T.R., Perryman, M., Andrady, A., Narayan, R. and Law, K.L., 2015. Plastic waste inputs from land into the ocean. *Science*, 347(6223), pp. 768–771.

Jephcote, C. and Mah, A., 2019. Regional inequalities in benzene exposures across the European petrochemical industry: A Bayesian multilevel modelling approach. *Environment international*, 132, p. 104812.

Johnston, J.E., Werder, E. and Sebastian, D., 2016. Wastewater disposal wells, fracking, and environmental injustice in Southern Texas. *American Journal of Public Health*, 106(3), pp. 550–556.

Kassotis, C.D., Tillitt, D.E., Lin, C.H., McElroy, J.A. and Nagel, S.C., 2016. Endocrine-disrupting chemicals and oil and natural gas operations: Potential environmental contamination and recommendations to assess complex environmental mixtures. *Environmental Health Perspectives*, 124(3), pp. 256–264.

Kelly, S., 2020. Shell's falcon pipeline dogged by issues with drilling and permit uncertainty during pandemic. *DeSmog*, June 15. Available at: www.desmogblog.com/2020/06/15/shell-falcon-pipeline-construction-pandemic-permits (Accessed August 5, 2020).

Kinchy, A., Parks, S. and Jalbert, K., 2016. Fractured knowledge: Mapping the gaps in public and private water monitoring efforts in areas affected by shale gas development. *Environment and Planning C: Government and Policy*, 34(5), pp. 879–899.

Klein, N., 2015. *This changes everything: Capitalism vs. the climate*. Simon and Schuster.

Korfmacher, K.S., Jones, W.A., Malone, S.L. and Vinci, L.F., 2013. Public health and high volume hydraulic fracturing. *New Solutions*, 23(1), pp. 13–31.

Krieg, E.J., 2005. Race and environmental justice in Buffalo, NY: A ZIP code and historical analysis of ecological hazards. *Society and Natural Resources*, 18(3), pp. 199–213.

Ladd, A.E., 2018. Conclusion: Standing at the energy policy crossroads. In Ladd, A.E. (ed.), *Fractured communities: Risk, impacts, and protest against hydraulic fracking in US shale regions*. Rutgers University Press, pp. 271–286.

Lerner, S., 2012. *Sacrifice zones: The front lines of toxic chemical exposure in the United States*. MIT Press.

Li, C. and Chen, J., 2019. The significance of restrictions on waste import in promoting green development in China. *American Journal of Environmental Protection*, 8(1), pp. 5–16.

Lips, J., 2018. *Debt and the oil industry-analysis on the firm and production level*. Available at: https://papers.ssrn.com/sol3/papers.cfm?abstract_id=3026063.

Logan, J.R. and Molotch, H.L., 2007. *Urban fortunes: The political economy of place*. University of California Press.

Lutz, B.D., Lewis, A.N. and Doyle, M.W., 2013. Generation, transport, and disposal of wastewater associated with Marcellus Shale gas development. *Water Resources Research*, 49(2), pp. 647–656.

Mah, A., 2015. Dangerous cargo and uneven toxic risks: Petrochemicals in the port of New Orleans. In Birtchnell, T. and Savitzky, S. and Urry, J. (eds.), *Cargomobilities*. Routledge, pp. 149–162.

Mah, A. and Wang, X., 2019. Accumulated injuries of environmental injustice: Living and working with petrochemical pollution in Nanjing, China. *Annals of the American Association of Geographers*, 109(6), pp. 1961–1977.

Malin, S.A., 2014. There's no real choice but to sign: Neoliberalization and normalization of hydraulic fracturing on Pennsylvania farmland. *Journal of Environmental Studies and Sciences*, 4(1), pp. 17–27.

Malin, S.A. and DeMaster, K.T., 2016. A devil's bargain: Rural environmental injustices and hydraulic fracturing on Pennsylvania's farms. *Journal of Rural Studies*, 47, pp. 278–290.

McAdam, D. and Boudet, H., 2012. *Putting social movements in their place: Explaining opposition to energy projects in the United States, 2000–2005*. Cambridge University Press.

McCormick, E., Murray, B., Fonbuena, C., Kijewski, L., Saraçoğlu, G., Fullerton, J., Gee, A. and Simmonds, C., 2019. Where does your plastic go? Global investigation reveals America's dirty secret. *The Guardian*, June 17. Available at: www.theguardian.com/us-news/2019/jun/17/recycled-plastic-america-global-crisis (Accessed January 6, 2020).

Mix, T.L. and Raynes, D.K.T., 2018. Denial, disinformation and delay: Recreancy and induced seismicity in Oklahoma's shale plays. In Ladd, A.E. (ed.), *Fractured communities: Risk, impacts, and protest against hydraulic fracking in U.S. shale regions*. Rutgers University Press, pp. 173–197.

Moore, J.R., 2015. Potential litigation over ethane economics: Who picks up the tab for the party that just ended? *Petroleum Accounting and Financial Management Journal*, 34(3), pp. 1–9.

Moore, J.W., 2011. Ecology, capital, and the nature of our times: Accumulation & crisis in the capitalist world-ecology. *Journal of World-Systems Research*, pp. 107–146.

Morello-Frosch, R., Zuk, M., Jerrett, M., Shamasunder, B. and Kyle, A.D., 2011. Understanding the cumulative impacts of inequalities in environmental health: Implications for policy. *Health Affairs*, 30(5), pp. 879–887.

Nace, T., Plant, L., and Browning, B., 2019. Pipeline bubble. *Global Energy Monitor*, April.

National Academies of Science Engineering and Medicine, 2016. *The changing landscape of hydrocarbon feedstocks for chemical production: Implications for catalysis: Proceedings of a workshop*. National Academies Press.

Nixon, R., 2011. *Slow violence and the environmentalism of the poor*. Harvard University Press.

North, E.J. and Halden, R.U., 2013. Plastics and environmental health: The road ahead. *Reviews on Environmental Health*, 28(1), pp. 1–8.

Pellow, D.N., 2004. *Garbage wars: The struggle for environmental justice in Chicago*. MIT Press.

Pellow, D.N., 2007. *Resisting global toxics: Transnational movements for environmental justice*. MIT Press.

Pellow, D.N., 2018. *What is critical environmental justice?* John Wiley & Sons.

Perrow, C., 2011. *Normal accidents: Living with high risk technologies*. Princeton University Press.

Perry, S.L., 2013. Using ethnography to monitor the community health implications of onshore unconventional oil and gas developments: Examples from Pennsylvania's Marcellus Shale. *New Solutions: A Journal of Environmental and Occupational Health Policy*, 23(1), pp. 33–53.

Phillips, S., 2021. PUC judge orders Sunoco to improve its public safety guidance, and pipeline safety, on Mariner East project. *StateImpact Pennsylvania, National Public Radio*, April 14.

PlasticsEurope, 2016. *Plastics – the facts 2016*. Available at: www.plasticseurope.org/application/files/4315/1310/4805/plastic-the-fact-2016.pdf (Accessed June 20, 2020).

PlasticsEurope, 2019. *Plastics – the facts 2019*. Available at: www.plasticseurope.org/application/files/9715/7129/9584/FINAL_web_version_Plastics_the_facts2019_14102019.pdf (Accessed August 1, 2020).

Pulido, L., 2017. Geographies of race and ethnicity II: Environmental racism, racial capitalism and state-sanctioned violence. *Progress in Human Geography*, 41(4), pp. 524–533.

Rabe, B.G. and Borick, C., 2013. Conventional politics for unconventional drilling? Lessons from Pennsylvania's early move into fracking policy development. *Review of Policy Research*, 30(3), pp. 321–340.

Rankin, J., 2019. European parliament votes to ban single-use plastics. *The Guardian*, March 27.

Shell Co., n.d. *Our growth projects*. Available at: www.shell.com/business-customers/chemicals/about-shell-chemicals/our-growth-projects.html (Accessed August 18, 2020).

Shonkoff, S.B., Hays, J. and Finkel, M.L., 2014. Environmental public health dimensions of shale and tight gas development. *Environmental Health Perspectives*, 122(8), pp. 787–795.

Sicotte, D.M., 2020. From cheap ethane to a plastic planet: Regulating an industrial global production network. *Energy Research & Social Science*, 66, p. 101479.

Silva, G.S., Warren, J.L. and Deziel, N.C., 2018. Spatial modeling to identify sociodemographic predictors of hydraulic fracturing wastewater injection wells in Ohio census block groups. *Environmental Health Perspectives*, 126(6), p. 067008.

Singer, M., 2011. Down cancer alley: The lived experience of health and environmental suffering in Louisiana's chemical corridor. *Medical Anthropology Quarterly*, 25(2), pp. 141–163.

Smith, M., Love, D.C., Rochman, C.M. and Neff, R.A., 2018. Microplastics in seafood and the implications for human health. *Current Environmental Health Reports*, 5(3), pp. 375–386.

Staggenborg, S., 2018. Marcellus Shale protest in Pittsburgh: Fractured communities: Risk, impacts, and protest against hydraulic fracking in US shale regions. In Ladd, A.E. (ed.), *Fractured communities: Risk, impacts, and protest against hydraulic fracking in U.S. shale regions*. Rutgers University Press, pp. 107–127.

Sze, J., 2006. *Noxious New York: The racial politics of urban health and environmental justice*. MIT Press.

Taylor, D., 2014. *Toxic communities: Environmental racism, industrial pollution, and residential mobility*. New York University Press.

Teuten, E.L., Saquing, J.M., Knappe, D.R., Barlaz, M.A., Jonsson, S., Björn, A., Rowland, S.J., Thompson, R.C., Galloway, T.S., Yamashita, R. and Ochi, D., 2009. Transport and release of chemicals from plastics to the environment and to wildlife. *Philosophical Transactions of the Royal Society B: Biological Sciences*, 364(1526), pp. 2027–2045.

Tishman Environment and Design Center, 2019. *U.S. municipal solid waste incinerators: An industry in decline*. Available at: https://static1.squarespace.com/static/5d14dab43967cc000179f3d2/t/5d5c4bea0d59ad00012d220e/1566329840732/CR_GaiaReportFinal_05.21.pdf (Accessed March 3, 2020).

U.S. Census Bureau, 2019. *Median housheold income, U.S. and Marcus Hook Borough, Pennsylvania. 2014–2018 American Community Survey 5-Year Estimates*. U.S. Census Bureau.

U.S. Department of Energy, 2018. Natural gas liquids primer, with a focus on the Appalachian Region. U.S. Department of Energy.

U.S. Energy Information Administration, 2018. *U.S. ethane consumption, exports to increase as new petrochemical plants come online*. Available at: www.eia.gov/todayinenergy/detail.php?id=35012 (Accessed August 4, 2020).

U.S. Energy Information Administration, 2019a. *EIA improves its propane and other hydrocarbon gas liquids data*. Available at: www.eia.gov/todayinenergy/detail.php?id=41935 (Accessed August 7, 2020).

U.S. Energy Information Administration, 2019b. *The United States expands its role as world's leading ethane exporter*, February 5. Available at: www.eia.gov/todayinenergy/detail.php?id=38232 (Accessed March 25, 2019).

U.S. Energy Information Administration, 2020. *U.S. imports to India of ethane*. Available at: www.eia.gov/dnav/pet/hist/LeafHandler.ashx?n=PET&s=M_EPLLEA_EEX_NUS-NIN_2&f=M (Accessed August 31, 2020).

U.S. Environmental Protection Agency, 2019. *Advancing Sustainable Materials Management: 2017 Fact Sheet*. Available at: https://www.epa.gov/facts-and-figures-about-materials-waste-and-recycling (Accessed January 20, 2020).

U.S. House Subcommittee on Energy, 2020. *Modernizing the natural gas act to ensure it works for everyone*. Testimony of Susan Tierney, p. 6, February 5. Available at: https://docs.house.gov/meetings/IF/IF03/20200205/110468/HHRG-116-IF03-Wstate-TierneyS-20200205.pdf (Accessed August 7, 2020).

Verma, R., Vinoda, K.S., Papireddy, M. and Gowda, A.N.S., 2016. Toxic pollutants from plastic waste-a review. *Procedia Environmental Sciences*, 35, pp. 701–708.

Villarosa, L., 2020. Pollution is killing Black Americans: This community fought back. *New York Times*, July 28.

Walker, G., 2012. *Environmental justice: Concepts, evidence and politics*. Routledge.

Warner, B. and Shapiro, J., 2013. Fractured, fragmented federalism: A study in fracking regulatory policy. *Publius: The Journal of Federalism*, 43(3), pp. 474–496.

World Economic Forum, 2016. *The new plastics economy: Rethinking the future of plastics*. Available at: www.weforum.org/press/2016/01/more-plastic-than-fish-in-the-ocean-by-2050-report-offers-blueprint-for-change/ (Accessed October 22, 2019).

Wright, B., 2005. Living and dying in Louisiana's 'cancer alley'. In Bullard, R.D. (ed.), *The quest for environmental justice*. Sierra Club Books, pp. 87–107.

Zwickl, K., 2019. The demographics of fracking: A spatial analysis for four US states. *Ecological Economics*, 161, pp. 202–215.

12

LATIN AMERICA CAUGHT BETWEEN INEQUALITY AND NATURAL CAPITAL DEGRADATION

A view from macro and micro data

Juan-Camilo Cardenas

Introduction: how does Latin America fare in terms of inequality and the environment

Latin America is, paradoxically, a region with one of the largest stocks of natural capital and with a large contribution in the provision of ecosystem services to humankind, while being the most unequal region on the planet with a substantial fraction of the current environmental conflicts worldwide. This snapshot of the current state of affairs also needs to be evaluated dynamically through the lens of the trends in these contrasting dimensions. While the region struggles – rather ineffectively – to reduce inequality, the threats to conservation of its natural capital continue growing. Deforestation of important natural areas in key ecosystems like the Amazon and tropical humid forests near the Pacific coast continues to threaten habitats of endemic species and key hot spots sustaining local and global biodiversity. Forest loss and soil and water pollution continue to happen because of mining at different scales and intensity in core Andean and Amazonian biomes. Water extraction for supplying cities and agro-industrial activities continues to increase, depleting the existing stocks of surface and groundwater sources in the region. Giant hydroelectric projects, depending on major watersheds, affect the possibility of maintaining the natural self-regulatory capacity of these basins to respond to climate changes associated with global warming as well as natural phenomena like El Niño and La Niña.

The Figure 12.1 compares a selection of inequality and environmental governance indicators in the Latin American and Caribbean region with some other regions of the world. While levels of inequalities expressed in Gini coefficients or quintiles income ratio[1] are relatively high, the region also exhibits higher percentages of protected lands for conservation. According to various sources, an important fraction of the top countries in terms of hosting biodiversity of species is located in the region. Strategies to declare protected areas in national parks systems and recognition to indigenous groups of their ancestral occupation of territories have contributed to exclude external forces from these biodiversity-rich areas (Cardenas, 2020). One example of such concentration of biodiversity in the region, provided by the Mongabay ranking,[2]

DOI: 10.4324/9780367814533-13

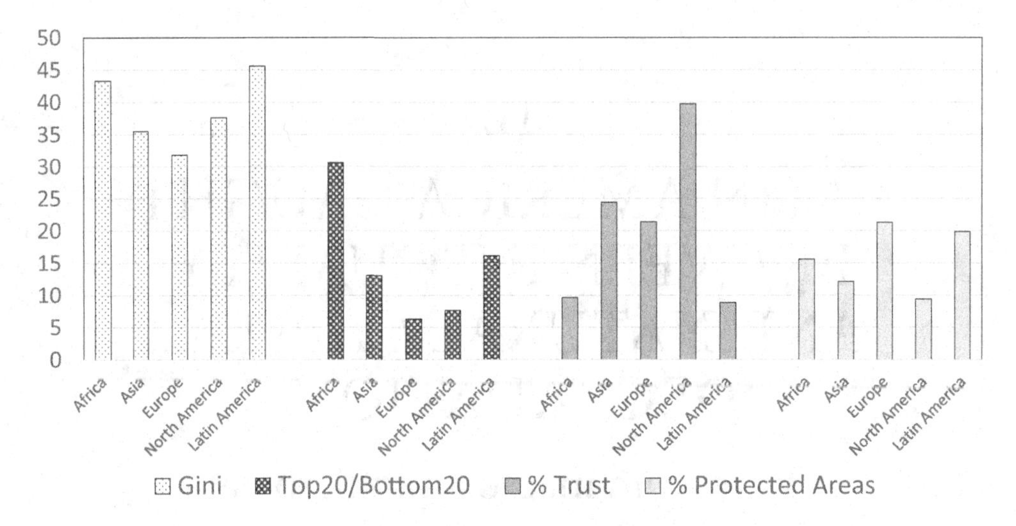

Figure 12.1 Inequality, trust and natural protected areas around the world

shows that in terms of birds, amphibians, reptiles, fish, mammals, and vascular plants, five out of the top 10 countries in the world are in South America. According to UNEP (2016), Latin America and the Caribbean countries have 57% of the world's primary forest and host almost 60% of the planet's terrestrial life with Brazil holding almost half of forest cover of the region. The Caribbean holds 12% of the world's mangrove forests. At the same time, freshwater stocks have declined between 1960 and 2015 from 60,000 to 20,000 cubic meters per capita (Leon and Cardenas, 2020).

This co-existence of a persistently unequal economic system with rich natural capital under threat creates a challenging future for the Latin American region. As noted in UNEP (2016), there is concern regarding the correlation between poverty and biodiversity decline in tropical regions and arguments in favor of connecting conservation actions along with poverty reduction. However, less mention is made of the role of inequality per se. Further, it might be the case that inequality harms the possibilities of preserving these ecological amenities for the present and future generations. It is important then to explore the mechanisms that could explain this relationship and, from those, evaluate the possible paths ahead to create a fair and sustainable future in the region.

In this chapter, I will summarize the possible mechanisms that may tie inequality with environmental degradation and explore the possibilities of preserving natural capital while taking a critical view of how inequality may hamper those possibilities. The focus will be mostly at the micro level and from the perspective of an active citizenship immersed in a mix of market, state and civic institutions.

The empirical support for the arguments will come from a set of experiments from the lab and the field, as well as from a very recent survey conducted in 13 Latin American cities in seven countries in the region. Using a behavioral approach to the micro problem, this chapter explores the connections between preferences for the environment and preferences for a fair economic system. These connections will have to pass through the analysis of what mechanisms may provide some possibilities for a more sustainable and fair social contract.

How inequality can affect the protection of natural capital

Solving environmental problems often requires coordination and collective action among parties. For two or more households sharing a space that provides ecosystem services for them to firms, cities and countries, the social dilemma is quite similar. Each agent can benefit from or be harmed by the actions of the other. Efforts to preserve those ecosystem services involve private costs for each, but the benefits accrued from the provision of those environmental amenities remain non-excludable. Creating social interactions to solve this social dilemma requires a set of clear property rights, but institutions must also create an environment of trust to make credible the commitments to maintain the environmental quality desired. Given the limitations of the state mechanism to enforce conservation measures and the indifference of most market mechanisms to these externalities, societies depend highly on self-governed mechanisms to reach collective action in these environmental dilemmas. Elinor Ostrom received the Nobel Prize in economics in 2009 because of her contributions in this regard. Her seminal book (Ostrom, 1990) provided a suitable framework for exploring how self-governance could succeed, emerging from reviewing field evidence of multiple cases around the world. In her presidential address to the American Political Science Association (Ostrom, 1998), she developed a micro-behavioral framework to explain how groups were able to cooperate through a virtuous cycle of trust, reciprocity and reputation.

Understanding the social interactions that determine if individual actions converge with the social goal of providing these ecosystem services is key. Within those interactions there are obviously costs and benefits, private and social, that will affect the decision to contribute to the preservation of nature or to reduce damages to the environment. Refraining from an environmentally harmful action or taking action to preserve an ecosystem function will imply private costs. Reducing the use of private transportation based on a combustion engine, recycling or forgoing the consumption of meat brings personal sacrifice. The effects of those actions will indeed contribute marginally in reducing the harms to the environment, and others in society will benefit from those actions. However, a significant and persistent number of actions of these kinds would be necessary to achieve, at the aggregate level, a result that is significant in terms of a sustainable path for present and future generations.

The possibilities for these actions to happen depend, indeed, on the valuation of the costs and benefits already mentioned, but also on the asymmetries between players in the game. The positions and action space available to each player will constrain the chances of the environmental outcome. Such positions and possibilities of action will also depend on the level of power each player has to affect the institutional setting towards personal or social goals. The literature on how inequality affects the possibilities for the environment is large and mixed, as reviewed in much of the volume compiled by Baland et al. (2007). The discussion on heterogeneity among players in collective action problems was discussed in the seminal book by Olson (1965), where he stated the idea that the privileged subgroup involved in the provision of public goods would be interested in contributing despite the free riding by the non-privileged group. Under that mechanism, Olson argued that greater heterogeneity would be beneficial for collective action. Bergstrom et al. (1986) also conjectured in this direction. Other authors later explored theoretically the mechanism further and qualified the result showing that depending on the structure of payoffs and technology would not hold in all cases (see Sandler, 1992). A number of empirical works coming from survey data, fieldwork and laboratory experiments, however, have also contested Olson's conjecture, and several of the papers in Baland et al. (2007) show evidence on the negative relationship between inequality and the possibility for collective action to provide environmental goods and nature conservation. Dayton-Johnson and Bardhan (2002) also provide theoretical mechanisms on how inequality affects the possibility of cooperation in local commons, supported by evidence from the Mexican ejidos and their irrigation systems (Dayton-Johnson, 2000).

In two series of experiments conducted with rural villagers in various sites in Colombia, I explored how two forms of inequalities, one based on people's actual socio-economic status in their villages and another induced within the laboratory setting, could affect the possibilities of conservation of a common pool resource (Cardenas, 2003, 2007; Cardenas et al., 2002). In all experimental games, groups of eight villagers made decisions about their level of extraction of a forest, and the monetary incentives within the experiment made clear that overexploiting the forest would give them personal gains, but collectively their payoffs would decrease if all players adopted such a strategy. In all cases, a number of rounds were played in which they had to make such decisions privately and simultaneously, and then we allowed them to have an open, non-binding, face-to-face conversation regarding the game before each round. Consistent with the literature from the field and the lab (Ostrom, Gardner & Walker, 1994), these conversations increased social efficiency and reduced free riding on average.

However, when unequal and equal situations were compared, interesting results were found. In the case of the inequality based on the real wealth that these people owned, which was assumed was common knowledge among the participants, being neighbors in the same village, a reduction in the effectiveness of the face-to-face communication stage to increase cooperation in the conservation of the forest was observed. Groups with larger heterogeneities in terms of the wealth possessed by the eight participants showed smaller increases in the cooperation levels. Based on the individual decisions and the relative wealth with respect to the other seven players, the data suggests that social distance, rather than absolute poverty, was what refrained people from trusting the others in their signals or intentions during the communication stage in the game. Inequality of wealth, which probably mapped inequality of power and social distance in the village, played a key role in eroding the willingness to cooperate with others within the game. In other words, trust, mediated by social distance and power asymmetries, was impeding the gains from cooperation that emerged within these experiments.

The second type of inequality, experimentally introduced, also provides some clues into the mechanisms that may interfere with the possibilities of collective action to protect natural capital and provide environmental goods. With the same experimental framework and in the same villages but with different participants, following a between-group design, we assigned randomly two of the players within one session to a situation in which they had enough private wealth that the returns from their private assets sustained their income without much dependence on their cooperation with the rest of the group. The other six players, however, had very little return from their private options and therefore depended much more on cooperation with the others. The results showed clearly that the two "rich" players could stick to their dominant strategies of exploiting the commons to the level where the marginal private gains matched their private marginal costs, while the six "poor" players moved away from the individual maximization of gains strategies and towards the cooperative choices. After communication was allowed, in these experiments, the strategy reinforced the need for the "poor" to reach consensus on a strategy to preserve the forest and sustain their income, which was much more dependent on their levels of collective action. Wade-Benzoni et al. (1996) show a very similar result in an experiment in which the participants represented companies that had to negotiate the fishing stocks in the northeastern United States. In the control treatment, all companies had the same stakes in the game, but in the asymmetric treatment some companies had long-term stakes while others short-term ones. The percentage of negotiations reaching a sustainable outcome changed from 64% in the control group to 10% in the unequal treatment. Moving from the lab to the field, Varughese and Ostrom (2001) present evidence from 18 user groups of forests in Nepal and show how heterogeneity makes it more difficult to achieve collective action. Similar evidence is found with respect to water.

From these experiments, a few lessons are worth highlighting here. One, inequality can erode the possibilities of creating trust-based mechanisms for collective action, key for conservation and the provision of ecosystem services. Trust seems to be key in creating such self-governed solutions, but inequality can erode trust by creating social distance among individuals. At a macro scale, Busso and Messina (2020) illustrate such a case for the world and the Latin American region in particular. Based on country-level data for Gini coefficients and the World Values Survey for the 1981–2014 period, they show a clear negative relationship between inequality and interpersonal trust. Furthermore, in their analysis, the region is located in the upper range of inequality and the lower end of trust. This is consistent with Figure 12.1 in which we also included the levels of trust for different regions in the world.

Causality is yet difficult to infer from the macro data, given that logical arguments can work in both directions, at this aggregate level. Countries that have reduced inequality probably have created a more inclusive economic system and therefore the possibilities of contact and sharing of the benefits from progress have reduced social distance. Likewise, it could be that more interpersonal trust also creates more trust in institutions, fewer rent-seeking activities and less corruption and therefore less friction for the state and markets to generate inclusion and distribution of benefits towards the poor. However, our micro-level experimental data may suggest that inequality is a causal effect in the erosion of willingness to commit to pro-environmental behaviors.

Aside from eroding trust within groups through larger social distance, other mechanisms can link inequality to environmental protection or degradation. Inequality can create the capture of the state by elites interested in rent-seeking activities based on the exploitation of natural capital, usually associated with extractive industries that create large short-run rents. Also, and related partially, unequal access to resources or benefits from the state can create conditions for illegal groups interested in exerting violence and promoting activities that are highly profitable, such as illegal mining and illicit crops. Ultimately, environmental conflicts increase in frequency and intensity as the powerful take on extractive enterprises (both illegal and legal), while the weak (indigenous, Afro-descendent and *campesino* groups) are excluded from the benefits of these short-run based activities while being exposed to the environmental downstream damages associated with the alteration or degradation of water sources or forest depletion resulting from them. All of these mechanisms are associated with the unequal distribution of power and are all consistent in various ways with Boyce (1994), who shows that as power is more unevenly distributed, the equilibrium between the marginal benefits for those creating an environmental harm and the marginal costs of those suffering it shifts in favor of the powerful.

How do Latin Americans perceive their environmental and social challenges?

Our argument so far leads to several questions regarding the state of affairs and the willingness of individuals to act upon the interaction between inequality, trust and the preservation of natural capital. A particular area of interest is how much citizens worry about these issues and if they see them as being related, a question which is explored here in the light of a new dataset from a survey conducted in the Latin American region.

During 2019, the Center of the Sustainable Development Goals for Latin America (CODS, 2020) conducted a large representative survey in the two main cities of Colombia, Argentina, Brazil, Mexico, Chile and Peru, and in the capital city of Costa Rica, covering a total of 13 cities[3] and 4,200 respondents. The survey collected information on the concerns, attitudes and behaviors regarding the environment and social issues related to sustainable development, and it provides rich and up-to-date empirical material related to the theoretical concerns of this chapter.

Overall, the data suggest a high concern for social and environmental issues in these countries. For instance, when asked how concerned the respondent was with respect to climate change and its consequences to her country, and inequalities in the access of education and health in her country, 70.0% and 76.7% respectively answered "very worried". Meanwhile, when asked about their concern that entrepreneurs do not have sufficient success in their businesses, only 42.7% manifested to be "very worried". More interestingly, when asked whether the government should prioritize preserving the environment over generating employment, 50.6% answered in favor of the environment. This percentage, however, ranged from 29.7% in Buenos Aires (Argentina) to 73.1% in Arequipa (Peru). Not surprisingly, the measured levels of air pollution in the cities for which there are available data predict these preferences for the environment over employment, as reported in Leon and Cardenas (2020).

In order to explore how attitudes and behaviors with respect to the environment and social issues are interrelated in the Latin American context, a set of three indices based on the questionnaire used in the CODS (2019) survey were created. Table 12.1 shows the questions used and the weights given to the different answers. Each index is composed of the sum of the points received by each of the questions according to the third column in the table, and then normalized to a 0–1 range to facilitate comparability, since each index has a different number of questions. The first index (sustainable behavior index) refers to reported recent behaviors by the respondent that may suggest a concrete action taken in favor of the environment. The second (pro-environmental index) refers to attitudes and concerns regarding nature and environmental quality but not necessarily associated with actual behaviors taken by the respondent. Third, the prosocial attitudes index is based on questions associated with concerns towards inequality and vulnerability of the poorest.

Based on the answers to these questions, the three indices offer a picture of the preferences and behaviors of Latin Americans in these cities with respect to the environmental and social issues associated with exclusion, poverty and the more vulnerable. The variation across cities and countries, but also by gender, age group or education level, also offers some potential for analyzing the data and builds a more interesting picture of the prosocial and pro-environmental preferences of citizens.

In a first interpretation, the data suggest that there is a correlation between prosocial and environmental concerns and actions. Figure 12.2 shows the scattered plots of the prosocial index (horizontal axis) against the sustainable behavior and pro-environmental attitudes indices (vertical axis) for each of the countries. The graphs also include the 95% confidence interval for both linear regression fitted lines. For all countries, these correlations are statistical significant ($p = 0.000$). Of particular interest is the fact that prosocial concerns predict the variation of the sustainable behaviors and pro-environmental attitudes and concerns, and at least suggest they may go hand in hand. It is hard, however, to attribute causality to these relationships, and instead they probably are part of a self-reinforcing mechanism that involves caring for others and caring for the environment.

Table 12.1 Behaviors and attitudes towards the environment and social issues (Based on CODS, 2019)

Index	*Questions used*	*Points given to the different answers*
Sustainable behavior index	In the last six (6) months, how often have you used or purchased products that reduce environmental damage?	Never (0) – Few times (0.33) – Frequently (0.67) – Always (1)
	In the last two (2) weeks, have you used reusable bags in your shopping?	No (0) – Yes (1)
	In the last two (2) weeks, have you turned off the lights and disconnected appliances that are not in use?	No (0) – Yes (1)
	In the last two (2) weeks, have you consciously limited the amount of water you use for bathing/brushing your teeth/washing dishes?	No (0) – Yes (1)
	In the last two (2) weeks, have you purchased products that have stamps or labels that certify environmental or social actions?	No (0) – Yes (1)
	In the last two (2) weeks, have you separated and recycled trash from your home?	No (0) – Yes (1)
	Which of the following is your main means of transportation?	Private motorbike or car (0) – Taxi, Uber, mototaxi, bicitaxi (0.33) – Mass public transportation (0.67) – Walk or bike (1)
	In the past year, have you participated in carpooling activities with friends, family or study/work colleagues?	Never (0) – A few times (0.25) – Frequently (0.75) – Always (1)
	In your last home purchase or rental decision, did you consider reducing the size of your home?	No (0) – Yes (1)
	In your last home purchase or rental decision, have you considered reducing time or distance to work or study?	No (0) – Yes (1)
	In your last home purchase or rental decision, have you considered less expensive public services?	No (0) – Yes (1)
	In your last home purchase or rental decision, have you considered the availability of environmentally friendly technologies?	No (0) – Yes (1)
	In the last two (2) weeks, have you consumed less meat?	No (0) – Yes (1)
	In the last two (2) weeks, have you consumed less sugar?	No (0) – Yes (1)
	In the last two (2) weeks, have you consumed more vegetables?	No (0) – Yes (1)

(*Continued*)

Table 12.1 (Continued)

Index	*Questions used*	*Points given to the different answers*
Pro-environmental attitudes index	How worried are you about climate change and its consequences in your country?	Not at all worried (0) – Little worried (0.33) – Somewhat worried (0.67) – Very worried (1)
	How worried are you about the future conservation of the seas and forests of your country?	Not at all worried (0) – Little worried (0.33) – Somewhat worried (0.67) – Very worried (1)
	On the effects of climate change, do you think that. . . ?	Will never happen (0) – It's a long way from happening (0.33) – They will happen soon (0.67) – Are already happening (1)
	Climate change will affect plants and animals the most (compared to affecting the poorest, future generations, economic development)	No (0) – Yes (1)
	Do you agree with this statement? "We should buy and use environmentally friendly products even if they are expensive"	Totally disagree (0) – Partially disagree (0.33) – Partially agree (0.67) – Totally agree (1)
	Do you agree with this statement? "We should buy and use useful products even if they are NOT environmentally friendly"	Totally disagree (0) – Partially disagree (0.33) – Partially agree (0.67) – Totally agree (1)
	When a new product or service appears on the market that is environmentally friendly, or supports a social cause,. . .	You will find out when a lot of people use it (0) – You will find out from people close to you (0.5) – You will be the first to use it (1)
Prosocial attitudes index	How worried are you that families do not have enough money in your country?	Not at all worried (0) – Very worried (1)
	How worried are you about the inequalities in access to education and health in your country?	Not at all worried (0) – Very worried (1)
	Climate change will affect the poorest more	No (0) – Yes (1)
	Climate change will affect the future generation more	No (0) – Yes (1)
	In the last two weeks, have you bought any products that have stamps or labels that certify environmental or social actions?	No (0) – Yes (1)
	How much would you be willing to pay extra for a coffee that supports a social cause, such as a fair payment to growers, support for a vulnerable community or group?	I would not be willing to pay more (0) – 5% additional (0.25) – 10% additional (0.5) – 15% additional (0.75) – I would pay more than 15% (1)

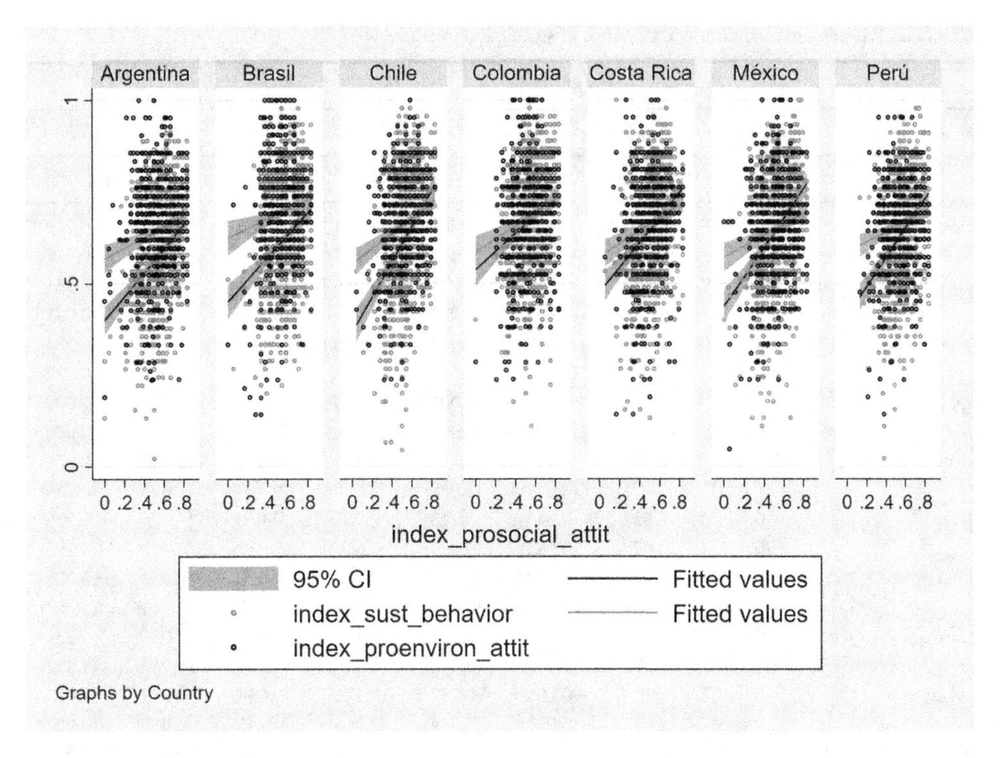

Figure 12.2 Correlations between prosocial attitudes and pro-environmental attitudes and behavior in Latin America

The next step is to explore if some of the demographic, contextual and attitudinal factors of these citizens can be explained by the variation in these preferences and behaviors. Because of space limitations, I will focus on explaining variation in the sustainable behaviors reported in the questionnaire as a function of the following variables: first, the prosocial and pro-environmental indices (see Table 12.1); then, demographic variables including gender, age and self-reported ethnic group,[4] socio-economic status and education. The regressions reported in Table 12.2 include one with pooled data of fixed effects for countries and then separate regressions for each country.

Several results are worth highlighting from the estimation results. Although not surprising at first glance, the pro-environmental attitudes strongly predict the variation in sustainable behaviors. As illustrated in Figure 12.2, the prosocial attitudes or concerns also help predict sustainable behaviors. Notice, further, that the coefficients for the prosocial and pro-environmental preferences are of similar magnitude, and they are also similar in magnitude and statistical significance for all countries. The fixed effects by country, however, do show that the index of sustainable behaviors shows differences in mean values that vary across nations. In particular, it seems that comparing against Colombia as the baseline, Argentina, Brazil, Chile and Peru have lower levels of these sustainable behaviors, and Mexico and Costa Rica similar in levels as Colombia.

In terms of gender, females seem to show a slightly higher report of sustainable behaviors than men for the overall sample, but such statistical significance only remains strong for Costa

Table 12.2 Regression estimates on an index of sustainable practices

	(1)	(2)	(3)	(4)	(5)	(6)	(7)	(8)
Variables	*Index of sustainable behaviors (15 questions, normalized 0–1)*							
	Pooled	*Colombia*	*Argentina*	*Brazil*	*Mexico*	*Costa Rica*	*Chile*	*Peru*
index_proenviron_attit	0.249***	0.189***	0.195***	0.268***	0.244***	0.306***	0.294***	0.270***
	(0.017)	(0.041)	(0.042)	(0.043)	(0.047)	(0.047)	(0.055)	(0.052)
index_prosocial_attit	0.324***	0.239***	0.295***	0.274***	0.356***	0.310***	0.423***	0.361***
	(0.015)	(0.040)	(0.035)	(0.042)	(0.039)	(0.044)	(0.044)	(0.042)
Gender (Fem=1)	0.027***	0.018	0.019	0.024*	0.018	0.046***	0.024*	0.033***
	(0.005)	(0.012)	(0.012)	(0.012)	(0.012)	(0.012)	(0.013)	(0.012)
Age (yrs)	0.001***	0.002***	0.001*	0.001	0.000	0.000	0.000	0.002***
	(0.000)	(0.000)	(0.000)	(0.000)	(0.000)	(0.000)	(0.000)	(0.000)
White	0.046*	0.057	0.026	0.119	0.126**	0.017	0.027	−0.031
	(0.024)	(0.071)	(0.048)	(0.088)	(0.050)	(0.038)	(0.043)	(0.059)
Black	0.044	0.032	−0.043	0.118	0.083	0.030		0.026
	(0.027)	(0.078)	(0.083)	(0.089)	(0.077)	(0.051)		(0.063)
Indigenous	0.070***	−0.023	0.069	0.143	0.165***	0.070	0.052	−0.020
	(0.026)	(0.092)	(0.064)	(0.097)	(0.051)	(0.049)	(0.044)	(0.057)
Mestizo	0.057**	0.065	0.039	0.147	0.153***	0.034	0.031	−0.033
	(0.024)	(0.071)	(0.048)	(0.090)	(0.048)	(0.039)	(0.043)	(0.052)
Mulatto	0.069***	0.111	−0.006	0.137	0.171***	0.034	0.099	0.012
	(0.025)	(0.075)	(0.057)	(0.088)	(0.061)	(0.043)	(0.066)	(0.067)
Other group	0.054**	0.016	−0.006	0.041	0.161**	0.117**	0.062	0.072
	(0.027)	(0.075)	(0.051)	(0.097)	(0.079)	(0.050)	(0.060)	(0.069)
Not answered	0.040	0.064	0.054	0.169*	0.088	−0.023	0.013	0.038
	(0.027)	(0.076)	(0.056)	(0.096)	(0.061)	(0.050)	(0.047)	(0.062)
Socio-econ status	0.004	−0.004	−0.019	0.019	−0.001	−0.003	0.024*	0.020
	(0.005)	(0.013)	(0.013)	(0.012)	(0.011)	(0.014)	(0.014)	(0.016)
Education (yrs)	−0.000	−0.008	0.005	−0.000	0.003	−0.003	−0.007	−0.005
	(0.002)	(0.005)	(0.005)	(0.005)	(0.003)	(0.005)	(0.007)	(0.007)
Argentina	−0.085***							
	(0.008)							
Brazil	−0.062***							
	(0.009)							
Mexico	−0.010							
	(0.010)							
Costa Rica	−0.012							
	(0.008)							
Chile	−0.040***							
	(0.009)							
Peru	−0.041***							
	(0.008)							
Constant	0.216***	0.324***	0.253***	0.070	0.116*	0.238***	0.097	0.161*
	(0.031)	(0.085)	(0.063)	(0.104)	(0.067)	(0.062)	(0.067)	(0.083)
Observations	4,164	600	591	591	591	591	598	602
R-squared	0.206	0.154	0.191	0.162	0.212	0.212	0.232	0.219

Source: Data from CODS Survey (2019).

Robust standard errors in parentheses: *** p<0.01, ** p<0.05, * p<0.1

Rica and Peru, and weak for Brazil and Chile. Age has a very small positive effect in the pooled sample, but its significance seems to be driven by Colombia and Peru.

As for the ethnic groups, and having "Asian" as the reference (omitted variable), there are no major patterns that suggest a strong effect across countries, although in the pooled regression "Indigenous" and "Mulattos" may show some positive correlation, although mostly explained by the Mexico sample. Finally, neither socio-economic status nor education seems to have a strong capacity to explain variation in these reported sustainable behaviors.

Conclusions: a prosocial and environmental agenda ahead

The evidence provided in this chapter, although limited in scope, suggests that addressing environmental challenges in Latin America will require considering the social issues associated with inequalities and prosociality of the social contract. Francesco's encyclical "Laudato Si" (2015) connects the two eloquently in this passage:

> it would also be mistaken to view other living beings as mere objects subjected to arbitrary human domination. When nature is viewed solely as a source of profit and gain, this has serious consequences for society. This vision of "might is right" has engendered immense inequality, injustice and acts of violence against the majority of humanity, since resources end up in the hands of the first comer or the most powerful: the winner takes all.

Injustices and inequalities can create and be altered by the degradation of natural capital.

However, the CODS (2019) survey suggests that people who are more motivated by and concerned with social inequalities and injustices are also more willing to engage in behaviors that are consistent with a pattern of consumption that is more sustainable.

Further, this does not seem to be a luxury of the well off. Neither the experimental evidence mentioned in the second section nor the data from the survey suggest that socio-economic status is an impediment or driver of more prosocial or pro-environmental attitudes. Instead, it is the coherence between a more inclusive social contract and a more sustainable idea of inhabiting earth that seems more compatible with the data.

The Latin American region is facing some challenging times. Its natural wealth is being threatened, the rise in conflicts from extractive activities has been particularly high in the region (Andrews et al., 2017), and its profound inequalities are most likely associated with these threats.

However, the data and analysis given earlier provide clues. I have argued that, rather than poverty, inequality is key for solving the collective action problem of protecting the environment and that inequality will hamper collective action because of the erosion of trust among unequal partners. Further, those more concerned with the environment are also more concerned with an unfair economic system. It seems, then, that policies that address both social inclusion and environmental goals have a better chance to find support among its citizens. This invites the design of mechanisms that reduce inequality and seek ecological goals at the same time.

Pro-poor strategies to protect the most vulnerable have been part of the social agenda in the region for some time now. These strategies have been rather successful in bringing the most vulnerable out of extreme poverty in many of the region's countries. Cash-transfer programs like *Bolsa Familia* in Brazil, *Oportunidades* in Mexico and *Familias en Acción* in Colombia have now shown decades of positive results. However, reducing inequality has shown to be elusive in the region. With the current low levels of interpersonal trust, correlated with low levels of trust in governments and politicians, the challenge is immense. One of the challenges for policies to be able to reduce inequality and not just poverty is the large gaps that persist in the region between territories (urban vs rural) and between ethnicities (blacks and indigenous vs mestizos and whites).

Fortunately, there are small seeds in policies and proposals coming from the region that offer such possibilities. For example, the "Buen Vivir" (Good Living) project emerging from indigenous groups in the Andes (Gudynas, 2011) offers a rewriting of the social contract with nature through a more inclusive economic system that requires the demotion of growth as a goal. Also, the recognition of rights for historically marginalized groups has also shown some promise. That is the case for a law in Colombia that recognized the ancestral rights to occupy tropical forest areas in the Pacific coast after the abolition of slavery. This Law 70 of 1993 has issued collective titles to around 6 million hectares to Afro-descendent communities that own and manage their territories with no private but communal property. In two studies by Vélez et al. (2020) and Peña et al. (2017), an impact evaluation of this law has shown that this collective titling both reduced deforestation and increased school attendance, per capita income and private investments in housing, when compared to similar geographies and demographics in neighboring areas. A policy that targeted both social inclusion and nature conservation seems to use the mechanisms discussed in this chapter.

Caring for others and caring for the biophysical environment seem to go hand in hand, and future actions that make each goal work for the other might prove more promising than projects that attempt conservation while deep inequalities persist around them. Designing prosocial and pro-environmental projects will require more equal distribution of burdens and benefits in terms of their social, economic and ecological outcomes, as well as a governance system that builds trust through participation.

Notes

1 Measured as income of the top 20%/income of the bottom 20% using the right horizontal axis.
2 See https://rainforests.mongabay.com/03highest_biodiversity.htm.
3 Bogotá, Medellín, Buenos Aires, Córdoba, Sao Paulo, Rio de Janeiro, Ciudad de México, Guadalajara, San José, Santiago de Chile, Concepción, Lima and Arequipa.
4 Asian, White, Black, Indigenous, Mestizo, Mulato, Other group, NotAnswered.

Bibliography

Andrews, T., Elizalde, B., Le Billon, P., Oh, C. H., Reyes, D., & Thomson, I. (2017). *The rise in conflict associated with mining operations: What lies beneath.* Canadian International Resources and Development Institute (CIRDI), 1–127.

Baland, J. M., Bardhan, P., & Bowles, S. (Eds.). (2007). *Inequality, cooperation, and environmental sustainability.* Princeton University Press.

Bergstrom, T., Blume, L., & Varian, H. (1986). On the private provision of public goods. *Journal of Public Economics*, 29(1), 25–49.

Boyce, J. K. (1994). Inequality as a cause of environmental degradation. *Ecological Economics*, 11(3), 169–178.

Busso, M., & Messina, J. (2020). *The inequality crisis: Latin America and the Caribbean at the crossroads.* Inter-American Bank. http://dx.doi.org/10.18235/0002629.

Cardenas, J. C. (2003). Real wealth and experimental cooperation: experiments in the field lab. *Journal of Development Economics*, 70(2), 263–289.

Cardenas, J. C. (2007). Wealth inequality and overexploitation of the commons: Field experiments in Colombia. *Inequality, Cooperation, and Environmental Sustainability*, 205–233.

Cardenas, J. C. (2020). Commons. In Kaltmeier, O., Tittor, A., Hawkins, D., & Rohland, E. (Eds.), *The Routledge handbook to the political economy and governance of the Americas.* Routledge.

Cardenas, J. C., Stranlund, J., & Willis, C. (2002). Economic inequality and burden-sharing in the provision of local environmental quality. *Ecological Economics*, 40(3), 379–395.

CODS. (2019). *Encuesta de consumo y medio ambiente*. Centro de los Objetivos de Desarrollo Sostenible para América Latina CODS – Universidad de los Andes.

CODS. (2020). *Índice ODS 2019 para América Latina y el Caribe*. Centro de los Objetivos de Desarrollo Sostenible para América Latina y el Caribe. https://cods.uniandes.edu.co/indice-ods/.

Dayton-Johnson, J. (2000). Determinants of collective action on the local commons: A model with evidence from Mexico. *Journal of Development Economics*, 62(1), 181–208.

Dayton-Johnson, J., & Bardhan, P. (2002). Inequality and conservation on the local commons: A theoretical exercise. *Economic Journal*, 112(481), 577–602.

Francesco, P. (2015). *Laudato si'*. Edizioni Piemme.

Gudynas, E. (2011). Buen Vivir: Today's tomorrow. *Development*, 54(4), 441–447.

Leon, D., & Cardenas, J. C. (2020). *Latin America and the Caribbean: Natural wealth and environmental degradation in the XXI century*. UNDP LAC COVID-19 Policy Documents Series. www.latinamerica.undp.org/content/rblac/en/home/library/crisis_prevention_and_recovery/latinoamerica-y-el-caribe-riqueza-natural-y-degradacion-ambient.html.

Olson, M. (1965). *The logic of collective action: Public goods and the theory of groups*. Harvard University Press.

Ostrom, E. (1990). *Governing the commons: The evolution of institutions for collective action*. Cambridge University Press.

Ostrom, E. (1998). A behavioral approach to the rational choice theory of collective action. *American Political Science Review*, 92(1), 1–22, March.

Ostrom, E., Gardner, R., Walker, J., Walker, J. M., & Walker, J. (1994). *Rules, games, and common-pool resources*. University of Michigan Press.

Peña, X., Vélez, M. A., Cárdenas, J. C., Matajira, C., & Perdomo, N. (2017). Collective property leads to household investments: Lessons from land titling in Afro-Colombian communities. *World Development*, 97, 27–48.

Sandler, T. (1992). *Collective action: Theory and applications*. University of Michigan Press.

UNEP. (2016). *The state of biodiversity in Latin America and the Caribbean: A mid-term review of progress towards the Aichi biodiversity targets*. United Nations Environmental Programme.

Varughese, G., & Ostrom, E. (2001). The contested role of heterogeneity in collective action: some evidence from community forestry in Nepal. *World Development*, 29(5), 747–765.

Vélez, M. A., Robalino, J., Cardenas, J. C., Paz, A., & Pacay, E. (2020). Is collective titling enough to protect forests? Evidence from Afro-descendant communities in the Colombian Pacific region. *World Development*, 128, 104837.

Wade-Benzoni, K. A., Tenbrunsel, A. E., & Bazerman, M. H. (1996). Egocentric interpretations of fairness in asymmetric, environmental social dilemmas: Explaining harvesting behavior and the role of communication. *Organizational Behavior and Human Decision Processes*, 67(2), 111–126.

13

AIR QUALITY CO-BENEFITS OF CLIMATE MITIGATION IN THE EUROPEAN UNION

Klara Zwickl and Simon Sturn[1]

Introduction

Climate policy is typically framed as distributional conflict between countries, with some countries free riding on the costly efforts of curbing carbon emissions of others, or between generations, with current generations living at the expense of future ones. These perspectives neglect that burning fossil fuels not only releases greenhouse gases into the atmosphere, but also emits several hazardous co-pollutants that negatively affect local air quality and harm the health of people living in proximity to the emitting sources. Climate policy thus not only reduces global climate change, but also improves local air quality.

Incorporating these air quality co-benefits into an assessment of climate policy is crucial to get a broader picture of its costs and benefits. A substantial number of studies show that the benefits of climate policy are vastly underestimated if co-benefits are omitted from the analysis. Co-benefits alone are typically found to justify a substantial increase in carbon prices, independent of any climate benefits. Incorporating co-benefits also highlights the importance of another distributional dimension that is often neglected; the one between rich and poor neighborhoods. Since socioeconomically disadvantaged neighborhoods are typically more exposed to co-pollutants – both within and across countries – they benefit more from air quality improvements through policies that reduce fossil fuel combustion. Climate policy can thus reduce environmental inequality in air pollution exposure.

This chapter discusses the role of air quality co-benefits of climate mitigation in the European Union. In section What are air quality co-benefits and why do they matter?, we describe what air quality co-benefits are and why they matter. In section Why are greenhouse gases and co-pollutants regulated separately?, we review the regulation of greenhouse gases and co-pollutants in the EU. In section How large are air quality co-benefits?, we give an overview of the literature on the magnitude of air quality co-benefits, and then in section What are implications of including air quality co-benefits in European climate policies?, we discuss implications of including co-benefits into EU climate policy. Conclusion section concludes the chapter.

What are air quality co-benefits and why do they matter?

Climate policy, or the lack thereof, is often viewed as a global coordination failure. Since a ton of carbon emissions has the same climate impact independent of the location of its release, in

DOI: 10.4324/9780367814533-14

the absence of binding international agreements every country or region has incentives to free ride on others' carbon mitigation efforts. As a public good, climate stabilization will therefore be underprovided and carbon leakage – emissions shifting from regulated to unregulated locations – can undermine emissions reductions of single countries or regions (Fowlie 2009; Kuik and Hofkes 2010; Eichner and Pethig 2011; Aldy and Stavins 2012). Additionally, the long timespan between emissions releases and their climate effects imply a trade-off between generations, which requires balancing the economic costs of carbon mitigation today with the environmental benefits arising in the future (Arrow et al. 1996; Nordhaus and Boyer 2000).

Framing the optimal level of climate policy as distributional struggle across countries and generations overlooks the local and immediate benefits of carbon mitigation here and now. A growing body of literature on co-benefits of climate policy has emphasized the existence of large positive spillovers from fossil fuel abatement, ranging from environmental health improvements to positive fiscal impacts due to carbon revenues and reductions in fossil fuel subsidies (for an overview, see Parry et al. 2014, 2015). The most widely studied co-benefit of carbon mitigation is improved air quality, since the combustion of fossil fuels releases not only greenhouse gases, but also several hazardous co-pollutants, such as particulate matter, sulfur dioxides, and nitrogen oxides. Carbon mitigation consequently can simultaneously provide global climate stabilization, as well as local health benefits, through improved air quality.

In contrast to greenhouse gases, co-pollutants of fossil fuel combustion have direct and short-term impacts on air quality at the location of their release. In fact, air pollution improvements can have such immediate health effects that variation in pollution exposure at the same school over time has been found to significantly affect students' absenteeism (Currie et al. 2009), and variation in outdoor particulate matter exposure around the workplace has been shown to significantly affect workers' productivity (Chang et al. 2016). WHO (2019, p. 15) considers air pollution "the biggest environmental risk to health". Since the environmental health damages of co-pollutants are large, air quality co-benefits could change the incentive structure of countries and regions to contribute to global carbon mitigation, also in the absence of binding international agreements (Parry et al. 2014, 2015; Zwickl et al. 2021).

Given current climate and air pollution challenges, air quality co-benefits of climate mitigation are a straightforward way for policymakers to simultaneously address both problems in Europe. With a 10% share of global greenhouse gas emissions in 2017, the EU-28 is currently the third biggest global carbon emitter, following China and the US. Considering cumulative, historical emissions up to 2017, the EU's share increases to 22% (Ritchie and Roser 2020). At the same time, the EU faces a severe public health crisis due to high levels of air pollution. While air quality has improved over the last decades, especially particulate matter and ground-level ozone levels still exceed the World Health Organization (WHO) safety thresholds in wide parts of Europe. According to the EU air quality report (EEA 2020), 48% of the urban population was exposed to levels above the WHO threshold for PM_{10}, 74% above the $PM_{2.5}$ threshold, and even 99% above the O_3 threshold in 2018. Moreover, the health impacts of specific levels of pollution exposure are continuously corrected upwards, especially as the harmful cardiovascular and respiratory health effects of very fine particulates are better understood. A recent study suggests that air pollution reduces average life expectancy by 2.2 years and accounts for 129 deaths per 100,000 inhabitants in the EU-28 (Lelieveld et al. 2019), which is substantially higher than previous European Environmental Agency (EEA) and WHO estimates.

Air pollution exposure is distributed very unequally, both within and between countries. Globally, low- and middle-income countries face a disproportionately high air pollution burden (WHO 2019). Within countries, air pollution exposure disparities are also widespread. For the US, where regionally disaggregated air pollution as well as socioeconomic data have existed for more than two decades, many studies show that poor and minority neighborhoods are disproportionately exposed to high and unsafe levels of air pollution (for an overview, see Ringquist 2005; Zwickl et al. 2014). For the EU, where geographically fine-scale socioeconomic data do not yet exist, the first evidence from country-case studies (e.g. Rüttenauer 2018 for Germany) and regional EU-wide analyses (e.g. EEA 2018) suggests that air pollution exposure disparities are also substantial.

Moreover, within most countries and regions, air pollution is spatially very concentrated and clustered in pollution hot spots (for the US, see Boyce et al. 2016). For the EU, an assessment of air pollution damages from large industrial point sources regulated under the European Pollutant Release and Transfer Register (E-PRTR) found that the top 147 emitting facilities – roughly the top 1% of facilities included in the registry – contributed to 50% of the total industrial air pollution damage costs (EEA 2014). The large majority of these are thermal power stations and mineral oil and gas refineries; most of them also reported high greenhouse gas emissions (based on own calculations using E-PRTR data). Replacing energy from these sources with renewable ones would generate substantial climate as well as air quality co-benefits. Since many of these facilities are located in proximity to cities in Eastern European countries with already higher pollution levels, pollution abatement at these locations would also narrow regional and socioeconomic environmental disparities. For the US, Boyce and Pastor (2013) have shown that air quality co-benefits are highest in poor and minority neighborhoods. Climate mitigation, especially when targeting pollution hot spots, can thus reduce not only overall air pollution levels, but also disparities in pollution exposure.

Why are greenhouse gases and co-pollutants regulated separately?

Both climate change and air pollution largely stem from the same activity, the combustion of fossil fuels. Yet carbon emissions and local air pollutants are regulated separately. While the role of air quality co-benefits is increasingly emphasized by policymakers, the majority of existing policies are not designed to maximize positive spillovers from one to the other.

Air pollution regulations have been established in many EU countries from the 1970s onwards and have been frequently updated and harmonized ever since. The "Ambient Air Quality Directive" currently in place is binding to its member states. It includes thresholds, limit values, and target values for each air pollutant, air quality plans, and short-term action plans for pollution hot spots, as well as the implementation of national bodies responsible for data collection and compliance. The Ambient Air Quality Directive is supplemented by the National Emissions Ceiling Directive, which sets country limits on air emissions, as well as additional regulations for vehicles and for large combustion plants (EEA 2016, 2020).

The EU's main climate policy is the EU Emissions Trading Scheme (EU ETS). Implemented in 2005, it is the first multinational emissions trading scheme, including all EU member states, as well as Iceland, Liechtenstein, and Norway. The EU ETS is a classical cap-and-trade program, where first an EU-wide cap is set to meet the EU's climate goals, and then the existing permits are allocated across countries, which in turn allocate them to the participating industrial firms and facilities. At the beginning, permits were allocated for free to the historical polluters; over time, the share of auctioned permits and thus also the carbon revenues are increasing (Ellerman et al. 2016; Carl and Fedor 2016). EU ETS installations are free to trade permits across participating countries and sectors. Trading with other emissions trading schemes (such as the Swiss ETS) and options for international credits and offsets are gradually introduced. Moreover, there currently exist binding national annual greenhouse gas emissions targets for sectors not covered under the ETS, as well as binding targets for an increase in the share of renewable energy sources. Some European countries, such as Sweden, additionally have more stringent national climate policies, such as CO_2 taxes on energy use (OECD 2018b).

Two differences between greenhouse gases and local air pollutants have shaped their regulation. The first is related to the spatial distribution of pollution damages. Greenhouse gases are uniformly mixed, implying that one ton of CO_2 has the same climate impact independent of the location of its release. Standard environmental economic models have suggested that pollution mitigation is most efficient where abatement costs are lowest. Since regulators do not know the exact abatement costs of each plant, market-based environmental policies, such as taxes or permit schemes, have been argued to be most efficient (Stavins 2003). Co-pollutants such as particulate matter, sulfur dioxide, and nitrogen oxides, by contrast, are non-uniformly mixed. The total environmental and human health damages strongly depend on the fate and transport of emissions, which include not only emission releases but also factors such as wind and weather patterns and the number of people exposed. Because location matters, for the latter type of pollutants, spatial variations in abatement benefits have to be considered along with abatement costs (Tietenberg 1995; Muller and Mendelsohn 2009; Boyce and Pastor 2013).

The second difference between greenhouse gases and co-pollutants is related to existing pollution control technology, especially the availability of end-of-pipe controls. The only way to currently abate CO_2 is to reduce fossil fuel use or to some extent to switch to other fossil fuels that emit less CO_2, such as natural gas (which however, might only be a temporary solution, and moreover its climate impact might be strongly underestimated since natural gas releases the powerful greenhouse gas methane, see Howarth 2014[2]). With the exception of technologies such as carbon capture and storage, which are currently still very expensive and cannot yet be implemented at large scale, no end-of-pipe technology for greenhouse gases currently exists. For co-pollutants such as nitrogen oxides and sulfur dioxide, by contrast, a variety of end-of-pipe controls are available that can filter out some of the pollutants during combustion. The two most well known are catalytic converters for cars and scrubbers for industrial power plants. Both have been mandated in most EU member countries for several decades. These policies have the distinct advantage that they are easy to implement and monitor. A disadvantage, however, is that they can only abate co-pollutants, while providing no positive spillovers on greenhouse gas mitigation. In fact, spillovers can even turn slightly negative when these pollution abatement technologies require energy, which in turn is generated through more fossil fuel use (Holland 2010).

Historically therefore, taxes, trading schemes, and attempts to achieve binding international agreements have been the main policy tools for greenhouse gas abatement, whereas emissions standards and technology requirements have been used for local pollution control. In the next section, we will discuss potential benefits when regulating the two together.

How large are air quality co-benefits?

Many studies find that air quality co-benefits are high enough to justify more stringent climate policies independent of their climate impacts. Most of these studies are simulation studies that compare a proposed or implemented climate policy to a business-as-usual scenario. They then compute air quality co-benefits per ton of carbon emissions and compare their magnitude to conventional social cost of carbon (SCC) estimates. If the monetized benefits of air quality improvements exceed those of the SCC, these studies conclude that carbon mitigation is beneficial independent of its climate benefits.

For example, Thompson et al. (2014), Driscoll et al. (2015), Shindell et al. (2016), and Buonocore et al. (2016) model different US climate policy scenarios and find that any of these climate policies is associated with air quality co-benefits of such a magnitude that their implementation would be beneficial to the US regardless of climate mitigation goals. West et al. (2013) model the global averted mortality resulting from a global carbon price limiting global temperature increase to 2.5 degrees until 2100 and find that air quality and health co-benefits are high enough to exceed carbon mitigation costs by far. In a global review of co-benefit studies, Nemet et al. (2010) find that co-benefit estimates range from 2–196 USD/ton of carbon with an average of 49 USD/ton of carbon. Moreover, with an average of 81 USD/ton of carbon, co-benefits are substantially higher in developing countries, compared to 44 USD/ton of carbon in developed countries.

Parry et al. (2014, 2015) analyze co-benefits of climate mitigation of the world's top 20 carbon-emitting countries for 2010. While they consider a broader set of co-benefits than most other studies, including reduced traffic accidents and government revenues through reduced fossil fuel subsidies, the largest co-benefits are found to be air quality improvements from reduced coal combustion. The average price across the 20 countries is USD 57.5 per ton of CO_2, which is substantially larger than the US social cost of carbon, estimated to lie at USD 35 in 2013. While there is strong heterogeneity in co-benefits across countries, depending on the fuel mix of the economy as well as existing environmental regulations, implementing 'nationally efficient' CO_2 prices (that correct for all of the negative externalities except for greenhouse gas mitigation) would result in a global reduction of greenhouse gas emissions by 10.8%.

A second important finding from the literature on air quality co-benefits is that they vary across place and across pollution sources. Buonocore et al. (2016) model health benefits of a proposed US environmental policy at the US county level. While they find positive co-benefits for every single US county, their magnitude varies substantially and depends largely on the share of the county's electricity coming from coal as well as the county's air quality prior to the policy. Muller (2012) assesses co-pollutant emissions per ton of carbon emissions for more than 10,000 point sources in the electric power generation and transportation sectors. The largest co-benefits of around 87 USD/ton of carbon (and carbon equivalent greenhouse gas) emissions can be found at bituminous coal-fired electric power generators, followed by other coal-fired power generators. Residual oil-fired electric power generators have co-benefits of around 49 USD/tCO_e, and natural gas-fired electric power generators have substantially lower co-benefits of 3 USD/tCO_e. In the transportation sector, co-benefits range from 11 to 23 USD/tCO_e, depending on the vehicle type (Muller 2012, p. 709f).

Population density also matters for the health impacts of co-pollutants. Muller and Mendelsohn (2009) find for the US that the damages per ton of SO_2 and $PM_{2.5}$ emissions vary by more than 100 times depending on whether the emissions are released in a densely populated city or in a rural area. Boyce and Pastor (2013) calculate co-pollutant intensity ratios (co-pollutants per ton of CO_2 emissions) for 1,540 large industrial point sources in eight industrial sectors and compare inter- and intra-sectoral variation in these ratios. In addition to the three most widely used co-pollutants, SO_2, NO_X, and particulate matter, they include toxic air pollutants from EPA's Risk Screening Environmental Indicators model (RSEI). Moreover, they report results with and without population weighting to illustrate the role of population density on total co-pollutant impacts. They find that ranks in co-pollutant ratios by sector vary strongly depending on the co-pollutant and depending on whether population weighting is applied. For example, although co-pollutants per ton of carbon emissions are higher for power plants than for petroleum refineries, population weighted damages are 3–10 times higher for refineries because they are located in areas with higher population density. Boyce and Pastor (2013) are also the first study to assess co-pollutant impacts by race/ethnicity and the level of income of the neighborhood. For air toxics, which have been found to disproportionately affect people of color (Ash and Fetter 2004; Zwickl et al. 2014), they find that petroleum refineries, chemical manufacturers, and power plants rank highest in disproportionately affecting people of color. Refineries and chemical manufacturers most disparately affect low-income people (Boyce and Pastor 2013). Overall, they find that co-pollutant damages are highest for facilities located in minority and low-income neighborhoods, and conclude that climate policy that incorporates air quality co-benefits could provide substantial equity – in addition to efficiency – improvements.

Finally, the literature discussed so far assesses co-benefits based on a unit elasticity assumption, where a 1% reduction or increase in CO_2 leads to a similar sized change in co-pollutants. However, this unit-elasticity assumption does not need to hold for various reasons. For example, end-of-pipe controls such as scrubbers can reduce co-pollutants, while at the same time requiring electricity. Increasing the combustion temperature in natural gas-fired power plants reduces CO_2 per unit of output but increases NO_X emissions. Since natural gas has lower sulfur content, fuel switching from coal to natural gas reduces both CO_2 and SO_2 emissions. Zwickl et al. (2021) empirically estimate co-pollutant elasticities for industrial point sources in Europe based on data from 2007 to 2015 and document a substantial heterogeneity across economic activities. Looking at climate policy–induced co-pollutant elasticities in the energy-producing sector they find elasticities of 1.2 to 1.8 for SO_X, 1.1 to 1.5 for NO_X, and 0.8 for PM_{10}, suggesting that previous studies might have been underestimating the magnitude of co-benefits. Monetizing the health effects of policy-induced co-pollutant emissions in the energy sector using co-pollutant damage costs from the EEA, they obtain air quality co-benefits of about 60 to 160 EUR per ton of CO_2 (converted into today's EUR), which is substantially higher than EEA or other conventional estimates of climate damage costs.

Summing up, the literature on air quality co-benefits has two main conclusions for EU climate policy, which will be explored in the next section. First, a substantially higher carbon price can be justified independent of its climate benefits. Second, variation in the magnitude of co-benefits across sectors and space suggests that a spatially differentiated carbon policy may provide benefits over a uniform one.

What are implications of including air quality co-benefits in European climate policies?

The literature on air quality co-benefits of climate mitigation has two important implications for current EU climate policy, especially the EU ETS. First, air quality co-benefits justify raising the price of burning fossil fuels substantially. The monetized damage costs of co-pollutants amount to more than 60 EUR/tCO_2 according to the mean estimate from various studies for all sectors in European countries reported by Nemet et al. (2010, Table A.1, converted into current EUR). Zwickl et al. (2021) quantify co-benefits in the European electricity sector to lie at about 60 to 160 EUR/tCO_2 in current euros. While the EU ETS price has been gradually increasing in the past years and has reached an all-time high of 31 EUR for a ton of carbon at the end of 2020 (Shepphard 2020), it is only at about 50% of the level that would internalize the costs of co-pollutants, completely ignoring any climate benefits. To fulfill the climate goals of the Paris Agreement of limiting the global average temperature increase to 1.5–2 °C above pre-industrial levels, a carbon price across countries and sectors of 33–66 EUR (40–80 USD) per ton of CO_2 in 2020 and 41–82 EUR (50–100 USD) per tCO2 by 2030 is required (Carbon Pricing Leadership Coalition 2017).[3] Also, this magnitude tends to be below the monetized co-benefits of improved air quality.

The overall carbon price could not only be increased by raising the price in existing carbon pricing schemes, such as the EU ETS, but also by extending regulatory coverage to previously un- or underpriced sectors. The carbon pricing gap in the EU, the percent of the country's emissions not priced at least 30 (60) EUR per ton of CO_2 in 2015, ranges from 30% (33%) in Luxembourg to 71% (80%) in Estonia (OECD 2018b). The EU ETS covers roughly 45% of the EU's emissions.

Finally, to increase the price of carbon, fossil fuel subsidies, which are still widespread, might be eliminated. In their quantification of global fossil fuel subsidies, Coady et al. (2017, 2019) distinguish between pre-tax and post-tax subsidies. The narrower indicator, pre-tax subsidies, measures the difference between consumer fuel prices and the opportunity cost of fuel supply (mostly applying to petroleum, natural gas, and electricity). Global pre-tax subsidies declined from 0.77% to 0.36% percent of global GDP from 2012 to 2016, which is in line with national and international efforts of fossil fuel subsidy reforms (Coady et al. 2019). While the majority of global pre-tax subsidies are provided in developing and emerging countries, a novel OECD (2018a) database on the Inventory of Support Measures of Fossil Fuels documents that various instruments, such as energy tax reliefs or exceptions for specific sectors or for heating fuels, are still widely in place across EU countries. Post-tax subsidies provide a broader measure, reflecting the difference between consumer fuel prices and prices necessary to internalize environmental costs and meet fiscal revenue requirements like the costs of climate change, local air pollution, broader vehicle externalities, and forgone consumption tax revenues. This broader measure thus also captures air quality co-benefits of carbon mitigation. In contrast to pre-tax subsidies, post-tax subsidies increased from 5.4% to 6.5% of global GDP from 2010 to 2016 (Coady et al. 2019). While post-tax subsidies are also substantially higher in most developing and emerging countries, they amount to relevant economic magnitudes also in European countries (for example, they amounted to 2.1% of GDP in Germany and 1.4% of GDP in France in 2015).

The second important implication of air quality co-benefits for current EU climate policy is that climate policy also has consequences on environmental inequality. As outlined earlier, especially poorer neighborhoods suffer from high air pollution. Evidence from the US further shows that the benefits from cleaner air due to fossil fuel abatement are more pronounced in poor and minority neighborhoods (Boyce and Pastor 2013). Environmental benefits from fossil fuel abatement also vary substantially across locations, activities, and sectors (see also Zwickl et al. 2021 for European evidence). Climate policy thus might be expected to have the largest benefits in poorer areas and areas with air pollution hot spots resulting from industrial activity. Without a consideration of co-benefits, carbon mitigation might not be prioritized in such areas. In fact, California's cap-and-trade program has not been found to reduce environmental disparities in co-pollutant exposure between 2011 and 2015 (Cushing et al. 2018). The latter is not surprising when emissions can be traded freely across locations and sectors, with no incentives for pollution abatement in places with the highest overall environmental damages.

For the EU, currently too little is known about the impact of permit trading in the ETS on co-pollutant emissions, as well as on the role of generating and reinforcing existing pollution hot spots. At the same time, however, various initiatives are set to increase the flexibility of EU ETS trading, such as allowing trading with other emissions trading schemes (such as the Swiss ETS as of 2020) and including carbon crediting mechanisms and offsetting (World Bank 2020). While the latter can bring some important benefits in minimizing the costs of climate policy, granting polluters more flexibility in how and where they want to abate emissions might also increase pollution hot spots. Moreover, sectoral coverage of the EU ETS has been steadily increasing, hereby also increasing flexibility of trading.

Monitoring the distributional effects of carbon trading and co-pollutant impacts is therefore of great importance. If regressive distributional effects and pollution hot spots from emissions trading and emissions offsetting can be identified, various policy instruments can be implemented within or in addition to the EU ETS. Within the ETS, spatially or sectorally differentiated pricing, trading zones, or trading ratios could be implemented if high co-pollutant costs can be found in specific locations or sectors (Boyce and Pastor 2013). Moreover, revenues from permit auctions could be used to specifically address co-pollutant damages and accelerate fossil fuel abatement in places with the highest overall environmental damages (e.g. coal-fired power plants in densely populated areas) or investment in renewable energy sources. Besides the ETS, conventional regulations, such as stricter air quality ceilings, can also target the adverse effects of co-pollutants and pollution hot spots.

The current discussion on whether climate policy hurts the poor neglects an important dimension if co-benefits are omitted. This discussion centers on the regressive effects of rising carbon prices through consumption effects (since low-income households spend a larger share of their incomes on energy, especially for transportation and heating), broader macroeconomic effects (such as a shift in factor prices and sectoral employment shifts), and the role of redistributing carbon revenues in ways to avoid harming the poor (Rausch et al. 2011; Boyce 2018; Metcalf 2019). Since climate policy tends to reduce co-pollutant exposure and increase air quality co-benefits especially in poorer neighborhoods, co-benefits provide an additional mechanism through which climate policy can benefit the poor.

Conclusion

Framing climate policy as a distributional conflict across countries and generations overlooks important reasons for carbon mitigation here and now. Fossil fuel combustion emits not only carbon, but also hazardous co-pollutants with large direct and immediate environmental and health effects. These air quality co-benefits of carbon mitigation are high enough that carbon mitigation should be "in a country's own interest" (Parry et al. 2015).

Considering air quality co-benefits in climate policy in the EU, which currently has the world's largest and only multinational emissions trading scheme, could justify a substantially higher carbon price, which also would increase the chances of meeting the Paris goals of limiting global temperature increase to 1.5–2 °C compared to pre-industrial levels.

Moreover, considering air quality co-benefits also adds a new dimension to debates on the distributional effects of climate policy, since socioeconomically disadvantaged neighborhoods are usually more exposed to co-pollutants, both within and across countries, and benefit more from air quality improvements caused by fossil fuel abatement. Variations in the magnitude of co-benefits across sectors, activities, and locations suggest that co-pollutant damages should be carefully monitored, especially as the European carbon market is getting more and more flexible due to an increase in the geographical and sectoral coverage, as well as new laws that allow carbon crediting mechanisms such as offsetting. The more flexible the market gets, the more likely it is that hot spots of co-pollutants are generated as a side effect. The latter, however, could be addressed by a variety of measures, including differentiated pricing, supplementary regulatory measures, and carbon revenue use that offsets adverse co-pollutant impacts.

Since high levels of co-pollutants and pollution hot spots tend to be disproportionately located in socioeconomically vulnerable areas, neglecting the role of air quality co-benefits while increasing the flexibility of trading in existing carbon pricing schemes could widen socioeconomic environmental and health disparities. Monitoring co-pollutant impacts of carbon trading and considering air quality co-benefits will therefore be of increasing importance to narrow environmental inequality.

Notes

1 The authors gratefully acknowledge funding from the Austrian Science Fund (FWF) single project P 31608-G31.
2 While natural gas releases substantially less carbon dioxide than coal or oil, it also releases another greenhouse gas, methane. Compared to carbon dioxide, methane is a far more effective greenhouse gas, but it also stays in the atmosphere for a shorter time span (12 years, compared to a century or more for CO_2). As a consequence, a comparison of the climate effect between the two has to consider a specific period. When choosing a short period, such as 20 years, which is the period currently most relevant to avert the most severe damages from climate change, and when considering the full life-cycle of greenhouse gas emissions of natural gas versus coal or oil (from mining/drilling to its end use), natural gas is not found to have a lower greenhouse gas impact (Howarth 2014).
3 The High-Level Commission on Carbon Prices emphasizes, however, that prices might have to be substantially lower in developing countries, implying conversely that they have to be higher in the EU.

References

Aldy, J. E., and Stavins, R. N. (2012). The promise and problems of pricing carbon: Theory and experience. *The Journal of Environment & Development*, 21(2), 152–180.
Arrow, K., Cline, W., Maler, K., Munasinghe, M., Squitieri, R., and Stiglitz, J. (1996). Intertemporal equity, discounting, and economic efficiency. In Bruce, J. P., Lee, H., and Haites, E. F., eds., *Climate change 1995 economic and social dimensions of climate change: Contribution of working group III to the second assessment report of the intergovernmental panel on climate change.* Cambridge: Cambridge University Press, pp. 130–144.
Ash, M., and Fetter, T. R. (2004). Who lives on the wrong side of the environmental tracks? Evidence from the EPA's risk-screening environmental indicators model. *Social Science Quarterly*, 85(2), 441–462.
Boyce, J. K. (2018). Carbon pricing: Effectiveness and equity. *Ecological Economics*, 150, 52–61.
Boyce, J. K., and Pastor, M. (2013). Clearing the air: Incorporating air quality and environmental justice into climate policy. *Climatic Change*, 120(4), 801–814.
Boyce, J. K., Zwickl, K., and Ash, M. (2016). Measuring environmental inequality. *Ecological Economics*, 124, 114–123.
Buonocore, J. J. et al. (2016). An analysis of costs and health co-benefits for a US power plant carbon standard. *PloS One*, 11(6), e0156308. https://doi.org/10.1371/journal.pone.0156308.
Carbon Pricing Leadership Coalition. (2017). *Report of the high-level commission on carbon prices.* https://static1.squarespace.com/static/54ff9c5ce4b0a53decccfb4c/t/59b7f2409f8dce5316811916/1505227332748/CarbonPricing_FullReport.pdf.
Carl, J., and Fedor, D. (2016). Tracking global carbon revenues: A survey of carbon taxes versus cap-and-trade in the real world. *Energy Policy*, 96, 50–77.
Chang, T., Graff Zivin, J., Gross, T., and Neidell, M. (2016). Particulate pollution and the productivity of pear packers. *American Economic Journal: Economic Policy*, 8(3), 141–69.
Coady, D., Parry, I., Sears, L., and Shang, B. (2017). How large are global fossil fuel subsidies? *World Development*, 91, 11–27.
Coady, D. et al. (2019). *Global fossil fuel subsidies remain large: An update based on country-level estimates.* Washington, DC: International Monetary Fund Working Paper 19/89.
Currie, J., Hanushek, E., Kahn, E. M., Neidell, M., and Rivkin, S. (2009). Does pollution increase school absences? *Review of Economics and Statistics*, 91(4), 682–694.
Cushing, L., Blaustein-Rejto, D., Wander, M., Pastor, M., Sadd, J., Zhu, A., and Morello-Frosch, R. (2018). Carbon trading, co-pollutants, and environmental equity: Evidence from California's cap-and-trade program (2011–2015). *PLoS Medicine*, 15(7), e1002604.
Driscoll, C. T. et al. (2015). US power plant carbon standards and clean air and health co-benefits. *Nature Climate Change*, 5(6), 535–540. https://doi.org/10.1038/nclimate2598.
EEA [European Environment Agency]. (2014). *Costs of air pollution from European industrial facilities 2008–2012 – an updated assessment.* Copenhagen: European Environmental Agency.
EEA[European Environmental Agency]. (2016). *National emissions ceilings directive.* www.eea.europa.eu/themes/air/air-pollution-sources-1/national-emission-ceilings (4.12.2020).
EEA [European Environmental Agency]. (2018). *Unequal exposure and unequal impacts: Social vulnerability to air pollution, noise and extreme temperatures in Europe*, EEA Report No. 22/2018. www.eea.europa.eu/publications/unequal-exposure-and-unequal-impacts (4.12.2020).

EEA [European Environmental Agency]. (2020) *Air quality in Europe 2020 report*. www.eea.europa.eu/publications/air-quality-in-europe-2020-report (4.12.2020).
Eichner, T., and Pethig, R. (2011). Carbon leakage, the green paradox, and perfect future markets. *International Economic Review*, 52(3), 767–805.
Ellerman, A. D., Marcantonini, C., and Zaklan, A. (2016). The European Union emissions trading system: Ten years and counting. *Review of Environmental Economics and Policy*, 10(1), 89–107.
Fowlie, M. (2009). Incomplete environmental regulation, imperfect competition, and emissions leakage. *American Economic Journal: Economic Policy*, 1(2), 72–112.
Holland, S. P. (2010). *Spillovers from climate policy*. Cambridge: National Bureau of Economic Research, Working Paper No. 16158. https://doi.org/10.3386/w16158.
Howarth, R. W. (2014). A bridge to nowhere: Methane emissions and the greenhouse gas footprint of natural gas. *Energy Science & Engineering*, 2(2), 47–60.
Kuik, O., and Hofkes, M. (2010). Border adjustment for European emissions trading: Competitiveness and carbon leakage. *Energy Policy*, 38(4), 1741–1748.
Lelieveld, J. et al. (2019). Cardiovascular disease burden from ambient air pollution in Europe reassessed using novel hazard ratio functions. *European Heart Journal*, 40(20), 1590–1596. https://doi.org/10.1093/eurheartj/ehz135.
Metcalf, G. E. (2019). The distributional impacts of US energy policy. *Energy Policy*, 129, 926–929.
Muller, N. Z. (2012). The design of optimal climate policy with air pollution co-benefits. *Resource and Energy Economics*, 34(4), 696–722. https://doi.org/10.1016/j.reseneeco.2012.07.002.
Muller, N. Z., and Mendelsohn, R. (2009). Efficient pollution regulation: Getting the prices right. *The American Economic Review*, 99(5), 1714–1739.
Nemet, G. F., Holloway, T., and Meier, P. (2010). Implications of incorporating air-quality co-benefits into climate change policymaking. *Environmental Research Letters*, 5, 1–9. https://doi.org/10.1088/1748-9326/5/1/014007.
Nordhaus, W., and Boyer, J. (2000). *Warming the world: Economic models of global warming*. Cambridge, MA: MIT Press.
OECD. (2018a). *OECD companion to the inventory of support measures for fossil fuels 2018*. https://read.oecd-ilibrary.org/energy/oecd-companion-to-the-inventory-of-support-measures-for-fossil-fuels-2018_9789264286061-en#page1.
OECD. (2018b). *Effective carbon rates 2018: Pricing carbon emissions through taxes and emissions trading*, www.oecd-ilibrary.org/sites/9789264305304-en/index.html?itemId=/content/publication/9789264305304-en (12.12.2020).
Parry, I., Heine, D., Li, S., and Lis, E. (2014). How should different countries tax fuels to correct environmental externalities? *Economics of Energy & Environmental Policy*, 3(2), 61–77. https://doi.org/10.5547/2160-5890.3.2.ipar.
Parry, I., Veung, C., and Heine, D. (2015). How much carbon pricing is in countries' own interests? The critical role of co-benefits. *Climate Change Economics*, 6(4), 1550019. https://doi.org/10.1142/S2010007815500190.
Rausch, S., Metcalf, G. E., and Reilly, J. M. (2011). Distributional impacts of carbon pricing: A general equilibrium approach with micro-data for households. *Energy Economics*, 33, S20–S33.
Ringquist, E. J. (2005). Assessing evidence of environmental inequities: A meta-analysis. *Journal of Policy Analysis and Management: The Journal of the Association for Public Policy Analysis and Management*, 24(2), 223–247.
Ritchie, H., and Roser, M. (2020). CO_2 and greenhouse gas emissions. *OurWorldInData.org*. https://ourworldindata.org/co2-and-other-greenhouse-gas-emissions.
Rüttenauer, T. (2018). Neighbours matter: A nation-wide small-area assessment of environmental inequality in Germany. *Social Science Research*, 70, 198–211.
Shepphard, D. (11.12.2020). Price of polluting in EU rises as carbon price hits record high. *Financial Times*. www.ft.com/content/11bd00ee-d3b5-4918-998e-9087fbcca3cd (4.12.2020).
Shindell, D. T., Lee, Y., and Faluvegi, G. (2016). Climate and health impacts of US emissions reductions consistent with 2 degree C. *Nature Climate Change*, 6(5), 503–507. https://doi.org/10.1038/nclimate2935.
Stavins, R. N. (2003). Experience with market-based environmental policy instruments. *Handbook of Environmental Economics*, 1, 355–435.
Thompson, T. M. et al. (2014). A systems approach to evaluating the air quality co-benefits of US carbon policies. *Nature Climate Change*, 4(10), 917–923. https://doi.org/10.1038/nclimate2342.

Tietenberg, T. (1995). Tradeable permits for pollution control when emission location matters: What have we learned? *Environmental and Resource Economics*, 5, 95–113.
West, J., Smith, S., Silva, R. et al. (2013). Co-benefits of mitigating global greenhouse gas emissions for future air quality and human health. *Nature Climate Change*, 3, 885–889. https://doi.org/10.1038/nclimate2009.
WHO. (2019). *Ambient air pollution: A global assessment of exposure and burden of disease* https://apps.who.int/iris/bitstream/handle/10665/250141/9789241511353-eng.pdf (4.12.2020).
World Bank. (2020). *State and trends of carbon pricing 2020.* https://openknowledge.worldbank.org/bitstream/handle/10986/33809/9781464815867.pdf?sequence=4&isAllowed=y.
Zwickl, K., Ash, M., and Boyce, J. K. (2014). Regional variation in environmental inequality: Industrial air toxics exposure in US cities. *Ecological Economics*, 107, 494–509.
Zwickl, K., Sturn, S., and Boyce, J. K. (2021). Effects of carbon mitigation on co-pollutants at industrial facilities in Europe. *The Energy Journal*, 42(5).

14

DESIGNING URBAN SUSTAINABILITY

Environmental justice in EU-funded projects

Ian M. Cook and Tamara Steger

Urban sustainability initiatives in European cities generally aim to promote ecological and social well-being and nurture a competitive edge. Still, urban sustainability efforts lag behind in their attention to social equality even as urban environmental disparities continue to run deep across race, class and gender lines. Low-income populations in European cities, for example, tend to be exposed to more air pollution, noise and extreme temperatures than are wealthier segments of the population (EEA, 2018; WHO, 2019) and have less access to green space (The Marmot Review, 2010). Immigrants and slum dwellers, in particular, generally tend to live in the most polluted areas (WHO, 2010). Women, children and the aged can face more hardships when it comes to certain environmental living conditions (WHO, 2019). Efforts to improve urban sustainability, furthermore, may backfire in cases, for example, where 'green' initiatives result in gentrification leading to further exclusion of disadvantaged communities (Anguelovski et al., 2018a).

Participatory approaches in urban sustainability initiatives can play a fundamental role in not only explicitly incorporating social equality or equity as an important concern or goal, but also building in the perspectives of struggling communities, for example, based on 'livability' (Trudeau, 2018) to overcome persistent, institutionalized inequalities. Based on a review of 94 articles on stakeholder participation in urban sustainability governance (2013–2016), Soma et al. (2018) affirm the importance of stakeholder participation in assuring green livable cities and legitimate, accountable and transparent decision-making but point out that not all participatory approaches sufficiently engage sustainability principles, such as social justice. Rather, there is an emphasis on the urban built environment (John et al., 2015), and "it remains unclear how to proceed in marginalised urban areas when the context is 'inequality', 'social exclusion', 'poverty', 'high conflict levels', 'high corruption levels', 'pollution', 'water shortage', and/or 'lack of institutional capacity', among others" (Soma et al., 2018, pp. 444–445).

In the last decades, the European Union (EU) has invited project proposals that can advance participatory or inclusive governance processes as a means to promote urban sustainability. We believe this offers an interesting analytical entry point for upscaling findings from a project-specific scale to the dominant project-led governance approach. By looking at the design of EU-funded urban governance-related projects, we can learn about the potential and limits of the wider trend for projectification, especially as it relates to environmental justice in urban

DOI: 10.4324/9780367814533-15

contexts. We thus explore the participatory mechanisms of select EU-funded urban sustainability projects based on an environmental justice framework.[1]

In this chapter, we articulate the conceptual and methodological boundaries of the research. Then, we provide an analytical discussion of projectified participatory mechanisms while evoking examples of select projects to demonstrate our emerging findings. In conclusion, we provide a summary of our insights for building urban sustainability based on participatory approaches from an environmental justice perspective.

Methodology: an 'optimistic' approach

Methodologically, we drew on our involvement in the EU-funded project "UrbanA – Urban Arenas for Sustainable and Just Cities" that aims to distil approaches from EU projects that focus on either sustainability or social justice in urban contexts.[2] For the most part, the projects were funded under the Framework Programmes for Research and Technological Development 4–7 (FP 4–7) and Horizon 2020 schemes, but also included other EU and non-EU projects that were relevant to the aims of the project. This involved a mapping of EU projects, as part of a wider consortium, before honing in on select projects and/or approaches. As a consortium the authors played a role in a multistage process that mapped projects advancing approaches that sought to tackle urban (un)sustainability and (in)justice. The process's wider goal was to broker knowledge across 'city-makers' in European cities, thus helping the cities transform into becoming sustainable and just places. As such, aside from drawing out approaches from the projects, the mapping also identified individuals and groups, some of whom we brought together at 'urban arenas'.

The mapping process involved exploring possibilities through a consortium-generated 'gut list' (n = 25–30), an open call on social media and a systematic scanning via the Cordis[3] database of EU-funded projects to create a basic long list (n = 100+), a 'quick scanning' of these projects to create a final long list (n = 100), a desk study of these projects and the approaches contained within them to produce a hotlist of approaches (n = 30–40), and finally a distilling of this research to produce a Wiki database of approaches.[4] These were then shared with city-makers at an arena event for validating and co-creation, leading to changes to the Wiki. In this chapter we draw from our role within this process and a further reanalysis of specific projects and approaches covered within UrbanA.[5]

The mapping was focused around four central themes – urban, sustainability, justice and transformative potential. Regarding (in)justice, we looked for exclusion and attempts for inclusion with an intersectional eye (i.e. ethnicity, age, gender, income, etc.) based on an environmental justice framework (see later). In our further review of the projects and approaches for this chapter, we reanalyzed the materials with a specific focus on participatory approaches. Based on this analysis of the project documentation of select urban sustainability projects, we identified several key themes: inclusivity, diverse roles, integration, networks and networking, and different kinds of knowledge and ways of knowing. In addition, we sought out reflections upon completion of various elements of the projects or the projects as a whole to identify 'lessons learned' from the perspective of the project partners who were responsible for implementing the project work packages.

As such, this is not a comprehensive (or quantitative) analysis of all potentially relevant projects. Such an analysis would require, we believe, a much more detailed ethnographic approach, including extensive interviews that capture a project's unfolding in real time and how it interacted with its target groups and the wider social, cultural and political context within which it is embedded. Rather, our aim here is tightly circumscribed to how project partners present their projects –in both their planning and their self-evaluations through the required documentation found on their respective websites and within the Cordis database. We treat the project documentation as a textual representation of an ideal project which, though it may not fully reflect how those involved or affected by the project experienced its activities, nevertheless reveals how environmental justice is written into project conceptualization and design.

Our 'optimistic' approach is one in which we emphasize elements that enhance environmental justice in the participatory approaches to promoting urban sustainability. We identified three main tenets through which the projectification of urban sustainability straddles an environmental justice framework. (1) Urban sustainability is historically based on the concept of sustainable development that asserts overlapping economic, ecological and social spheres. However, researchers and practitioners alike have been increasingly compelled to add 'just' or 'justice' to the term sustainability to overcome the emerging predominance of the economic and ecological spheres. This further coincides with the long-standing debates and discussions around the conceptualization of the 'environment' and the relationships across nature and culture manifest in a projectified context, in which culture may be either central or peripheral to urban sustainability initiatives rather than assumed as an integrated and mutually informed dynamic. (2) Projects may design and/or test participatory mechanisms that attempt to build in or emphasize one or more of the following: inclusivity, diversity, integration, networking and different approaches to knowledge building that may or may not deliberately address the discriminatory, racist and/or classist basis that environmental justice remedies. (3) Finally, environmental justice is challenged by the perpetuation of structural inequalities that take time to transform. In a projectified context in which many EU projects last three years, the rigidity or limited elasticity of such structures can only yield to environmental justice over the long haul with an enduring persistence.

Urban sustainability, environmental justice and projectification

Urban sustainability, from the standpoint of sustainable development,[6] is configured as three separate but overlapping spheres: economic growth, environmental protection and social equity. However, from an environmental justice perspective, economic, ecological and social equity do not constitute separate, overlapping spheres; rather, they are integrated fundamental aspects of everyday life and survival (Di Chiro, 1996). Di Chiro refers to "living environmentalisms" (2008) to assert a more integrated and dynamic understanding of sustainability from an environmental justice perspective reasoning that the separation of economy, ecology and society does not reflect the reality of our existence in place, and such a separation sets the stage for the forces that assert economic growth as central.[7] Inequalities are thus manifested by the forces that preclude or deny this interrelationship. Green city initiatives that compartmentalize environmental and economic issues, for example, tend to emphasize energy efficiency inspired predominantly by climate change and the economic principles of competitiveness with little or tangential attention to equity or justice (see, for example, Winter, 2016; Pearsall and Pierce, 2010). Thus while the greening of cities can benefit all, in practice as some cities become greener, their residents can become whiter and wealthier. Greening cities have thus been increasingly challenged to assure proportionate access to green spaces while avoiding the displacement of poorer subordinated communities (Anguelovski et al., 2018a; Gould and Lewis, 2017; Winter, 2016; Wolch et al., 2014; Curran and Hamilton, 2012; Checker, 2011), which raises questions about the distribution of green space and if greening could be done in a more even and holistic manner.

In response to this, environmental justice scholars assert the need for "just sustainabilities" (Agyeman et al., 2003; Agyeman, 2013) or sustainability as a focus on social justice, community, environment and economic security "that is meaningful for all communities" (Di Chiro, 2013). While environmental justice from a social movement standpoint is a call for distributive justice, inclusive participation and recognition (Schlosberg, 2004), the environmental justice movement has a pluralistic nature embracing diversity in modes of organization, foci and action (Schlosberg, 1999) and is associated with multiple spatialities (Walker, 2009). The forces for sustainability should be diverse and contextualized based on sharing (Agyeman et al., 2016, refer to Agyeman et al., 2013). Thus fundamental to urban sustainability from an environmental justice viewpoint are the perspectives, experiences and meaningful participation of struggling communities.

Institutional and structural forces are instrumental in the promotion of environmental justice, raising questions about how to overcome deeply rooted structural inequalities and trajectories (Pellow, 2000; Schlosberg, 2004, refers to I.M. Young, 1990, 2000) and engage the importance of history (Escobar, 1996). Thus, urban sustainability from an environmental justice perspective nurtures the transformation of the structural and institutional mechanisms that perpetuate the subordination and vulnerable-ization of certain communities. This is no small task and has significant temporal considerations. It is a long-term 'project' central to urban sustainability from an environmental justice perspective.

In general, EU funding mechanisms are characterized by projectification – a preference for specially created ephemeral organizational forms tasked with meeting set goals within a limited period of time (Sjöblom et al., 2013). The appeal of projects, an organizational framework inspired perhaps initially from the private sector, lay in their perceived ability to both respond to specific needs in a timely manner and offer more efficiency, flexibility and innovation than permanent structures do. Projects dealing with urban sustainability are flexible coordinated efforts (in contradistinction to standing organizations) using specified tools applied toward a specific goal within a certain time frame, "delivering an idea" and solving problems in new ways (Cerne and Jansson, 2019, p. 357). Diverse and multiple ideas and goals emerging in and across different localities is methodologically well aligned with the pluralistic nature of environmental justice.

Despite the perpetuation of embedded historical structural inequalities and trajectories, a good deal of sustainability work occurs in the form of projects that are temporally challenged by the need for efficiency, deadlines and finality. The challenge is thus how to address a long-standing dynamic through multiple short-term project processes. In such projectified contexts, a delicate balance that bridges project initiatives with more permanent organizations is called for so as to benefit from the former's flexibility and innovative capacities and the latter's coordination and continuity (see Godenhjelm et al., 2015).

Another key to overcoming the challenges inherent in projectified work is to assure continuous and long-term advances in urban sustainability through the narratives of their participants and learning from project "mistakes and downfalls" (Vifell and Soneryd, 2012, p. 26). However, one scholar notes that dissensus, "[u]nlike methods assembled out of consensus-building exercises performed amongst the usual suspects in the comfort of well-funded frameworks", reveals more about the needed pathways toward urban sustainability that is socially just (Kaika, 2017, p. 11). Moreover, on an EU-level this has led to varying degrees of exclusion as an expert-populated 'project world' has emerged with its own conventions, rhetoric, regulations and standards (Büttner and Leopold, 2016). Whilst these experts can 'scale hop' between the local, national and EU levels, local inhabitants are often left without the project world skills and language needed to effect change through project funding mechanisms (Szőke, 2013).

As such, in the following analysis of EU-funded projects, we explore inclusivity and participation from an environmental justice perspective. For example, in what ways, if at all, are disadvantaged or vulnerable communities a part of the participatory narrative schema of urban sustainability projects and what challenges and opportunities have been garnered in that process?

Participatory processes within urban sustainability and justice projects

Urban sustainability projects supported by the EU's funding schemes straddle different assumptions and conceptualizations of sustainability that have implications for environmental justice. Environmental justice activists, for example, raise questions about the purpose of sustainability and its beneficiaries (Di Chiro, 2013). Accordingly, we should ask who benefits from sustainability initiatives, and what participatory or recognition support mechanisms exist within them. Our reading of EU-funded project documentation suggests that guiding principles are asserted when it comes to particular participatory processes and methodologies in relationship to urban sustainability that include, among others:

- Inclusivity (e.g. youths)
- Diverse roles (e.g. citizen science, citizen or participatory budgeting)
- Different kinds of knowledge and ways of knowing: participatory research methodologies (e.g. action research)
- Networks and networking or partnership building
- Integration (e.g. co-construction of knowledge across experts and policymakers)

These projectified points of engagement, however, reveal different assumptions about what constitutes inclusivity, diverse roles, integration, networks or partnership building, and different kinds of knowledge or ways of knowing. Further, EU project calls provide guidance on these matters that may incorporate social involvement in the form of enhancing environmental stewardship and, more recently, environmental citizenship.

Inclusivity

Inclusivity in a projectified context widely emphasizes organizational or sectoral representation (e.g. civil society, government, private industry, scientists, environmental issues such as energy), while an explicit environmental justice focus might target 'disadvantaged', 'vulnerable' and/or immigrant, youth and elderly communities. However, inclusivity is also indicated in a project's conceptualization of urban sustainability, including the extent to which social equity is engaged. Inclusivity can also be a driver of project-based research. For example, GREENLULUS (Green Locally Unwanted Land Uses) involved a review and analysis of 99 cities across Europe, the United States and Canada focused on environmental justice and gentrification. They concluded that stronger urban social policies on affordability and 'staying in place' protect community access to the benefits of green initiatives while preventing gentrification that can further exclude socio-economically disadvantaged groups (Anguelovski et al., 2018b).

Other projects have acknowledged the importance of garnering disadvantaged youth perspectives in policymaking and decision-making (e.g. SocIEty – Social Innovation | Empowering the Young for the Common Good). Meanwhile, ProGIreg (Productive Green Infrastructure for post-industrial urban regeneration) focuses on "urban areas that face the challenge of post-industrial regeneration. These areas suffer from social and economic disadvantages, inequality and related crime and security problems."[8] One of their project's participatory forums that took place in Zagreb, Croatia, also specifically included a non-governmental organization ('ISKRA') that represents vulnerable groups and a representative from a Roma ethnic minority association. Projectified work can further build inclusivity along the lines of balanced representation across government, academia, private enterprise and civil society actors. Additional efforts from an environmental justice perspective build in representation from the perspective of community diversity in which poor and struggling communities are targeted to address liveability and the critical issue of survival.

A recent trendy iteration of an urban sustainability methodology is 'nature-based solutions' (NBS). While it is not defined clearly or consistently across the projects that we reviewed, it generally draws on environmental services in urban economies focusing on the built environment. NATURVATION asserts that it is "the use of nature properties of ecosystems" to "limit impacts of climate change, enhance biodiversity and improve environmental quality while contributing to economic activities and social well-being".[9] URBAN GreenUP describes a main part of their project: "Through the implementation of very technical nature-based solutions in large-scale districts, URBAN GreenUP aims at achieving a variety of impacts related with both environmental and socio-economic aspects."[10] GrowGreen "aims to create climate and water resilient, healthy and livable cities by investing in nature-based solutions (NBS). Making nature part of the urban living environment improves quality of life for all citizens and will help business to prosper."[11] Some examples of nature-based solutions include work to build in bike paths, green roof gardens, parks that reduce the urban heat island effect and water flow management through varied porous surfaces and encatchments.

While nature-based solutions tend to emphasize technical, economic and environmental aspects, a 'nature-based solutions' methodology does not necessarily preclude issues of equality. NATURVATION included a literature review based on the "social and cultural benefits" of nature-based solutions and found that 11 out of 65 identified relevant articles focused on "enhancement of equality" (Maia da Rocha et al., 2017). The discourses surrounding such solutions, furthermore, have incorporated the articulation of demands for open, participatory green spaces in cities, thus exceeding the ongoing marketization of urban nature that such an approach often engenders (Kotsila et al., 2020). The project NATURVATION[12] provides guidance on how marginalized groups can be reached and included and describes what works in terms of approaches and methods to promote citizen engagement for nature-based solutions. These include citizen panels (a "large, demographically representative group of citizens that is surveyed regularly"); the use of appreciative inquiry to assess the valuable aspects of an idea or topic and see what resources already exist; advocacy planning, where specialists assist and advise citizens; and community organizing through which citizens can assert their interests. URBANAT, another NBS project, specifically engages deprived communities in their efforts by focusing on the "regeneration and integration of deprived social housing urban developments through an innovative and inclusive catalogue of NBS, ensuring sustainability and mobilising driving forces for social cohesion".[13]

In general, however, projectified urban sustainability efforts driven by a predominance of market and ecological values proliferate, in part, because social equity is not sufficiently rooted in urban sustainability (Gould and Lewis, 2017). From an environmental justice standpoint, this integration is central from the perspective of not only articulating environmental problems, but also asserting solutions.

Diverse roles

Indeed, recognition of different ways of knowing and their role in promoting urban sustainability is critical from an environmental justice standpoint. Citizen observatories, for example, create the possibility "to learn from citizen experience and perception and enable citizenship co-participation in community decision making and co-operative planning" (see, for example, CITI-SENSE). Citizens or "civic scientists" collect environmental data or may participate in the development of a public municipal budget. For example, the neighbourhood citizen budgeting process introduced in Amsterdam in 2010 emphasized human rights and transparency and culminated in a budgetary output in 2014. In such efforts, "municipal bureaucracy stands next to the citizen instead of in opposite him/her" (Wittmayer and Rach, 2016, p. 8 quote Engbersen et al., 2010), "bridge the commitment between citizen and government" (interview cited in Wittmayer and Rach 2016, p. 9) which can be done, for example, in a citizens' perspectives paper. This effort spanned years, and the budgeting process was part of an ongoing cycle involving a significant learning curve on the part of community participants. It is unclear, however, the extent to which these civic-based initiatives recognized the views and concerns of disadvantaged community members. And, a general appreciation for diverse epistemologies can be critical.

Different kinds of knowledge and ways of knowing

Indeed, there is a tacit recognition in some projects that different kinds of knowledge or ways of knowing are related to inclusivity. This is another way of saying that for projects to be inclusive, they must not only acknowledge but also meaningfully utilize the knowledge brought and generated by diverse individuals and groups of people.

URBANSELF and other projects have focused on an actor-centred approach when engaging local forms of communication, knowledge and modes of survival. This is opposed to placing technical expertise at the forefront of urban change. As such, URBANSELF notes that there is a difference between, on the one hand, training inhabitants in ways that allow them to appreciate urban planning, and on the other, recognizing local citizens' competences, knowledge and capacities. Ultimately, this is to suggest that shared visions of the future and planning possibilities should be brought forth by integrating citizens into urban governance models. URBAN-SELF concluded,

> Sustainable urban development therefore requires a broader societal dialogue. A transition is needed from a rational, technocratic policy process towards a participative process of 'messy governance' in pursuit of 'clumsy solutions'; one that recognises the validity of multiple, subjective stakeholder viewpoints.[14]

Different ways of engaging different knowledge forms may also require a reconfiguration of actor roles. IN-CONTEXT noted, "When faced with this type of initiatives, local governments can take the role of bystanders, passengers, or drivers" and "[i]n the community arena people meet as individuals, including their values, emotions and hopes on one side and their institutional environments on the other."[15] However, they note a challenge with taking on different roles for government actors in participatory processes,

> Elected leaders or public servants may experience local initiatives in an ambivalent way: on the one hand they might want to support engaged citizens. On the other hand, they might see local initiatives as a threat to the self-concept of being a representative body that holds special expertise and is elected to fulfil public duties.[16]

However, it is further noted that local government representatives' roles need to be asserted based on a case-by-case basis.

Still, IN-CONTEXT's approach using "out-of-town-hall" participatory forums are challenged by the organizing principles of more traditional or formal public participation processes, expressing,

> [i]t is not easy to organise successful public participation processes. Rows of empty seats in public meetings and 'the usual suspects' bringing up the same issues over and over again are familiar to all those responsible for implementing participation on the ground.[17]

Including people with diverse roles and recognizing different ways of knowing was seen as only the first step in some projects, with also the desired need for more robust networks.

Inclusivity and integration emphasize meaningful social involvement in the production of knowledge, if predominantly originating from assumptions about common divisions made between scientists and policymakers and layperson and expert knowledge. Participatory research methodologies (e.g. action research) might serve as inspiration for both affirming and revisiting projectified governance in order to better promote environmental justice in urban sustainability initiatives. Such methodologies can convene networks that link scientists, policymakers and civil society and/or communities in producing knowledge while also recognizing different ways of knowing.

Networks and networking or partnership building

Bringing together a combination of scientists and policymakers, and possibly civic entities, is a common methodology to facilitate science-policy networks in the co-construction or implementation of knowledge in policy making. Foodlinks, for example, introduces itself as "a collaborative project . . . with the purpose of evaluating knowledge brokerage activities to promote sustainable food consumption and production: linking scientists, policymakers and civil society organizations".[18] URBAN-NEXUS[19] sought to engage those from different "geographical, cultural and professional backgrounds" and "different types of actors, such as municipalities, universities, national ministries, knowledge and lobby institutes and civil society organisations" in a

> structured dialogue approach . . . aimed to promote a culture change in the stakeholders working in urban areas across Europe. It will introduce them, via Dialogue Cafés, to individuals, organisations and disciplines that they would not have previously encountered in their roles.[20]

The creation of networks is one of the ways in which novel forms of knowledge are brokered. For these networks of knowledge to build in more environmental justice, however, additional participatory targets may be considered. For instance, the project URBAN-SELF's main goal was to identify working solutions for cities through different forms of self- organization and then to take these findings and spread them to various stakeholders.[21] Self-organization can draw on pre-existing strengths and initiative and could be complemented by support mechanisms for self-organizing that attempt to overcome pre-existing barriers to participation. Such an actor-centred approach can draw on local knowledge/survival strategies first and foremost, rather than starting from technical expertise. In this particular case, for example, there was the extension of consideration "for those who are not fully integrated into a market economy".[22]

A much more solidified network has been created through the three iterations project-turned-programme URBACT. The idea of the programme is to allow cities to work together to find solutions to shared challenges through co-learning and networking. It describes itself as a

> network of 26 associations in 23 European countries enhancing the competences of local actors that are engaged in the social and solidarity economy, supporting a new paradigm of economic development in order to fight poverty and to diffuse an equitable and sustainable way of living.[23]

There is further evidence, within the project documentation, of an understanding of the need to integrate the knowledge co-constructed from within these networks.

Integration

One way such an integration is realized is through community of practice (CoP), a concept that has gained popularity within project documentation as a specific means of promoting the kinds of diverse, inclusive and integrated networks needed to build urban sustainability. However, generally the concept has not been historically strictly defined or consistently applied (Li et al., 2009). Rather, it has evolved from initially focusing on social learning among practitioners and the professionalization of novice participants in interaction with experts to a wider emphasis on individual development in group dynamics and ultimately considered in the managerial facilitation of corporate competitiveness (Li et al., 2009, refer to the works of Lave and Wenger, 1991; Wenger, 1998; Wenger et al., 2002).

The project Foodlinks defines CoP as

> a group that evolves or is created around a common interest in a particular field with the goal of expanding knowledge related to that field. It is through the process of sharing information and experiences with the group that the members learn from each other and have an opportunity to broaden their understanding of the matter
>
> (Wenger et al., 2002).[24]

Project documentation noted, however, that some participants found it difficult to bring in different ways of knowing in contradistinction to traditional scientific knowledge, acknowledging the importance of balanced representation across academics and practitioners and understanding links between government, research and civil society organizations.[25]

Other projects such as IN-CONTEXT noted that in transition management approaches, the

> cooperation of the engaged citizens and the local government needs to be on equal footing, each bringing in their knowledge, values and ideas. The nature of the relationship in transition management approaches is thus fundamentally different from the traditional forms of governance dominated by expert-knowledge and top-down policy making. It is characterised by respect, non-directivity, openness of agenda and creativity.[26]

Some projects combine these aspects; in CITIES4PEOPLE,[27] for example, develop flexible 'citizen mobility toolkits' to be used in different localities when determining needs and wants for people-led transport solutions.

Calling time on projects?

A common thread found throughout all of the projects is that bottom-up participation in processes aimed at urban (un)sustainability and (in)justice requires time (and money). As reflected upon within the INCONTEXT[28] project, "aiming for a sustainability transition tends to be more time-consuming and therefore more expensive (at least at face value) than implementing ordinary top-down approaches". What is less readily apparent within project documentation, however, is the different types of temporal mediation and contradictions that arise when different groups of people engage in democratic participatory processes (see Abram, 2014). This is to suggest that different groups of people have very different temporal frames through which they might imagine a particular issue and how it relates to other aspects of their lives and their city.

A project, by its very definition, is a temporally constrained mechanism that, once its initially agreed upon time frame has expired, ceases to operate formally. This is evident with even mundane matters. For instance, when conducting research for the UrbanA project, we struggled with this as most projects create websites that cease to exist once the project ends (thus any outcomings or learnings from the project are much harder to track down). Still, collective efforts in the context of project work can build relationships and experiences that are not forgotten or lost even after the project comes to a close.

Additional temporal issues that challenge projectified work and other urban sustainability initiatives are related to the election cycle. For example, in URBAN-NEXUS, it was noted, "[k]ey constraints for integrating urban governance include the democratic electoral cycle that reinforces a political culture of short termism and the proliferation of sector-specific funding regimes, conceived and implemented separately, which compound further the challenges of coordination."[29]

When dealing with issues of urban environmental justice, however, the finality of projects poses a significant challenge. It is particularly problematic because transforming structural inequalities takes time, more than is allowed for within the standard project time frames. By distilling approaches from past EU-funded projects, UrbanA seeks to alleviate this by providing the knowledge and translocal connections citymakers need to instigate transformation. This assumption about the long-term sustainability of projects through launching enduring networks is also shared by URBAN-NEXUS. As URBAN-NEXUS notes, "the strong focus of URBAN-NEXUS on integrated combinations between practices between various themes within urban sustainability will lead to long term, integrated partnerships with participants from all thematic themes."[30] They conclude further, "[t]he traditional focus on 'big bang' policies is inappropriate. A 'progressive incremental' approach that focuses on how small steps cumulatively produce significant returns over time is more appropriate."[31] Internal and external projectified networks play an important role in assuring that efforts continue. More broadly, this points to the need to engage a longer-term effort to understand the effects of (and then integrate) the project form with more stable institutional arrangements (Munck af Rosenschöld and Wolf, 2017).

Conclusion

The project-based participatory mechanisms reviewed and discussed here draw attention to inclusivity, diverse actors and roles, different kinds of knowledge and ways of knowing, networks and integration that, to varying agrees, are able to align with environmental justice principles along the lines of conceptualizing sustainability more holistically, targeting social inequalities and navigating short-term efforts to address enduring, institutionalized problems. Individual projects recognize the need for decision-making to be more inclusive, with some specifically targeting struggling communities in a way that integrates different forms of knowledge and ways of knowing that can, if done right, instigate longer-term networks of projectified notions of co-creation and co-decision-making. Following an optimistic approach in this chapter, we have highlighted how EU-funded projects have promoted approaches to inclusivity that nurture new networks and the co-construction of integrated knowledge across different actors and roles, some of which explicitly engage struggling and deprived urban communities. In the latter process, justice takes its rightful place in urban sustainability as the communities for whom urban sustainability is the most critical for survival come to the fore.

Notes

1 This chapter emerges from our work within the UrbanA project, which received funding from the European Union's Horizon 2020 Research and Innovation Programme under grant agreement No. 822357.
2 https://urban-arena.eu/
3 https://cordis.europa.eu/
4 https://wiki.urban-arena.eu/
5 Mapping guidelines and UrbanA arena design details can be found at https://urban-arena.eu/resources/
6 See the 1987 Brundtland Report; the 1992 United Nations Conference "Earth Summit" on Environment and Development and the United Nations Sustainable Development Goals (SDGs) (2015–2030).
7 See, for example, Harvey, D. 2007. *A Brief History of Neoliberalism*. Oxford: Oxford University Press, 247.
8 https://progireg.eu
9 https://naturvation.eu/about
10 www.urbangreenup.eu/solutions/
11 http://growgreenproject.eu/about/project/
12 https://naturvation.eu/
13 https://urbinat.eu
14 https://cordis.europa.eu/project/id/282679/reporting
15 www.incontext-fp7.eu/sites/default/files/D5.3_Out%20of%20the%20townhall-final.pdf
16 www.incontext-fp7.eu/sites/default/files/D5.3_Out%20of%20the%20townhall-final.pdf
17 www.incontext-fp7.eu/sites/default/files/D5.3_Out%20of%20the%20townhall-final.pdf
18 See www.foodlinkscommunity.net/foodlinks-home.html
19 https://cordis.europa.eu/project/id/282679
20 See https://cordis.europa.eu/project/id/282679/reporting
21 https://cordis.europa.eu/project/id/268931
22 https://cordis.europa.eu/project/id/268931
23 https://urbact.eu/urbact-glance
24 www.foodlinkscommunity.net/fileadmin/documents_organicresearch/foodlinks/publications/Foodlink-broschuere-knowledge-brokerage.pdf
25 www.foodlinkscommunity.net/fileadmin/documents_organicresearch/foodlinks/publications/Foodlink-broschuere-knowledge-brokerage.pdf
26 www.incontext-fp7.eu/sites/default/files/D5.3_Out%20of%20the%20townhall-final.pdf
27 https://cities4people.eu/hu/
28 www.incontext-fp7.eu/sites/default/files/D5.3_Out%20of%20the%20townhall-final.pdf
29 https://cordis.europa.eu/project/id/282679/reporting
30 www.asde-bg.org/docs/projects/bg/31_Urban-Nexus_SDPF_22.03.2012_BG.pdf
31 https://cordis.europa.eu/project/id/282679/reporting

Bibliography

Abram, S. 2014. The Time It Takes: Temporalities of Planning. *Journal of the Royal Anthropological Institute* 20 (April): 129–147. https://doi.org/10.1111/1467-9655.12097.

Agyeman, J. 2013. *Introducing Just Sustainabilities: Policy, Planning and Practice*. London: Zed Books.

Agyeman, J., R. Bullard, and B. Evans (Eds.). 2003. *Just Sustainabilities: Development in an Unequal World.* Cambridge: MIT Press.

Agyeman, J., D. McLaren, and A. Shaeffer-Borrego. 2013. *Sharing Cities.* London: Friends of the Earth.

Agyeman, J., D. Schlosberg, L. Craven, and C. Matthews. 2016. Trends and Directions in Environmental Justice: From Inequity to Everyday Life, Community, and Just Sustainabilities. *Annual Review of Environment and Resources* 41: 321–340. https://doi.org/10.1146/annurev-environ-110615-090052.

Anguelovski, I., L. Argüelles, F. Baró, H. Cole, J. Connolly, M. Lamarca, S. Loveless, C. Pérez del Pulgar, G. Shokry, T. Trebic, and E. Wood. 2018b. *Green Trajectories: Municipal Policy Trends and Strategies for Greening in Europe, Canada and United States (1990–2016) Barcelona Laboratory for Urban Environmental Justice and Sustainability.* www.bcnuej.org/wp-content/uploads/2018/06/Green-Trajectories.pdf. Last visited January 14, 2020.

Anguelovski, I., J. Connolly, L. Masip, and H. Pearsall. 2018a. Assessing Green Gentrification in Historically Disenfranchised Neighborhoods: A Longitudinal and Spatial Analysis of Barcelona. *Urban Geography*, 39 (3): 458–491. https://doi.org/10.1080/02723638.2017.1349987.

Büttner, S.M., and L. Leopold. 2016. A "New Spirit" of Public Policy? The Project World of EU Funding. *European Journal of Cultural and Political Sociology* 3 (1): 41–71. https://doi.org/10.1080/23254823.2016.1183503.

Cerne, A., and J. Jansson. 2019. Projectification of Sustainable Development: Implications from a Critical Review. *International Journal of Managing Projects in Business* 12 (2): 356–376. https://doi.org/10.1108/IJMPB-04-2018-0079.

Checker, M. 2011. Wiped Out by the "Greenwave": Environmental Gentrification and the Paradoxical Politics of Urban Sustainability. *City & Society* 23 (2): 210–229.

Curran, W., and T. Hamilton. 2012. Just Green Enough: Contesting Environmental Gentrification in Greenpoint, Brooklyn. *Local Environment* 17: 1027–1042.

Di Chiro, G. 1996. Nature as Community: The Convergence of Environment and Social Justice. In *Uncommon Ground: Rethinking the Human Place in Nature.* Ed. William Cronon. New York: Norton, 298–320.

Di Chiro, G. 2008. Living Environmentalisms: Coalition Politics, Social Reproduction, and Environmental Justice. *Environmental Politics* 17 (2): 276–298.

Di Chiro, G. 2013. *Connecting Sustainability and Environmental Justice.* Swarthmore College: News and Events. Audio Transcript. www.swarthmore.edu/news-events/giovanna-di-chiro-connecting-sustainability-and-environmental-justice. Last visited January 31, 2020.

Engbersen, R., K. Fortuin, and J. Hofman. 2010. *Bewonersbudgetten, wat schuift het? Ervaringen van gemeenteambtenaren met bewonersbudgetten. Publicatiereeks over burgerparticipatie.* Den Haag: Ministerie van Binnenlandse Zaken en Koninkrijksrelaties.

Escobar, A. 1996. Construction Nature: Elements for a Post-Structuralist Political Ecology. *Futures* 28 (4): 325–343.

European Environment Agency (EEA). 2018. *Unequal Exposure and Unequal Impacts: Social Vulnerability to Air Pollution, Noise and Extreme Temperatures in Europe.* Report No. 22/2018. www.eea.europa.eu/publications/unequal-exposure-and-unequal-impacts/. Last visited October 18, 2020.

Godenhjelm, S., R.A. Lundin, and S. Sjöblom. 2015. Projectification in the Public Sector – the Case of the European Union. *International Journal of Managing Projects in Business* 8 (2): 324–348. https://doi.org/10.1108/IJMPB-05-2014-0049.

Gould, K., and T. Lewis. 2017. *Green Gentrification: Urban Sustainability and the Struggle for Environmental Justice.* London and New York: Routledge, Taylor & Francis Group.

John, B., L.W. Keeler, A. Wiek, and D.J. Lang. 2015. How Much Sustainability Substance Is in Urban Visions? −An analysis of Visioning Projects in Urban Planning. *Cities*, 48: 86–98.

Kaika, M. 2017. "Don't call me resilient again!": The new urban agenda as immunology . . . or . . . what happens when communities refuse to be vaccinated with "smart cities" and indicators. *Environment and Urbanization.* https://doi.org/10.1177/0956247816684763.

Kotsila, P., I. Anguelovski, F. Baró, J. Langemeyer, F. Sekulova, and J.J.T. Connolly. 2020. Nature-Based Solutions as Discursive Tools and Contested Practices in Urban Nature's Neoliberalisation Processes. *Environment and Planning E: Nature and Space.* https://doi.org/10.1177/2514848620901437.

Lave, J., and E. Wenger. 1991. *Legitimate Peripheral Participation in Communities of Practice. Situated Learning: Legitimate Peripheral Participation.* Cambridge: Cambridge University Press.

Li, L.C. et al. 2009. Evolution of Wenger's Concept of Community of Practice. *Implementation of Science* 4: 11. https://doi.org/10.1186/1748-5908-4-11.

Maia da Rocha, S., D. Almassy, and L. Pinter. 2017. *Social and Cultural Values and Impacts of Nature-Based Solutions and Natural Areas*. https://naturvation.eu/sites/default/files/result/files/naturvation_social_and_cultural_values_and_impacts_of_nature-based_solutions_and_natural_areas.pdf. Last visited January 29, 2020.

The Marmot Review. 2010. *Fair Society, Healthy Lives: Strategic Review of Health Inequalities in England Post-2010*. London: UCL IHE.

Munck af Rosenschöld, J., and S. A. Wolf. 2017. Toward projectified environmental governance? *Environment and Planning A* 49 (2): 273–292.

Pearsall, H., and J. Pierce. 2010. Urban Sustainability and Environmental Justice: Evaluating the Linkages in Public Planning/Policy Discourse. *Local Environment* 15 (6): 569–580. https://doi.org/10.1080/13549839.2010.487528.

Pellow, D. 2000. Environmental Inequality Formation Toward a Theory of Environmental Injustice. *American Behavioral Scientist* 43 (4): 581–601.

Schlosberg, D. 1999. *Environmental Justice and the New Pluralism*. Oxford: Oxford University Press.

Schlosberg, D. 2004. Reconceiving Environmental Justice: Global Movements and Political Theories. *Environmental Politics* 13 (3): 517–540.

Sjöblom, S., K. Löfgren, and S. Godenhjelm. 2013. Projectified Politics – Temporary Organisations in a Public Context. *Scandinavian Journal of Public Administration* 17 (2): 3–12.

Soma, K., M.W.C. Dijkshoorn-Dekker, and N.B. P. Polman. 2018. Stakeholder Contributions Through Transitions Towards Urban Sustainability. *Sustainable Cities and Society* 37: 438–450.

Szőke, A. 2013. Projecting the "Disadvantaged": Project Class, Scale Hopping and the Creation of Ruralities. In *Shaping Rural Areas in Europe*. Eds. Luís Silva and Elisabete Figueiredo. Dordrecht: Springer, 75–92.

Trudeau, D. 2018. Integrating Social Equity in Sustainable Development Practice: Institutional Commitments and Patient Capital. *Sustainable Cities and Society* 41: 601–610.

Vifell, A.C., and L. Soneryd. 2012. Organizing Matters: How "the Social Dimension" Gets Lost in Sustainability Projects. *Sustainable Development* 20: 18–27. https://doi.org/10.1002/sd.461.

Walker, G. 2009. Beyond Distribution and Proximity: Exploring the Multiple Spatialities of Environmental Justice. *Antipode* 41 (4): 614–636.

Wenger, E. 1998. *Communities of Practice: Learning, Meaning, and Identity*. New York: Cambridge University Press.

Wenger, E., R.A. McDermott, and W. Snyder. 2002. *Cultivating Communities of Practice*. Boston, MA: Harvard Business School Press.

Winter, A.K. 2016. *Contested Sustainability and the Environmental Politics of Green City Making*. Doctoral Dissertation. Central European University. Budapest, Hungary.

Wittmayer, J.M., and S. Rach. 2016. *Participatory Budgeting in the Indische Buurt; Chapter 5 of TRANSIT Case Study Report Participatory Budgeting*. TRANSIT: EU SSH.2013.3.2-1 Grant Agreement No: 613169. www.transitsocialinnovation.eu/content/original/Book%20covers/Local%20PDFs/185%20Participatory%20budgeting%20in%20the%20Indische%20Buurt%202015.pdf.

Wolch, J., J. Byrne, and J.P. Newell. 2014. Urban Green Space, Public Health, and Environmental Justice: The Challenge of Making Cities "Just Green Enough". *Landscape and Urban Planning* 125: 234–244. ISSN 0169-2046. https://doi.org/10.1016/j.landurbplan.2014.01.017.

World Health Organization (WHO). 2010. *Social and Gender Inequalities in Environment and Health*. Copenhagen. www.euro.who.int/__data/assets/pdf_file/0010/76519/Parma_EH_Conf_pb1.pdf. Last visited October 18, 2020.

World Health Organization (WHO) Regional Office for Europe. 2019. *Environmental Health Inequalities in Europe*. Second Assessment Report. Copenhagen: WHO.

Young, I.M. 1990. *Justice and the Politics of Difference*. Princeton, NJ: Princeton University Press.

Young, I.M. 2000. *Inclusion and Democracy*. Oxford: Oxford University Press.

PART 2

From analysis to modelling and policy

From analysis to policy

15

FROM THE WELFARE STATE TO THE SOCIAL-ECOLOGICAL STATE

Éloi Laurent

Introduction: a welfare state for our era

In a landmark book, which for the first time defined, clarified and detailed the notion of social policy, English sociologist Richard Titmuss writes, "We do not have policies about the weather because, as yet, we are powerless to do anything about the weather" (Titmuss 1976: 1). Times have changed. Powerful as we have become, we can now do something about storms, droughts and floods that affect and sometimes ravage human communities all over the planet because of the climate change and more generally ecological crises for which we are responsible. The Anthropocene is, in the geological sense as in the meteorological sense, the era of the time on our watch. But we also have power over the social consequences of ecological crises we have put in motion.

In its simplest form, the welfare state opposes a right to a risk. The combination of the industrial revolution and the "first globalization" at the end of the 19th century, by increasing economic vulnerabilities, in turn reinforced demands for collective protection. More precisely, increased social volatility of human existences has given rise to the need for a double or better joint protection, of human well-being and the biosphere. Remarkably and too rarely underlined, ecology has emerged at the end of the 19th century as a domain of knowledge, only a few years before the birth of social protection as a public policy (the welfare state was born in 1883 with the law on labor accidents granted by Bismarck to German workers tempted by socialism, while in 1868, Ernst Haeckel defined ecology as the science of the relations between living organisms and their organic and inorganic environment).

The modern welfare state was devised in the 1880s in unified Germany to forge a new alliance between labor and capital (out of the influence of what Bismarck called "socialists") and was built later in most of Europe's countries upon the idea that human beings are entitled to receive protection against the hazards of social life. "Social security" – currently guaranteed to fewer than 30% of the world's population in about half of the planet's countries – is in fact a considerable extension of the "civil security" that Hobbes entrusted to the Leviathan in the mid-1600s. This chapter essentially argues that our era calls for a deep transformation of this very successful institution into a social-ecological state, able to provide social-ecological protection of human well-being in the face of climate change, the destruction of biodiversity and the degradation of ecosystems visible and tangible everywhere on the planet. As the brutal "yellow

DOI: 10.4324/9780367814533-17

vests" crisis in France has shown in 2018–2019, to be fair, accepted and therefore effective, ecological transition policies must take into account the social dimension of ecological issues. The transition will be just or just not be (Laurent 2020). As the COVID-19 pandemic has shown a year later, the welfare state is therefore a strategic institution in the century of environmental crises.

In fact, studies arguing for the advent of a "sustainable welfare state" are now developing, as evidenced by a special issue of the journal *Sustainability* published in early 2020 (Hirvilammi and Koch 2020) that extends the pioneering work of James Meadowcroft, Ian Gough and others on the link between social policies and climate change (Gough et al. 2008). Social-ecological analysis and policy are gaining ground.

This chapter first retraces the genealogy of the social-ecological state (first section) to then clarify its philosophy (second section), its perceived foundational need of economic growth (third section), its functions (fourth section) and, as an illustration of its necessary generalization to all levels of governance, its application to urban policy (final section).

A genealogy of the social-ecological state

Why begin the metamorphosis of our welfare state, thought in the 19th century to overcome the conflict between work and capital, into a social-ecological state calibrated for the 21st century, designed to reconcile social issues and environmental challenges? How do we build robust institutions capable of guaranteeing social-ecological progress, defined as the democratic progress of human development in the Anthropocene age? We can take two different paths to establish a philosophical continuity between the welfare state and the social-ecological state: that of social risk and that of individual well-being.

If we retain the risk approach, it now appears that social risk includes an important environmental dimension (embodied in the floods, heat waves, storms, etc. that individuals, groups, localities and nations around the world are experiencing with increasing frequency and intensity). Citizens are therefore entitled to expect public authorities to develop new means of risk pooling and protection policy. If we chose to consider not the risk facing individuals and groups but the sources of their well-being, it also appears that it is in great part determined by the environmental conditions of existence (climate, air pollution, water quality, access to energy, etc.). There is thus every reason for social policy to include the environmental dimension. This integration is still in its infancy in the beginning of the 21st century, but it can be traced back to Europe's 14th century.

In fact, one could argue that social-ecological policy preceded social policy in Europe. If the first social welfare law dates back to 1883, the first social-ecological decree appeared in 1306, when the English king Edward I attempted to ban the use of coal in London for sanitary reasons (his own mother having fallen sick because of the thick pollution of sulfur engulfing the city). It was not until 1956, some 650 years later, that the British Parliament voted for the Clean Air Act, approved in the aftermath of the 1952 "Great Smog" that killed 4,000 Londoners by air poisoning.

In the same post-war period in the United Kingdom, at a time when the welfare state was taking off thanks to pioneering figures like William Beveridge, researchers rediscovered the importance of environmental factors in the state of health of populations, a link at the origin of the hygienist policies in the 18th and the 19th century that had been gradually neglected in the first half of the 20th century. Social policy, brought to light as an academic discipline by Richard Titmuss, was extended as a public domain to environmental issues, most notably by François Laffite, Titmuss's co-author. Laffite (Titmuss and Lafitte 1963) implicitly conceptualized social-ecological policy when he defined social policy as a policy of the local environment, encompassing not only social conditions of life (family, work, leisure), but also access to environmental amenities, control of urban pollution and all the environmental factors likely to influence the well-being of individuals. In doing so, he extended the realm of protection granted by the welfare state and before the welfare state by the state.

The philosophy of the social-ecological state: from unequal uncertainty to mutualized risk

The fundamental goal of the state, as theorized in the 16th and the 17th century by Machiavelli, Bodin and Hobbes, is indeed to produce security, a civil security extended to social security with the advent of the welfare state in the 19th century. Social protection is not only about risk recognition and accident insurance: it aims to turn uncertainty into risk in order to pool and reduce uncertainty and thus reduce social inequality. In this perspective, social protection in the 21st century should return to its primary purpose to enlarge it to environmental protection. Let's briefly review the three historical stages that led to public mutualization of social risk (Laurent 2018a).

The first stage, before the 19th century and the advent of collective protections, is that of social uncertainty answered by solitary prayer. François Ewald (1996) puts it well: "For a long time, we shared the risks through the Church and religion: we responded to the risk through Providence. The more the divine was incomprehensible, the more it was necessary to have faith." But as Ewald adds,

> this response was exhausted in 1755 with the Lisbon earthquake. Through a phenomenon of sudden dissolution, a common way of experiencing events had become impossible: that of Leibnizian optimism. From then on, it will be said that it is human nature and short-sightedness that are the causes of our misfortunes and not the divine will. Risk became both an individual problem through the suffering it implies and a social problem through the responsibilities it involves
>
> (1996: 54)

This thinking lays the way for the second stage, where foresight (or on the contrary improvidence) becomes the expression of moral responsibility. Foresight, first individual, each member of society assuming and being accountable for his or her risks, becomes collective throughout the 19th century. In France, for instance, mutual benefit societies are created outside the state and then recognized by the state in 1835 and then encouraged by the state with the law of April 1, 1898, on labor accidents. This law in fact tipped the French system towards social protection and the third stage, where it is public solidarity, embodied in the welfare state, that addresses social risk.

The function of security production and risk reduction is at the root of such a welfare state, which intends to measure, supervise and predict society. It is based on the distinction established by Franck Knight (2006) between risk (measurable unknowns) and uncertainty (unmeasurable ones). If social life is uncertain in the sense of Knight, then the welfare state will not be able to protect human well-being from unforeseen events. But if social accidents can be normalized, in the statistical sense of the term, then the apparent fatality can be standardized and domesticated through insurance mechanisms. Unpredictable individual risks become manageable social risk that can be calculated, pooled and mitigated.

To go back to the French case, one can see this metamorphosis of uncertainty into risk when comparing the opening statement and Article 1 of the ordinance of October 4, 1945, creating

Social Security in France. Uncertainty is the social problem explicitly identified in the opening statement:

> Social security is the guarantee given to everyone that in all circumstances he will have the necessary means to ensure his subsistence and that of his family in decent conditions. Finding its justification in a basic concern for social justice, it responds to the concern to rid the workers of the uncertainty of tomorrow, of this constant uncertainty which creates a sense of inferiority and which is the real and profound basis of the distinction of classes between the self-assured and their future self-possessed and the workers on whom weighs, at any moment, the threat of misery.

Article 1 of the ordinance provides the solution to this problem, public risk sharing:

> A social security organization is set up to guarantee workers and their families against risks of any kind likely to reduce or eliminate their earning power, to cover maternity expenses and the family expenses they bear.

Social protection is here clearly defined as the mutualization of social risk with a view to reducing it and more equitably distributing its burden among citizens. At the start of the 21st century, social risk cannot be thought of (and mutualized) without taking into account the ecological crises that increasingly threaten human well-being in all corners of the world.

While attempting to understand the fundamental principle underlying the welfare state, Danish sociologist Gøsta Esping-Andersen proposed, in line with Richard Titmuss's founding work and after Karl Polanyi, the notion of "de-commodification" of labor, that is to say the protection of work from market logic by means of social law aiming at an ethically superior value, namely human well-being. For Esping-Andersen, social protection is founded on the idea that work is not a commodity. In his view, this common principle is embodied throughout the world in distinct institutional logics, which lead to giving it more or less strength.[1]

The guiding principle of the social-ecological state is de-naturalization (or, positively, socialization), that is to say the social domestication of environmental crises. De-naturalization consists more precisely of transforming ecological uncertainty into social risk, by means of guarantees and public insurance aimed at making the social consequences of the environmental crises of the 21st century as fair as possible, therefore mitigating their natural violence. As the next sections show, like the welfare state, the social-ecological state has to assume a function of allocation, redistribution and stabilization but without relying as much as the welfare state on economic growth that is accelerating ecological crises it intends on mitigating.

In the emerging literature on social-ecological analysis and policy, the term "eco-social state" is sometimes found (see for instance Koch and Fritz 2014). Yet, eco-social could as well refer to "economic-social" as to "ecological-social" given the meaning and historical use of the Greek radical *oikos* (meaning household) by Aristotle and Xenophon to define economics, well before it was used to define ecology. For this reason, *social-ecological state* seems a preferable concept (Laurent 2021).

Is the social-ecological state a growth state?

Hirvilammi (2020)[2] reminds us that the system of full employment was conceptualized in the form of a "virtuous circle" by the Swedish economist Gunnar Myrdal, thinker and architect of the social protections of his country and theorist of the welfare state. Myrdal's "virtuous circle" aims at formalizing the alliance between social protection and economic growth. This circle is virtuous because of two feedback nodes – full employment on the one hand and education and training policies on the other – that link wage levels and labor productivity. The social-economic alliance, typical of the second half of the 20th century in Europe, is cumulative: economic growth fueled by the increase in labor productivity and employment in turn feeds social progress through the reduction of inequalities and extension of social protection to all areas of the life cycle (education, housing, employment, pensions). Attitudes and behaviors (political confidence, aspirations for social progress, etc.) propagate the structural dynamic. In this balance resides a good part of what has been called social democracy.

Can the social-ecological state do without growth? We might want to do without growth for reasons of ecological sobriety but could find it an impossible task because we need to "finance" (or afford in a more conservative view) our social policies. We therefore face in fact a formidable dilemma between two contradictory demands: ecological transition and social progress. This allegedly consubstantial link between the welfare state and economic growth is in reality not robust, and it appears even less convincing in a social-ecological approach to public policies.

First, from a historical perspective, it is important to recognize that the welfare state was born and developed in a context of weak and unstable economic growth, the end of the 19th century in Europe, its financial scope being considerably extended in Europe in an equally weak and unstable growth regime, the 1940s. Conversely, the social austerity policies that have greatly weakened European welfare states since the early 1990s (with disastrous consequences on the human level, as shown by the health record of the United Kingdom, Italy, Spain and France in the face of the COVID-19 pandemic) were deployed in a context of moderate but sustained growth. The case of the United Kingdom in particular shows that it is not the level of economic growth that governs the choice of whether to "finance" social policy, but political considerations. The case of Japan shows, on the contrary, that the brutal collapse and the prolonged absence of economic growth can be accompanied by a very strong increase in social spending (in this case a doubling of their share as a percentage of GDP between years 1980 and the end of the 2010s). It can also be shown empirically, by widening the historical lens, that the unprecedented acceleration of human development in the 20th century depends much more on the meteoric improvement in health and education than on the increase in income by inhabitant (for a comprehensive argument on the perceived dependency of the welfare state on growth, see Laurent 2021).

More fundamentally, the argument that, without GDP growth, redistribution policies become impossible, is empirically very fragile. Such an assertion first forgets that the state of primary and secondary inequalities has a major influence on production capacities and therefore generation of growth (inequalities are not only unjust socially, they are also economically ineffective). Above all, however, it ignores the fact that GDP growth weighs almost nothing against the structural parameters that determine social spending, as the French case shows, whose net social spending (according to the definition of the OECD) is today the highest in the world (see Laurent 2021).

In a pay-as-you-go system like the French system, the financial equilibrium depends fundamentally on a demographic equilibrium between the number of contributors and the number of retirees, which itself depends on the age pyramid, the lengthening of life expectancy at 60 years, workforce activity and the average age of retirement, all parameters that are only marginally influenced by the growth rate of the GDP. The equilibrium is also based on the increase, not of GDP, but of activity income, that is, mainly wages, which itself depends on the distribution of added value and therefore, again, on choices of distribution and redistribution and not of production (i.e. equity and not efficiency issues). Even with weak growth in the aftermath of the great recession of 2009, social systems were able to return to equilibrium in France (a fact often overlooked; the general pension system had returned there to equilibrium in 2017).

Similarly, health spending depends essentially on the speed of demographic aging, the growing influence of environment-related diseases on chronic and transmissible pathologies (pollution, quality of food, etc.) and the cost of medical technologies. Expenditure on family, education, housing or poverty policies also depends fundamentally on demographic structures and the state of social inequality. Finally, the outlook for social spending depends on structural parameters such as demography (net migration, fertility, mortality) and activity and employment behaviors much more than on the increase in GDP. It is therefore these fundamental parameters that must be improved to guarantee the sustainability of social spending, if that is indeed the objective pursued.

What is more, the idea that growth is necessary to "finance" social policies is an archaic way of conceiving these policies in the age of environmental challenges: it is important today, in social as in energy matters, to move from a logic of spending to a logic of sobriety. Indeed, the ecological extension of the welfare state – imposed by the social risks engendered by environmental crises – is based on a logic of savings rather than spending pledged on taxes, themselves based on income. The financing of the social-ecological state can thus be ensured by the colossal savings of social expenditure allowed by the mitigation of ecological crises.

Let us think of the savings made possible by rational, that is to say non-self-destructive, treatment of ecosystems and biodiversity which would have made it possible to avoid the epidemics of AIDS, Ebola, MERS and SARS and of course COVID-19. Consider the savings in social spending made possible by the gradual alleviation of the ozone layer crisis, which has started to regenerate as a result of effective global governance and has thus avoided tens of millions of skin cancer cases on the planet. Consider the potential savings in social spending by mitigating climate change or air pollution, not to mention the health and therefore financial consequences of improving eating habits, sports practices or urban mobility.

Basically, the COVID-19 pandemic and its treatment by public authorities around the world signals the relegation of economic growth to the rank of a third-rate indicator in the 21st century: ecosystems determine health which determine economic possibilities. In other words, increasing economic growth while degrading ecosystems and therefore human health is a counterproductive development strategy. From this point of view, the social-ecological state freed from growth is not a post-materialist luxury: it is an economic necessity. What exactly does it consist of?

The three functions of the social-ecological state

The great strength of the best typologies of economic thought is to regain their relevance in radically new historical contexts on the condition of minor updates. This is the case with the tripartite analysis of the functions of public spending proposed by Richard Musgrave 60 years ago (Musgrave and Peacock 1958). Musgrave distinguished three "branches" of public finance, akin to the three powers that the American Constitution separates.

The first, called allocation, aims at the supply of public goods (or the demarcation of the border between public goods and private goods by the state); the second, distribution (and not just redistribution) aims to put public finances at the service of the collective preferences of citizens; the third, called stabilization, uses public finance as an instrument to maximize the "magic square" in its Kaldorian version.[3]

Going further, what could be the purpose of a social-ecological state? It would be no different from that fulfilled by the welfare state through its functions of allocation, redistribution and stabilization, but these functions would be applied to environmental issues. In this respect, there is no fundamental difference between social and environmental policy: both aim at correcting market economy failures such as imperfect information, incomplete markets, externalities, and so on that justify public intervention.

Allocation: a sober hence economical social-ecological state

We must begin here by emphasizing, to better get around it, the main defect in Musgrave's typology of separating issues of social justice and issues of economic efficiency. Ecological crises, such as the COVID-19 pandemic, show how these issues are in fact intertwined, inseparable, inextricable. The allocation function therefore obviously has powerful distributive effects which must precisely be the subject of social compensation (via redistribution), as in the case of the regulation of carbon emissions.

In this regard, it should be noted that while Musgrave takes care to specify that regulatory policies of this type are not included in the allocation function, on the contrary, they must be integrated into them: in a social-ecological approach, regulation policies are indeed the major component of the allocation function and the central reason for its economic and social positive impact on human well-being.

As has been said, the social-ecological state is mostly financed by savings, not by taxes. Even when new taxes have to be introduced, such as carbon taxation, this can easily lead, if properly calibrated, to double savings in terms of quality of life and income for the majority of the population (Berry and Laurent 2019; Boyce in this volume; Malliet and Haalebos in this volume). Note that to measure these benefits, there is no need to resort to fragile and ethically questionable methods of monetization of human life or growth points gained or lost by environmental policies. There are indeed many reliable environmental health indicators (Laurent 2018b).

The allocation function of the social-ecological state essentially means revealing the hidden social costs of ecological crises – such as respiratory diseases, strokes, and so on caused by air pollution in European urban centers – in order to reduce them as well as mitigating the inequality that they compound. Numerous reports indeed stress the beneficial effect of environmental regulations on health and well-being (such as the Clean Air Act in the United States and the Montreal Protocol on the ozone layer at the global level). The social cost of ecological crises must be made visible in order to reveal the misguided allocation of resources to which the current economic systems lead.

Distribution and redistribution

Scientific advances in the understanding of ecological crises bring us closer to the moment when, as with the social phenomena of the end of the 19th century and of the post-war period, collective responsibility will replace fatality and environmental uncertainty will make way for social-ecological risk. The IPCC indeed takes great care, in its reports, to probabilize climate risks in four distinct categories: which is almost certainty (the substantial increase in extreme temperatures by the end of the 21st century); which is very likely (the average increase in sea and ocean levels, which will contribute to an increased risk of coastal submersion), which is likely (the greater frequency of heavy precipitation and the strength of most tropical cyclones) and finally which is quite likely (heat waves and retreat of the glaciers). Often questioned by its detractors for the apocalyptic nature of its predictions, the IPCC actually helps us, by means of reasoned forecasts, to familiarize us on the matter of ecological crises with the language and logic of risk.

However, so-called natural disasters (which in fact arise more and more from human action and are better understood as social-ecological disasters, Laurent 2011) are still not fully "insurable" for three main reasons: they have a highly variable gravity, which prevents modeling and evaluating their cost; they are not entirely random (in particular due to the human factor described earlier); and they are subject to an anti-geographic selection (certain regions, for example coastal, are much more subject than others to disasters, which leads to very high insurance costs that cannot be spread evenly over the population). Solidarity must therefore overcome the limits of private insurance, which in certain years only covers half of the growing financial costs linked to the multiplication and intensification of ecological disturbances, particularly climatic. For instance, out of the 158 billion dollars in cost of so-called natural disasters estimated by the Swiss reinsurance group Swiss Re for the year 2016 worldwide (compared to 94 billion dollars in 2015), only approximately 49 billion dollars in damage were covered by insurance companies.

A social-ecological state must therefore pool these costs, reduce them and more fairly distribute them, just as the welfare state has done for social risks for more than a century. Faced with ecological crises for which we are fully responsible, we must in short again rely on the equalizing power of the welfare state, which can transform uncertainty into risk, hazard into protection, chance into justice. This may imply, institutionally, in a country like France, the creation of a new "social-ecological" branch of Social Security or the integration, within each of the existing branches, of environmental risks.

When it comes to the fiscal underpinning of redistribution, a direction of reform could be to shift tax systems toward the penalization of the excessive use of natural resources, starting with fossil fuels. Countries of the world, chief among them the OECD group, should embark on a third tax revolution, after the taxation of income at the beginning of the 20th century and of consumption in the 1950s. Currently, on average, environmental taxation in the European Union, for instance, the most advanced region of the world on the matter, only represents 6% of total taxation and has actually declined since 2002 (when it represented 7%). In the rest of the developed world, it remains much too low to influence behaviors and shift production and consumption systems toward greater sustainability, with only a few countries being an exception, such as Denmark.

Stabilization: a social-ecological state that preserves essential well-being

Let us now consider the function of stabilization. In its traditional meaning, this consists of governments' bringing into play automatic stabilizers and discretionary policy in order to cushion an economic shock and prevent a recession from degenerating into a depression. The stabilization function thus increases resilience. The social-ecological stabilization function is, by the same logic, aimed at enabling individuals to deal with ecological shocks (e.g. the heat wave of 2003 in Europe that killed 70,000 people) by preserving their well-being.

The macroeconomic objectives that justify Musgrave's function of stabilizing are clearly outdated and should therefore be updated (for the central banks themselves, the inflation target is gradually becoming obsolete). Essential well-being appears to be a relevant stabilization target, in particular the protection of the essential components of human well-being from ecological shocks (pandemics, heat waves, etc.).

Understood in this broad framework, what would a social-ecological policy look like? The development of a social-ecological policy requires prior identification and analysis of the associated and sometimes inextricable character of the social and environmental dimensions: there is a need to recognize the ecological stakes within social issues, as well as to reveal the social stakes of ecological issues, at the national as well as European level (the social-ecological dimension of carbon taxation,[4] for instance, is a national and European policy matter). This approach can be formalized using a social-ecological matrix (see Figure 15.1).[5]

How can we represent the mutually beneficial social-ecological interactions that sustain the functions of the social-ecological state in a dynamic rather than the static way used in Figure 15.1? Since the key feature of social-ecological policies is to dovetail social issues and ecological challenges, we can sketch a social-ecological feedback loop (Figure 15.2) reproducing the mathematical symbol of infinity but also evoking a Möbius strip (the shape that inspired the recycling logo since the early 1970s and by extension the circular economy).

This image, which depicts dynamic social-ecological synergies, clarifies the background of the cumulative social-ecological loop argument by emphasizing two essential nodes: the link between inequalities and ecological crises (the sustainability-justice nexus, see chapter 3) and the link between ecosystem health and human health (the full health nexus). This is an essential change compared to 20th-century welfare states: the transition from full employment to

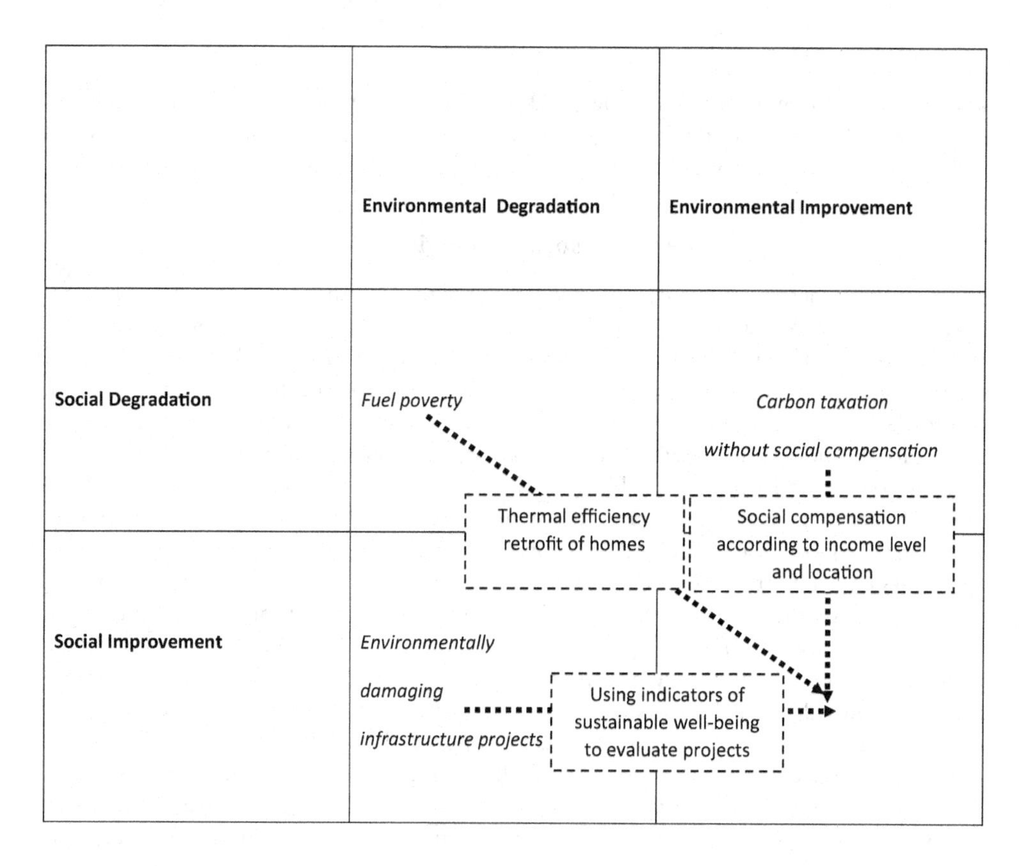

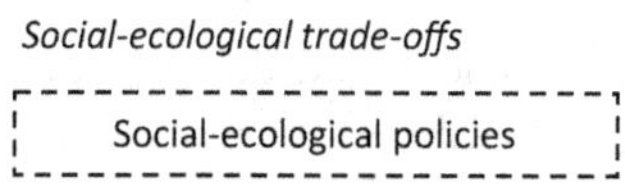

Figure 15.1 Social-ecological trade-offs and synergies

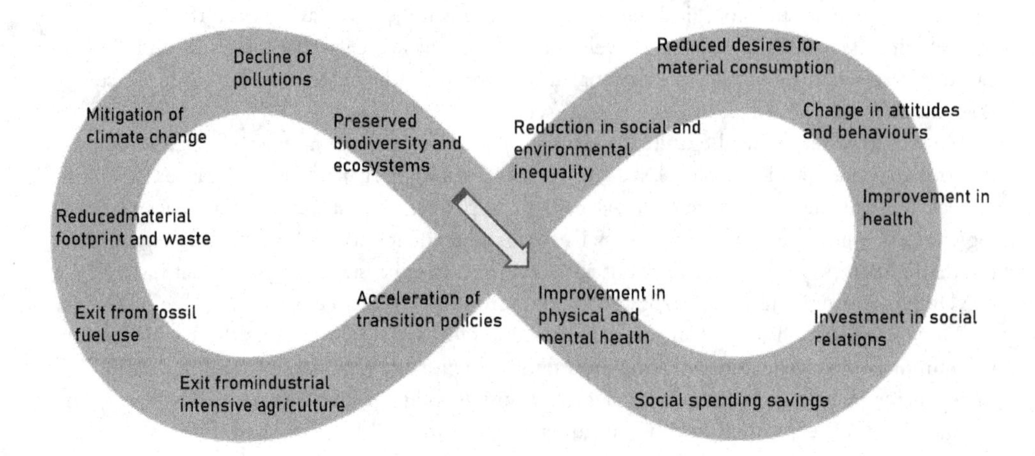

Figure 15.2 The social-ecological feedback loop

full health – that is, human health understood in all its ramifications and implications (physical health, mental health, social links, happiness, health inequalities, environmental health, social and environmental inequalities).

The urban social-ecological state

The shift toward a social-ecological state able to carry through the just transition involves not only central or national authorities but all levels of governance, from regional powers such as the European Union to municipalities. The mobilization of urban spaces in this social-ecological transition is in fact crucial, for cities now concentrate a majority of inhabitants (a proportion that reaches 75%–80% in North America and Europe). While occupying only 5% of the surface of the planet, cities account for 66% of the energy consumed and 75% of CO_2 emissions.

Four approaches to the sustainability of cities can be defined using four different disciplines. A city is first of all an administrative and geographical place. Cities can be defined, in this first approach, as dense and interconnected collective living spaces. The notion of "urban systems" reflects these spatial and human interrelations. The question of the hierarchy between spaces (especially in view of the contemporary logic of metropolitanization, which sees the connection of an urban pole to one or more peripheries), inducing more or less controlled mobility between them, appears critical from the point of view of environmental sustainability. It naturally leads to the second approach to urban systems, the economic approach, centered on the concepts of agglomeration and sprawl. From this perspective of urban economy, a city is essentially a place of efficient agglomeration (of jobs, goods, services, people, institutions and ideas). But, in a perspective of sustainability, this agglomeration can also prove to be inefficient, generating considerable environmental and social costs, like those attached to air pollution.

The third definition of urban spaces or systems comes from sociology and defines cities as spaces of social cooperation. Urban space must be shared to fulfill its essential vocation. According to this third approach, a city is the product of human density and social diversity and in fact embodies a certain vision of social justice. Finally, the city, an enterprise of human cooperation, is subject to the conditions of its environment and affects it in return, locally and globally. The key concepts here are those of urban metabolism and urban adaptation, including climate change. Urban metabolism considers the city as a living organism or an ecosystem and focuses on the quantity of resources it needs to function (water, energy, etc.) and the waste it rejects. Urban adaptation refers to the process of adjusting urban systems to global environmental change (climate change, destruction of biodiversity, degradation of ecosystems), taking into account its observed or expected effects.

These four approaches outline four main axes of urban social-ecological policy (Figure 15.3):

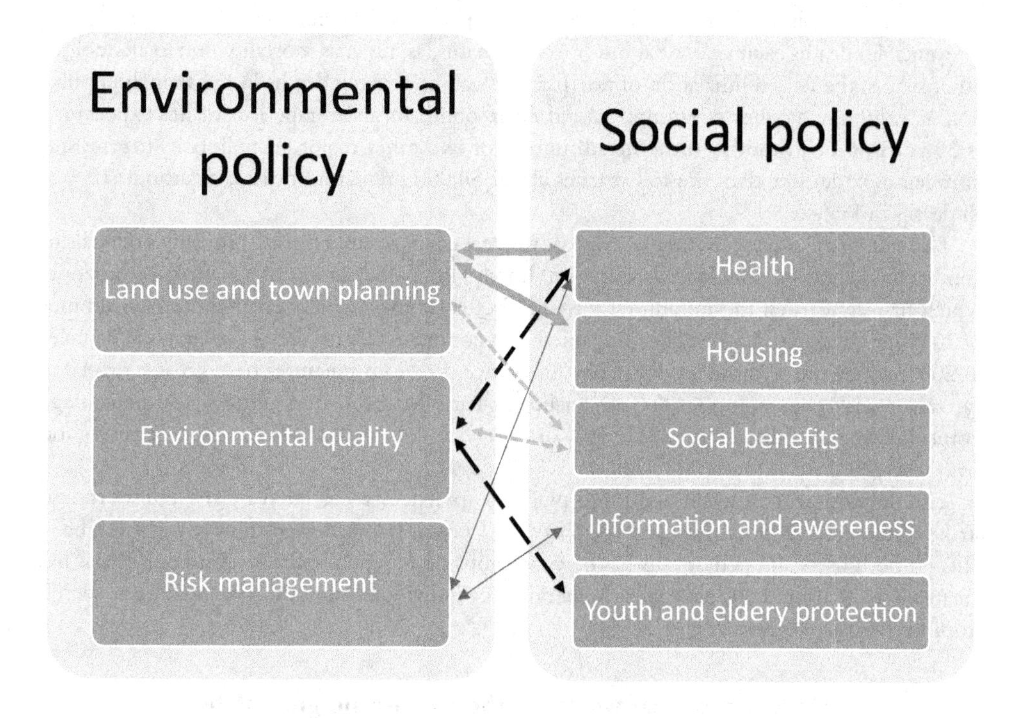

: Axis 1 - traffic speed regulation to limit urban pollution; development of non-polluting public transport to facilitate access to housing.

: Axis 2 - uniform collection of garbage in all areas of the city; social aid for non-polluting mobility.

: Axis 3 - monitoring and evaluation of environmental health indicators such as air pollution; quality of food in school canteens.

: Axis 4 - alert and social-ecological measures in the event of heat waves in the direction of isolated elderly people; containment measures when threatened with the spread of animal viruses such as COVID-19.

Figure 15.3 Urban social-ecological policies in France

- Axis 1: Mobility-environment-health: mobility for access to housing, employment, leisure and public services in connection with associated pollution and its effects on human health (environmental health);
- Axis 2: Social and environmental justice: environmental inequalities in access to amenities and exposure to risks; social policy aimed at easing ecological transition;
- Axis 3: Quality of life: improvement of human well-being and integration of well-being indicators into public policies;
- Axis 4: Footprint and vulnerability: urban metabolism (consumption of natural resources, pollution and waste generated) and adaptation to climate change.

One can illustrate this approach with the case of air pollution regulation and mitigation in French urban areas. The vast majority of French cities exceed the WHO safety thresholds for fine particle pollution, 17 of the 20 largest French cities exceeding the standards for $PM_{2.5}$ particles in 2016. In mainland France, pollution by fine particles alone represents more than 48,000 (preventable) deaths each year, or about 8% of all deaths, as much as mortality due to alcohol. In other words, the forced inhalation of fine particles causes as many deaths in the French population as voluntary alcohol consumption and corresponds to an average loss of life expectancy at 30 years of nine months. If the health impact of two other major air pollutants (ozone and nitrogen dioxide) is added, the toll reaches about 58,000 premature deaths, or around 10% of all deaths in France.

The case of Paris is interesting because of the recent successful efforts of the city's officials to combat air pollution along social-ecological lines. The establishment of a low-emission zone, which first concerned the intramural territory before being partially enlarged in the summer of 2019 to the greater has been accompanied by a complete ban on diesel and gasoline vehicles in 2030, a measure without an equivalent in France. Previous measures have greatly regulated car traffic with convincing results: air quality in Paris has been improving by 30 percentage points in less than a decade (in 2019, 70% of days were considered to be of good or very good air quality).

The development of cycle paths and practices (practices accelerated by the transportation strike of the 2019–2020 winter and the COVID-19 crisis that followed) providing health benefits for both users and pedestrians has been accompanied by public financial support: Paris has created a set of around 30 financial aids intended for individuals and businesses willing to switch from fossil fuels to electric vehicles.

Conclusion: four worlds of the social-ecological state?

When Esping-Andersen identified his three worlds of welfare capitalism in the early 1990s, the OECD undertook a long-term work to measure the impact of "structural rigidities" – at the forefront of which were social protections now strongly developed in Europe and beyond – on "labor market performance" assessed using unemployment and growth rates. The perspective of these studies was radically opposed to that of Esping-Andersen on two counts: work was relegated to its economic utility, and the convergence towards a single social model was promoted, a model considered from the almost exclusive angle of its cost-benefit "optimality."

Thirty years later, it is clear that the debate on the welfare state has largely turned to the advantage of the proponents of economic efficiency, who have succeeded in convincing those in power, especially in Europe where it was born, that social protection is a burden rather than a lever. This does not mean that the principles of the welfare state have become obsolete or that the resulting public policies have ceased to be effective and just, but rather that a simplistic vision of the functioning of the economy, which opposes a predator state to a liberating market, has come to dominate public debate.

French president Emmanuel Macron's speech on March 12, 2020, delivered under the shock of the COVID-19 health crisis, sounds from this point of view as an epiphany as radical as it is late: "What this pandemic is already revealing, is that free health care, without condition of income, course or profession, our welfare state, are not costs or burdens, but precious goods, essential assets when fate strikes"; "There are goods and services which must be placed outside the laws of the market."

All of this is true. It is also diametrically opposed to the policy conducted in France since the 2017 presidential election and during the previous mandate, when Emmanuel Macron exerted a considerable influence on the Holland presidency. It is also not precise enough. If "fate" "strikes" humanity today, it does not fall from heaven: humans, in the age of the environmental crises of the Anthropocene, have become their own fatality.

The 2020 decade, which opens the 21st century, is indeed that of the ecological challenge: faced with climate change, the destruction of biodiversity and the degradation of ecosystems, human communities must initiate a profound transformation of attitudes and behaviors of their members to prevent the 21st century from being one of self-destruction of human well-being. The first months of the first year of this decisive decade leave little doubt about the urgency of this collective effort.

First, Australia was ravaged by a succession of giant fires that only rain eventually put out. Then it was the COVID-19 pandemic that led to the arrest of almost half of humanity and, with it, the entire global economy. Then, giant fires erupted again, on the West Coast of the United States.

It is hardly debatable that states around the world are ill-equipped to deal with these ecological shocks. Hence the need to make progress in the field of social-ecological analysis and policy that this chapter has explored and to build a social-ecological state calibrated for our time. In fact, different social-ecological states are emerging in the world according to different criteria: vulnerability (exposure to risks, state of health of the population, etc.), protection (development of social protections, the degree of social inequality, etc.) and resilience (social cohesion, trust, the quality of institutions, etc.). Using these three criteria, four different regimes appear on the planet. Four worlds of the social-ecological state are currently visible.

Bio-techno power is the first world of the social-ecological state. What Michel Foucault called half a century ago "the power over life" is today combined with digital control tools whose omnipotence he could not imagine. In the light of the management of the COVID-19 crisis, a mode of socialization of environmental crises becomes clearer, which combines strong exposure to risk, authoritarian power, civil discipline and digital surveillance. South Korea is the most emblematic country of this model, but China has prefigured and applied it on a larger scale. The admiration for this social-ecological regime, palpable in European countries in 2020 whose populations were considered less reliable and whose governments are deemed to be too lax, disregards what ecological authoritarianism has cost the whole world when the first alerts on what was only a regional epidemic were fiercely repressed by the Chinese power, in the fall of 2019. The "effectiveness" of bio-techno power is thus doubly doubtful, from both the factual and the ethical points of view.

The second world is that of *ecological neoliberalism*. In Brazil, the United States and Australia, economic fundamentalism takes the place of social-ecological policy. Environmental regulations as well as health protections are weakened in favor of a small minority who have captured political power and exploited it as a rent to extract huge profits from activities of health privatization and environmental degradation. Yet in these countries, exposure to environmental risks is high and collective protection is already weak and fragile, as the COVID-19 health tragedy in the US makes clear.

Economic naturalism appears as the third world of the social-ecological state, and it is the prerogative of European countries. Unable to define together a new social-ecological regime calibrated for the 21st century, they opted for a naturalization of the economic system they have built in common since the 1950s, notions borrowed from the living world such as growth and competition, ending up governing human societies and social systems. We can see today how secondary these superficial economic realities are destructive of the social cooperation which underlies it. The health crisis triggered by COVID-19, for instance, hit the French health care system at the exact moment when political power – neither "globalization" nor "demographic aging" – was pushing it, knowingly, to its breaking point. The national madness of the so-called budgetary rationalization of the social system is the reflection of European rules that seem to have as their objective collective ill-being.

The fourth and last world of the social-ecological state is that of *natural regulations*. Even if the welfare state continued its global expansion in the contemporary period, it still concerns only 30% of humanity today. In most of Africa and Asia, human communities simultaneously face very high exposure to environmental risk while enjoying very little social protection. The case of India, where health spending per capita is around $60 (70 times lower than that of OECD countries), is significant from this point of view. Humans therefore need to rely mostly if not solely on natural protections, such as the heat, varying with seasons, that has the power to destroy many viruses. More generally, it is the regulatory services provided by ecosystems that protect humans: climate regulation, purification of air and water, tsunami mitigation, destruction of parasites and pathogens, and so on. These natural regulations, more or less degraded by humans since the industrial revolution, are for them both enemies and allies, the heat waves appearing when viruses disappear, the mangroves protecting marine submersions caused by human-induced climate change.

The major difference between this rudimentary typology and that, much more sophisticated, of Esping-Andersen is due to its temporality: Esping-Andersen conceived its ideal-types after a century of evolution of the welfare state, while a strong path dependency had helped stabilize its different regimes. The four worlds of the social-ecological state as we can see them today are still in their infancy. Far from being crystallized, their internal contradictions will make them evolve rapidly.

In fact, as with the nascent welfare state of the late 19th century, the social-ecological state remains largely to be invented in the years to come. We are now called to a double revolution: putting health back at the heart of our public policies, and putting the environment at the heart of our health policies.

Notes

1 The Esping-Andersen typology, which has become classic, is, as was Titmuss's, a tripartition that contrasts the "corporatist" (as in Germany), "social democratic" (as in Sweden) and "liberal" (as in the United States) models, each characterized by a purpose, a funding method and different governance principles. At the end of the 20th century, Esping-Andersen therefore perceived three worlds of what he called "welfare capitalism."

2 "Sustainable Welfare beyond Growth," *Special Issue of Sustainability*, 2020, www.mdpi.com/journal/sustainability/special_issues/sustainable_welfare_beyond_growth

3 The magic square is a graphic representation of the economic health of a country. It summarizes the four main objectives of a country's short-term economic policy, namely growth, full employment of factors of production, the external balance of the trade balance and price stability.

4 Gough (2017) has very compelling arguments as to the respective merits (or comparative advantages) of social-ecological policies, in particular he argues for social investment (such as subsidies for home retrofitting) rather than social compensations (such as transfers in cash).
5 This figure is adapted from Laurent 2015.

Bibliography

Berry, Audrey, and Eloi Laurent. 2019. "Taxe carbone, le retour, à quelles conditions?". *Sciences Po*. https://ideas.repec.org/p/fce/doctra/1906.html.

Ewald, François. 1996. *Histoire de l'État providence: les origines de la solidarité*. Paris: Grasset.

Gough, Ian. 2017. *Heat, Greed and Human Need: Climate Change, Capitalism and Sustainable Wellbeing*. London: Edward Elgar.

Gough, Ian, James Meadowcroft, John Dryzek, Jürgen Gerhards, Holger Lengfeld, Anil Markandya, and Ramon Ortiz. 2008. "JESP Symposium: Climate Change and Social Policy". *Journal of European Social Policy* 18 (4): 325–344. doi:10.1177/0958928708094890.

Hirvilammi, Tuuli. 2020. "The Virtuous Circle of Sustainable Welfare as a Transformative Policy Idea". *Sustainability* 12 (1): 391. doi:10.3390/su12010391.

Hirvilammi, Tuuli, and Max Koch. 2020. "Sustainable Welfare Beyond Growth". *Sustainability* 12 (5): 1824. doi:10.3390/su12051824.

Knight, Frank H. 2006. *Risk, Uncertainty and Profit*. Mineola, NY: Dover Publications. http://catalog.hathitrust.org/api/volumes/oclc/62282701.html.

Koch, Max, and Martin Fritz. 2014. "Building the Eco-Social State: Do Welfare Regimes Matter?" *Journal of Social Policy* 43 (4): 679–703. doi:10.1017/S004727941400035X.

Laurent, Éloi. 2011. *Social-écologie*. Paris: Flammarion.

Laurent, Éloi. 2015. *Social-Ecology: Exploring the Missing Link in Sustainable Development*. Documents de Travail de l'OFCE 2015-07, Observatoire Francais des Conjonctures Economiques (OFCE). https://ideas.repec.org/p/fce/doctra/1507.html.

Laurent, Éloi. 2018a. "La protection sociale: de l'incertitude au risque, de l'État Providence à l'État social-écologique". *Revue Française de Socio-Économie* 20 (1): 191. doi:10.3917/rfse.020.0191.

Laurent, Éloi. 2018b. *Measuring Tomorrow: Accounting for Well-Being, Resilience, and Sustainability in the Twenty-First Century*. Princeton and Oxford: Princeton University Press.

Laurent, Éloi. 2021. *From Welfare to Farewell: The European Social-ecological State Beyond Economic Growth*. Working Paper 2021.04, Brussels: ETUI, https://www.etui.org/publications/welfare-farewell.

Musgrave, Richard A., and Alan T. Peacock, éds. 1958. *Classics in the Theory of Public Finance*. London: Palgrave Macmillan. doi:10.1007/978-1-349-23426-4.

Titmuss, Richard Morris. 1976. *Essays on "the Welfare State"*, 3rd ed. London: Allen & Unwin.

Titmuss, Richard M., and Francois Lafitte. 1963. "Social Policy in a Free Society". *The British Journal of Sociology* 14 (3): 288. doi:10.2307/587745.

16

PROMOTING JUSTICE IN GLOBAL CLIMATE POLICIES

Michel Bourban

Introduction

Philosophy and political economy play a crucial role in the climate change discourse. Normative and empirical perspectives overlap because climate change is structured by radical inequalities. One of the most striking characteristics of climate change is that people and populations most vulnerable to climate impacts, due to geographic location and socio-economic situation, have contributed least to global greenhouse gas (GHG) emissions. The 57 small island developing states (SIDS) are responsible for less than 0.003% of total GHG emissions, but they are among the most vulnerable to climate impacts (de Águeda Corneloup and Mol 2014: 282). The 48 least developed countries (LDCs) have only contributed 0.34% of cumulative emissions of CO_2, but people there are five times more likely to die from climate-related disasters than people in other countries (Ciplet, Roberts, and Khan 2015: 7). The US and EU, on the other hand, are *each* responsible for about 25% of cumulative fossil fuel CO_2 emissions since 1751 (Hansen and Sato 2016: 6–7). The richest (high and upper-middle income) countries in the world account for 86% of CO_2 emissions but only half of the global population (Ritchie and Roser 2019). On a per capita basis, global carbon inequalities are even more pronounced. While the richest 10% of the world's population are responsible for around 50% of global emissions, the poorest 50% (around 3.5 billion people), most of whom live in countries most vulnerable to climate change, are responsible for only around 10% of global emissions (Oxfam 2015).

This spatial inequality is compounded by a temporal one. While past and present generations have contributed substantially to exhausting the global carbon budget, future generations will suffer severe and probably catastrophic climate impacts. Present generations play a dominant role in this radically unequal situation. More than half of cumulative anthropogenic CO_2 emissions between 1750 and 2010 have occurred since 1970, with larger increases toward the end of this period. GHG emissions grew on average by 1.3% per year from 1970 to 2000, and by 2.2% per year from 2000 to 2010 (IPCC 2014: 6). After a three-year period in which emissions remained largely steady, global fossil CO_2 emissions rose by approximately 1.6% in 2017 and 2.7% in 2018 (Le Quéré et al. 2018). Climate change is therefore characterized by a strong spatial and temporal dispersion of causes and effects, which makes it a social-ecological challenge representative of the entry into the era of environmental inequalities (Gardiner 2011: 19–48; Laurent 2011: 102–111; Guivarch and Taconet this volume).

DOI: 10.4324/9780367814533-18

The Intergovernmental Panel on Climate Change (IPCC) warns that for a 66% probability of limiting global warming to 1.5°C, humanity has a remaining carbon budget of no more than 420 $GtCO_2$ (Allen et al. 2018: 12). At current emission rates (42 $GtCO_2$ per year), this budget is expected to be exhausted in about 10 years. This situation is urgent and raises serious ethical issues, especially in terms of equity, as the IPCC highlights in its report on Global Warming of 1.5°C:

> Ethical considerations, and the principle of equity in particular, are central to this report, recognizing that many of the impacts of warming up to and beyond 1.5°C, and some potential impacts of mitigation actions required to limit warming to 1.5°C, fall disproportionately on the poor and vulnerable (*high confidence*). Equity has procedural and distributive dimensions and requires fairness in burden sharing both between generations and between and within nations.
>
> (Allen et al. 2018: 51; original emphasis)

The fact that these lines have been written by climate scientists shows how crucial normative research on climate change is. In fact, due to the great inequalities that structure the climate problem, neither natural scientists nor political actors can avoid using normative notions, such as "equity", "responsibility", and "capacity". The international climate regime, which is at the intersection of politics and science, at the crossroads between Conference of the Parties (COPs) and IPCC reports, frequently mentions ethical considerations, starting with the United Nations Framework Convention on Climate Change (UNFCCC 1992). One of the original roles of climate ethics has been to investigate the normative issues raised by scientific, political, and economic discussions on climate change (Gardiner et al. 2010). As Henry Shue (1992: 381) stressed in one of the first papers written in this field, climate negotiations are "a process constrained all along by some considerations of justice", because questions of climate justice are "unavoidable".

This chapter assesses climate policies from a climate justice perspective, with a particular interest in international negotiations. The first section explains how the relationship between the philosophy and the economics of climate change evolved from opposition to complementarity. The rest of the chapter explores promising institutional reform proposals made possible by this recent overlap between the two fields. From a climate justice perspective, states have three main kinds of duty: mitigating GHG emissions, helping vulnerable populations to adapt to climate impacts, and compensating the victims of climate impacts that have not been avoided (Bourban 2018: 94–100). The second section focuses on mitigation policies and explains how carbon pricing schemes could contribute to a just energy transition. The third section moves on to adaptation policies and focuses on the prioritization criteria that could contribute to a just allocation of adaptation finance. This last section also deals with questions of compensatory justice raised by the topic of adaptation finance.[1]

The philosophy and the economics of climate change

In a seminal contribution to climate ethics, Dale Jamieson (1992: 144–146) writes that since climate impacts are so broad, diverse, and uncertain, "conventional economic analysis is practically useless", and "[t]he tools of economic evaluation are not up to the task". This is due, in large part, to the influence of cost-benefit analyses (CBA) on the economics of climate change. Stephen Gardiner (2004: 571) stresses that influential CBA undertaken by statistician Bjørn Lomborg and economist William Nordhaus are indeed "extremely controversial", especially as they set sharp limits on intergenerational concerns by underestimating the costs imposed on future people by climate change. The social discount rates used to calculate future costs (4%–6% for Lomborg, 3%–6% for Nordhaus) make even the most catastrophic costs disappear after a few decades in economic models, which has serious consequences from an intergenerational ethics perspective. Likewise, many costs to nonhumans and all noneconomic costs to humans cannot be integrated into economic models, leading to a massive underestimation of the human and ecological damage of climate change.[2]

In its initial stages, in the 1990s and early 2000s, climate ethics therefore mainly developed as a critical reaction to economic analyses of climate change, leading to a clash between the two disciplines. The first generation of climate ethicists, led by Jamieson and Gardiner, stood largely in opposition to the first generation of climate change economists, led by Nordhaus. However, with the rise of the second wave of debates between philosophers and economists working on climate change since the end of the 2000s, things have evolved in both fields, and more interdisciplinary approaches have emerged. There are two main reasons for this.

First, more and more economists are questioning the relevance of CBA of climate change; they have also started to address normative issues, such as questions of intergenerational ethics. Simon Dietz et al. (2018: 456) explain, "CBA of climate change requires a series of methodological choices to be made, some of which have an ethical or otherwise philosophical character, where economics can provide limited guidance." More generally, Simon Dietz, Cameron Hepburn, and Nicholas Stern (2008: 366) highlight, "by its very nature climate change demands that a number of ethical perspectives be considered, of which standard welfare economics is just one." Second, philosophers have also made efforts to integrate the results of economic analyses into their research. Jamieson (2014: 143, 37) is more nuanced today and believes that although "economic analysis of climate change rests on normative assumptions that it does not have the resources to justify", the "economic model, thinking, and considerations are important and helpful".

Interdisciplinary projects are a good indication of the mutually enriching discussions on climate policies taking place today between philosophers and economists. Simon Caney and Cameron Hepburn (2011) have jointly published a paper on the ethical dimensions of market mechanisms in response to normative and empirical objections to carbon markets. More recently, Ravi Kanbur and Henry Shue (2019: 2, 14) have edited a volume integrating economics and philosophy in order to strengthen the "overlap and mutual enforcement between the economic and philosophical discourses on climate justice" and to call for more "interplay and interaction" between the two disciplines. According to them, "a major issue on which a joint effort by economists and philosophers is necessary and possible [is] the sustainability of the Paris Agreement" (Kanbur and Shue 2019: 14).

Just mitigation policies

The Paris Agreement

The Paris Agreement was adopted in 2015 at COP21 and came into force in 2016. It is the most recent – and perhaps, the major – outcome of the international climate regime. It is a long-term framework agreement, with a periodic review of what has been promised and realized, which has moved from the top-down approach of the Kyoto Protocol to a bottom-up architecture in which countries propose voluntary mitigation targets. The idea of this pledge-and-review approach is no longer to set absolute targets in a legally binding international treaty; it is to let each state decide the content of its nationally determined contribution (NDC) in an agreement based on the promises of the different countries. The Paris Agreement has the legal status of an international treaty, which implies that it creates legal obligations for all countries that have ratified it; however, many parts of this agreement are not binding, such as the level of ambition of NDCs or respecting their content (Bodansky 2016).

This evolving structure can be explained by the fact that the geopolitics of climate change can be conceived as a geopolitics of energy led by a trio of fossil fuel providers and consumers that renders any ambitious and binding international mitigation policy difficult: China, which owns the world's largest reserves of coal and consumes nearly half of world production on its own; Russia, which uses gas for half of its primary energy and exports it massively; and the US, the largest consumer of oil in the world and a major producer of shale gas (Aykut and Dahan 2015: 430–433). To complete the picture, one should add the alliance between the major oil-exporting countries, the Gulf Cooperation Council (GCC), headed by Saudi Arabia, which, together with the US Congress, has played a decisive role in constantly slowing or blocking progress in international cooperation. This geopolitical balance of power helps us understand why the goal of climate negotiations is not so much a legally binding climate treaty as an agreement made up of voluntary national contributions.[3]

At the same time, international negotiations are necessary to face the global collective action problem of climate change. Cooperation between states is needed to mitigate global emissions, to help vulnerable countries adapt to climate impacts, and, if possible, to compensate for harm resulting from climate impacts. Only international negotiations can bring together representatives of the world's countries around a process mutually accepted by all in order to face a problem that concerns humanity as a whole.

Distributing the costs of mitigation

Ethical considerations, especially in terms of justice or fairness, can have a real and observable influence on climate negotiations. They play a crucial role in the creation, maintenance, and evolution of influential international norms. While other factors such as economic and political interests also matter, the role of ethics in the formation, alteration, and application of norms should not be underestimated (Pickering 2010).

Climate justice is a field of normative political theory composed of two related major branches: burden-sharing justice, which focuses mainly on the allocation of the costs of climate policies (mitigation, adaptation, compensation), and harm-avoidance justice, which aims at ensuring that the most vulnerable to climate impacts are protected (both by maintaining global warming below a certain threshold and by helping them adapt to unavoidable climate impacts) and that climate policies do not result in further harming the global poor.[4] The justification and institutionalization of principles of justice play a major role in this field, from both the distributive perspective (the polluter pays, ability to pay, and beneficiary-pays principles) and the harm-avoidance perspective (the harm principle).

While climate justice scholars have developed their own theories (for instance, cosmopolitan, utilitarian, or Rawlsian), it is interesting to note that considerations of justice are raised both by climate scientists, as we saw earlier with the IPCC, and by climate negotiators and other policymakers. Climate change has been framed as a problem of distributive justice since the start of international climate negotiations, about three decades ago. The parties to the UNFCCC have indeed agreed that their core objective would be to "protect the climate system for the benefit of present and future generations of humankind, on the basis of equity and in accordance with their common but differentiated responsibilities and respective capabilities" (UNFCCC 1992: art. 3.1). This norm of common but differentiated responsibilities and respective capabilities (CBDR-RC) plays a major role in all the main outcomes of climate negotiations, including in the Paris Agreement, which states that NDCs should be set according to this principle (UNFCCC 2015: art. 4.3).

The interpretation of the norm of CBDR-RC has been the object of endless political and academic debates. There has been substantial philosophical debate on distributive principles of climate justice.[5] One possible way to understand this debate is to see it as an attempt to clarify the normative implications of CBDR-RC. While the polluter pays principle explains why some states are more responsible than others (because they have emitted and are still emitting the most), the ability to pay and the beneficiary pays principles explain why some are more capable than others (because they have higher financial and technological ability and because they have benefited more from past emitting activities). But how can such principles be operationalized?

Here, climate justice scholars must rely on the research of social and political scientists to explain how to measure or operationalize the abstract principles they support. This operationalization process allows clear identification of which countries are meeting the expectations of climate justice and which are not. Sivan Kartha et al. (2009) and Paul Baer (2013) have developed a highly relevant instrument to this end: the "greenhouse development rights (GDR) framework". This is a formula for global burden-sharing based on an index composed of a responsibility indicator (cumulative national emissions since 1990) and a capacity indicator (per capita annual income above a development threshold of $8,500). This responsibility–capacity index determines the percentage of total global obligation for each country by giving each indicator the same weight: in 2010, the US held 29.4% of global obligation, the EU 26%, Japan 7.6%, Russia 5.8%, China 5.1%, Brazil 2.8%, and India and South Africa 0.9% each. In total, high-income countries held 73.7% of global obligation and LDCs 0.3%, with the remaining 26% being held by new emitting countries, such as BASIC countries (Brazil, South Africa, India, and China).[6]

Recognizing that their choice of indicators is debatable according to the different countries' interpretation of responsibility and capacity, some authors of the GDR have developed a second index, the "climate equity reference framework" (CERF), which proposes a "climate equity reference calculator" to assign each country and region its fair share of mitigation efforts (Holz, Kartha, and Athanasiou 2018). In contrast to the GDR, the CERF includes an "equity range" or an "equity band" that allows more or less demanding effort-sharing parameters to be chosen, giving countries enough flexibility to calculate their level of global obligation.[7] Among high-capacity and high-responsibility countries, such as the US, EU members, and Japan, "NDCs fall far short of the fair-share contributions as bounded by the equity band" (Holz, Kartha, and Athanasiou 2018: 127). This means, whatever the choice of parameters, these countries fail to assume their global responsibilities. The US pledged only 16%–24% of its fair-share contributions, the EU 21%–23%, and Japan about 10%. In addition, fair shares of the global effort to mitigate emissions are higher than plausible domestic reductions in these countries. This means that, to fulfill their mitigation efforts, they must contribute, through climate finance, to mitigation opportunities in other countries where mitigation potential exceeds domestic obligations.[8]

The major (and perhaps unsurprising) finding is that, if we take the norm of CBDR-RC seriously, most developed countries' pledges are largely insufficient so far and, therefore, unfair. This remains true even if we choose undemanding indicators of responsibility and capacity. Hence, more ambitious climate policies are urgently needed. But how can fairer mitigation policies be reached?

A just energy transition

Reinforcing the pledge-and-review approach that has dominated climate negotiations for a decade with a price-signal approach is a promising way forward. The logic of promises and verifications currently guiding international negotiations is both insufficient (because most NDCs are currently unfair) and ineffective (because global GHG emissions keep rising). To remedy these two major weaknesses, a price-signal approach could be integrated into the climate regime in order to incentivize producers and consumers to reduce their emissions. This might contribute to closing the gap between current emission levels and a trajectory that would avoid dangerous anthropogenic interference with the climate system.

The goal of the Paris Agreement is to hold "the increase in the global average temperature to well below 2°C above pre-industrial levels" and to pursue efforts "to limit the temperature increase to 1.5°C above pre-industrial levels" (UNFCCC 2015: art. 2.1). Even if all countries were to fulfil their current NDCs, temperature increases would still reach (at least[9]) between 2.6°C and 3.1°C by 2100 (Rogelj et al. 2016). To give NDCs the necessary boost they need to keep global warming "well below 2°C", new promises by high-emitting countries are not enough: if no problem of collective action has so far been solved by the voluntary actions of different countries, a problem as profoundly global and intergenerational as climate change will probably not be an exception (Stiglitz 2015). A new framework to guide mitigation policies is therefore needed. A carbon pricing regime, which aims to attribute a price rationale to countries' pledges, can help facilitate the transition from intentions to action and put an end to the waiting game of high-emitting countries (Gollier and Tirole 2015). In other words, "[t]he current weaknesses of the existing climate negotiation framework could be remedied in large part by introducing into climate negotiations the goal to develop a robust carbon pricing system" (Laurent 2020b: 105).

To meet the demands of climate justice, a carbon pricing regime has to integrate both burden-sharing considerations (how can GHG emissions be fairly priced?) and harm-avoidance considerations (how can the regressive effects of carbon price policies be avoided?). I propose here a roadmap for a just energy transition in four main steps.[10]

The first step is to rapidly and radically reduce government subsidies for fossil fuels. Global subsidies for fossil fuels reached the mind-boggling amount of $4.9 trillion in 2013 and $5.3 trillion in 2015 (6.5% of global GDP for both years) (Coady et al. 2017). The first reason why these subsidies must be reduced directly and rapidly is that they are stimulating the overexploitation of the energy sources at the root of the climate problem and air pollution. Eliminating such subsidies would have reduced global emissions by 21% and fossil fuel air pollution deaths by 55% in 2013 (Coady et al. 2017). The second reason is that such subsidies make carbon pricing meaningless: putting a price on carbon emissions while massively subsidizing fossil fuels would be contradictory.

This leads to the second step: putting an increasing price on GHG emissions. This would allow coordination between the parties to the UNFCCC's efforts to achieve their mitigation targets and gradually increase their ambition to reduce the gap between NDCs and the objective of the Paris Agreement. Implementing a price signal is precisely the objective of cap-and-trade systems and the carbon tax. These two market mechanisms provide an incentive to reduce emissions by including the cost of environmental externalities in the cost of energy use. Raising the price of energy-intensive products and services encourages producers and consumers to change their behavior and to use low-emitting sources of energy.

Although the two instruments have the same objective and very similar effects, they differ in how they operate. Cap-and-trade systems primarily concern large industries, such as electricity, aluminum, and steel production, and ensure, through their cap, that a specific emission reduction target is achieved. The tax can cover emissions from other sectors, such as housing, transport, and agriculture, but does not guarantee a specific mitigation goal. What is important is that these two mechanisms are largely complementary (Criqui, Faraco, and Grandjean 2009: 179). For emissions from large industrial companies, it is desirable to maintain and develop carbon markets with strengthened rules governing them; for diffuse emissions not covered by these markets, carbon taxes allow all households and companies not subject to the carbon market to adjust to an increasing carbon price. It is therefore preferable to set up a system where the two instruments coexist in hybrid schemes (Hepburn 2009).

Since the carbon price should be set according to the principle of CBDR-RC, it is not identical for all countries. The idea is to define a price trajectory according to the share of global responsibility of different countries as set by the CERF's equity calculator, not a single, global price. Many carbon pricing policies are already implemented, with a total of 57 initiatives in 2019. However, only 20% of global GHG emissions are currently covered by a carbon price policy and less than 5% are priced at a level consistent with reaching the goal of the Paris Agreement (WB 2019). The triple objective of complementing the current pledge-and-review approach with a price-signal approach is (1) to allow coordination between existing carbon pricing policies, (2) to extend them to countries that are not yet participating, and (3) to set carbon prices that are consistent with the goal of keeping global warming well below 2°C.

The amount levied on GHG emissions by the tax and the auctioning of emission rights is recovered by public authorities, which creates significant revenue. The existing incentive structure can only be effectively transformed if alternative energy sources are available and affordable for individuals and companies that decide to change their source of energy due to the rising price of oil, coal, and gas. States can play a decisive role by investing massively in the research, development, and deployment of renewable energies such as solar, hydro, and wind power. This is the third major step. This investment can be mainly funded by the money raised through tax and auctioning of emission rights, but also by subsidies that are no longer invested in fossil fuels. If the $5.3 trillion per year currently used to subsidize fossil energy were invested instead in renewable energies, a rapid energy transition would become possible.

According to the IPCC (2011: 20), renewable energies could cover up to 43% of global energy demand by 2030 and up to 77% by 2050, provided the right policy decisions are taken. Wind, water, and sun could cover 80% of energy demand by 2030 and 100% of this demand by 2050 in 139 countries (Jacobson et al. 2017). Subsidizing the transition from fossil fuels to renewables would also lead to significant co-benefits, such as the creation of 24.3 million net long-term, full-time jobs, the avoidance of millions of annual air pollution deaths, and increased worldwide access to energy (Jacobson et al. 2017).

As the IPCC stresses, however, "some potential impacts of mitigation actions" can also "fall disproportionately on the poor and vulnerable" (Allen et al. 2018: 51). Carbon markets and carbon taxes can have a regressive effect: any increase in the cost of fossil fuels affects low-income individuals and households, who spend a larger proportion of their budget on energy expenditure. Since livelihood and health depend largely on access to affordable energy sources, the energy transition could contribute to maintaining and even worsening energy poverty, thereby widening environmental inequalities.

To avoid this potential clash between climate justice and social justice, compensatory measures ought to be implemented as a fourth step. For instance, governments could set incentives to improve energy efficiency in poor households: low-energy infrastructures would cost them less, and large savings on their energy budgets would be possible due to their very low level of consumption. Similarly, public transportation would need to get more subsidies to extend networks, improve efficiency, and reduce prices (or make it free), since increasing the cost of hydrocarbons has a significant impact on the travel of the working and middle classes. Another mechanism would be to redistribute part of the revenue collected through abolishing fossil fuel subsidies and through the carbon market and tax in the form of an energy dividend to offset the increased cost of energy.[11]

What about countries where such measures cannot be implemented? As part of their responsibility to mitigate global emissions, developed countries should help developing countries secure access to low-emitting sources of energy. States like Bangladesh, India, and many African countries all have very favorable geographical provisions for the deployment of alternative energy sources, especially solar energy. If these countries were to succeed in exploiting this gigantic source of energy, through both clean technology transfer and climate finance, the challenge would be largely met.

The renewable energy market is now a profitable option for multiple investors and industrialists.[12] In many parts of the world, wind and solar energy are now competitive with fossil fuels in terms of cost. These market opportunities show that the costs of the energy transition are declining, while the direct benefits associated with reducing GHG emissions are becoming more apparent (Moellendorf 2015: 83). The collective structure of the climate problem, according to which it is collectively rational for states, corporations, and individuals to cooperate to reduce global emissions but individually irrational to do so, is therefore partially eroding. This gives us reason to hope that dangerous climate change can still be avoided without jeopardizing the fight against global poverty.

Just adaptation policies

The adaptation finance gap

The design of adaptation policies would also benefit from the overlap between philosophy and political economy. Adaptation is today at the top of the UNFCCC's agenda because of the visible increase in climate-related disasters such as droughts, floods, heat waves, and hurricanes and because of the rising awareness that current and future mitigation measures will not be sufficient to avoid dangerous climate impacts (Ciplet, Roberts, and Khan 2013: 52).

Developed countries have committed to help developing countries mitigate and adapt through monetary transfers and technical assistance. By providing "fast-start" finance pledges approaching $30 billion between 2010 and 2012 and "medium-term" finance pledges scaling up to $100 billion a year by 2020, the 2009 Copenhagen COP generated strong political momentum in climate finance (UNFCCC 2009: para. 8). The Paris Agreement adds that parties to the UNFCCC "shall set a new collective quantified goal from a floor of USD 100 billion per year, taking into account the needs and priorities of developing countries" (UNFCCC 2015: para. 54), meaning that the parties have to agree on a more ambitious goal before 2025.

The major problem is that there is insufficient funding for the number of adaptation projects. There is a persistent and growing "adaptation finance gap" between funds needed and those promised and delivered, for two main reasons. First, pledged amounts of climate finance are much lower than estimates of adaptation costs in developing countries. The costs of adaptation in developing countries could reach up to $300 billion a year by 2030, and between $500 billion and $1 trillion a year by 2050, even if global temperatures remain below 2°C (Puig et al. 2016: 6; Oxfam 2018: 6). Although this wide range of estimates shows that there are significant uncertainties involved in putting a precise cost on adaptation, it is fairly certain that tens and probably hundreds of billions of dollars will be needed every year to deal with climate impacts in the coming decades (Betzold and Weiler 2018: 8–9). Second, OECD countries have not respected their financial commitments, having for instance only provided about $2.35 billion in 2012 for genuine adaptation projects, even though they claimed that $10.1 billion was adaptation-related (Weikmans et al. 2017: 466).

It is therefore not surprising that, just as in the case of mitigation policies, adaptation finance policies so far have proved insufficient and mostly unfair. In a recent evaluation of climate finance through a climate justice lens, Mizan Khan et al. (2020: 265) conclude, "Ambiguity in key areas of climate finance governance related to distributive, procedural, recognition, and compensatory forms of justice still plague the UNFCCC regime." From a climate justice perspective, adaptation finance is a redistribution of wealth premised on, or at least partially justified by, prior or ongoing injustices that ought to be compensated (Duus-Otterström 2016: 659; Baatz 2018: 74). Developed countries have a duty of compensatory justice to contribute their fair share to adaptation efforts in developing countries, based on the principle of CBDR-RC. The persistent and growing adaptation finance gap shows that existing efforts should be substantially scaled up to match the demands of climate justice.

Sharing the costs of adaptation equitably

The adaptation finance gap implies that funding for adaptation projects is scarce. To ensure that funding is allocated to those who need it the most, prioritization criteria are necessary. Following the categories of recipient need, recipient merit, and donor interest influential in the development aid literature, Weiler, Klöck, and Dornan (2018) found that physical vulnerability, good governance, and economic interests were major prioritization criteria in the distribution of bilateral adaptation finance between 2010 and 2015.

Vulnerability is the most-discussed criterion in prioritizing the allocation of adaptation finance, in both academic and political circles. Scholars agree that adaptation finance ought to benefit those who are most vulnerable to climate change (Ciplet, Roberts, and Khan 2013: 60), and all major UNFCCC agreements reflect this consensus (Klein and Möhner 2011: 16–17). The current allocation of adaptation finance seems to take into account this crucial criterion, but three issues need to be stressed.

First, many studies on the distribution of adaptation finance do not find that the most vulnerable countries have received adequate amounts of money (Khan et al. 2020). And even if the countries that are physically vulnerable to climate change, that is, more exposed and sensitive to climate risks, did receive more adaptation money, this would still disregard crucial vulnerability factors. Vulnerability is also influenced by social factors, such as levels of inequality, marginalization, and social injustice, which determine people's adaptive capacity (IPCC 2014: 21), and by political factors, such as government effectiveness, levels of democracy, and levels of corruption (Mikulewicz 2018). An allocation of adaptation finance based on physical vulnerability may therefore not prioritize the most vulnerable, all things considered.

Second, despite broad agreement in scientific literature and political discussions that adaptation finance is owed as a priority to those who are the most vulnerable to climate change, Jonathan Pickering (2012: 5) stresses that this agreement "masks two important areas of disagreement, namely how vulnerability should be conceived and measured, and whether (and which) other principles could also inform prioritization".

We therefore come back here to the crucial question of the operationalization of principles of distributive justice. How should vulnerability be measured? Several studies have developed complex aggregate global vulnerability indices. For example, the World Risk Index (UNU-EHS 2019) uses several indicators for each of the main factor of physical and social vulnerability: exposure, susceptibility, coping capacities, and adaptive capacities. There are, however, major problems with aggregate indices of global vulnerability: the rankings of countries diverge greatly between different indices (Mathy and Blanchard 2016: 757), there is no agreed way to compare countries' vulnerability (Klein and Möhner 2011: 16), and "all attempts to allocate adaptation funding based on aggregate national-level indices of vulnerability to climate change have been deeply unsatisfying" (Füssel, Hallegatte, and Reder 2012: 323). Data aggregation requires many empirical and normative assumptions, some of which are controversial. As long as no agreed methodology exists, aggregating many heterodox factors that influence vulnerability into a single number will "not reveal more but rather disguise[s] what is known" (Hinkel 2011: 205).

This raises the other question mentioned by Pickering: which additional criteria should then inform prioritization? This points toward a multi-criteria approach to fairly allocate adaptation finance, and to the third issue related to the current allocation of adaptation finance: trade-offs might arise between different criteria.

Avoiding trade-offs

Among the three prioritization criteria mentioned earlier, a possible trade-off could arise between vulnerability and good governance. Many vulnerable people in LDCs and SIDS are governed by authoritarian, rights-violating regimes. If "better governed" countries receive higher amounts of adaptation finance, then vulnerable individuals who live in "poorly governed" countries, and who greatly need adaptation assistance, receive less or no funding. This trade-off seems already to be happening to some extent in the current practice of bilateral adaptation finance:

> Donors seem concerned with aid effectiveness and tend to give more adaptation aid to better governed countries which are (seen as) better able to use resources in an efficient and effective manner, even if these countries are also better able to cope with climate change and hence are less vulnerable.
>
> (Betzold and Weiler 2018: 166)

This trade-off between vulnerability and good governance raises the following question: how can we increase the likelihood of highly vulnerable individuals who live in poorly governed states and who ought to be protected benefiting from adaptation finance?

Göran Duus-Otterström (2016: 668) rightly points out,

> vulnerable individuals should not be penalized for living under nondemocratic and ill-functioning governments, and so if poor domestic institutions stand in the way of their protection, it would seem that the right conclusion is not to withhold finance, but to try to sidestep the national government in various ways.

But how can this be done? A relevant option would be to help populations living in authoritarian regimes to claim their basic social and political rights by helping them gain control over the design and implementation of adaptation projects supported by climate finance. There are different and complementary ways to achieve this goal, for instance by supporting local, regional, or national resistance in authoritarian regimes from the outside (Caney 2015) or by implementing new trade policies that reduce corruption in authoritarian states with rich natural resources (Wenar 2016).

A third measure would be to reduce corruption by supporting ongoing democratization processes in vulnerable societies. This would directly address the political factors of vulnerability. Democratization processes can play a key role in reducing high levels of corruption and reducing vulnerability by limiting the misappropriation of adaptation finance. There is a consensus in the adaptation literature that local adaptation governance should be democratized in order to prioritize the predicament of the poor and the marginalized (Mikulewicz 2018: 26). Thus, to avoid the trade-off between vulnerability and good governance, adaptation finance should, in part, support the development of democratic institutions in authoritarian countries with populations highly vulnerable to climate change.

Democratizing the local decision-making process within vulnerable countries with high levels of corruption would achieve a more equal distribution of power within the community, and therefore a more equitable use of adaptation finance. From this perspective, the good governance criterion and the vulnerability criterion support each other: democratization is a basis for tackling the political nature of vulnerability. One possible way to support democratization processes in poorly governed countries is to help civil society with financial support to local NGOs dealing with climate impacts in their community.

In many SIDS, for instance, NGOs provide key community-level development and adaptation services but are often heavily underfunded: "Frequently, the few paid staff in an organization end up spending most of their time chasing grants just to keep a handful of core staff on the payroll and end up with little time for program work" (Johnson 2015: 68). In the context of the Marshall Islands, Giff Johnson (2015: 69) suggests implementing specific funding for core NGO staff that would ensure existing NGOs can provide their community services and won't spend most of their time looking for new grants or simply disappear. Silja Klepp comes to a similar conclusion in the case of Kiribati (Klepp 2013: 316; Klepp and Chavez-Rodriguez 2018: 3–4). In this context, joint efforts by civil society and government officials have led to remittances from migration programs and increased financial support for adaptation projects, but most of the available resources are used for external consultants who are not familiar with the context and the needs of the population, and local knowledge and circumstances are often neglected in the implementation of adaptation projects. These examples show the need for more financial support to civil society, especially NGOs employing or working closely with the local population, to ensure that vulnerable people living in poorly governed countries get the financial assistance they are entitled to and take part in the design and implementation of adaptation projects whose *raison d'être* is to protect them.

How do we measure the progresses made by vulnerable countries in their democratization process and how do we avoid similar problems to those mentioned previously regarding the operationalization of the vulnerability criterion? The Varieties of Democracy (V-Dem) Indicators seem to represent a promising reply to this question (Lührmann, Mechkova, et al. 2018; Lührmann, Dahlum, et al. 2018). On the one hand, the V-Dem project allows to measure key core elements of democracy by relying on 450 indicators and indices, such as the egalitarian principle (via the V-Dem Egalitarian Component Index) and the participatory principle (via the V-Dem Participatory Component Index). On the other hand, strong correlations within and between indices show that there are good grounds for the robustness of the data. While notions such as equality and participation are not easy to measure, they are less heterogeneous than the ones of physical, social, and political vulnerability. The point here is not to turn indices measuring democracy into a distributive formula but rather to consider the information they provide in decision-making regarding the distribution of adaptation finance, together with the information provided by other criteria, such as vulnerability, effectiveness, and sustainability.[13]

Conclusion

There are two main topics on which political economy and philosophy overlap in discourses on climate change: first, in discussions on principles of distributive justice, such as the norm of CBDR-RC in mitigation policies and the criteria of vulnerability and good governance in adaptation policies. While philosophers are more focused on justifying the responsibility and capacity of countries based on the polluter pays, ability to pay, and beneficiary pays principles, political economists have developed useful indices to operationalize these principles and compare the different levels of responsibility and capacity between regions and countries. Both tasks are essential and complementary, and fulfilling them together could lead to an "ex datis philosophy", a philosophy that strongly draws on empirical data.

Second, the two disciplines also complement each other on the crucial task of finding institutional reforms to make the international climate regime more just and more effective. They can both provide crucial resources to reduce the mitigation gap and the adaptation finance gap. Climate justice, as an interdisciplinary field, is interested in both identifying climate injustices and finding ways to reduce such injustices. The institutional reforms explored earlier, complementing the pledge-and-review approach with a price-signal approach and supporting democratization processes in poorly governed vulnerable countries, raise both empirical and normative challenges. These challenges sketch an agenda for future interdisciplinary research on climate change from a climate justice perspective.[14]

Notes

1 On the related subfield of loss and damage ethics, see Wallimann-Helmer et al. (2019).

2 The influence of CBA in the economics of climate change is clearly illustrated by William Nordhaus being awarded the Nobel Prize in Economic Sciences in 2018 for his integrated assessment models (IAMs), which rely extensively on CBA. According to his "dynamic integrated climate-economy" (DICE) model, an optimal climate policy would result in global warming of about 3.5°C by 2100 (Nordhaus 2018: 348). Ironically, the Nobel announcement was made on the same day as the release of the IPCC report emphasizing the crucial importance of keeping global warming below 1.5°C (IPCC 2018). For critical reactions by political economists to the attribution of this prize to Nordhaus, see Gareth (2018) and Laurent (2020b: 133–134). For a recent criticism of Nordhaus's approach, which integrates both effectiveness and equity concerns, see Boyce (2018).

3 The failure of the 2009 Copenhagen Summit to produce a long-expected new climate treaty to replace the Kyoto Protocol also played an important role in this evolution of the architecture of the international climate regime. The Cancun Agreements, which incorporated most of the outcome of the Copenhagen conference, achieved almost unanimity in 2010. The new bottom-up structure was then solidified at successive meetings in Durban, Doha, Warsaw, and, finally, Lima, where the common framework and guidelines for the 2015 negotiations were set by the "Lima Call to Action". For more details on the Paris Agreement from a climate justice perspective, see Bourban (2017) and Light (2017).

4 This is an adaptation of a distinction first introduced by Caney (2014).

5 For the major contributions to burden-sharing climate justice, see Caney (2010), Page (2011), and Shue (2015).

6 For alternative indices combining other indicators and leading (in general) to complementary results, see Dellink et al. (2009), Müller, Höhne, and Ellermann (2009), and Laurent (2020b: 106–10). For instance, in his model of fair allocation of the carbon budget, Éloi Laurent proposes to rely on per capita emissions rather than national emissions and to include levels of human development (according to the Human Development Index) and projected population increase. His index is complementary to the GDR framework in the sense that he also finds high-income countries, especially the US, Canada, Germany, and Japan, as the most responsible for bearing the burden of mitigation policies.

7 For a politically feasible proposal to include the CERF equity calculator into the climate regime, see Bourban (forthcoming).

8 Here again, other indices lead to similar results. For instance, Laurent (2020b: 108) finds that the US, Canada, Germany, and Japan owe respectively 17, 9, 2, and 1 billion(s) of tons of CO_2 to other countries, a "negative carbon budget" they have to pay "by investing in carbon sinks or by transferring technology and/or financing to accelerate emission reductions in carbon positive carbon budget countries".

9 "At least", because if we take into account possible tipping points in the climate system, crossing the planetary threshold of 2°C could lead the entire earth system into a "hothouse earth" pathway, even if

GHG emissions are subsequently reduced (Steffen et al. 2018). In this scenario of cascade of positive feedbacks, global temperature would raise much higher than 3.1°C.

10 I draw here on Bourban (2018: 283–295). Note that the idea of a "just transition" was promoted in the early 1990s by US labor leader Tony Mazzocchi to resolve the conflict between jobs and the environment. It has found an important echo recently, in both the academic and the political world, as part of the "Green New Deal" promoted by the UN-DESA (2009), and then supported in early 2019 by politician and activist Alexandria Ocasio-Cortez (as well as economists Joseph Stiglitz and Paul Krugman), and then by the European Commission at the end of 2019. For a critical examination of the concepts framing the "European Green Deal", including the one of "just transition", see Laurent (2020a); for a development of normative considerations of the original UN-DESA strategy, see Shue (2013).

11 For instance, British Columbia has designed a highly progressive carbon tax in which about 40% of revenue is reimbursed to households through income tax cuts and lump-sum payments (Beck et al. 2015).

12 Global investment in renewable energy capacity reached $272.9 billion in 2018, which was the fifth successive year in which renewables capacity investment exceeded $250 billion. From 2010 to 2019, a total of $2.6 trillion was invested in renewable capacity (UNEP 2019).

13 For more information on the democracy criterion and its relationship with other prioritization criteria, see Baatz and Bourban (2019).

14 I am grateful to Lisa Broussois and Éloi Laurent for their very helpful comments and suggestions on how to improve this chapter. I would also like to thank Christian Baatz and Konrad Ott for our very enriching discussions on adaptation finance justice. Finally, I gratefully acknowledge financial support provided by the Deutsche Forschungsgemeinschaft's (DFG) Cluster of Excellence 80 "The Future Ocean" (Project CP1771) and by the Swiss National Science Foundation (SNSF) (grant P400PG_190981).

Bibliography

Allen, M.R., O.P. Dube, W. Solecki, F. Aragón-Durand, W. Cramer, S. Humphreys, M. Kainuma, J. Kala, N. Mahowald, Y. Mulugetta, R. Perez, M. Wairiu, and K. Zickfeld. 2018. 'Framing and Context.' In V. Masson-Delmotte, P. Zhai, H.-O. Pörtner, D. Roberts, J. Skea, P.R. Shukla, A. Pirani, W. Moufouma-Okia, C. Péan, R. Pidcock, S. Connors, J.B.R. Matthews, Y. Chen, X. Zhou, M.I. Gomis, E. Lonnoy, T. Maycock, M. Tignor, and T. Waterfield (eds.), *Global Warming of 1.5°C. An IPCC Special Report on the Impacts of Global Warming of 1.5°C above Pre-industrial Levels and Related Global Greenhouse Gas Emission Pathways, in the Context of Strengthening the Global Response to the Threat of Climate Change, Sustainable Development, and Efforts to Eradicate Poverty* (World Meteorological Organization: Geneva).

Aykut, Stefan C., and Amy Dahan. 2015. *Gouverner le climat? Vingt ans de négociations internationales* (Presses de Sciences Po: Paris).

Baatz, Christian. 2018. 'Climate Adaptation Finance and Justice. A Criteria-Based Assessment of Policy Instruments', *Analyse & Kritik*, 40: 73–105.

Baatz, Christian, and Michel Bourban. 2019. 'Distributing Scarce Climate Adaptation Finance Across SIDS: Effectiveness, not Efficiency.' In Carola Klöck and Michael Fink (eds.), *Dealing with Climate Change on Small Islands: Towards Effective and Sustainable Adaptation?* (Göttingen University Press: Göttingen).

Baer, Paul. 2013. 'The Greenhouse Development Rights Framework for Global Burden Sharing: Reflection on Principles and Prospects', *Wiley Interdisciplinary Reviews: Climate Change*, 4: 61–71.

Beck, Marisa, Nicholas Rivers, Randall Wigle, and Hidemichi Yonezawa. 2015. 'Carbon Tax and Revenue Recycling: Impacts on Households in British Columbia', *Resource and Energy Economics*, 41: 40–69.

Betzold, Carola, and Florian Weiler. 2018. *Development Aid and Adaptation to Climate Change in Developing Countries* (Palgrave Macmillan: Cham).

Bodansky, Daniel. 2016. 'The Legal Character of the Paris Agreement', *Review of European, Comparative & International Environmental Law*, 25: 142–50.

Bourban, Michel. 2017. 'Justice climatique et négociations internationales', *Négociations*, 27: 7–22.

Bourban, Michel. 2018. *Penser la justice climatique. Devoirs et politiques* (PUF: Paris).

Bourban, Michel. Forthcoming. 'Climate Justice in the Non-Ideal Circumstances of International Negotiations.' In Sarah Kenehan and Corey Katz (eds.), *Applying Principles of Justice and Feasibility to Climate Politics* (Rowman & Littlefield: London/New York).

Boyce, James K. 2018. 'Carbon Pricing: Effectiveness and Equity', *Ecological Economics*, 150: 52–61.
Caney, Simon. 2010. 'Climate Change and the Duties of the Advantaged', *Critical Review of International Social and Political Philosophy*, 13: 203–28.
Caney, Simon. 2014. 'Two Kinds of Climate Justice: Avoiding Harm and Sharing Burdens', *Journal of Political Philosophy*, 22: 125–49.
Caney, Simon. 2015. 'Responding to Global Injustice: On the Right of Resistance', *Social Philosophy and Policy*, 32: 51–73.
Caney, Simon, and Cameron Hepburn. 2011. 'Carbon Trading: Unethical, Unjust and Ineffective?', *Royal Institute of Philosophy Supplement*, 69: 201–234.
Ciplet, David, J. Timmons Roberts, and Mizan R. Khan. 2013. 'The Politics of International Climate Adaptation Funding: Justice and Divisions in the Greenhouse', *Global environmental politics*, 13: 49–68.
Ciplet, David, J. Timmons Roberts, and Mizan R. Khan. 2015. *Power in a Warming World. The New Global Politics of Climate Change and the Remaking of Environmental Inequality* (MIT Press: Cambridge/London).
Coady, David, Ian Parry, Louis Sears, and Baoping Shang. 2017. 'How Large Are Global Fossil Fuel Subsidies?', *World Development*, 91: 11–27.
Criqui, Patrick, Benoît Faraco, and Alain Grandjean. 2009. *Les États et le carbone* (PUF: Paris).
de Águeda Corneloup, Inés, and Arthur P. J. Mol. 2014. 'Small Island Developing States and International Climate Change Negotiations: The Power of Moral "Leadership"', *International Environmental Agreements: Politics, Law and Economics*, 14: 281–297.
Dellink, Rob, Michel den Elzen, Harry Aiking, Emmy Bergsma, Frans Berkhout, Thijs Dekker, and Joyeeta Gupta. 2009. 'Sharing the Burden of Financing Adaptation to Climate Change', *Global Environmental Change*, 19: 411–21.
Dietz, Simon, Alex Bowen, Baran Doda, Ajay Gambhir, and Rachel Warren. 2018. 'The Economics of 1.5°C Climate Change', *Annual Review of Environment and Resources*, 43: 455–480.
Dietz, Simon, Cameron Hepburn, and Nicholas Stern. 2008. 'Economics, Ethics and Climate Change.' In Kaushik Basu and Ravi Kanbur (eds.), *Arguments for a Better World: Essays in Honor of Amartya Sen, Volume 2: Society, Institutions, and Development* (Oxford University Press: Oxford).
Duus-Otterström, Göran. 2016. 'Allocating Climate Adaptation Finance: Examining Three Ethical Arguments for Recipient Control', *International Environmental Agreements: Politics, Law and Economics*, 16: 655–670.
Füssel, H.-M., S. Hallegatte, and M. Reder. 2012. 'International Adaptation Funding.' In O. Edenhofer (ed.), *Climate Change, Justice and Sustainability: Linking Climate and Development Policy* (Springer: Dordrecht).
Gardiner, Stephen M. 2004. 'Ethics and Global Climate Change', *Ethics*, 114: 555–600.
Gardiner, Stephen M. 2011. *A Perfect Moral Storm: The Ethical Tragedy of Climate Change* (Oxford University Press: Oxford).
Gardiner, Stephen M., Simon Caney, Dale Jamieson, and Henry Shue. 2010. *Climate Ethics: Essential Readings* (Oxford University Press: Oxford).
Gareth, Dale. 2018. 'Climat: deux Nobels pyromanes', *L'écologiste*, 19: 42–44.
Gollier, Christian, and Jean Tirole. 2015. 'Negotiating Effective Institutions Against Climate Change', *Economics of Energy & Environmental Policy*, 4: 5–28.
Guivarch, Céline, and Nicolas Taconet. this volume. 'Global Inequalities and Climate Change.' In Éloi Laurent, Rebecca Ray and Klara Zwickl (eds.), *The Routledge Handbook of the Political Economy of the Environment* (Routledge: New York).
Hansen, James, and Makiko Sato. 2016. 'Regional Climate Change and National Responsibilities', *Environmental Research Letters*, 11: 1–9.
Hepburn, Cameron. 2009. 'Carbon Taxes, Emissions Trading and Hybrid Schemes.' In Dieter Helm and Cameron Hepburn (eds.), *The Economics and Politics of Climate Change* (Oxford University Press: Oxford).
Hinkel, Jochen. 2011. '"Indicators of Vulnerability and Adaptive Capacity": Towards a Clarification of the Science – Policy Interface', *Global Environmental Change*, 21: 198–208.
Holz, Christian, Sivan Kartha, and Tom Athanasiou. 2018. 'Fairly Sharing 1.5: National Fair Shares of a 1.5 °C-compliant Global Mitigation Effort', *International Environmental Agreements: Politics, Law and Economics*, 18: 117–134.
IPCC. 2011. 'Summary for Policymakers.' In O. Edenhofer, R. Pichs-Madruga, Y. Sokona, Seyboth, K.P. Matschoss, S. Kadner, T. Zwickel, P. Eickemeier, G. Hansen, S. Schlömer and C. von Stechow (eds.),

IPCC Special Report on Renewable Energy Sources and Climate Change Mitigation (Cambridge University Press: Cambridge and New York).

IPCC. 2014. 'Summary for Policymakers.' In O. Edenhofer, R. Pichs-Madruga, Y. Sokona, E. Farahani, S. Kadner, K. Seyboth, A. Adler, I. Baum, S. Brunner, P. Eickemeier, B. Kriemann, J. Savolainen, S. Schlömer, C. von Stechow, T. Zwickel and J.C. Minx (eds.), *Climate Change 2014: Mitigation of Climate Change. Contribution of Working Group III to the Fifth Assessment Report of the Intergovernmental Panel on Climate Change* (Cambridge University Press: Cambridge).

IPCC. 2018. 'Summary for Policymakers.' In V. Masson-Delmotte, P. Zhai, H.-O. Pörtner, D. Roberts, J. Skea, P.R. Shukla, A. Pirani, W. Moufouma-Okia, C. Péan, R. Pidcock, S. Connors, J.B.R. Matthews, Y. Chen, X. Zhou, M.I. Gomis, E. Lonnoy, T. Maycock, M. Tignor, and T. Waterfield (eds.), *Global Warming of 1.5°C. An IPCC Special Report on the Impacts of Global Warming Of 1.5°C above Pre-industrial Levels and Related Global Greenhouse Gas Emission Pathways, in the Context of Strengthening the Global Response to the Threat of Climate Change, Sustainable Development, and Efforts to Eradicate Poverty* (World Meteorological Organization: Geneva).

Jacobson, Mark Z., Mark A. Delucchi, Zack A.F. Bauer, Savannah C. Goodman, William E. Chapman, Mary A. Cameron, Cedric Bozonnat, Liat Chobadi, Hailey A. Clonts, Peter Enevoldsen, Jenny R. Erwin, Simone N. Fobi, Owen K. Goldstrom, Eleanor M. Hennessy, Jingyi Liu, Jonathan Lo, Clayton B. Meyer, Sean B. Morris, Kevin R. Moy, Patrick L. O'Neill, Ivalin Petkov, Stephanie Redfern, Robin Schucker, Michael A. Sontag, Jingfan Wang, Eric Weiner, and Alexander S. Yachanin. 2017. '100% Clean and Renewable Wind, Water, and Sunlight All-Sector Energy Roadmaps for 139 Countries of the World', *Joule*, 1: 108–121.

Jamieson, Dale. 1992. 'Ethics, Public Policy, and Global Warming', *Science, Technology, & Human Values*, 17: 139–153.

Jamieson, Dale. 2014. *Reason in a Dark Time: Why the Struggle Against Climate Change Failed – and What It Means for Our Future* (Oxford University Press: Oxford).

Johnson, Giff. 2015. *Idyllic No More: Pacific Island Climate, Corruption and Development Dilemmas* (CreateSpace: North Charleston).

Kanbur, Ravi, and Henry Shue. 2019. *Climate Justice: Integrating Economics and Philosophy* (Oxford University Press: Oxford).

Kartha, Sivan, Paul Baer, Tom Athanasiou, and Eric Kemp-Benedict. 2009. 'The Greenhouse Development Rights Framework', *Climate and Development*, 1: 147–165.

Khan, Mizan, Stacy-ann Robinson, Romain Weikmans, David Ciplet, and J. Timmons Roberts. 2020. 'Twenty-five Years of Adaptation Finance through a Climate Justice Lens', *Climatic Change*, 161: 251–269.

Klein, Richard J.T., and Annett Möhner. 2011. 'The Political Dimension of Vulnerability: Implications for the Green Climate Fund', *IDS Bulletin*, 42: 15–22.

Klepp, Silja. 2013. 'Small Island States And the New Climate Change Movement.' In Matthias Dietz and Heiko Garrelts (eds.), *Routledge Handbook of the Climate Change Movement* (Routledge: New York).

Klepp, Silja, and Libertad Chavez-Rodriguez. 2018. 'Governing Climate Change: The Power of Adaptation Discourses, Policies, and Practices.' In Silja Klepp and Libertad Chavez-Rodriguez (eds.), *A Critical Approach to Climate Change Adaptation: Discourses, Policies, and Practices* (Routledge: London/New York).

Laurent, Éloi. 2011. *Social-écologie* (Flammarion: Paris).

Laurent, Éloi. 2020a. 'The European Green Deal: Bring back the New', *OFCE Policy Brief*, 63: 1–10.

Laurent, Éloi. 2020b. *The New Environmental Economics: Sustainability and Justice* (Polity: Cambridge/Medford).

Le Quéré, C., R.M. Andrew, P. Friedlingstein, S. Sitch, J. Hauck, J. Pongratz, P. A. Pickers, J.I. Korsbakken, G.P. Peters, J.G. Canadell, A. Arneth, V.K. Arora, L. Barbero, A. Bastos, L. Bopp, F. Chevallier, L.P. Chini, P. Ciais, S.C. Doney, T. Gkritzalis, D.S. Goll, I. Harris, V. Haverd, F.M. Hoffman, M. Hoppema, R.A. Houghton, G. Hurtt, T. Ilyina, A.K. Jain, T. Johannessen, C.D. Jones, E. Kato, R.F. Keeling, K.K. Goldewijk, P. Landschützer, N. Lefèvre, S. Lienert, Z. Liu, D. Lombardozzi, N. Metzl, D.R. Munro, J.E.M.S. Nabel, S. Nakaoka, C. Neill, A. Olsen, T. Ono, P. Patra, A. Peregon, W. Peters, P. Peylin, B. Pfeil, D. Pierrot, B. Poulter, G. Rehder, L. Resplandy, E. Robertson, M. Rocher, C. Rödenbeck, U. Schuster, J. Schwinger, R. Séférian, I. Skjelvan, T. Steinhoff, A. Sutton, P.P. Tans, H. Tian, B. Tilbrook, F.N. Tubiello, I.T. van der Laan-Luijkx, G.R. van der Werf, N. Viovy, A.P. Walker, A. J. Wiltshire, R. Wright, S. Zaehle, and B. Zheng. 2018. 'Global Carbon Budget 2018', *Earth System Science Data*, 10: 2141–2194.

Light, Andrew. 2017. 'Climate Diplomacy.' In Stephen Gardiner and Allen Thompson (eds.), *The Oxford Handbook of Environmental Ethics* (Oxford University Press: Oxford).

Lührmann, Anna, Sirianne Dahlum, Staffan I. Lindberg, Laura Maxwell, Valeriya Mechkova, Moa Olin, Shreeya Pillai, Constanza Sanhueza Petrarca, Rachel Sigman, and Natalia Stepanova. 2018. 'V-Dem Annual Democracy Report 2018. Democracy Facing Global Challenges.' In University of Gothenburg: V-Dem Institute.

Lührmann, Anna, Valeriya Mechkova, Sirianne Dahlum, Laura Maxwell, Moa Olin, Constanza Sanhueza Petrarca, Rachel Sigman, Matthew C. Wilson, and Staffan I. Lindberg. 2018. 'State of the world 2017: Autocratization and exclusion?', *Democratization*, 25: 1321–1340.

Mathy, Sandrine, and Odile Blanchard. 2016. 'Proposal for a poverty-adaptation-mitigation window within the Green Climate Fund', *Climate Policy*, 16: 752–767.

Mikulewicz, Michael. 2018. 'Politicizing Vulnerability and Adaptation: On the Need to Democratize Local Responses to Climate Impacts in Developing Countries', *Climate and Development*, 10: 18–34.

Moellendorf, Darrel. 2015. 'Can Dangerous Climate Change Be Avoided?', *Global Justice: Theory Practice Rhetoric*, 8: 66–85.

Müller, Benito, Niklas Höhne, and Christian Ellermann. 2009. 'Differentiating (Historic) Responsibilities for Climate Change', *Climate Policy*, 9: 593–611.

Nordhaus, William. 2018. 'Projections and Uncertainties about Climate Change in an Era of Minimal Climate Policies', *American Economic Journal: Economic Policy*, 10: 333–360.

Oxfam. 2015. Extreme Carbon Inequality.' In Oxford: Oxfam International.

Oxfam. 2018. 'Climate Finance Shadow Report 2018: Assessing Progress Towards the $100 Billion Commitment.' In Oxford: Oxfam International.

Page, Edward A. 2011. 'Climatic Justice and the Fair Distribution of Atmospheric Burdens: A Conjunctive Account', *The Monist*, 94: 412–432.

Pickering, Jonathan. 2010. 'Global Negotiations on Climate Finance: What Role can Fairness Play?' In *Berlin Conference on the Human Dimensions of Global Environmental Change*, 1–23. Berlin.

Pickering, Jonathan. 2012. 'Adaptation Finance in the Asia – Pacific Region: Strengthening Fairness, Effectiveness and Transparency in Allocation.' In *Workshop on Climate Change Governance in the Asia-Pacific Region: Agency and Adaptiveness*. Australian National University (ANU).

Puig, Daniel, Anne Olhoff, Skylar Bee, Barney Dickson, and Keith Alverson. 2016. 'The Adaptation Finance Gap Report.' In Nairobi: United Nations Environment Programme.

Ritchie, Hannah, and Max Roser. 2019. 'CO_2 and Greenhouse Gas Emissions'. https://ourworldindata.org/co2-and-other-greenhouse-gas-emissions.

Rogelj, Joeri, Michel den Elzen, Niklas Höhne, Taryn Fransen, Hanna Fekete, Harald Winkler, Roberto Schaeffer, Fu Sha, Keywan Riahi, and Malte Meinshausen. 2016. 'Paris Agreement climate proposals need a boost to keep warming well below 2 °C', *Nature*, 534: 631–639.

Shue, Henry. 1992. 'The Unavoidability of Justice.' In Andrew Hurrell and Benedict Kingsbury (eds.), *The International Politics of the Environment: Actors, Interests and Institutions* (Clarendon Press: Oxford).

Shue, Henry. 2013. 'Climate Hope: Implementing the Exit Strategy', *Chicago Journal of International Law*, 13: 381–402.

Shue, Henry. 2015. 'Historical Responsibility, Harm Prohibition, and Preservation Requirement: Core Practical Convergence on Climate Change', *Moral Philosophy and Politics*, 2: 7–31.

Steffen, Will, Johan Rockström, Katherine Richardson, Timothy M. Lenton, Carl Folke, Diana Liverman, Colin P. Summerhayes, Anthony D. Barnosky, Sarah E. Cornell, Michel Crucifix, Jonathan F. Donges, Ingo Fetzer, Steven J. Lade, Marten Scheffer, Ricarda Winkelmann, and Hans Joachim Schellnhuber. 2018. 'Trajectories of the Earth System in the Anthropocene', *Proceedings of the National Academy of Sciences*, 115: 8252–8259.

Stiglitz, Joseph. 2015. 'Overcoming the Copenhagen Failure with Flexible Commitments', *Economics of Energy & Environmental Policy*, 4: 29–36.

UN-DESA. 2009. 'A Global Green New Deal for Climate, Energy, and Development.' In New York.

UNEP. 2019. 'Global Trends in Renewable Energy Investment 2019.' In Frankfurt: United Nations Environment Programme.

UNFCCC. 1992. 'United Nations Framework Convention on Climate Change. Document FCCC/INFORMAL/84.' In New York.

UNFCCC. 2009. 'Report of the Conference of the Parties on Its Fifteenth Session, Held in Copenhagen from 7 to 19 December 2009. Addendum. Part Two: Action Taken by the Conference of the Parties at its Fifteenth Session. Document FCCC/CP/2009/11/Add.1.' In Copenhagen.

UNFCCC. 2015. 'Adoption of the Paris Agreement. Decision 1/CP.21. Document FCCC/CP/2015/10/Add.1.' In Paris.

UNU-EHS. 2019. 'World Risk Report.' In Alliance Development Works and United Nations Universtiy.

Wallimann-Helmer, Ivo, Lukas Meyer, Kian Mintz-Woo, Thomas Schinko, and Olivia Serdeczny. 2019. 'The Ethical Challenges in the Context of Climate Loss and Damage.' In Reinhard Mechler, Laurens M. Bouwer, Thomas Schinko, Swenja Surminski, and JoAnne Linnerooth-Bayer (eds.), *Loss and Damage from Climate Change. Concepts, Methods and Policy Options* (Springer: Cham).

WB. 2019. 'State and Trends of Carbon Pricing 2019.' In Washington, DC: World Bank.

Weikmans, Romain, J. Timmons Roberts, Jeffrey Baum, Maria Camila Bustos, and Alexis Durand. 2017. 'Assessing the Credibility of How Climate Adaptation Aid Projects are Categorised', *Development in Practice*, 27: 458–471.

Weiler, Florian, Carola Klöck, and Matthew Dornan. 2018. 'Vulnerability, Good Governance, or Donor Interests? The Allocation of Aid for Climate Change Adaptation', *World Development*, 104: 65–77.

Wenar, Leif. 2016. *Blood Oil: Tyrants, Violence, and the Rules that Run the World* (Oxford University Press: Oxford).

17

CARBON PRICING AND CLIMATE JUSTICE

James K. Boyce

Many economists believe that the best way to curb the use of fossil fuels for climate change mitigation is to put a price on carbon emissions. If our use of the biosphere's limited capacity to absorb emissions carries a price tag, rather than being free of charge, consumers, firms, and governments alike would have an incentive to use it less and to invest more in energy efficiency and renewable energy.

Carbon pricing proposals have encountered opposition, however, from some climate justice advocates. These opponents have leveled three major criticisms at the strategy: first, that carbon pricing is a 'false solution' that will not really arrest climate change; second, that it is a regressive policy, hitting the poor harder than the rich; and third, that it does not adequately protect vulnerable communities that suffer the greatest environmental harms from the extraction, processing, and burning of fossil fuels.

None of these criticisms can be dismissed lightly. But none of them is insurmountable.

To promote climate justice, carbon pricing must be effective in curbing emissions; it must be equitable; and it must address the needs of vulnerable communities. The key to effectiveness is to anchor the carbon price firmly to an ambitious emissions reduction trajectory. The price then emerges as a consequence of keeping the fossil fuels in the ground.

The key to equitable carbon pricing is to recycle the revenue to the people, based on the principle that the limited ability of the biosphere to absorb emissions safely belongs equally to all. Equal per capita dividends transform carbon pricing from a regressive policy to a progressive one.

The key to protecting vulnerable communities overburdened with pollution is to embed carbon pricing within a broader policy mix that ensures emissions reductions and environmental health improvements at the locations that need them most. This chapter discusses how these requirements can be met.

Effectiveness

Merely instituting a carbon price does not guarantee that climate policy goals will be met: the price must be adequate to ensure that emissions reduction goals are met. Actually existing carbon prices have largely failed to achieve this, leading some to conclude that carbon pricing is a

DOI: 10.4324/9780367814533-19

false solution. The remedy, however, is not to make emissions free, but to implement a carbon price that is up to the job.

The climate policy litmus test

The litmus test for effective climate policy is whether it keeps enough fossil fuel in the ground to prevent global temperatures from rising more than 1.5–2 °C above their pre-industrial level, the target set forth in the 2016 Paris Agreement. Many policies can contribute to this, but there is only one way to be absolutely certain that we achieve emission reductions commensurate with this target: put a hard ceiling on the amount of fossil carbon we allow to enter the economy and ratchet the ceiling steadily down over time.

The most straightforward way to do this is to issue carbon permits up to the level set by the ceiling. If the target is to cut emissions by 85% in 30 years, for example, this means cutting the number of permits by about 6% each year. At every tanker port, pipeline terminal, and coal mine head, fossil fuel corporations would be required to surrender one permit for each ton of carbon they bring into the economy. When these permits are auctioned, the firms will bid what they expect to recoup from higher prices paid by consumers. The carbon price is a direct result of this limit on supply.

If other climate mitigation policies dramatically reduce demand for fossil fuels, the price will be lower than would otherwise be the case. Indeed, if other policies are so successful that they achieve the targeted emissions reduction on their own, the supply limit would turn out to be redundant and the permit price would fall to zero. Like fire insurance, in an optimistic scenario the carbon price would turn out to be unnecessary – but optimism is not a good reason to forgo insurance.

To guarantee that we meet the target, it is not enough to set a carbon price and hope it will do the job: the price must be anchored to the emissions trajectory. Likewise, just investing in mass transit and hoping for the best, or passing fuel economy standards and hoping for the best, is not enough. We know these measures will help, but we cannot know exactly how much.

Today the world is past the stage where just hoping for the best is good enough. We need to make absolutely certain that we cut emissions decisively in the coming years. And we need to face up to the reality that is almost certain to come with this objective: higher prices on fossil fuels.

Existing carbon prices

Today, carbon pricing systems cover approximately 11 gigatons of carbon dioxide emissions worldwide, equivalent to roughly 20% of total greenhouse gas emissions. In 2019 the prices ranged from US$1 to $127 per metric ton of carbon dioxide (mt CO_2). In more than half the cases, emissions were priced at less than $10, equivalent to less than 10 US cents on a gallon of gasoline or 3 euro cents on a liter of petrol (World Bank 2019, p. 12).

Indeed, many countries actively subsidize the use of fossil fuels. Direct subsidies by governments amounted to $333 billion/year worldwide in 2015 (Coady et al. 2017).[1] The average subsidy worldwide was roughly five times larger than the average carbon price. In effect, therefore, the average net carbon price was *negative*.

Even those jurisdictions with the most robust carbon pricing policies implemented so far have failed to cut their emissions along anything close to the 6% per year trajectory that would be needed globally to meet the target established by the Paris Agreement. Sweden, the country with the world's highest carbon price ($127/mt CO_2), barely changed its greenhouse gas emissions from 2014 to 2019, not only because its price arguably remains too low but also because it applies to less than half of the country's total emissions.[2] British Columbia, with the highest carbon tax in Canada ($26/mt CO_2), applying to more than 70% of its greenhouse gas emissions, registered an overall decline of only 0.5% between 2007 and 2017.[3] The lesson is not that carbon prices 'don't work.' It is that carbon prices not securely anchored to hard emissions targets don't work.

Carbon prices based on neoclassical efficiency: the 'social cost of carbon'

Prescriptions for the 'right' carbon price necessarily rest on an ethical foundation. One candidate for such a foundation is the efficiency criterion of neoclassical economics, where the optimal price – termed the 'social cost of carbon' (SCC) – maximizes the net present value of the benefits of emission reductions minus their cost.

Measuring the benefits of emission reductions and translating them into monetary terms is notoriously difficult.[4] The equations used in integrated assessment models (IAMs) to estimate GDP losses as a function of increases in global temperature are, as Pindyck (2013, p. 870) observes, 'completely made up, with no theoretical or empirical foundation.' The treatment of uncertainty in IAMs is deeply problematic in the presence of catastrophic risks of unknown and unknowable probability and magnitude (Weitzman 2011; Ackerman 2017). The use of discount rates, based on the dubious premise that the time-preference logic that individual mortals use in thinking about their own futures ought to tell us how to think about the well-being of future generations, causes the monetary value assigned to future damages to melt away, much as polar ice today is melting with climate change.[5] The benefits of reducing emissions of hazardous co-pollutants emitted by fossil fuel combustion, such as sulfur dioxide and nitrogen oxides, are excluded from SCC calculations, despite evidence that their monetized value often is comparable to or even greater than conventional estimates of climate damages.[6] For these among other reasons, a review on the *Journal of Economic Literature* concludes that the models used to compute the SCC are 'so deeply flawed as to be close to useless as tools for policy analysis' (Pindyck 2013, pp. 861–862).

Further difficulties arise in the measurement of marginal abatement costs, which are compared to marginal damages in order to arrive at the ostensibly 'optimal' carbon price and emissions trajectory. Future abatement costs, in particular, are uncertain: the cost curves shift downward over time, and public policies, including carbon pricing, can accelerate this shift. For both damage functions and abatement costs, extrapolations outside the range of past experience are especially problematic.

Carbon prices based on science: target-consistent prices

An alternative way to prescribe the carbon price is to anchor it to emission targets based on climate science, such as the objective of holding the rise in global mean surface temperature to 1.5 °C above its pre-industrial level. Here the normative criterion is safety rather than neoclassical efficiency.[7]

The safety criterion is the foundation of much environmental law. The U.S. Clean Air Act, for example, directs the U.S. Environmental Protection Agency to set air quality standards for 'the protection of public health and welfare' while 'allowing an adequate margin of safety' – not to determine clean air standards by weighing marginal benefits of protecting public health against its marginal costs.[8]

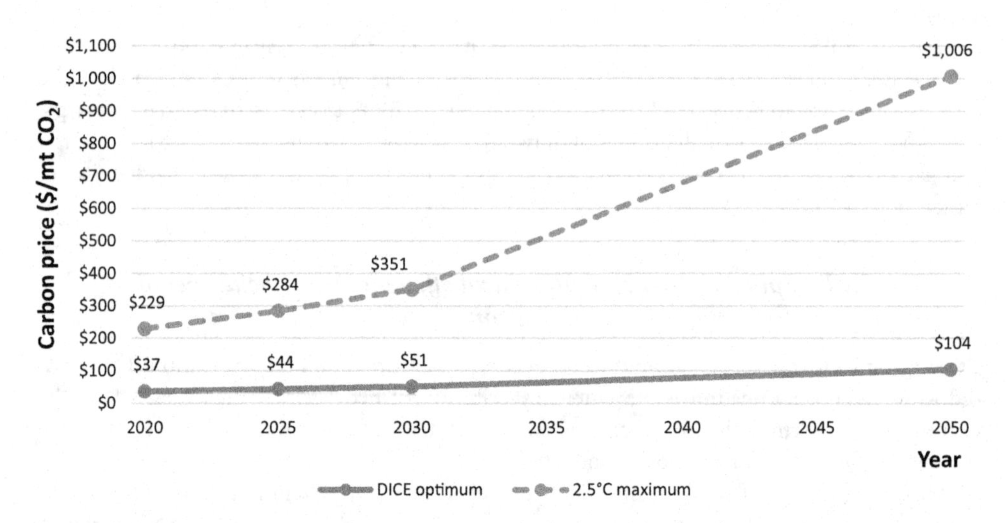

Figure 17.1 Carbon price paths

Note: Global CO_2 price in 2010 US dollars.

Source: Boyce (2018), based on data from Nordhaus (2017a, Table 17.1).

To be sure, there is always some degree of arbitrariness in delineating what qualifies as 'safe,' but scientists are better placed than economists to address this question. Unlike neoclassical efficiency, the safety criterion does not require the estimation of marginal damages and marginal abatement costs across the range of potential outcomes. Economists who accept this approach frame their policy advice more modestly, in terms of choosing the most cost-effective means for reaching the climate policy goal rather than choosing the goal itself.[9]

Figure 17.1 illustrates the potentially huge differences between prices prescribed on the bases of neoclassical efficiency and safety. The estimates shown here are drawn from a paper by William Nordhaus (2017a), who compared the SCC derived from an IAM called DICE (the dynamic integrated model of climate and the economy) to the price that the model estimates would be needed to hold the global mean temperature increase to 2.5 °C.[10] The 'welfare-optimizing' SCC prescribed by the neoclassical efficiency criterion rises from $37/mt CO_2 in 2020 to about $100 in 2050. The temperature increase accompanying this trajectory would be 3.5 °C by the turn of the century, rising further thereafter.[11] The price required to meet the 2.5 °C maximum starts more than six times higher at about $230/mt CO_2 in 2020, and rises to about $1,000 in 2050. The gap between the two trajectories would be even wider if the safety constraint were taken to be the 1.5–2 °C target set out in the Paris Agreement.

Mechanics of effective carbon pricing

The cost of meeting emissions targets cannot be known with much precision in advance, since it will depend on how abatement costs shift over time. If there are certain carbon price thresholds beyond which fossil fuels will be replaced quickly by alternative energy, the cost will be lower than one might conclude by merely extrapolating from past experience. On the other hand, it is plausible that marginal costs will rise more sharply as emissions are cut more deeply. This is one reason to set the quantity of emissions and the let the carbon price adjust accordingly, rather than merely setting a price and hoping it turns out to be right.

Carbon taxes vs carbon caps

A carbon price can be instituted via a tax or an emissions cap. A tax sets the price and lets the quantity of emissions vary; a cap sets the quantity of emissions and lets the price vary. A third option is to combine the two, via a hybrid system in which the tax serves as a floor price in permit auctions.

Although the future relationship between carbon prices and emission quantities cannot be known in advance, past experience may offer a first approximation as to what to expect. A meta-analysis of estimates of the price elasticity of energy demand, based on hundreds of empirical studies published between 1990 and 2016, found that a 10% increase in energy prices results on average in a 2.1% decline in the quantity consumed in the short run and in a 6.1% decline in the long run (Labandeira et al. 2017). The inelastic response, with the fall in quantity being less than the rise in price, stems from the fact that energy to a large extent is a necessity rather than a luxury.

Different studies reported a wide range of estimates, however, reflecting variations across energy products, locations, time, and estimation techniques as well as differences in public policies. The central importance of meeting emissions targets, coupled with considerable uncertainty as to the relationship between quantity and price, provides a strong argument for setting the trajectory for reducing carbon emissions and letting the price adjust, rather than vice versa.

One way to anchor the carbon prices to the emissions targets is to place a cap on total emissions. The annual quantity declines over time; each year the number of permits is set by the cap. During a recession, when energy demand is weak, the permit price will be lower than during an economic boom. If energy-saving technological change proceeds rapidly, the permit price will be lower than if technological change proves to be slow. Regardless of these uncertainties, the cap guarantees that the target is met.

A second option is a carbon tax with a rate that automatically adjusts in response to the distance between present emissions and targets. Switzerland has done this in its CO_2 levy on power plants. Hafstead et al. (2017) recommend adjusting the tax rate annually or biennially, with the extent of adjustment depending on the difference between actual emissions and targets. Metcalf (2018) proposes a carbon price that rises at an annual real rate of 5% when emissions targets are met and 10% when they are not, but if the base is too low, the percentage adjustments may prove inadequate for meeting the targets.[12]

A third option is a tax-and-cap combination, in which the tax serves as the floor price in permit auctions. The cap sets the ceiling on the total number of available permits. If the tax turns out to be high enough to keep demand within this limit, it is the carbon price. But if the tax alone proves too low to meet the target for reductions in emissions, then the cap kicks in, and permit auctions let the carbon price rise accordingly. Compared to a cap alone, this policy has the merit of providing price certainty on the downside, which could be helpful in guiding investment decisions.

Where to implement the carbon price?

Carbon pricing is most easily implemented upstream – at the ports, pipeline terminals, and mine heads where fossil fuels enter the economy. For each ton of CO_2 that eventually will be emitted when the fuel is burned, the supplier must surrender one permit or pay the tax. In the U.S., an upstream system would involve roughly 2,000 collection points nationwide (U.S. Congressional Budget Office 2001). If, instead, the compliance entities were the consumers of fossil fuels, the administrative costs would be much larger.

CO_2 emissions can be calculated simply from the carbon content of fossil fuels prior to their combustion, eliminating any need for end-of-pipe monitoring. In this respect, CO_2 differs from conventional pollutants like sulfur dioxide, where emissions per ton of fuel vary depending on fuel quality and pollution-control equipment. The predictability of CO_2 emissions makes a low-cost upstream pricing system feasible.

Existing carbon pricing systems often have midstream compliance entities – power plants, large industrial facilities, or fuel distributors – who are located between the firms that first bring fossil fuels into the economy and to the final consumer. When these entities are few in number, the administrative costs are tractable. But midstream systems typically are less comprehensive than an upstream system, since they do not cover all sectors of the economy.

Wherever implemented, the carbon price in the end will be passed through to final consumers. When the cost of coal goes up, for example, so does the cost of electricity. In other words, it is not the upstream or midstream compliance entities who ultimately pay the carbon price. This is a feature of any carbon pricing system, not a bug: it is the cost pass-through that transmits price signals to consumers, firms, and governments to curtail their carbon footprints.

To trade or not to trade?

A cap-and-permit system is not necessarily a cap-and-trade system. Most permits in society, such as parking permits and driving permits, are not tradeable. There is no intrinsic reason that carbon permits should be different. 'Cap-and-trade' became a part of the climate policy lexicon – so much so that sometimes it is incorrectly assumed to be synonymous with a cap-and-permit system – only because early pollution permit systems, like the U.S. sulfur dioxide program for power plants and the European Union's emissions trading system for CO_2, gave away free permits to firms by means of formulae based on historic emissions. These permits were tradeable so that firms with higher abatement costs could purchase them from firms with lower abatement costs.

If permits are auctioned rather than given away, there is no need whatsoever for them to be tradeable. Each firm simply buys as many permits as it wants at the auctions, which are held quarterly or annually. If a firm buys more permits than it needs, it can save the extra ones to use in a subsequent compliance period.

Permit trading has several drawbacks. First, it introduces possibilities for market manipulation and speculation. Second, it multiplies administrative costs, since permit trades must be tracked. Third, it diverts part of the money that consumers pay in higher fuel prices into trader profits, at the opportunity cost of putting these funds to other uses. Finally, permit giveaways confer windfall profits on the recipients, effectively rewarding them for past pollution.

'Offsets' are a variant of permit trading whereby firms can pay for emissions reduction (or carbon sequestration) elsewhere as a substitute for buying a permit. Although appealing to economists on cost-effectiveness grounds, offsets are beset by the formidable practical difficulties of verification and additionality.[13] Moreover, they can create perverse incentives for polluters to increase baseline emissions in order to garner more payments.[14] In effect, offsets risk turning the emissions cap into a sieve. A better strategy is to pursue emissions reduction elsewhere and carbon sequestration separately, so that their benefits come in addition to, rather than instead of, the emissions reductions mandated by the cap.

A uniform international price?

An international agreement on a uniform world carbon price is unlikely, and perhaps undesirable. Experience suggests that it is far more likely that individual nations (or subnational units) will continue to establish carbon pricing policies independently, with prices that vary across jurisdictions.

Apart from the practical and political impediments to international agreement on a carbon price, different countries may have sound reasons to want different prices. In effect, a uniform international price would allocate the earth's remaining carbon space based on ability to pay: high-income countries would be able to afford more emissions than would low-income countries. Yet the United Nations Framework Convention on Climate Change provides that countries will reduce emissions according to their 'common but differentiated responsibilities and respective capacities,' a formulation that implies that higher-income countries should do more, not less, to curb their emissions.

Cross-country differences in the air quality benefits of reduced fossil fuel use may provide a further motive for price differentiation (Boyce 2020). Insofar as the public health impacts of fossil fuel combustion are more severe in some low- and middle-income countries, they ultimately may prefer higher carbon prices.

In any case, the prospects for effective carbon pricing do not hinge on international action. Air quality co-benefits alone may be sufficient for countries to decide to adopt the policy. The economic and employment benefits brought by the clean energy transition may provide a further motive. And, as discussed in the next section, if revenues from carbon pricing are recycled to the public as dividends, the net financial impact can be positive for the majority of each country's residents. Together, these attractions may well be sufficiently compelling for countries to adopt carbon pricing policies without awaiting an international accord.

Distributional equity

The revenue generated through carbon pricing is likely to be substantial, especially if the price is robust enough to be effective in curbing emissions. A simple calculation will illustrate the potential order of magnitude. CO_2 emissions from fossil fuel combustion in the U.S. currently amount to about 5.2 billion mt/yr. At $230/mt CO_2 (the initial carbon price in the safety-based trajectory depicted in Figure 17.1), total carbon revenue would amount to about $1 trillion/year, the exact amount depending on the extent of the associated change in quantity. If the demand for fossil fuels remains price-inelastic, total revenue will rise as the cap tightens in future years. Who pays and who receives the carbon revenue will pose critical distributional questions.

Carbon rent

When the supply of fossil fuels entering an economy is curtailed, their price goes up. This can be expected to occur regardless of the means and motive for the supply restriction (see Table 17.1). When OPEC cuts oil production, for example, the world prices go up. In this case, the extra money that consumers pay ultimately flows back to the producer cartel. Similarly, if a country were to simply 'keep the oil in the soil,' a slogan popular among climate justice activists, prices at the pump would rise. In this case, the extra money would flow to those producers who continue to extract oil. Carbon pricing differs in that it opens other possibilities for allocating the extra money that is paid by consumers.

The higher cost of fossil fuels as a result of carbon pricing – here termed 'carbon rent' – is a transfer, not a resource cost. This money is not spent to abate emissions; rather, it is paid for fossil fuel use that is not abated. It is not used to produce more fossil fuels. Nor does it magically disappear. Instead, it is transferred to the recipients of the carbon rent, whoever these may be.

Table 17.1 Three ways to cut the supply of fossil fuels

Strategy	*Motive*	*Effect*
Supply restriction by cartel	Market manipulation	Higher profits for producers
Just say no	Climate stabilization	Higher profits for producers
Carbon pricing	Climate stabilization	Decided by policy design

Source: Boyce (2019b, p. 64).

Who are these recipients? Broadly, there are three options. One is to give free permits to firms in the policy called cap-and-trade (or, more precisely, cap-and-giveaway-and-trade). In this option, the transfer leads to windfall profits.[15] The second is to auction the carbon permits (or, equivalently, levy a carbon tax) and let the government retain the revenue. In this option, the final distributional impact depends on how the government uses the revenue. The third is to auction the permits (or levy the tax) and return the money to the people as equal per-person dividends. In this option, the result is a net transfer from those with bigger-than-average carbon footprints to those with smaller-than-average carbon footprints.

Incidence of carbon pricing

The amount that any given household pays in higher fuel bills (and higher costs for other goods and services that use fossil fuels in their production and distribution) as a result of carbon pricing depends upon the size of its carbon footprint. Those who consume more, pay more; those who consume less, pay less. Local, state, and federal governments likewise pay in proportion to their use of fossil fuels.[16] Firms, on the other hand, behave as intermediaries, passing on the costs on to these final consumers.

In general, households at the upper end of the income distribution have the biggest carbon footprints, for the simple reason that they consume more of just about everything, including fossil fuels. They heat and cool bigger homes, they travel more often in airplanes, and so on. So in absolute amounts, they will pay more as a result of carbon pricing than low- and middle-income households.

Relative to their incomes, however, upper-income households often pay less.[17] Figure 17.2 shows the distributional incidence of a \$50/mt CO_2 tax in the U.S. In the lowest expenditure quintile, the tax claims 2.8% of household expenditure; in the top quintile, 1.9%. The impact is thus regressive.[18] It also would be quite evident to the public, as fuel prices are among the most visible in the economy.[19]

Sharp increases in the prices of fossil fuels could generate a backlash from consumers in general, and from low- and middle-income consumers in particular, jeopardizing the policy's political sustainability. Whether this happens, however, may depend crucially on where the money goes.

Carbon dividends

Returning carbon revenue to the people as equal per-person dividends would transform the policy into one with a progressive effect on income distribution (Boyce 2019b). Upper-income households generally would pay more in higher prices than they receive back as dividends. Lower-income households generally would receive more than they pay. Middle-income

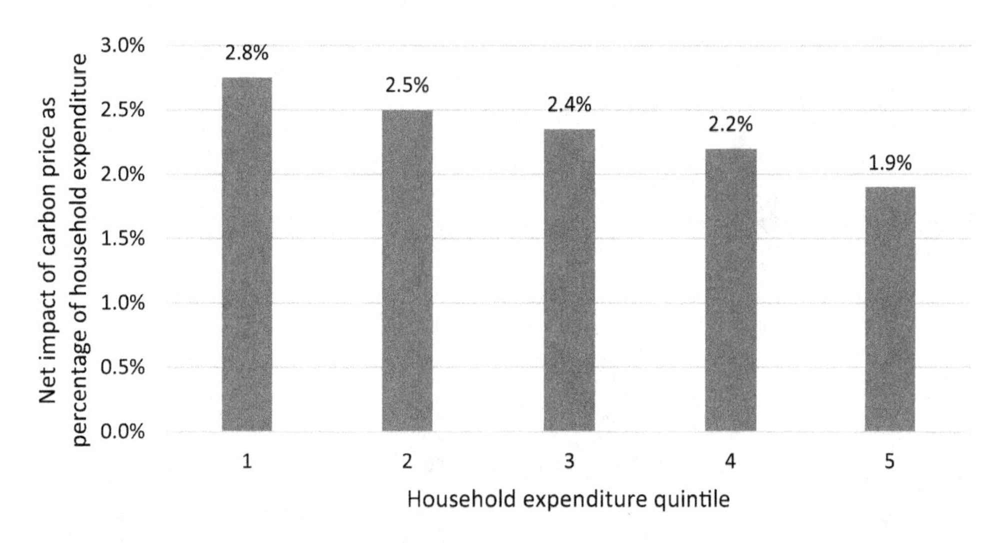

Figure 17.2 Incidence of \$50/ton CO_2 tax in U.S.

Note: Based on consumer expenditure survey data for 2012–2014.

Source: Calculated from data in Fremstad and Paul (2019, Table 17.2).

households generally would more or less break even, thus being protected from negative impacts on their net incomes. The result would be a decrease in income inequality.

Carbon dividends are an example of a 'feebate': individuals pay fees in proportion to their use of a commonly owned resource, and the money collected in fees is returned in equal rebates to all co-owners. The idea can be illustrated with an analogy. Imagine that 1,000 people work in an office building whose parking lot has only 300 spaces. If everyone could park for free, the result would be chronic congestion and excess demand. To avoid this outcome, a parking fee is charged that limits demand to fit the lot's capacity. Every month the proceeds from the fee are distributed in equal payments to everyone who works in the building. Those who take public transportation or bicycle to work come out well ahead: they pay nothing and get their share of the revenue. Those who carpool to work roughly break even. And those who commute every day in a single-occupancy vehicle pay more into the revenue pot than they get back.

Carbon dividends apply the same idea to the atmospheric parking lot. The incentive for a household to reduce its use of fossil fuels is not diminished by the rebate, since its individual use only affects what it pays, not what it receives.

Net impact of a carbon price-and-dividend policy

If a substantial share of the carbon rent is returned to the public in equal per-person dividends, the net distributional impact of carbon pricing becomes sharply progressive. This is illustrated in Figure 17.3, which shows the net impact of the \$50/mt CO_2 price with dividends. The bottom quintile receives a positive transfer, net of what they spend as a result of the carbon price, that is equivalent to 3.9% of their household expenditure. The top quintile sees a negative net impact equivalent to 0.8% of theirs.

Although the net impact of a carbon dividend policy on the vertical distribution of income is progressive, there will be variations within any given income stratum. Fremstad and Paul (2019), from whose data Figures 17.2 and 17.3 are derived, calculate that 95% of households in the

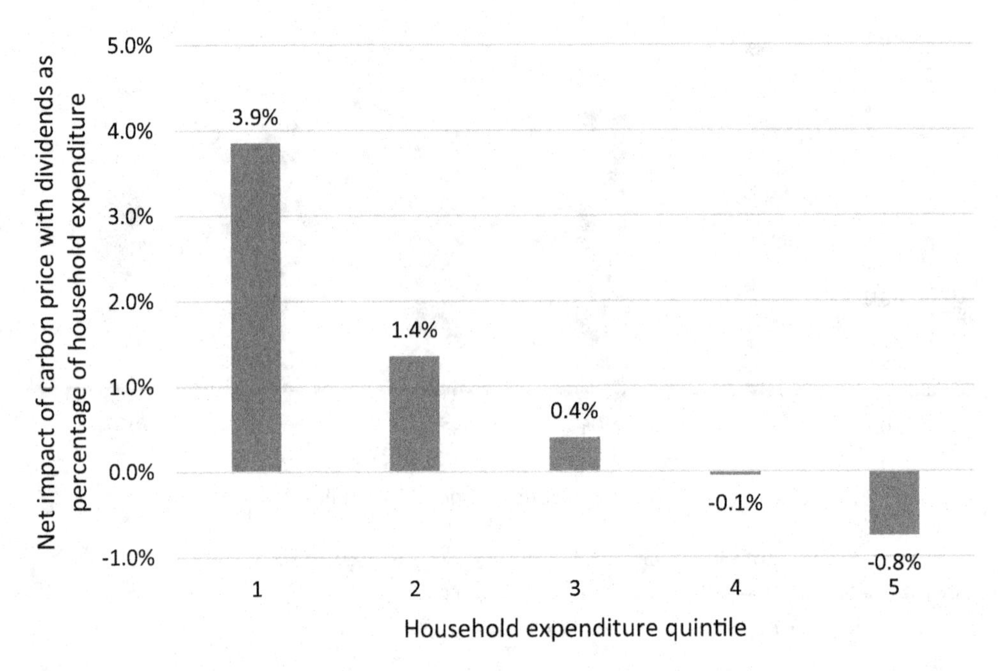

Figure 17.3 Net incidence of \$50/ton CO_2 tax coupled with dividends in U.S.

Source: Boyce (2018), calculated from data in Fremstad and Paul (2019, Table 17.2).

bottom quintile come out ahead – their dividends exceeding what they pay as the result of carbon pricing – and that in the top quintile, 93% pay more than they get back, while in the middle quintiles the results are more mixed. Some of the reasons for horizontal variations within the same income stratum include circumstances that are largely beyond the control of individual households, such as rural-urban differences in vehicle miles driven and regional differences in home heating and air conditioning needs. On grounds of horizontal equity, policymakers may want to take such factors into account in allocating part of the carbon rent (Boyce and Riddle 2011; Cronin et al. 2017).

Dividends and political sustainability

Distributional considerations are important in climate policy in general, and carbon pricing in particular, not only for reasons of equity but also for reasons of political sustainability. While most people are distressed by the possibility (and, increasingly, the reality) of climate destabilization, this prospect often is overshadowed by the more immediate demands of day-to-day survival. Significantly higher fuel prices, unless counterbalanced by carbon dividends or comparably visible return flows of money, would cut into living standards and would not be accepted with equanimity.

This may be the main reason why existing carbon prices to date have been too low, and too incomplete in their coverage, to make a bigger dent in total emissions. Many politicians are prepared to support weak measures to address climate change, accompanied by a great deal of handwaving, but their appetite for strict policies to curb the use of fossil fuels is suppressed by their fear of a backlash from the public.

The political perils of carbon pricing were illustrated vividly in France in November 2018, when the 'yellow vest' protests erupted in response to President Macron's announcement of an increase in taxes on gasoline and diesel as a step to address climate change. Macron 'talks about the end of the world,' demonstrators complained, 'while we are talking about the end of the month' (Rubin 2018). Understandable worries about paying bills at the end of the month are shared by many workers across the world.

Carbon pricing in the broader policy mix

If the carbon price is securely anchored to the emissions reduction trajectory, the policy can be effective. If the carbon revenue is returned to the public as equal per-person dividends, it can be equitable. But carbon pricing is by no means the only desirable climate policy in the toolkit. Smart regulations, too, can help to 'bend the cost curve,' accelerating technological changes and reducing the costs of energy efficiency and clean energy. Smart public investment likewise can save money and spark key innovations.[20] And while carbon emissions from fossil fuels is the proverbial elephant in the climate change room, this is not the only important task: we also need to reduce emissions of other greenhouse gases, develop smart land management practices for carbon sequestration, and invest in adaptation to deal with the climate changes we have failed to prevent.

For climate justice, important complements to effective and equitable carbon pricing include 'hot spot' remediation measures, public investment to ensure a just transition, and adaptation grounded on the bedrock principle that access to a clean and safe environment is a human right shared by everyone.

Hot spot remediation

As already mentioned, carbon pollution accompanies a number of hazardous co-pollutants, including sulfur dioxide, nitrogen oxides, particulate matter, and air toxics. Far from being distributed uniformly across the population, co-pollutant impacts often are concentrated in communities that are at a disadvantage by virtue of their relative lack of purchasing power, political power, or both. In the United States, for example, there is abundant evidence that low-income and predominantly African-American and Latino communities are disproportionately exposed to these and other air pollutants (Boyce and Pastor 2013; Zwickl et al. 2014).

Carbon pricing on its own cannot ensure that disparities in co-pollutant exposure are reduced. Indeed, in the absence of safeguards in the policy mix, environmental disparities could be exacerbated. This concern led many environmental justice advocates in California to oppose the state's cap-and-trade program. Initial evidence suggests that these worries were not misplaced. A study of the program's first three years revealed that even as overall emissions went down, emissions actually went up in some socioeconomically disadvantaged neighborhoods (Cushing et al. 2018). To ensure emissions reductions in the places where they matter most, carbon pricing policies must be accompanied by mandates to (i) identify vulnerable communities that are disproportionately burdened by pollution from the production or combustion of fossil fuels; (ii) monitor emission trends in these communities; and (iii) take whatever regulatory or other measures are needed to guarantee that emissions reductions in these communities at least match those mandated by the policy's overall trajectory (Boyce and Pastor 2013).

Public investment in a just transition

Some revenue from carbon pricing may be dedicated to public investment as well as dividends.[21] Because spending by local, state, and federal governments accounts for a substantial fraction of total carbon emissions and hence of the carbon rent, devoting a commensurate share to public investment would not reverse the progressive net impact of dividends. It would be regrettable from an equity standpoint, however, if public investment in clean energy were financed primarily by regressive taxation.

The phrase 'just transition' is often used to refer to policies designed to safeguard the well-being of workers and communities that currently depend on employment in the fossil fuel industry during the clean energy transition. More broadly, it can be used to refer to policies that channel a fair share of public investment toward communities that have suffered disproportionate environmental harms from fossil fuels. These include urban neighborhoods overburdened with pollution as well as rural communities saddled with the toxic legacies of fossil fuel extraction.

California law today mandates that a substantial share of carbon auction revenues goes to investments in disadvantaged communities that have borne disproportionate environmental burdens.[22] Similar provisions are warranted for all public investments in ecological restoration and environmental health, whether funded from carbon revenues or otherwise.

Adaptation for all

Because it is too late to prevent climate change altogether, investments in adaptation will be an important item on the climate policy agenda in the decades ahead. Funds for this purpose are likely to be scarce relative to needs, raising the question of how adaptation resources should be allocated. Neoclassical economics prescribes that investments ought to be guided by cost-benefit analysis. But conventional measures of costs and benefits rest on willingness to pay, which in turn reflects ability to pay. The implications of this approach were spelled out with brutal clarity more than two decades ago in a World Bank memorandum signed by then-chief economist Lawrence Summers, in which he asked whether the Bank should encourage more migration of dirty industries to developing countries, and concluded that 'the economic logic of dumping a load of toxic waste in the lowest-wage country is impeccable and we should face up to that.'[23] Climate change can be understood as a new kind of toxic waste.

A radically different criterion for the allocation of adaptation resources would be to count each human being equally, rather than each dollar. The ethical basis for this alternative approach is the principle that a clean and safe environment is a human right, rather than a commodity to be distributed based on purchasing power, or a privilege to be distributed based on political power.

For example, in evaluating where and how best to safeguard coastal populations against sea-level rise and storm surges, the wealth-based approach would prioritize construction of sea walls to protect expensive real estate while flooding less valuable adjoining properties. The rights-based approach would prioritize protecting human lives, regardless of whether the people in question happen to be rich or poor. Again, the normative criteria of neoclassical efficiency and safety generate contrasting prescriptions. The implications of this difference for climate justice are evident.

Conclusions

To recap succinctly, carbon pricing can advance the transition to a clean energy economy in a manner that is both effective and equitable.

The key to effectiveness is to anchor the price to an emissions-reduction trajectory that is consistent with the goal of climate stabilization. This can be done by means of a cap that tightens along the emissions-reduction trajectory, a tax with a rate indexed to this trajectory, or a combination in which the tax serves as the floor price and permits are auctioned when the demand for permits at the price set by the tax exceeds the supply of permits set by the cap.

The key to equity is to recycle most or all of the carbon revenue to the public in the form of equal per-person dividends. This converts what would otherwise be a regressive policy into a progressive one, and may help to safeguard the political durability of the policy. At the same time, carbon pricing can and should be coupled with complementary policies to advance the goal of climate justice, including measures to protect environmental quality and public health in those communities that suffered the greatest harm from the fossil-fueled economy of the past.

Notes

1 For more on fossil fuel subsidies, see Sovacool (2017).
2 Price and coverage from (World Bank 2019, p. 28); emission trends from Statistics Sweden (2019).
3 Price and coverage from (World Bank 2019, p. 28); emission trends Environmental Reporting BC (2019).
4 For discussion, see for example, Azar 1998; Ackerman et al. 2009; Pindyck 2013, 2017; van den Bergh and Botzen 2014; Howard and Sterner 2017.
5 For discussion, see National Academies of Sciences, Engineering, and Medicine (2017), chapter 6.
6 See, for example, Shindell et al. 2016.
7 For discussion of safety versus neoclassical efficiency as a criterion for policy making, see chapter 1 in this handbook.
8 42 U.S. Code § 7409 – National primary and secondary ambient air quality standards, section (b)(1).
9 See, for example, Stiglitz and Stern 2017.
10 Nordhaus dismisses a more ambitious target as 'infeasible.' However, the Intergovernmental Panel on Climate Change reported in 2018 that exceeding 1.5 °C would entail serious risks of heightened damages. The IPCC concluded that meeting the 1.5 °C target will 'require rapid and far-reaching transitions in energy, land and infrastructure (including transport and buildings), and industrial systems' that are 'unprecedented in terms of scale, but not necessarily in terms of speed' (IPCC 2018, p. 17).
11 Nordhaus (2017b, Figure 4 and Table A-5). To put this number in perspective, the last time the earth experienced mean temperatures 3.5 °C above pre-industrial levels was about 125,000 years ago, long before the advent of cave painting (about 40,000 years ago) or agriculture (about 10,000 years ago). Global sea levels were about 6 meters higher than at present.
12 For example, with an initial price set at $40 and a maximum annual increase of 10%, the carbon price in 15 years could be no higher than $167.
13 For example, an analysis of the Clean Development Mechanism, an international offset program established under the Kyoto Protocol, found that 85% of the projects analyzed had 'a low likelihood that emissions reductions are additional and are not over-estimated' (Cames et al. 2016, p. 11).
14 For discussion and proposals for potential remedies, see Bushnell (2012) and Bento et al. (2016).
15 Countervailing policies could limit or eliminate windfall profits to the firms. For example, government regulators may prevent electric utilities from raising prices to consumers, albeit with the side effect of weakening or eliminating the price signal to end users of electricity. Alternatively, governments can tax the windfall profits.
16 In the U.S., for example, federal, state, and local governments account for roughly one-fourth of total fossil fuel use. An important issue in carbon pricing is whether, and if so, how, some of the carbon rent will be recycled to 'keep government whole' (Boyce 2019a, ch. 25).
17 The picture may differ in low-income countries where fossil fuels are more a luxury than a necessity. In such settings, the incidence of carbon pricing may be progressive. For examples, see Brenner et al. (2007) on China in the 1990s, and Datta (2010) on India in the early 2000s.
18 The measured extent of regressivity depends, among other things, on whether household income or expenditure is taken as the base for calculations (Hassett et al. 2009). It also may depend on whether inflation-indexed changes in government transfer payments are taken into account (Cronin et al. 2017).
19 For evidence on the keen awareness of fuel prices among the U.S. public, for example, see Ansolabehere et al. (2013).
20 On the role of regulatory standards in innovation, see Ashford and Hall (2011). On the role of public investment in innovation, see Mazzucato (2013).

21 A 2009 U.S. Senate bill proposed, for example, that 75% of carbon permit auction revenue be allocated to dividends and 25% to a clean energy trust fund for public investment. For discussion, see Boyce and Riddle (2011).
22 For an analysis of the California policy, see Callahan and DeShazo (2014).
23 The memorandum was leaked to *The Economist,* which published the relevant excerpt on 8 February 1992 under the headline, 'Let Them Eat Pollution.'

References

Ackerman, F. 2017. *Worst-Case Economics: Extreme Events in Climate and Finance.* London: Anthem Press.

Ackerman, F., S.J. DeCanio, R.B. Howarth and K. Sheeran. 2009. Limitations of integrated assessment models of climate change. *Climatic Change* 95, 297–315.

Ansolabehere, S., M. Meredith and E. Snowberg. 2013. Asking about numbers: Why and how. *Political Analysis* 21(1), 48–69.

Ashford, N.A. and R.P. Hall. 2011. The importance of regulation-induced innovation for sustainable development. *Sustainability* 3, 270–292.

Azar, C. 1998. Are optimal CO_2 emissions really optimal? *Environmental and Resource Economics* 11, 301–315.

Bento, A., R. Kanbur and B. Leard. 2016. On the importance of baseline setting in carbon offsets markets. *Climatic Change* 137, 625–637.

Boyce, J.K. 2018. Carbon pricing: Effectiveness and equity. *Ecological Economics* 150, 52–61.

Boyce, J.K. 2019a. *Economics for People and the Planet: Inequality in the Era of Climate Change.* London: Anthem.

Boyce, J.K. 2019b. *The Case for Carbon Dividends.* London: Polity.

Boyce, J.K. 2020. Distributional issues in climate policy: Air quality co-benefits and carbon rent. In G. Chichilnisky and A. Rezai, eds., *Handbook on the Economics of Climate Change.* Cheltenham: Edward Elgar Press, forthcoming.

Boyce, J.K. and M. Pastor. 2013. Clearing the air: Incorporating air quality and environmental justice into climate policy. *Climatic Change* 102(4), 801–814.

Boyce, J.K. and M.E. Riddle. 2011. *CLEAR Economics: State-Level Impacts of the Carbon Limits and Energy for America's Renewal Act on Family Incomes and Jobs.* Amherst, MA: Political Economy Research Institute.

Brenner, M. M.E. Riddle and J.K. Boyce. 2007. A Chinese sky trust? distributional Impacts of carbon charges and revenue recycling in China. *Energy Policy* 35(3), 1771–1784.

Bushnell, J.B. 2012. The economics of carbon offsets. In D. Fullerton and C. Wolfram, eds., *The Design and Implementation of U.S. Carbon Policy.* Chicago: University of Chicago Press, pp. 197–209.

Callahan, C. and J.R. DeShazo. 2014. *Investment Justice Through the Greenhouse Gas Reduction Fund.* Los Angeles: UCLA Luskin Center for Innovation.

Cames, M. et al. 2016. *How Additional is the Clean Development Mechanism?* Berlin: Institute for Applied Ecology.

Coady, D., I. Parry, L. Sears and B. Shang. 2017. How large are global fossil fuel subsidies? *World Development* 91, 11–27.

Cronin, J.A., D. Fullerton and S.E. Sexton. 2017. Vertical and horizontal redistributions from a carbon tax and rebate. Cambridge, MA: National Bureau of Economic Research, Working Paper 23250. March.

Cushing, L., D. Blaustein-Rejto, M. Wander, M. Pastor, J. Sadd, A. Zhu, and R. Morello-Frosch. 2018. Carbon trading, co-pollutants, and environmental equity: Evidence from California's cap-and-trade program (2011–2015). *PLoS Medicine*, https://doi.org/10.1371/journal.pmed.1002604.

Datta, A. 2010. The incidence of fuel taxation in India. *Energy Economics* 32, S26–S33.

Environmental Reporting BC. 2019. Trends in Greenhouse Gas Emissions in B.C. (1990–2017). State of Environment Reporting, Ministry of Environment and Climate Change Strategy, British Columbia, Canada. Available at www.env.gov.bc.ca/soe/indicators/sustainability/ghg-emissions.html, accessed 15 September 2020.

Fremstad, A. and M. Paul. 2019. A impact of a carbon tax on inequality. *Ecological Economics* 163, 88–97.

Hafstead, M., G.E. Metcalf and R.C. Williams III. 2017. Adding quantity certainty to a carbon tax through a Tax Adjustment Mechanism for Policy Pre-Commitment. *Harvard Environmental Law Review* 41, 41–57.

Hassett, K., A. Mathur and G. Metcalf. 2009. The incidence of a US carbon pollution tax: A lifetime and regional analysis. *Energy Journal* 30(2), 155–178.

Howard, P.H. and T. Sterner. 2017. Few and not so far between: A meta-analysis of climate damage estimates. *Environmental and Resource Economics* 68(1), 197–225.
Intergovernmental Panel on Climate Change (IPCC). 2018. *Global Warming of 1.5 °C: Summary for Policymakers*. Geneva: IPCC.
Labandeira, X., J.M. Labeaga and X. López-Otero. 2017. A meta-analysis on the price elasticity of energy demand. *Energy Policy* 102, 549–568.
Mazzucato, M. 2013. *The Entrepreneurial State: Debunking Public vs. Private Sector Myths*. London: Anthem Press.
Metcalf, G.E. 2018. *An Emissions Assurance Mechanism: Adding Environmental Certainty to a Carbon Tax*. Washington, DC: Resources for the Future, June.
National Academies of Sciences, Engineering, and Medicine. 2017. *Valuing Climate Damages: Updating Estimation of the Social Cost of Carbon Dioxide*. Washington, DC: National Academies Press.
Nordhaus, W.D. 2017a. Revisiting the social cost of carbon. *Proceedings of the National Academy of Sciences* 114(7), 1518–1523.
Nordhaus, W.D. 2017b. Projections and uncertainties about climate change in an era of minimal climate policies. Cambridge, MA: National Bureau for Economic Research, Working Paper 22933. September.
Pindyck, R.S. 2013. Climate change policy: What do the models tell us? *Journal of Economic Literature* 51(3), 860–872.
Pindyck, R.S. 2017. Coase Lecture – Taxes, targets and the social cost of carbon. *Economica* 84, 345–364.
Rubin, A. 2018. Macron inspects damage after 'yellow vest' protests as France weighs state of emergency. *New York Times*, 1 December.
Shindell, D.T., Y. Lee and G. Faluvegi. 2016. Climate and health impacts of US emissions reductions consistent with 2° C. *Nature Climate Change* 6(5), 503–507.
Sovacool, B.K. 2017. Reviewing, reforming, and rethinking global energy subsidies: Towards a political economy research agenda. *Ecological Economics* 135, 150–163.
Statistics Sweden. 2019. Greenhouse gas emissions by the Swedish economy unchanged in 2018. Available at www.scb.se/en/finding-statistics/statistics-by-subject-area/environment/environmental-accounts-and-sustainable-development/system-of-environmental-and-economic-accounts/pong/statistical-news/environmental-accounts – emissions-to-air-fourth-quarter-of-2018/, accessed 15 September 2020.
Stiglitz, J. and N. Stern. 2017. *Report of the High-Level Commission on Carbon Prices*. Washington, DC: Carbon Pricing Leadership Coalition.
U.S. Congressional Budget Office. 2001. An evaluation of cap-and-trade programs for reducing U.S. Carbon Emissions. June.
Van den Bergh, J.C.J.M. and W.J.W. Botzen. 2014. A lower bound to the social cost of CO_2 emissions. *Nature Climate Change* 4, 253–258.
Weitzman, M.L. 2011. Fat-tailed uncertainty in the economics of catastrophic climate change. *Review of Environmental Economics and Policy* 5(2), 275–292.
World Bank. 2019. *State and Trends of Carbon Pricing 2019*. Washington, DC: World Bank. June.
Zwickl, K., M. Ash and J.K. Boyce. 2014. Regional variation in environmental inequality: Industrial air toxics exposure in U.S. cities. *Ecological Economics* 107, 494–509.

18

POLITICAL ECONOMY OF BORDER CARBON ADJUSTMENT

Paul Malliet and Ruben Haalebos

Introduction

The planet's global warming threat induced by the increase of greenhouse gas emissions concentration in the atmosphere calls for urgent actions to curb them. From a theoretical economics perspective, global warming is a public bad, in the sense that it is a negative externality from human activities that is not addressed by a market. Giving a price to this externality would allow internalizing its effects in the economic agent's trade-off, leading to a reduction of the public bad up to the point that would be optimal regarding the agent's preferences. The global nature of the problem makes it even more challenging to address since there is a lack of incentive for countries to reduce their emissions, which could adopt free-riding behaviors. The withdrawal of the United States from the Paris Accord in 2016 illustrates this issue. It is hard to reach a global agreement without any system of penalty and legal authority to enforce it (Nordhaus, 2015). Moreover, the unilateral implementation of carbon pricing at a regional or national level can lead to several adverse effects that could weaken the competitiveness of domestic firms or lead to carbon leakage, the outsourcing of polluting activities in regions where the environmental regulation is less stringent than the one implementing carbon pricing. Carbon taxation is still at an early stage since only 15% of global GHG emissions are covered by a pricing mechanism[1] (Ramstein et al., 2019), either through a carbon tax or an emissions trade scheme mechanism (ETS).

The international global value chain integration process that has emerged from the early 1990s makes this challenge of carbon pricing even more difficult. It is estimated that 22% of the world's GHG emissions stem from the production of goods and services consumed in a different country (Peters et al., 2012), which tends to make the question of the sovereign responsibility in emissions more entangled. This dilemma (implementing actions to curb emissions and ensuring a fair share of each country in the reduction effort) could be partly solved through the adoption of border carbon adjustment (BCA). If there are various ways to implement it (final consumer tax, production tax, tariffs), its aim remains the same and is to impose imported carbon at a particular tax rate. The present European Commission under the presidency of Ursula von der Leyen has already announced its will to implement such a mechanism at the EU border, which would need some insight into how to compute this carbon tax rate to internalize the cost of imported carbon with efficiency and equity.

DOI: 10.4324/9780367814533-20

Carbon pricing: theory and reality

The economic theory of climate change relies on the internalization of any externality's social cost into the price system, as stated by Pigou (1933), and is the cornerstone of climate change economic policy. From this perspective, this principle is sufficient to address the climate change issue by imposing a common carbon price at the world scale, since the marginal damage of a ton of CO_2 is the same wherever it is emitted.[2] This propriety can be relaxed in some situations, though, for instance when international transfers of carbon receipts are restricted and therefore cannot compensate each country up to its social cost (Chichilnisky and Heal, 2017). Sandmo (2007) also pointed out that the carbon price must be uniform if perfect international transfers are possible but should be differentiated when they do not reflect redistributive concerns. More recently, D'Autume et al. (2016) have shown that even in a second-best-case scenario (induced by the existence of constraints on the international payment), an equilibrium with a unique carbon price can be reached. In the same vein, carbon pricing should not be differentiated across sectors (Hoel, 1996). However, in practice, the dynamic of carbon pricing in the world has not really followed a synchronized path. Nowadays, there are 76 various carbon pricing jurisdictions in the world (46 national and 28 subnational), with carbon pricing starting from $1 in Ukraine and Poland to $127 in Sweden (Ramstein et al., 2019).

The consequences of a fragmented carbon pricing regime

The unilateral implementation of carbon pricing at a regional or national level can lead to pervasive effects (Lockwood and Whalley, 2010; Horn and Sapir, 2013) that could hamper its effectiveness in emissions abatement. The first effect is the price-competitiveness impact of such a policy, which would raise the production price of the companies within the region where the carbon pricing is implemented. Concerns initially focused on energy-intensive, trade-exposed firms (EITE), typically from the cement (Demailly and Quirion, 2006) and steel and aluminum sectors (Sartor, 2013). A second effect, called *carbon leakage*, is the outsourcing of polluting activities in a region where environmental regulations are less stringent than the one implementing carbon pricing.[3]

The range of this carbon leakage effect has been estimated by several studies that can be segmented into ex-ante studies using computable general equilibrium (cge) or partial equilibrium (pe) models and ex-post studies relying on econometric estimations. Whereas the latest don't find empirical evidence of carbon leakage (Ellerman et al., 2010; Quirion, 2011; Reinaud, 2008; Sartor, 2013), the ex-ante studies provide more balanced results. On the one hand, Böhringer et al. (2012) compared results from 12 different models. Under various assumptions,[4] the authors found that the global leakage rate[5] ranged from 5% to 19%, with a mean value of 12% in the reference case[6] against 2% to 12% with a mean value of 8% in the case where a BCA is implemented. On the other hand, McKibbin (2012) doesn't find a significant effect. With a pe analysis, Monjon and Quirion (2010) find a leakage rate of 10%. Another study by Branger et al. (2014) performed a meta-analysis of various economic studies and found, under certain assumptions,[7] leakage rate ranges from 5% to 25%, with a mean of 14%, in the absence of a BCA against a leakage ratio ranging between −5% and 15% (mean 6%) with BCA.[8]

Furthermore, the study concludes by stating, "BCA does reduce the leakage rate with robust statistical significance." These results provide a theoretical justification of the use of a BCA. However, as they are obtained using various models and assumptions, one cannot expect the same results in an empirical situation.

If this question has been widely discussed, mainly in the context of the implementation of the European Union ETS, there is no clear evidence that its development has led to carbon leakage from the European EITE sectors (Naegele and Zaklan, 2019). The great amount of free-allocation allowances received by European firms largely explains this result (Schmidt and Heitzig, 2014). The continuous under-optimal price of the carbon quotas might also have played a role in the absence of significant carbon leakage from European firms.

Border carbon adjustment as a solution to fragmentation

Introducing a BCA can be an efficient way to reduce leakage issues resulting from a national carbon tax. However, from the consumers' perspectives, imposing such an adjustment is not price neutral, as it could imply an additional tax on goods. This depends on the type of products concerned and on how the carbon adjustment is passed on to the final consumer. The effect on prices will vary according to the form of the adjustment that is implemented, more specifically which goods and what emissions from these goods will be affected. This technical question is key to the final success of the BCA and can be summarized into three questions:

- (i) Which goods are being targeted?
- (ii) Which emissions are accounted for?
- (iii) Which countries are being taxed?

The targeted goods

Computing the total emissions from the production of a good (or embedded emissions) could prove quite challenging as they can evolve each year and vary greatly according to the country of production for the final good. From a theoretical point of view, this information can be derived from input-output matrices, where both the carbon content and the origin of the product are available. However, products inside these input-output databases rarely correspond to final goods, and while this data is essential for theoretical research about the BCA, it can prove quite impractical for policymakers. Moreover, energy costs for most goods only represent a small part of the total cost, so computing the embedded emissions for all imported goods would vastly exceed the benefits. A way to overcome this difficulty is to precisely select products on which to impose the BCA. They should have a high energy intensity (to ensure that their price would be modified by the border adjustment) and should be exposed to trade (to ensure that the border adjustment limits the leakage). Only these goods, also called "energy-intensive, trade-exposed" goods (Kortum and Weisbach, 2017), would be subject to the BCA. For example, Houser et al. (2008) aggregated them into five categories: cement, chemicals, steel, aluminum, and paper. Whether they all are exposed to trade can be subject to debate, especially in the case of cement. Nevertheless, this classification provides a sensible way to identify goods on which the BCA will be the most useful.

The accounted emissions

The question of how to compute emissions associated with each good remains. Emissions can be divided into two categories: (1) emissions from energy use and (2) emissions from the process. For example, for cement production, both are substantial as the chemical process during the production of cement releases carbon dioxide, and the energy needed for this process is also essential. These two factors raise various accounting issues:

Emissions from the energy use

GHG emissions from energy use depend on the type of source to produce it. While it is relatively easy to compute the amount of carbon dioxide emitted through fossil fuels' combustion, knowing precisely the quantity and the type needed for a specific good can prove challenging. A practical solution can be to use a proxy such as a country average. If a company can demonstrate that its carbon intensity of energy is lower than the country average, it can document it to lower its BCA.

Emissions from the process

The production process could be hard to determine by country precisely. Cosbey et al. (2019) listed four possibilities: (1) the average emission intensity for each product by the exporting country, (2) the average emission intensity for each product by importing country, (3) the emission intensity for best-in-class technology, and (4) the emission intensity of the worst technology of the exporting or importing country. All of these benchmarks have advantages and disadvantages. A benchmark based on information provided by the exporting country will bear more incentives for the exporting country to reduce emissions. Still, it can prove very costly in terms of information collection.

In practice, countries have to submit national emission inventories (NEI) under the United National Framework Convention of Climate Change (UNFCCC) to assess and compare emissions reductions. However, there is no legal framework under which consumption-based accounting is performed, and several methodologies and statistical sources are co-existing. We distinguish two main approaches to calculate imported emissions accounting:

- The first one determines the emissions embodied in bilateral trade (EEBT) flows using the monetary bilateral trade statistics. These methods use NEI in order to determine the carbon intensity of production by sectors for a country *r* of exported products to a country *s*. The main drawback is that it does not distinguish between intermediary and final consumptions.
- The second approach determines the emissions embodied in every economic flow, within and between nations, and has flourished thanks to the development of environmentally extended multi-regional input-output (EE MRIO). EE MRIOs are a way to determine these emissions by offering a global framework constructed on a supply-use table by distinguishing the countries, activities, and products. These databases allow for accounting the carbon footprint by acknowledging the structure of the global value chain and therefore could break down imported emissions by geographical origin, type of product, and consumer's use. Several MRIO projects have been implemented in the past decade, the most popular being EXIOBASE (Stadler et al., 2018; Tukker et al., 2009), WIOD (Timmer et al., 2015), EORA (Lenzen et al., 2013), and GTAP (Andrew and Peters, 2013). These databases differ in their sectoral decomposition, environmental satellite stressors, base year, and geographical coverage. However, some flaws prevent it from being a sound and consensual tool yet, even if we currently observe convergences and ongoing harmonization (Moran and Wood, 2014).

The estimation of imported emissions through these databases is currently the best available emissions accounting system but cannot sufficiently assess the carbon intensity of exporting foreign firms' products. There is a need for a firm-based accounting system in order to ensure compliance with a BCA. The extension of the monitoring, reporting,[9] and verification[10] regulations of the EU ETS emissions to foreign firms exporting their products could be extended relatively quickly, but the question of emissions verification in another jurisdiction is still raised. The Copernicus program is planning a monitoring capacity for CO_2 anthropogenic emissions by 2025[11] and could be used to fulfill this purpose, at least for energy-intensive industries. Nevertheless, the technical aspects of a robust, transparent, and enforcing emissions monitoring system are crucial in the design of a BCA and should be precisely scrutinized during an implementation.

The taxed countries

The final question is which countries should be impacted by the BCA: all countries or only those without a comparable carbon price? If a BCA is imposed on imports from and exports to all countries, the countries with a current carbon price should impose the same adjustment in a similar design, or various problems could arise.[12] This would require significant coordination between the countries, the lack of which is precisely why a country imposes a BCA in the first place. If a BCA is imposed solely on countries without a comparable carbon price, other issues appear. For example, the definition of a comparable carbon price can be unclear given the number of pricing schemes. However, as EITE goods mainly come from a few countries, the option to select countries could prove useful. It would also simplify the administrative issues that emerge from emissions accounting.

Any BCA project should carefully look into these issues as they are decisive in the success of such a policy. Though there is no one-size-fits-all strategy for all countries, attaining a sense of harmonization between the country's policies could greatly help make the case for BCA and simplify international discussions.

Distributive impacts of carbon taxation

In imposing a BCA, several issues in its design must be considered in order to overcome implementation barriers. A quite significant branch of the literature has questioned the compatibility of BCA with the international legal framework of the World Trade Organization, which states, "the products imported in the territory of any contracting party shall not be subject to internal taxes in excess of those applied to like domestic products." For some authors, this legal obstacle either can be addressed by imposing a consumer tax rather than duties (Trachtman, 2016) or has been opposed by article XX[13] (Bureau et al., 2017). Nevertheless, viewing it from only a technical point of view overlooks what is maybe the greatest issue: public acceptance, and more specifically, redistributive effects.

Redistributive effects induced by trade have been addressed in several papers. (Galle et al., 2017) estimate the effects of international trade on heterogeneous groups of workers, for instance through a supply shock in China on US heterogeneous groups of workers, whereas Bas and Paunov (2019) extend their analysis to distributional effects on firms and consumers of trade liberalization in Ecuador. These studies find both a negative impact on the low-skilled workers and, for the latest, a negative effect on the least competitive firms and a positive impact on consumer prices. BCA can be related to this literature since it would also imply redistributive effects through the aforementioned different channels (effects on firms, workers, and consumers). As the BCA is associated with a local carbon tax, we could expect it is facing the same scrutiny from the public (Klenert et al., 2018). Carattini et al. (2017) listed five general reasons for public hostility against a carbon tax:

- The effect on prices: people associate carbon tax with a high personal cost, and in general, people prefer subsidies to a carbon tax (Alberini et al., 2018);
- The regressive nature of carbon taxes: without appropriate redistributive schemes, carbon taxes are regressive, meaning that households with lower incomes will pay a higher share of their total income;
- The effect on the general economy: a carbon tax can be perceived as negative for the broader economy. People think that it can lower the competitiveness of local companies and lead to job losses. Though it is hard to prove the scientific validity of these statements, they should be addressed;
- The marginal effect of carbon taxes on behavior: the price elasticity of demand for carbon-intensive goods is perceived as being low;
- The carbon tax is a way to increase the government's revenue: a carbon tax can be perceived as a fiscal instrument solely devoted to improving the government's revenues, especially if it is perceived as useless (see the previous point).

In France, the plan to augment the carbon part in the general tax on car fuel highlighted some of these concerns and launched the "yellow vest" protest. People were especially afraid about the effect on prices, the regressive nature, and global inefficiency to change behaviors. Indeed, as this tax only concerned car fuel, people could not lower their demand as there are, in the short term, few alternatives available. This shows how these concerns are often intertwined and can lead to public anger against ecological measures.

Theoretically, the border carbon tax helps to mitigate the issue of carbon leakage. However, within the broader debate on the carbon tax, it may also help alleviate some of the previous concerns, mainly those about the competitiveness of local companies and the regressive nature of the carbon price. These aspects will be studied in the following paragraphs, where we will first look at the effect on prices on the border carbon tax as well as redistributive issues. Next, we will illustrate these concepts using the case of France. Finally, we will explore various redistributive schemes and the political constraints associated with them.

Specific distributive impacts of BCAs

The BCA will influence prices. However, as we have seen in the previous paragraphs, this impact will vary according to the compelling form that the BCA will take. In the following sections, we will describe these effects by adopting a macro view by focusing on the efficiency of a BCA and the possible implications on gross domestic product (GDP) and industry competitiveness. Then, we will zoom in on the specific effect on price and the redistributive issues that can appear.

Between-countries effects

The BCA seems to be an efficient way to reduce leakage issues that can arise from a carbon tax at a limited cost for global GDP. Indeed, Fouré et al. (2016) studied the effect of an adjustment targeting the import by the European Union of EITE goods from 2013. They found that such a measure would lower the real income in 2020 by 0.28% compared to a baseline, and global

GDP (in volume) would diminish by 0.07% in 2020, while emissions would reduce by 9.6%. Change in real income and GDP must be compared to the benefits of lowering global emissions, as we now know that climate change will have a negative impact on global wealth. An appealing feature of their research is the study of differences in export due to retaliation from other countries. Indeed, some countries subject to the BCA can choose to limit imports from the imposing country or region to rebalance the flows. They found that partners most vulnerable to a BCA imposed by the European Union were China, the US, and Russia. Moreover, the agricultural sector in the EU would be mainly harmed by trade retaliation, leading to a difficult choice by the Europeans between reducing CO_2 emissions and protecting its own agriculture.[14] These results are in line with those of Böhringer et al., where a reduction in emissions of roughly 10% could lead to GDP losses ranging between 0.13% and 0.63%, with a mean of 0.35%.

However, a BCA could raise the cost of imported products that were unmoved by a domestic carbon price. The magnitude of this effect depends entirely on the type of goods subject to the BCA as well as which countries would be affected. As we have seen, for practical reasons, a BCA would mainly concern EITE products, which are generally raw materials. It is unclear how this supplementary cost would reverberate to the final good price. A plausible assumption is that the additional cost resulting from the BCA is simply passed on along the value chain until the final consumer price, partially or in its totality. The share of the original price surplus in the final good would depend on the degree of sophistication of this value chain and the willingness of the different actors to transfer the additional cost as well as the potential substitution effects. This aspect is studied by Sager (2019), in which a carbon price of 30 USD/tCO_2 is implemented in the EU, with a BCA. Overall, the cost of the BCA is limited compared to the overall cost of the carbon tax. Surprisingly, the distribution of the additional fee within the income groups follows an inverted U-shape, meaning that both the people with the largest and the smallest income would face the highest cost. An explanation by the author is that both these categories consume a high share of imported goods, though presumably not at the same level.

With the question of price comes the question of redistribution. The between-country effect will depend on which countries implement a BCA and for which trade partners. Should it be implemented by developed countries,[15] it could lead to a price increase, thus, to some extent, a decrease in consumption and imports. As stated in Bueb et al. (2016), this could profit developing countries as it could shift some of the consumption to developing countries, favoring their development. Böhringer et al. (2012) find an opposite impact, where countries subject to the BCA could face a deterioration of the competitiveness of their exports, and could thus suffer a deterioration in their terms of trade. Energy-exporting countries would be particularly adversely affected because the lowering in demand for fossil fuels could lower the global price and alter their economic activity. Overall, as one of the purposes of the BCA is to reduce the loss of competitiveness resulting from a national carbon price, one can expect exporting countries to be worse off with the instauration of a BCA.

Within-country effects

Within-country effects of a BCA are mainly related to the local price increase and the redistribution schemes implemented. Theoretically, a BCA could help reduce within-country inequalities resulting from a carbon tax,[16] as all imported goods would bear a carbon price. Generally, people with higher income tend to consume more imported products,[17] and the BCA can marginally rebalance the regressive effect of the carbon tax. Moreover, the proceeds of the former are relatively small compared to those of the latter. The redistribution effects of a BCA would depend on the general policies of the imposing country. Rebates of subsidies could be established for people with the lowest income, which could largely rebalance the regressive effect of the carbon tax. These effects will be studied at more length within the next section, using a concrete example from France.

The BCA could help alleviate some concerns arising from the implementation of a carbon tax, such as the loss of competitiveness of local companies and the carbon leakage phenomenon. On the latter, we notice that the consensus is that it is an efficient tool to limit leakage effect; thus, it helps further limit GHG emissions. On the former, the outcome would depend on the categories of goods included, but it would have a generally positive effect for EITE goods. This BCA, however, is not neutral in terms of influence on the global economy or local prices, and efficient redistribution schemes could be implemented to counter some adverse effects.

Price impact and redistribution strategies: the case of France

The within-country implication of a BCA is less studied than its overall effect on the economy. However, as we have seen in the previous paragraphs, they may not be negligible and can have an influence on the general acceptability of such a policy. As part of a project for the *Agence De l'Environnement et de la Maîtrise de l'Energie* (ADEME), we studied several consequences on prices and redistribution schemes of the implementation of a BCA for French households (Malliet et al., 2020). In the following paragraphs, some results of this research will be presented to complement the elements shown in the previous section.

First, we need to differentiate four different types of emissions for a specific good consumed in France:

- Direct emissions: they refer to the direct emissions resulting from fossil fuel combustion.
- Indirect emissions: they refer to emissions associated with the production process of the final good.
- Gray emissions: they refer to the total emissions related to the value chain of the final good.
- Emissions from the public administration: they refer to the emissions associated with the services provided by the public administration.

The total carbon footprint is the sum of these four types of emissions. For example, we take the example of a car directly imported to France from Germany. Direct emissions are the emissions in France from the use of the vehicle. Indirect emissions are the emissions in Germany resulting from the construction of the car. Gray emissions are the emissions in other countries needed to produce car parts.

In developed countries, direct emissions are often lower than indirect and gray emissions. This is also the case in France, where imported emissions account for approximately 60% of total emissions. This is also represented in Figure 18.1, where gray and indirect emissions represent an important part of the total carbon footprint of households. A carbon tax on direct emissions would not concern a large number of total emissions and target emissions with low elasticity to demand. A BCA could extend the tax base, with practical limits on goods, emissions, and countries targeted presented in the last part.

To represent this, we compared the cost of a 44.6 €/tCO_2 tax on direct emissions versus a 25 €/tCO_2 tax on imported goods,[18] considering consumption modifications. The result we

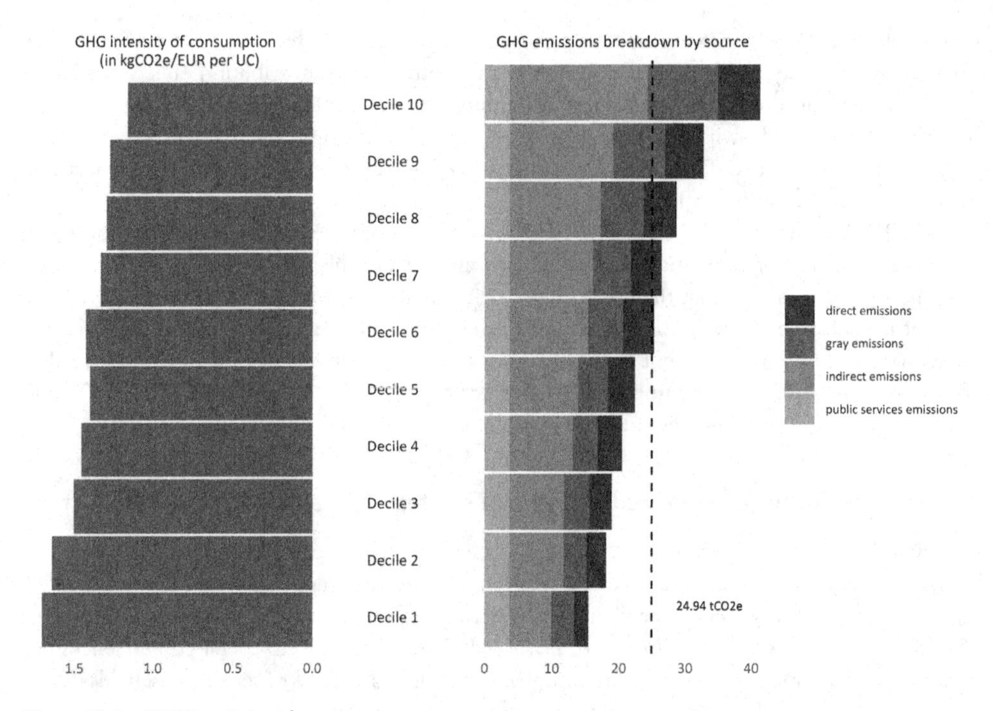

Figure 18.1 GHG emissions by type

Source: EXIOBASE 3, INSEE BDF 2011, authors' own computations.

found is presented in Figure 18.2, where we see that the overall cost is reduced with the BCA, whereas the cost is very similar when consumption modifications are not accounted for. This is due to the diverse nature of goods that are targeted in each scheme:

- A carbon tax on direct emissions applies to GHG emissions that are unresponsive to price changes in the short term, such as car fuel consumption and heating.
- A carbon tax on imported emissions, in our model, concerns final goods, more elastic to price.

Our example is schematic and only presents a hypothetical situation: a carbon tax on domestic emissions can be extended to final goods and not only to direct emissions, though it is not the case of the French "*Contribution Climate Energie*." Moreover, as we have seen, a BCA is unlikely to apply to all imported products, only to EITE goods. In the case of France, three categories of products would make the bulk of the tax revenue: steel products, products from refineries, and coking and chemical products. However, it shows the difference of effects two carbon taxation schemes can have, how they can complement each other, and how redistribution schemes should adapt to account for these differences.

To further analyze various redistribution schemes, we use the same two scenarios and compare the effects of two other redistribution schemes:

- A lump-sum redistribution: each household collects the same tax rebate.
- A degressive redistribution: each household collects a tax rebate that is inversely proportional to its total income (households with lower income get a higher tax rebate).

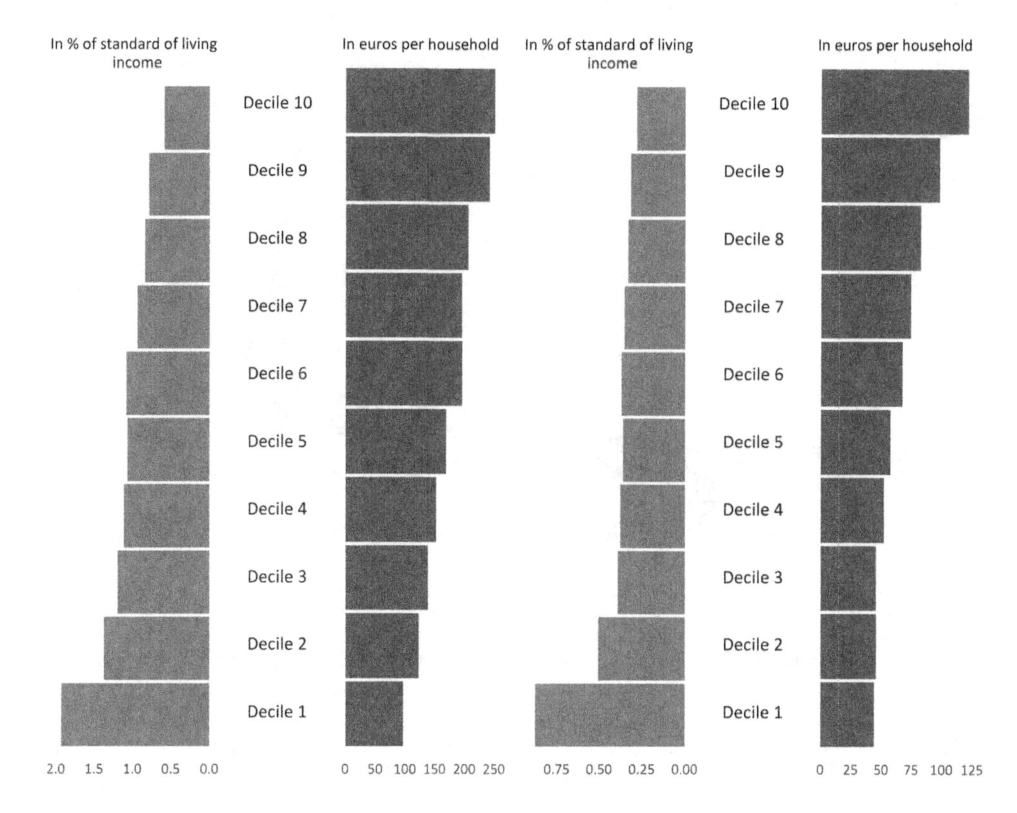

Figure 18.2 Cost per decile of income of a carbon tax on direct emissions (left) and imported emissions (right)

Source: EXIOBASE 3, INSEE BDF 2011, authors' own computations.

We compare the difference in points of percentage for *winners* between a carbon tax on direct emissions and a BCA for the two redistribution schemes. A winner is a household for which the rebate is higher than the tax. The results are presented in Figure 18.3.

These results endorse what was previously observed: the BCA is less regressive than a carbon tax on direct emissions. Indeed, for both redistribution schemes, there are more winners in the case of a carbon tax on direct emissions than for a BCA for deciles from 7 to 10, meaning that households with higher income could benefit more from those measures. Indirect emissions seem more elastic to price than direct emissions do, meaning that emissions gains in the case of a BCA could prove more substantial than do those in the case of a carbon tax on direct emissions. Hence, the taxation of indirect emissions (domestic and/or at the border) in a carbon tax scheme associated with a degressive redistribution scheme could allow going beyond the issue of social justice and acceptability of this tax.

Finally, we combine a carbon tax on direct emissions with a BCA to see the overall effect of such a scheme on French households. A hybrid redistribution scheme is implemented and is composed of (1) a general lump-sum rebate and (2) a rebate ensuring that at least 90% of the first three deciles and 66% of the following three deciles are totally reimbursed. The results of this scheme are presented in Figure 18.4, where the left part represents the mean rebates per

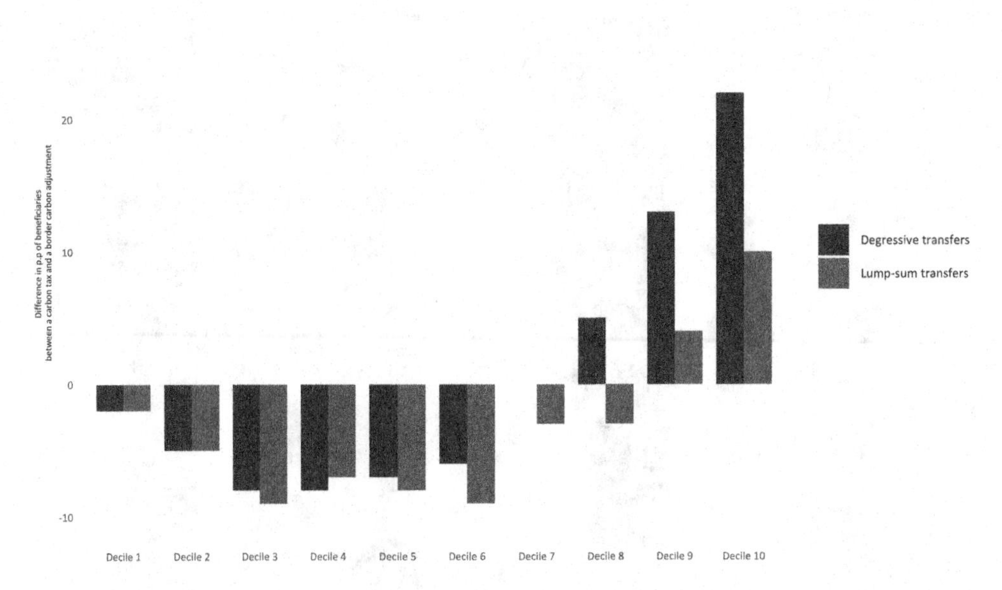

Figure 18.3 Difference in points of percentages between the share of winners between a carbon tax on direct emissions and a BCA

Note: An interpretation of this figure is the following: A BCA with a degressive redistribution leads to an increase of the share of winners in the fourth decile of 8 points compared to a carbon tax on direct emissions with a degressive redistribution.

Source: EXIOBASE 3, INSEE BDF 2011, authors' own computations.

household and per decile of income, and the right amount displays the mean total cost per household.

The overall tax revenue is approximately €7.5 billion, with two-thirds deriving from the carbon tax on direct emissions and one-third emanating from the BCA. With such a scheme, the five less favored deciles would not bear the weight of the implementation of a carbon scheme, and contribution is consequential from the seventh decile. This would favor acceptability, as less favored households would not carry the weight of such a policy.

Key takeaways from the study of the BCA in the case of France are the following:

- It would increase the tax base and the total tax revenue, allowing for the financing of targeted measures.
- It would lessen the regressive feature of a carbon tax on direct emissions.
- It would strengthen the carbon price signal by targeting more price elastic goods.

We are facing an urgency to reduce our GHG emissions, and implementing carbon taxes could help the transition of our production and consumption structures towards a low carbon economy. This, however, should not reinforce inequalities and lead part of a population to resent climate actions. With this in mind, the BCA should be considered as an option worth exploring by policymakers to overcome some of the hostility against carbon pricing.

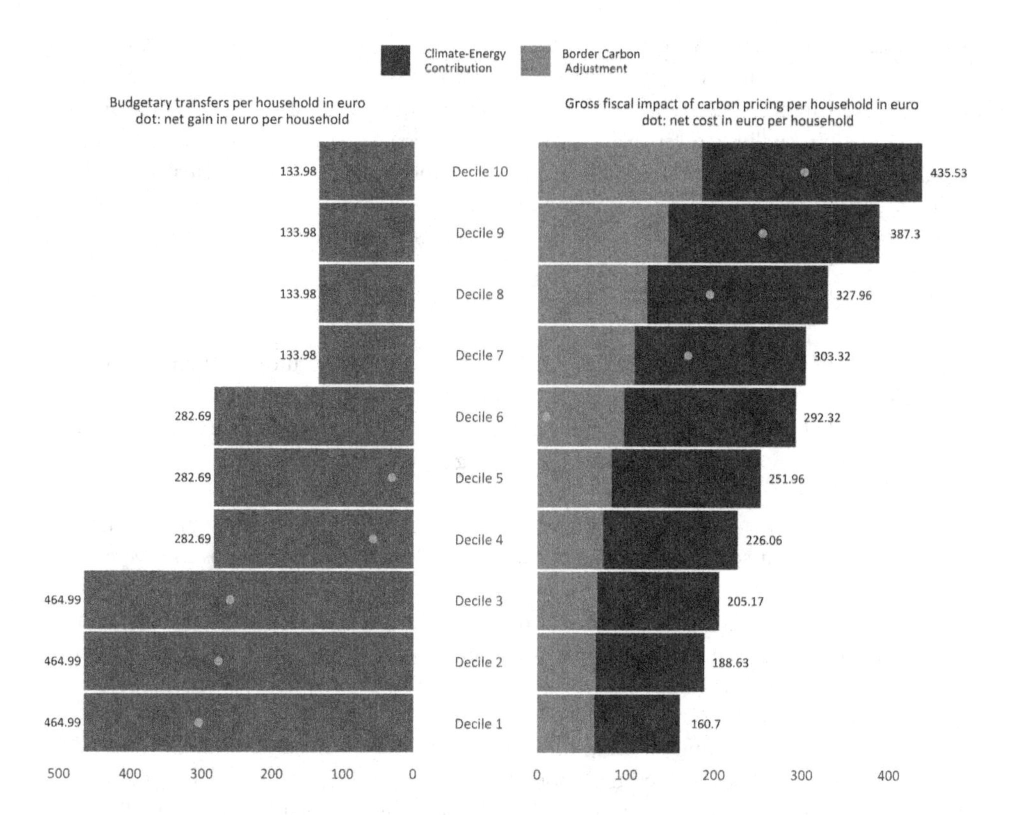

Figure 18.4 Rebates (left) and cost of the carbon price (right) per decile of income

Source: EXIOBASE 3, INSEE BDF 2011, authors' own computations.

Use of the tax and the political constraints

A carbon taxation scheme implies two different ways of reducing GHG emissions: through (1) the lowering of consumption due to the increase in price or the switch to low carbon goods and (2) using tax income to finance the transition to low carbon production methods. The former is an effect on consumption, while the latter is an effect on the producer and is essential, as it allows low carbon goods to be put on the market, and both effects are widely intertwined. The issue of redistribution of the income of a carbon taxation scheme towards the consumer is crucial as it can foster its adoption. In our previous example, the total tax income was used for redistribution. It can also be reversed in the general government revenue, as fiscal experts would recommend, to optimize the overall tax system (Bowen, 2015). However, public acceptance of the carbon tax depends on the use of its revenues by the government, and a plan aiming directly at redistribution and financing emission reduction projects could foster public approval. One can imagine a scheme where revenues from the domestic carbon tax are used for internal redistribution, whereas revenues from the BCA are used for investment.

Within this paragraph, we will explore different financing strategies through the use of marginal abatement cost curves (MAC curves). These curves are useful for this study as they allow one to measure the total GHG abatement that is attainable given an entire investment.

In practice, they measure the cost of reducing one unit of GHG emissions and thus give the marginal cost of carbon. Given that different countries have different levels of development, marginal costs can vary between regions. For example, the marginal cost of abating one unit of carbon in China will be lower than the same cost in France because fewer low carbon technologies have been implemented in China. Hence, in theory, it would be optimal to invest the totality of the tax revenue in countries with the lowest marginal carbon cost. In practice, such an option could prove unpopular as it could be perceived as an illegitimate use of the tax revenue because it is not reinvested in the local economy.

For this study, we have implemented four different scenarios:

- Scenario 1: financing of low carbon technologies within the European Union. This scenario would maximize acceptability. It will be referred to as *Reference Case scenario (ref)*.
- Scenario 2: financing of low carbon technology in China. This scenario represents the case where investments are made in a producing country. It will be referred to as *Producing country scenario (prod)*.
- Scenario 3: financing of low carbon technology in Africa. This scenario represents the case where investment serves a development purpose. It will be referred to as the *Development scenario (dev)*.
- Scenario 4: a mix of the previous scenario where one-third of total financing is in the European Union, one-third in China, and one-third in Africa. It will be referred to as the *Hybrid scenario (hybrid)*.

The total amount is computed using the yearly revenue from the BCA. For the sake of this example, we suppose that this income is constant between 2015 and 2030. This assumption is optimistic as it neglects behavioral changes ensuring the implementation of the BCA, but we think it gives a reasonable proxy of the actual amount that could be collected. The results are presented in Table 18.1.

The *Producing country scenario* is the one allowing for the most CO_2 reductions (3.65 times higher than in the *reference case scenario*). Surprisingly, reductions in the *Development scenario* are nearly equal to those in the *ref* case. However, considering that emissions are lower in Africa than in the European Union, abatement potential is lower, leading to the quick growth of the marginal carbon cost. Another interesting feature is that the abatement potential in the *Hybrid scenario* is close to that in the *prod* case. The bulk of reductions in China is realized using only a third of the total amount from the BCA, and it seems inefficient to invest the total revenue in a single country. In the end, the amounts allocated to each country should result from arbitration between within-country acceptability and maximum abatement that can be reached.

Table 18.1 CO_2 abatement for the different scenarios in respect to the reference case

Scenario	*Ratio*
Ref	1
Prod	3.65
Dev	0.99
Hybrid	3.55

Conclusion

Besides technical issues, one of the major obstacles to the implementation of carbon pricing policies are public concerns about its overall efficiency to reduce emissions, the harm to the local economy, and the effects on prices. A BCA can ease some of these concerns by limiting the leakage effect and reduce the loss of competitiveness of local companies. Moreover, it can enhance the efficiency of a carbon tax on direct emissions by widening the goods affected and seems to be less regressive. Nevertheless, as we saw in the case of France, an efficient redistribution policy needs to be implemented to limit the cost for the most vulnerable households.

A BCA could also be allocated to the financing of low carbon technologies on the producer side. It is a form of indirect rebate, as it would support new technologies and allow for a lowering of their price. It would be the most efficient in producing countries, where most of the emissions for developed countries are emitted. However, these investment decisions would be a compromise between local acceptability and maximum emission reduction attainable.

More generally, we consider that there is an implicit trade-off regarding the tax revenues recycling scheme determined by the general acceptability of these taxes, which can be decomposed into two primary sources. At first, an external set of constraints, coming from trade partners to accept without retaliation the imposition of a BCA, and second, an internal set of constraints determined by the distribution of the tax effect on domestic consumers. The BCA policy design should therefore also rely on these multiple acceptability dimensions.

Notes

1 The introduction of the Chinese ETS in 2020 will extend this coverage up to 20% of global emissions.
2 This also assumes that the marginal cost of abatement is equalized across countries.
3 IPCC provides a broader definition of carbon leakage and defines it as a "phenomenon whereby the reduction in emissions (relative to a baseline) in a jurisdiction/sector associated with the implementation of mitigation policy is offset to some degree by an increase outside the jurisdiction/sector through induced changes in consumption, production, prices, land use and/or trade across the jurisdictions/sectors" (IPCC Synthesis Report of the Fifth Asessment Report – Annex II, p. 125).
4 The assumptions are the following:

- The BCA is limited to EITE industries.
- Only direct emissions from the combustion of fossil fuel and indirect emissions associated with the production of electricity are considered.
- The BCA includes import tariffs and export rebates.
- The countries imposing a carbon tax are all Annex 1 regions (including the USA but without Russia).For a complete description of all assumptions, please refer to Böhringer et al. (2012).

5 Which is defined as the volume of emissions being shifted from domestic production activities to foreign countries concerning the decrease of emissions induced by the unilateral carbon pricing policy.
6 The reference case scenario is a scenario where only domestic emissions are subject to a carbon price (either a tax or a cap-and-trade scheme).
7 For a complete description of all assumptions, please refer to Branger et al. (2014).
8 It is noted that these rates are sensitive to several assumptions such as the elasticity of fossil fuel supply, the substitutability between foreign and domestic goods, and the access to mitigation measures for firms (Burniaux et al., 2013).
9 Commission Regulation (EU) No 601/2012 of 21 June 2012 on the monitoring and reporting of greenhouse gas emissions pursuant to Directive 2003/87/EC of the European Parliament and of the Council.
10 Commission Implementing Regulation (EU) 2018/2067 of 19 December 2018 on the verification of data and on the accreditation of verifiers pursuant to Directive 2003/87/EC of the European Parliament and of the Council.
11 With a resolution of 2 km, it would be sufficient to track major industrial installations. To be noted, an associated issue politically is whether foreign countries will accept continuous monitoring (Source: www.che-project.eu/).

12 See, for example, Kortum et al. (2017) for a description of these issues.
13 Article XX of General Agreement on Tariffs and Trade (GATT), "providing exceptions from the above restrictions for measures necessary to protect human, animal, or plant life or health (XX(b)) and for actions relating to conservation of exhaustible natural resources (XX(g))."
14 The recent example of the Dutch nitrogen crisis, where the government's plan to reduce nitrogen emissions for environmental reasons provoked massive strikes, demonstrates the difficulty of this choice.
15 This assumption seems the most plausible as of today because developed countries seem more eager to impose a carbon price. This, however, can change over the years.
16 A carbon tax is commonly assumed to be a regressive tax, meaning that those with lower incomes would pay a higher share of their total revenue.
17 This is a very general assumption but is partially confirmed by Sager (2019), in which we see that people with higher incomes could lose more with a BCA. However, we see that people with the lowest incomes would have a higher cost, so the effect is not unilateral.
18 We only consider imported goods outside of the EU. The tax rate on direct emissions is equivalent to the actual *Contribution Climate Energie* in France, while the second is equivalent to the carbon price on the EU ETS market. We do not consider rebates on exports.

Bibliography

Alberini, A., Ščasný, M., and Bigano, A. (2018) Policy- v. individual heterogeneity in the benefits of climate change mitigation: Evidence from a stated-preference survey. *Energy Policy* 121(November 2017): 565–575.

Andrew, R. M., and Peters, G. P. (2013) A multi-region input – output table based on the global trade analysis project database (GTAP-MRIO). *Economic Systems Research* 25(1): 99–121.

Bas, M., and Paunov, C. (2019) *What gains and distributional implications result from trade liberalization ?* (No. 2019.03).

Böhringer, C., Rutherford, T. F., Balistreri, E. J., and Weyant, J. (2012) Introduction to the EMF 29 special issue on the role of border carbon adjustment in unilateral climate policy. *Energy Economics* 34: S95–S96.

Bowen, A. (2015) Carbon pricing : How best to use the revenue? *Policy Brief November 2015* (November).

Branger, F., and Quirion, P. (2014) Would border carbon adjustments prevent carbon leakage and heavy industry competitiveness losses ? Insights from a meta-analysis of recent economic studies. *Ecological Economics* 99: 29–39.

Bueb, J., Hanania, L. R., and Clézio, A. Le (2016) *Border adjustment mechanisms Elements for economic, legal, and political analysis* (No. 2016/20).

Bureau, D., Fontagné, L., and Schubert, K. (2017) Commerce et climat: pour une réconciliation. *Les notes du conseil d'analyse économique* (No. n°37), Paris.

Burniaux, J.-M., Chateau, J., and Duval, R. (2013) Is there a case for carbon-based border tax adjustment? An applied general equilibrium analysis. *Applied Economics* 45(16): 2231–2240.

Carattini, S., Carvalho, M., and Fankhauser, S. (2017) *How to make carbon taxes more acceptable.* London: Grantham Research Institute on Climate Change and the Environment and Centre for Climate Change Economics and Policy, London School of Economics and Political Science.

Chichilnisky, G., and Heal, G. (2017) *Energy-Capital Substitution: A General Equilibrium Analysis MPRA Paper 8329.* Germany: University Library of Munich.

Cosbey, A., Droege, S., Fischer, C., and Munnings, C. (2019) Developing guidance for implementing border carbon adjustments: Lessons, cautions, and research needs from the literature. *Review of Environmental Economics and Policy* 13(1): 3–22.

D'Autume, A., Schubert, K., and Withagen, C. (2016) Should the carbon price be the same in all countries? *Journal of Public Economic Theory* 18(269788): 709–725.

Demailly, D., and Quirion, P. (2006) CO 2 abatement, competitiveness and leakage in the European cement industry under the EU ETS: Grandfathering versus output-based allocation. *Climate Policy* 6(November): 93–113.

Ellerman, D., Convery, F., and de Perthuis, C. (2010) *Pricing carbon: The European Union emissions trading scheme.* Cambridge, UK: Cambridge University Press Edition.

Fouré, J., Guimbard, H., and Monjon, S. (2016) Border carbon adjustment and trade retaliation: What would be the cost for the European Union? *Energy Economics* 54: 349–362.

Galle, S., Rodríguez-Clare, A., and Yi, M. (2017) *Slicing the Pie: Quantifying the Aggregate and Distributional Effects of Trade* (No. w23737). Cambridge, MA: National Bureau of Economic Research.

Hoel, M. (1996) Should a carbon tax be differentiated across sectors? *Journal of Public Economics* 59(1): 17–32.

Horn, H., and Sapir, A. (2013) *Can Border Carbon Taxes Fit Into The Global Trade Regime ?* Bruegel Policy Brief 2013/06, 9. Bruxelles.

Houser, T., Bradley, R., Childs, B., Werksman, J., and Heilmayr, R. (2008) *Leveling the carbon playing field: International competition and US climate policy design*. Washington, DC: Peterson Institute for International Economics and World Resources Institute.

Klenert, D., Mattauch, L., Combet, E., Edenhofer, O., Hepburn, C. Rafaty, R., and Stern, N. (2018) Making carbon pricing work for citizens. *Nature Climate Change* 8(8): 669–677.

Kortum, S., and Weisbach, D. (2017) The design of border adjustments for carbon prices. *National Tax Journal* 70(2): 421–446.

Lenzen, M., Moran, D., Kanemoto, K., and Geschke, A. (2013) Building Eora: A global multi-region input–output database at high country and sector resolution. *Economic Systems Research* 25(1): 20–49.

Lockwood, B., and Whalley, J. (2010) Carbon-motivated border tax adjustments: Old wine in green bottles? *World Economy* 33(6): 810–819.

Malliet, P., Haalebeos, R., and Nicolas, E. (2020) La fiscalité carbone aux frontières et ses effets redistributifs. *Rapport Expertise ADEME.*

McKibbin, W. J. (2012) A new climate strategy beyond 2012: Lessons from monetary history. *The Singapore Economic Review* 57(03): 1250016.

Monjon, S., and Quirion, P. (2010) How to design a border adjustment for the European Union Emissions Trading System? *Energy Policy* 38(9): 5199–5207.

Moran, D., and Wood, R. (2014) Convergence between the EORA, WIOD, EXIOBASE, and OPENEU's Consumption-based carbon accounts. *Economic Systems Research* 26(3): 245–261.

Naegele, H., and Zaklan, A. (2019) Does the EU ETS cause carbon leakage in European manufacturing? *Journal of Environmental Economics and Management* 93: 125–147.

Nordhaus, W. D. (2015) Climate clubs : Overcoming free-riding in. *The American Economic Review* 105(4): 1339–1370.

Peters, G. P., Davis, S. J., and Andrew, R. (2012) A synthesis of carbon in international trade. *Biogeosciences* 9(8): 3247–3276.

Pigou, A. C. (1933) *The economics of welfare*. Palgrave Macmillan.

Quirion, P. (2011) Les quotas échangeables d'émissions de gaz à effet de serre: éléments d'analyse économique. mémoire d'habilitation à diriger des recherches. *Technical report,* Paris: EHESS.

Ramstein, C., Dominioni, G., Ettehad, S., Lam, L., Quant, M., Zhang, J., Mark, L., Nierop, S., Berg, T., Leuschner, P., Merusi, C., Klein, N., and Trim, I. (2019) *State and Trends of Carbon Pricing 2019.* Washington, DC.

Reinaud, J. (2008) *Issues Behind Competitiveness and Carbon Leakage Focus on Heavy Industrys, IEA Information Paper.* Paris: International Energy Agency, OECD/IEA.

Sager, L. (2019) *The Global Consumer Incidence of Carbon Pricing: Evidence from Trade.* London: Grantham Research Institute on Climate Change and the Environment.

Sandmo, A. (2007) Global public economics: Public goods and externatilities. *Économie Publique/Public Economics* (18–19).

Sartor, O. (2013) *Carbon Leakage in the Primary Aluminium Sector: What Evidence after 6 ½ years of the EU ETS.* USAEE Working Paper. Paris.

Schmidt, R. C., and Heitzig, J. (2014) Carbon leakage: Grandfathering as an incentive device to avert firm relocation. *Journal of Environmental Economics and Management* 67(2): 209–223.

Stadler, K., Wood, R., Bulavskaya, T., Södersten, C. J., Simas, M., Schmidt, S., et al. (2018) EXIOBASE 3: Developing a time series of detailed environmentally extended multi-regional input-output tables. *Journal of Industrial Ecology* 22(3): 502–515.

Timmer, M. P., Dietzenbacher, E., Los, B., Stehrer, R., and de Vries, G. J. (2015) An illustrated user guide to the world input-output database: The case of global automotive production. *Review of International Economics* 23(3): 575–605.

Trachtman, J. P. (2016) *WTO laws constraints on Border Tax Adjustement and Tax Credit Mechanisms to Reduce the Competitive Effects of Carbon Taxes* (No. RFF DP 16–03). *RFF Discussion Papers*. Washington, DC.
Tukker, A., Poliakov, E., Heijungs, R., Hawkins, T., Neuwahl, F., Rueda-Cantuche, J. M., Giljum, S., Moll, S., Oosterhaven, J., and Bouwmeester, M. (2009) Towards a global multi-regional environmentally extended input-output database. *Ecological Economics* 68(7): 1928–1937.

19

POLITICAL ECONOMY OF FOREST PROTECTION

Alain Karsenty

Introduction

Conserving forests by halting deforestation and reducing their degradation has become one of the major global issues of the 21st century, given the importance of forest ecosystems for climate change, biodiversity conservation and human well-being. While initially seen as a technical issue devoted to forestry experts, the discourse was extended to other stakeholders as it became increasingly evident that collective choices and governance issues have to be addressed as a priority. Far from being mere technical exercises, simply defining what a forest is and measuring deforestation turned out to be full of political stakes and geopolitical implications. After the "mega-fires" that hit the Amazon and other forests throughout the world in 2019, the call to grant large tropical forests a "global common goods" status was formulated by prominent Western figures, sparking fierce nationalist reactions from Southern politicians who recalled the ecological debt contracted by industrialized countries.

Several major international initiatives have emerged over the past 20 years to try to curb deforestation and forest degradation. Independent certification came first from the world of the NGOs. Certification has become an institution, in the sociological sense of an established social form, and is no longer exclusively an instrument of private governance, as some governments have decided to use it in their public policies. Another initiative came from industrial countries, with public regulations criminalizing imports of illegally cut timber and proposals for bilateral trade partnerships with producing countries in order to "certify the country" (Karsenty 2019). In the meantime, a number of agribusiness firms, followed by governments, have focused on voluntary approaches to remove deforestation from commodity supply chains. Some Western governments have designed national strategies against "imported deforestation" embedded in agricultural product imports.

The best-known initiative, Reducing Emissions from Deforestation and forest Degradation (REDD+), came from an alliance of researchers, economists and NGOs based on a proposal to remunerate countries for conserving their forests. This proposal, formulated in a context of climate change negotiations and the development of carbon markets, reflects the influence of the economic framing of the deforestation issue (conservation of forests is an issue of opportunity cost), the attraction for results-based payments as an effective management tool to deliver international transfers, and the trust in economic incentives for transforming public governance.

DOI: 10.4324/9780367814533-21

So far, none of these initiatives has succeeded in curbing deforestation and the conversion of natural ecosystems to artificialized areas, either for food, urban settlements or energy crops. Diagnoses of the "forest crisis" are generally correct, but they often overlook major political economy issues, such as the fact that governments are not benevolent institutions acting for the common welfare of their people and that urban elites have little interest in the fate of forest-dependent people, who are not that numerous and often voiceless. More importantly, endeavours to tackle the forest crisis without questioning the unabated global demand for biomass, energy and agricultural land, and the rules of international trade, seem illusory. The same can be said about population growth: this major underlying cause of ecosystem conversion has long been occulted from international organizations' statements and was, until recently, not popular in academic circles, as pointing out the issue was assimilated to an attempt to curtail "the right to develop" of Southern countries.

This chapter will attempt to take stock of some of the most prominent international initiatives taken and instruments adopted over the last two decades for reducing deforestation and conserving forests, and their ambition and limitations will be examined from a political economy perspective.

The first part of the chapter analyzes the stakes involved in qualifying forest, deforestation and its causes. In this section, an attempt will also be made to clarify the discussion about the hypothetical "global public goods" nature of the world's forests and provide some considerations about the potential of "local commons", often institutionalized as community forests.

The second part of the chapter analyzes forest-related market-based instruments, designed to provide incentives to stakeholders for changing their practices or policies. Certifications and corporate voluntary commitments for sustainable forest management or deforestation-free agricultural production are coexisting with attempts by Western governments to ban illegal timber and "imported deforestation" from international trade. Lastly, the most ambitious initiative, REDD+, which aims to create an international regime of results-based payments, illustrates how a well-intentioned but short-sighted mechanism is likely to generate perverse incentives and favours strategic behaviours of rent-seeking governments.

In the third and last part of the chapter, an attempt is made to propose a rethink of results-based payments, since the principle has become the cornerstone of the international discourse on environmental protection. Effective results-based payments must give priority to investment and a joint agenda merging food security and forest protection. Finally, modifying consumption patterns and changing the rules of international trade are necessary crossing points for resolving the forest crisis.

Part I: Forest, deforestation and its causes

Political economy of defining forests and deforestation

The critical ecological role of forests

"Trees: they can save us." This was the front page headline of a popular French science magazine in 2019.[1] This statement testifies to the place held by forests in the imagination of the climate crisis. While deforestation, which accounts for the vast majority of land use changes, accounts for 14% of annual anthropogenic CO_2 emissions, terrestrial ecosystems, including forests, account for 29% of total human-made emissions (Global Carbon Project 2019). The reservoir they represent is immense, since forests store more than half of the carbon in the world's land (1,120 GtC). Thus, as the climate issue occupies a growing place on the international relations agenda, the fate of forests has become a political issue in its own right. The forest fires in the summer of 2019 in the Amazon and the controversies with the Bolsonaro government, accused of "letting the planet's lung be destroyed", demonstrate this.

Beyond their importance in the carbon cycle, forests are estimated to support 80% of the world's terrestrial biodiversity, and two-thirds of it is found in natural tropical forests. As for the global water cycle, the role of forests has been refined. Not only do large forests, such as the Amazon, make their own rain through tree transpiration (Staal et al. 2018), but the long-distance transportation of moisture through "rivers in the sky" gives rise to rainfall thousands of kilometres away (Ellison et al. 2017).

Role in human well-being

Forests also contribute to human well-being in several ways. Timber and wood-energy activities provide millions of jobs. According to the Food and Agriculture Organization of the United Nations (FAO 2014), in employment terms the formal sector provided 13.2 million jobs, while an estimated 41 million livelihoods were dependent on the informal sector. The FAO report also outlined the importance of forest contributions to food and nutrition security. About 11 kg of edible non-wood forest products were consumed per capita globally.

It is also estimated that a third of the world's population depends on woodfuel for cooking, while 764 million people use woodfuel to boil drinking water. Herders in arid and semi-arid lands depend on trees as a source of fodder for their livestock. In this respect, forests provide a "safety net" for poor rural people.

Forests also contribute to clean water by protecting watersheds: three-quarters of the globe's accessible fresh water comes from forested watersheds. Forests provide habitats for an estimated 80% of the world's biodiversity.

The disruption of forested ecosystems causes a multiplication of interactions with humans. It thus creates new gateways for microorganisms. The destruction of primary forests not only reduces these benefits for people, but also creates contacts with animals that are potentially vectors of zoonoses. Deforestation and the artificialization of soils, by depriving many wild animals of food, are driving many species to move closer to inhabited centres. And the populations living near relatively wild areas hunt and – more or less legally – bring bushmeat into the cities, where it is coveted by city dwellers. In both cases, the viruses carried by these wild animals are close to a particularly fertile environment: the massive crop farms and livestock farms on the outskirts of many towns and cities.

Defining a forest

The FAO defines forests as land with trees capable of reaching a height of at least 5 metres at maturity, and whose cover occupies at least 10% of an area of more than half a hectare. Here, "forest" includes natural afforestation and forest plantations but excludes rows of trees established for agricultural production (such as fruit trees) and trees planted in agroforestry systems.

From there, FAO defines deforestation as the conversion of forest to a different land use or the long-term reduction of tree cover below the minimum 10% threshold (FAO 2012). Forests that are entirely cut down but intended for natural or artificial reforestation are not counted as deforestation. Neither are forests destroyed by fires when the land is intended to become forested again. In most cases, deforestation is the transition from an already degraded forest to an area for agriculture, livestock or infrastructure activity.

The FAO has a quasi-monopoly on the production of data on deforestation at the global level, and the conventions it adopts are mandatory gateways, although other international organizations have adopted less extensive definitions of forest. The FAO has amended its conventions over time. While the 10% threshold has not changed, there have been significant changes about what should compose a forest. Thus, rubber trees were considered as part of agricultural production before the 2000 Forest Resource Assessment and not as planted forests. This was reversed from 2000 onwards (Penna 2010), to the great satisfaction of several nations, which were thus able to see their "official" rate of deforestation reduced. Since then, several countries, including Indonesia, have tried – without success so far – to have oil palm fields accepted as forests, which would again reduce deforestation in the statistics. However, the most significant change has been to choose "net" deforestation as the preferred indicator for reporting on the evolution of individual countries (Hoare 2005). Net deforestation represents "gross" deforestation, minus reforestation and natural regeneration. This puts natural ecosystems and artificial afforestation on an equal footing. It reflects the lesser importance given to the biological diversity of ecosystems compared to wood production or carbon storage functions. It also expresses the political clout of major emerging countries, such as China, India and Brazil, which have seen a significant increase in the industrial planting of fast-growing trees for pulp production, while their natural forests continue to decline.

Deforestation has generally worsened since the beginning of the century. Data provided by Global Forest Watch (www.globalforestwatch.org/), which does not indicate deforestation in the sense of the FAO (land use change) but lists annual losses of tree cover (that can be used as proxies for deforestation), suggest average annual losses of around 15 million hectares in the first decade of the century, increasing to about 20 million hectares from 2010 onwards. Most of the deforestation takes place in developing countries, and the fate of the Amazonian forest is anchored in the mind of billions of citizens, not only in Western countries. Nonetheless, deforestation and the conversion of natural ecosystems are also alarming in developed countries. Australia and Canada have recorded significant net deforestation between 2000 and 2015, as has Russia, even though this is gross deforestation (destruction of ancient forests) "compensated" for by natural expansion of forests with climate warming. Recent large fires occurring in these countries have aggravated these trends.

The multiple and embedded causes of deforestation

Deforestation is rarely due to a single cause. Gross deforestation has direct or indirect drivers, but above all underlying causes, which can be identified in public policies, governance and the cultural representations of societies. It seems that the most powerful cause is the attractiveness of the growth pathway based on the large-scale conversion of renewable natural resources into agro-industrial assets. This cause can be analyzed from a political economy perspective, highlighting the role of the enrichment of national elites through the development of "crony capitalism" relationships.

Underlying causes

The first underlying causes: demographic dynamics

The increase in rural density linked to high population growth, particularly in Africa, is strongly correlated with deforestation rates. In DR Congo, Defourny et al. (2011) highlighted this relationship by identifying "rural complexes" composed of more or less grouped habitats, fields and fallows. This increase in population density is related to the lack of capital of poor farmers (material, inputs, mastery of techniques allowing intensification, etc.), who tend to shorten fallow times and no longer allow the forest enough time to regenerate. The resulting progressive impoverishment of the soil leads to different types of responses: rural exodus, deforestation of new land, or evolution "à la Boserup".

In the scheme developed by E. Boserup (1965), rural societies respond to high human density through land individualization (hedged farmlands), livestock integration and intensification (through agroforestry, for example). However, this phenomenon is not always enough to stop the dynamics of deforestation. Sometimes, the individualization of land is socially impossible and the process of intensification cannot be initiated (cf. Marchal 1985, for Burkina Faso). But, above all, access to certain cash crops (cocoa in West Africa, oil palm in South-East Asia, etc.), combined with migration movements, leads to continued high deforestation, even though the forest cover has shrunk. Thus, in the case of Côte d'Ivoire, Ruf and Varlet (2017) wrote, "deforestation seems likely to continue until the last hectare is consumed. Zero deforestation cocoa does exist, but only when and where all the forest has already disappeared".

Another underlying cause: the continuous increase in food and non-food demand

The growing demand for land for urbanization, energy (dams, hydrocarbon deposits, etc.), minerals and, above all, agricultural products, be they food or non-food (rubber, paper, cosmetics and, increasingly, biofuels), is the second main underlying cause of forest conversion. International trade is playing an increasing role. In Brazil, about 30% of deforestation is linked to agricultural exports (Karstensen et al. 2013).

Intensification, previously based on chemical inputs but nowadays more on agro-ecological techniques (conservation agriculture, plot rotations, agroforestry, etc.), is frequently proposed as the ultimate solution to the problem of deforestation. Angelsen and Kaimowitz (2001: 3) called this vision "the Borlaug hypothesis", named after a famous agronomist considered as the father of the green revolution. They formulated it as follows: "With food demand expected to grow steadily over the next decades, one could argue that using new technologies to make agriculture more intensive is the only way to avoid rising pressure on tropical forests."

Although formally correct, this argument fails to distinguish between necessary and sufficient conditions (Phalan et al. 2016; Pirard and Belna 2012). Nevertheless, it is now used by most agronomists and agribusiness companies to promote, for example, the planting of oil palms, whose productivity is much higher than other oilseeds (soya, in particular), which would save space to satisfy a demand that it is assumed will increase. However, high productivity often also means higher profitability, which increases what can be called the "profitability perimeter" of deforestation. Just as the increase in the price of beef for Brazilian producers is correlated with the increase in deforestation in the Amazon (Chomitz et al. 2007), the increase in the profit margins of a handful of tropical crops (soya, cocoa, palm oil, rubber, sugar cane, etc.) leads, all other things being equal, to an increase in deforestation.

This new deforestation can be direct or indirect, the latter being generally misunderstood. To understand the concept of indirect land-use change (ILUC), it is necessary to consider the possibility that a sharp increase in demand (through, among other things, large-scale incorporation of vegetable oils in biofuels) will lead to an increase in the price of that vegetable oil, which will mean increased profitability for a number of oil palm producers. To take advantage of this, rubber, coffee, cocoa and livestock producers will convert all or part of their farms to oil palm. As a result, it will be necessary to produce rubber, cocoa, beef, and so on elsewhere (for example in forest areas) to meet unchanged demand (EEA 2011; Gawel and Ludwig 2011).

While intensification and the increase in agricultural yields is a necessary crossing point in developing countries, it should not be expected to provide a straightforward solution to the problem of deforestation. It is first and foremost to curbing or controlling demand for agricultural and non-agricultural products (in this respect the development of first-generation biofuels is an obvious problem) that political priority should be given.

An underlying socio-cultural cause: the attraction of the "Asian model" of accumulation

Most developing countries are ruled by urban elites fascinated by Chinese-like economic growth successes, who disregard natural environment potentials, apart from a handful of countries, such as Costa Rica, which has favoured nature-based tourism. In several South-East Asian countries, an accumulation model can be found that is based on the unregulated exploitation of natural forests, which has led to the development of an industrial base around wood processing and a powerful agro-industry based mainly on palm oil and pulp. The overexploitation of natural forests in Indonesia and Malaysia, combined with the ban on log exports that has fostered the development of a powerful timber industry based on a low price for wood resources, was the first step in this accumulation (Barr 2002). In this context of overexploitation, cutting cycles of 35 years are too short to allow commercial volume recovery (Sist et al. 1998). Such an immobilization of natural capital is considered as far too long by companies. Therefore, most of these degraded forests, instead of being left for natural or assisted regeneration, have been converted into plantations. As Indonesian or Malaysian forestry companies often belong to large conglomerates that also produce palm oil, pulp or rubber, the same economic interests often follow one another in these forest areas converted to agriculture (Casson 2002). This "accumulation pattern" largely inspires many governments aiming to attain economic "emergence", particularly on the African continent.

For the national elites who took over political power after the colonial eras, the control of the state has been an opportunity for personal enrichment and the development of a "crony capitalism" (Haber 2002) based on the looting of natural resources. The political economy of deforestation, involving top politicians monopolizing powers and business clients, was analyzed by C. Barr (1998) for Indonesia during the Suharto era. However, still in Indonesia, the decentralization of powers through subdivision of existing jurisdictions has aggravated illegal logging and subsequent deforestation, as more policy actors obtain rents from allowing illegal logging (Burgess et al. 2012). This political economy is embedded in an array of demographic, social and economic causes that combine in various ways in the different countries.

Indirect causes

Local agrarian systems in crisis

The role of poverty in deforestation is a subject of debate. Leaving aside the growing role of agribusiness, which is far from being present in all tropical areas, Geist and Lambin (2002) emphasized interactions between factors, such as population displacement, loss of access to part of their land and environmental degradation. Angelsen and Kaimowitz (1999) insisted on investment as a condition for converting forests to other uses, suggesting that it is not the poorest who deforest but those who achieve a certain level of accumulation. This idea was taken up by Moonen et al. (2016) for DR Congo, where it is the rural populations marketing their agricultural products, and also the most educated people, who are most active in land conversion processes (see also Pacheco 2009 for the Amazon).

Cultural factors should not be neglected. Forests as a "development frontier" that must be pushed back is a representation that can be found in both Latin America – particularly Brazil – and South-East Asia. On the island of Borneo, in the Indonesian sector (Kalimantan), people see the "exit from the forest" as an exit from poverty, including the Penan "indigenous" people (Feintrenie et al. 2010). The same can be said in Africa, where Bantu farmers in Cameroon commonly talk about "breaking the forest" to expand their fields.

Ambiguous land rights

Many developing countries endorsed the colonial conception of forests as spaces "without masters". It was a way of denying tenure rights to local people and communities in order to allocate timber or agricultural concessions to provide rents for their political clients.

The phenomenon of land grabbing is particularly prevalent in wooded areas (Gibbs et al. 2010; Messerli et al. 2014). Governments use a "presumption of State ownership" over forests to allocate large areas to companies, hoping to avoid land conflicts that could result from allocations in more narrowly appropriated agricultural areas (Karsenty 2018). This is one of the driving forces behind the phenomenon of "land grabbing" (Karsenty and Assembe 2011).

More generally, uncertainty about land rights can lead actors to "develop" (i.e. clear) land in an attempt to assert an individual right of ownership: many laws allow this, and it is also reflected in customary systems where ownership is achieved through an "axe right" or a "fire right".

Direct drivers

The role of agricultural dynamics

Agriculture accounts for about 80% of the direct causes of deforestation (Hosonuma et al. 2012). A distinction must be made between agribusiness (e.g. large soybean farms in the Amazon), family cash crops (e.g. small cocoa plantations in Côte d'Ivoire) and food-oriented agriculture (cassava or rainfed rice plantations). Commercial agriculture as a whole accounts for 68% of deforestation in Latin America, but only about 35% in Africa and Asia (Hosonuma et al. 2012). Overall, a third is attributable to food-oriented agriculture, with this percentage rising to 40%–50% and above in sub-Saharan Africa.

Outside Africa, the trend is towards an increase in the share of commercial agriculture, especially agribusiness (Boucher et al. 2011; Rudel et al. 2009). The development of oil palm at the expense of tropical forests is the most significant dynamic. This highly profitable crop is favoured by both agribusiness and smallholders. The installation of these plantations on "degraded" forests (ambiguous term, which most often refers to regenerating forests) often finds its rationale with the profits made by companies from the sale of wood resulting from conversion.

The "tandem" between logging and agriculture

Conversions of forests to pasture, oil palm or soybean fields are direct, immediate drivers of deforestation. However, quite often this radical change has been preceded by phases of degradation, which have themselves been encouraged by the opening up of roads, an indirect cause of future deforestation.

Different combinations of factors have been identified (Geist and Lambin 2002). The "selective logging–agriculture" pair is the most well known. In tropical countries, where most of the deforestation is concentrated, timber exploitation (logging) does not generally lead directly to deforestation, as only a few trees are extracted per hectare during selective harvesting. Nevertheless, the opening up of roads, the establishment of wood processing industries and the economic opening up caused by new activities may attract populations in search of agricultural land. This is often accompanied by the development of unregulated charcoal and hunting activities, which can gradually lead to deforestation. However, this is not inevitable and demographic characteristics matter: in Gabon, forest concessions cover three-quarters of the country, but deforestation is very low, due to the very limited presence of farmers outside peri-urban areas in a mostly urban country of around two million inhabitants.

Are forests commons?

Tropical forests are frequently presented by policymakers as a common, common good or common heritage, global or local, or sometimes as "global public goods" (see Humphreys 2014; Smouts 2003).

This one-dimensional vision inevitably leads to geopolitical misunderstandings. A twofold dimension of the object "tropical forest" should be considered for attempting to establish a shared international regime for its conservation. This double dimension refers, on the one hand, to forests as resources and, on the other hand, to forests as a support and condition for ecosystem services.

The "global public good" issue

The classic definition of a public good is that given by Samuelson (1954), which explicitly identifies the property of non-rivalry (consumption of the good by one agent does not reduce consumption by other agents) and implicitly the property of non-excludability (it is impossible to exclude an agent from consuming the good). This definition is theoretical and can only rarely be applied to the letter, hence the frequent use of a broader class of types of goods (see Table 19.1).

Awareness of global issues in the 1980s encouraged the extension of the concept of public goods on a global scale. The term global public good (GPG) was coined by the report entitled "Global Public Goods: International Cooperation in the 21st Century" (Kaul et al. 1999), and it is worth noting that the definition of GPG differs significantly from that of a simple public good on a broader scale. It is moving from a good characterized by its *consumption* (e.g. non-rivalrous and non-exclusive use, e.g. use of road infrastructure), to a good whose *effects* matter (e.g. less air pollution improves health). For example, Stiglitz (1999) refers to goods whose "benefits are for the entire world population". While GPGs are, therefore, generally intangible (climate,

Table 19.1 Different types of goods: rivalry and excludability

	Excludable	*Non-excludable*
Rivalry	Private goods	Common pool resources
Non-rivalry	Club goods	Public or collective goods

economic stability, knowledge, etc.), they depend on a material support. Thus, the production of these global services justifies the conservation of the support: the stabilization of the climate and the maintenance of services linked to biodiversity justify the protection of forests.

How do forests fit into these categories? A 2013 World Bank document starts with this statement, "The Amazon Rainforest is a global public good" (Navrud and Strand 2013). During the "mega-fires" that destroyed millions of hectares of the Brazilian Amazon in 2019, calls were made, notably by President Macron in France, to protect this "common good", triggering a stinging reply from J. Bolsonaro about neocolonialist temptations.

Resources versus services

The hesitation in qualifying tropical forests as a whole (excluding private forests) is indicative of a one-dimensional view of the object itself. In reality, this question can only be addressed by considering the two-dimensional nature of the object "tropical forest", which is at stake in international agreements, in an attempt to set up a shared regime for its conservation. This double dimension refers, on the one hand, to forests as resources and, on the other hand, to forests as a support and condition for services.

Tropical forests provide ecosystem *services* to the entire planet but depend on sovereign states and local actors with rights, who use them primarily as economic *resources* (timber, land, etc.). In this sense, they do not fall within the scope of "global public goods", insofar as they do not meet the classic characteristics of public goods in terms of the impossibility of exclusion of third parties and non-rivalry in consumption. For example, successive Brazilian governments, always highly suspicious of anything that might constitute, in their eyes, an attempt to "internationalize" the Amazon, do not intend to let the Brazilian natural forest be qualified as a global public good.

When dealing with *services*, we refer here to the notion of ecosystem services, popularized by the Millennium Ecosystem Assessment (MEA 2005), a collective expertise exercise conducted in the 2000s, whose report serves as a reference to guide international discussions on these subjects.

These services are defined as "the benefits that humans derive from ecosystems". They include "provisioning services" (wood, agricultural products, fibre, but also genetic resources), which refer to potential marketable goods. Regulatory services refer to what economists consider to be positive externalities (carbon fixation through photosynthesis, the capacity of a medium to filter water and regulate excess flows, biological diversity, pollination by certain insects, etc.). Other (cultural, etc.) services refer to intangible elements of a heritage nature (beauty of landscapes, spiritual inspiration contributing to a collective identity, etc.).

Local commons? The diverse fortunes of "community forests"

A large proportion of NGOs consider that the key to forest conservation lies in the recognition of communities' land rights and their empowerment in the management of their resources. This view is endorsed by IPBES (2019) in its "Global Assessment Report on Biodiversity and Ecosystem Services".[2] This hypothesis is based on the idea that "communities" would have a different, non-market relationship with nature and would be closely dependent on forest resources for their well-being. Language hesitates between "indigenous peoples" and "local communities", the use of the former being more frequent, and politically acceptable, in Latin America than in Africa, where the category of indigenous was, until recently, not considered relevant by anthropologists (Bahuchet and de Maret 2000). In West Africa, those who call themselves indigenous simply want to inform migrants that they have prior rights to land.

In reality, the relationships of the different human groups encompassed by the expression "local communities" with their natural environment are not homogeneous. Above all, these relationships are not static. In the 1990s, Indonesian agroforests, complex agroforestry systems that met the monetary needs of farmers while maintaining high biodiversity and forest-type cover, were frequently mentioned in the literature as a way forward for the future of forestry in populated rural areas. However, these agroforests have gradually been converted by the farmers themselves into less diverse, but more profitable, oil palm plantations. In Madagascar, "sacred forests", once no-go zones, were devastated by local communities who saw migrants take over the resources of these forests for themselves (Fauroux, 2001).

More generally, Robinson et al. (2011) conducted a meta-analysis of the "forest outcomes" of land tenure arrangements, and noticed contrasting results between Central America (with, globally, rather positive outcomes), South America (mitigated results) and Africa. They especially found an "association between negative forest outcomes and communal land in Africa". This suggests that tenure is only one factor among many others (local traditions and history, way of life, economic systems, governance context, etc.) that shapes the outcome of a given tenure system, a point too often overlooked.

Part II: Forest-related market-based instruments

The attempts to use international trade to tackle deforestation

The rise of independent certification of forests

In the 1990s, some "institutionalized NGOs", such as the World Wildlife Fund (WWF), agreed on the counterproductive nature of boycotting tropical timber. The hypothesis was that a large-scale boycott would reduce timber market value and could lead governments to withdraw from forest management efforts, and further encourage the conversion of forests into agricultural land.

An independent standard of certification, the Forest Stewardship Council (FSC), with a pluralistic governance (NGOs, scientists and the timber industry) saw the light of day in 1994. The standard proposed principles and criteria for "good management" of forests in order to reassure consumers and encourage them to pay more for wood bearing the label. A competing "global" label, the PEFC was launched some years after, first in Europe, then with a worldwide ambition. It was based on the mutual recognition of national certification standards. Twenty years later, more than 200 million ha were certified by the FSC (and even much more by the PEFC, thanks to mutual recognition). The vast majority of areas were certified in temperate countries rather than in tropical forests, which are the most threatened.

Certification, a market instrument supposed to express "consumer power", has often been greeted with some scepticism (see Karsenty 2019). This is because of the gradual South-South shift in tropical timber trade, the fragility of an instrument based exclusively on trust, the lack of a scientific consensus on "criteria and indicators" of sustainability, or because it does not address extra-sectoral dynamics or public governance, and it bypasses governments (Smouts 2003). Certification has not curbed deforestation, but its spread is indicative of the attractiveness of the idea of economic incentives and private governance over the traditional reliance on public regulations for forest management (Cashore et al. 2004).

Beyond timber, certification schemes have been introduced to distinguish between agricultural products, based on either social, safety or environmental criteria. This is particularly true for oil palm, with the Roundtable on Sustainable Palm Oil, and cocoa production, with the Rainforest Alliance. Currently, food commodity certification schemes are evolving to incorporate "zero deforestation" criteria in their standards.

Tackling illegal logging through bilateral trade agreements

Western countries have expressed concern about the degree of illegally harvested timber in international trade and have adopted regulations to criminalize imports of illegal timber (Brack et al. 2004). In the USA, it was a revision of the Lacey Act; in the EU, in 2013, it was the EU timber regulation (EUTR), which requires importers to carry out due diligence before marketing wood on the European market. Other countries, such as Japan and Australia, have adopted similar measures. Even China adopted, by late 2019, a new Forest Law that sanctions the purchase of timber known to be illegally harvested, whatever the country of origin.

The EU combines this regulation with a support programme for producer countries, the FLEGT (Forest Law Enforcement, Governance and Trade) initiative launched in 2003, which takes the form of voluntary partnership agreements (VPAs) aimed at helping countries to set up legality and traceability systems. This would mean allowing a kind of "country certification" through "FLEGT licences" that would cover timber exports to the EU, dispensing importers from carrying out the burdensome due diligence.

FLEGT efforts are clearly targeting governance through the reinforcement of public institutions and the participation of civil society, while certification is seeking to improve the forest management practices of the private sector. However, the FLEGT process greatly relies on technical instrumentation (databases, tracking technologies, etc.), while obstacles are often rooted in corruption and vested interests within the public administrations, and there is a lack of political will for improving transparency.

The EU points out as evidence of governance progress the increased participation of NGOs in several forums and projects. However, to date, and despite hundreds of millions of euros invested, only Indonesia has managed to issue FLEGT licences. In Africa, the continent with the largest number of countries involved, the process is lagging behind, especially in the Congo Basin. Despite initial ambitions of both national governments and the EU, timber sold on domestic markets by small companies or small-scale sawyers is not covered by national legality verification systems. In countries such as Cameroon, Ghana, DR Congo and Côte d'Ivoire, all involved in VPA processes, domestic markets are often larger than exports.

These disappointing results, so far, illustrate the overestimation by the EU of the potential of "incentives" (obtain better access to the lucrative EU market) targeted at countries with "limited statehood" (Krasner and Risse 2014). As for REDD+ (see later), the VPA proponents overestimated both the willingness and the capacity to reform of governments plagued by endemic corruption and ill-equipped administrations.

Private and public policies for deforestation-free commodities

Nowadays, industrial commodity production for export and trade is the largest driver of tropical deforestation, outpacing forest clearance for local consumption by subsistence farmers (Austin et al. 2017). A study for the European Commission (2013) found that forests – mainly in tropical countries – lost 127 million hectares between 1990 and 2008, of which 29 million hectares can be attributed, according to the report, to land conversions to meet demand from third-party countries. The EU contributed 8.4 million hectares to this assessment. For instance, the EU imports soybean meal from Brazil (20% of total soybean imports) to feed European livestock. In total, the production linked to meat consumption accounts for 60% of the deforestation imported by the EU, depending on the indicator used.

Roundtables and certifications

Because of this growing awareness of the responsibility of the global demand for food and biofuel in deforestation, large international corporations under pressure from Environmental Non Governmental Organizations (ENGOs) decided, from the 2000s, to draw up deforestation-free policies for "cleaning" their sourcing of agricultural products. These are called "zero deforestation" private policies.

The first initiative was the Roundtable on Sustainable Palm Oil (RSPO) in 2004, and the Round Table on Responsible Soy (RTRS) in 2006. Many others followed, for various commodities. Promoted by the WWF, these are international voluntary initiatives that deliver certifications (Garrett et al. 2019). RSPO has been the most widely adopted by corporations, but it has been criticized for its lack of effectiveness in terms of deforestation and biodiversity protection (Gatti et al. 2019; Ruysschaert and Salles 2014).

These certification schemes are also criticized by several scientists, as they are said to favour large-scale companies at the expense of smallholders (Lemeilleur and Allaire 2018; Napitupulu and Rafiq 2018; Saadun et al. 2018), given the cost of independent third-party verification and, for the most stringent ones, strict segregation of products and traceability systems. Some go further and question the market-oriented and utilitarian dimension of certifications, reducing "the expression of pluralism to a pluralism of interests" when defining sustainability (Cheyns et al. 2016).

Private initiatives, as such, rely on varying definitions of forests and deforestation. Certification is sometimes granted to recently converted areas, depending on the "cut-off date" adopted by the standard. This is the last date after which, for being certified today, forest clearing should not have taken place (e.g. "no forest on the plot after 31 December 2014"). Another voluntary initiative, the Brazilian Soy Moratorium, a private sector initiative established in 2006 by the largest soy companies in response to a damaging Greenpeace campaign, has led to a significant decline in soy-based deforestation in the Amazon (Gibbs et al. 2015). However, the pressure has been displaced from the Amazon to the Cerrado Biome, a biodiversity-rich woody savannah. This is a typical example of "leakage" and the limitation of acting only on the supply side, as global demand for soy – associated with the need to provide food for industrial livestock farming – remains unaddressed.

Emerging public policies in producer and consumer countries

Public policies were lagging behind these private initiatives until a 2014 UN Summit in NYC. Several governments and companies made commitments to promoting "zero-deforestation" agricultural commodity production. However, as for deforestation definition, the concept is subject to interpretation: some interpret it as "zero net deforestation", that is, with the possibility of "offsetting" for the loss of natural forests by planting trees on the least productive land. Garrett et al. (2019) show that the commitments made by nearly half of the companies allow agricultural expansion in the forest to continue without significantly deviating from "business-as-usual" practices, in particular by adopting "zero net deforestation" policies to "offset" losses.

For environmentalists, only a "zero gross deforestation" approach can conserve most of the ecological services of a natural ecosystem. But, as Brown and Zarin (2013) noticed, in tropical countries that have little non-forested land either suitable or available for agriculture, zero deforestation would essentially mean halting agricultural expansion, which may prove difficult to sustain. In Gabon, the government is considering the adoption of a new definition of "forest" based on a high threshold of carbon content per hectare (118 t C/ha, i.e. a dense forest) in order to keep the possibility to expand agriculture on wooded lands no more considered as forests. Compared to the forested area under the FAO definition, the new definition, if adopted, would allow a conversion of around 2.2 million hectares and markets the products cultivated on these lands as "zero deforestation" commodities.

In Europe, the Amsterdam Declarations were launched in 2015 by seven countries, with the ambition to preserve forests through "responsible supply chain management". Oil palm was the first commodity targeted, but soy is also considered. France prepared a national strategy against "imported deforestation", launched in 2018.

However, most of the actions proposed in these initiatives rely on information and voluntary commitments, not on coercive regulation, such as banning products from risky areas. A gap remains between the rhetoric of "imported deforestation" and specific policy measures to tackle EU consumption fuelling tropical deforestation (Weatherley-Singh and Gupta 2018). European countries, in particular, do not want to jeopardize their diplomatic relations with commercial partners such as Indonesia and Malaysia, which were already strained since the EU decided to remove palm oil from the list of "sustainable sources" for biofuels under the revision of the Renewable Energy Directive (RED II), and to phase it out by 2030. The aforementioned South-East Asian countries reacted fiercely, threatening European countries with trade retaliation – especially aircraft purchases.

Towards an international regime based on results-based payments?

Alongside private and public initiatives seeking to curb deforestation through forest-related products traded internationally, multilateral organizations have attempted to address the issue of forest conservation and sustainable management through the traditional instruments of international conventions, programmes and planning. Gradually, the approach evolved from a technical and sectoral approach to a broader and economic perspective: from forest policies to forest-related policies (i.e. all the public and private policies that affect forests, cf. Singer 2008), and the issue of economic incentives became central in constructing an international regime for protecting forests. REDD+ became emblematic of this incentives-based architecture currently being actively promoted by influential donors and organizations.

The initial influence of forest experts: planners and managers

The fight against deforestation dates back to the mid-1980s with the launch of the Tropical Forest Action Plan (TFAP) on the joint initiative of the FAO, United Nations Development Programme (UNDP), the World Bank and the World Resources Institute, and its national variations in the form of programming exercises and project "shopping lists".

The key words are "sustainable production of goods and services", "forest management plans" and "land use plans at national level". The perspectives were narrowly sectoral, and the debate was dominated by forest experts. The first forest-only international agreement, concluded under the auspices of the United Nations Conference on Trade and Development (UNCTAD) in 1994 (and updated in 2006), dealt with tropical timber, which indicates the importance given to the productive function. An international organization was set up (ITTO – International Tropical Timber Organization), inspired more by the logic of producer and consumer forums around commodities – which were flourishing at the time – than by a desire to collectively manage multifunctional ecosystems.

The results obtained were disappointing: deforestation continued and unregulated logging remained the most common practice. But the TFAP process succeeded in getting forest policy reviewed in several countries, particularly in Central Africa. This was followed by National Forest Action Plans (NFAP) or National Forest Programmes (NFP), national exercises that remained within the perspective of the TFAP and focused on field projects (management, reforestation, etc.).

The growing influence of theoretical framing by economics

At the end of the 1980s and the beginning of the following decade, many developing countries were in a difficult economic situation, linked in particular to the low prices of raw materials, which acted as an indicator of the weaknesses in the management of governments installed in rent-seeking habits.

The international financial institutions, led by the World Bank, concluded support agreements with countries experiencing budgetary difficulties and intervened directly in national policies. This was the time of structural adjustment and conditionalities (Seymour and Dubash 2000). The forestry sector was thereafter viewed from a different angle: the experts were, above all, economists who were moving away from the technical approach that characterized the "FAO period" to introduce new considerations, such as (woodland) prices, (good) governance and environmental externalities (Grut et al. 1991). Intervention in forest policy making was much more marked, with intrusion in sensitive areas (with regard to state sovereignty), such as forest taxation, forest title allocation policies and decentralization of forest management. The "laboratory" countries were Bolivia and Cameroon, with a very marked focus on the issue of forest concessions and tax reform (Bojanic and Bulte 2002; Brunner and Ekoko 2000). A global alliance between WWF and the World Bank was concluded and institutionalized in 1998. Its aim was to increase the surface area of protected areas by 50 million hectares and to promote forest certification. While the first objective was achieved, the second was a semi-failure. Only 22 million of the 200 million hectares were certified in the Bank's client countries, or just over 10%. This alliance also reflects the major shift in the 1990s and the rise of environmental issues in international discussions on forests in the wake of the United Nations Conference on Environment and Development in Rio (Humphreys 2014).

New instruments emerged in the late 1990s, such as "payments for environmental services" (Landell-Mills and Porras 2002) and "conservation concessions" (Niesten and Rice 2004). In the absence of a specific international convention on forests, the Convention on Biological Diversity became the focal point where concerns about forest decline were expressed and rhetoric in favour of local forest management developed.

A fragmented and ineffective international regime

While there is an international forest regime based on a number of international principles, networks and institutions (Humphreys 2012), it is fragmented and incomplete. Fragmented, because the various principles and institutions around which the international debate is organized cover only limited themes of the global forest issue. The lack of an international convention on forests is one aspect of this problem, but it is not the only one. Tropical forests are territorialized resources that support multiple activities and do not lend themselves well to a unified regime. The difficulty lies essentially in the limits of what are known as "forest policies" in developing countries, which have only a limited scope compared to other public policies: land, agricultural and social policies in particular. These other policies themselves express a number of collective choices of societies based on representations of economic growth, justice and sovereignty.

While these representations are not immutable and are subject to challenges between social and political forces within nations, it is striking to see how far tropical forest representations differ between the North and the South. Undoubtedly, the industrialized countries favour a conservation agenda justified by global changes. Developing countries are differently sensitive to such an agenda. Some, such as Brazil and Malaysia, fear that it will lead to limitation of their sovereignty and what they consider to be their right to develop using their natural resources – as did the industrialized countries. Others, such as African countries and Indonesia, consider that there are financial opportunities to be seized, without necessarily having the means or the will to transform their public policies (land, agriculture, etc.) and governance practices that affect forests.

The uncertain REDD+ process

Since the late 1990s and the first steps towards establishing international mechanisms for trading CO_2 emission permits, the issue of forests has been one of the most difficult issues in the negotiations on the Climate Convention. The Clean Development Mechanism (CDM), a project-based scheme which started in the early 2000s, aimed to reduce the emission reduction costs faced by companies in industrialized countries signing up to the Kyoto Protocol (UNFCCC 1997). The principle was to offset emissions in industrialized countries by carrying out projects in developing countries, where marginal emission reduction costs are lower. However, a specific difficulty interfered with an inclusion of forests on an equal footing: the non-permanence of stored carbon (whereas in other sectors, avoided emissions are considered "permanent").

The thorny issue of non-permanence

The IPCC has conventionally fixed the residence time in the atmosphere (i.e. the duration of its "radiative forcing" effect) of a CO_2 molecule at 100 years (standard for measuring the radiative powers of other gases). It is often suggested that complete neutralization of the climate effect from the emission of a CO_2 unit implies an equivalent sequestration of this gas in biomass for a period of a century. This is more a necessary convention for comparison with other greenhouse gases than atmospheric chemistry data (the IPCC indicates that the residence time varies between 5 and 200 years), since a fraction of the CO_2 emitted remains in the atmosphere for a very long time. What counts in global radiative forcing is the amount of excess CO_2 that remains in the atmosphere (related to the fact that there are more emissions than absorption at global level) (Möller 2010). The higher the stock of CO_2 in the atmosphere, the longer the average residence time increases. In fact, a complete neutralization of emissions by forest projects for the protection of forests or plantations would require almost perpetual storage, which no project can of course guarantee.

The solutions proposed by the experts to solve this problem, such as time-limited expiring credits for plantations, have never been judged satisfactory in terms of environmental integrity. Very few "expiring carbon credits" have been sold. As for forest conservation projects, they have never been eligible for the CDM because of the risk of "leakage", that is, the displacement of deforestation pressures from one place to another. Lastly, the UN CDM authority made a fairly strict application of the financial additionality criterion for forest plantation projects (projects that are theoretically profitable are not allowed to issue carbon credits). This also explains the very small proportion of forest projects in the total number of CDM projects. As of April 2012, fewer than 40 projects had been registered (less than 1% of the total). The CDM has been dormant for many years now and is to be replaced by an upcoming mechanism provided for by Article 6 of the 2015 Paris Agreement.

In response to this failure, a proposal to compensate countries for the reduction in emissions from deforestation, eventually extended to particularly include the increase in forest carbon stocks, was put forward in 2005 (Santilli et al. 2005). The objective of the researchers and NGOs making the proposal (which would later be called REDD+) was twofold: to have an instrument to remunerate forest conservation and to introduce an incentive instrument for encouraging forest-friendly public policies. The transition from the CDM "project" scale to the "country" scale made it possible to avoid the objection of leakage risks, even though the question of the displacement of deforestation between countries adopting divergent policies persisted. The financing of the mechanism remained an open question. Lula's Brazil, in particular, opposed the use of carbon credits, considering that it was primarily up to the industrialized countries, which had a historical responsibility, to make the necessary efforts to reduce their own emissions before "buying" reductions from the countries of the South. Evo Morales' Bolivia, and many NGOs, opposed carbon credits, assimilated to a "commodification of nature".

In the meantime, alongside the initial reduction of deforestation activity (combining the "biodiversity" and "carbon" agendas), other activities have been deemed eligible, namely plantations, forest management and the "conservation of carbon stocks". This last "activity", advocated by "high forest cover, low deforestation countries" such as Guyana, was designed to remunerate the "past efforts" having, supposedly, led to the conservation of forests. This inclusion indicates that the additionality principle, a key feature of the CDM, is only implicit in REDD+. Countries must design baseline scenarios of emissions from deforestation and degradation ("what would have been the emissions without the REDD+ measures?"), but allowing the payment for "past efforts" (often a rhetorical figure to take advantage of a low population and limited pressure on forests) shows how many governments intend to bypass the additionality principle.

National sovereignty and business-as-usual scenario

It took 10 years of negotiation for the instrument to be enshrined in the Paris Agreement in 2015. The latter, in the spirit of voluntary commitments, allowed for sales of "emission reduction units", but also left the possibility for the Green Climate Fund to remunerate countries for "result payments" based on REDD+ activities. The Green Climate Fund is the only multilateral window that can make payments for results, although several bilateral processes, with Norway or the World Bank, are also making payments.

In international commitment mechanisms, results are measured against a "baseline", that is, a past or anticipated level of emissions (a *business-as-usual* anticipation, or BAU). In fact, the use of a past reference implicitly suggests that the past is a predictor of a future BAU, that is, unchanged economic policies and conditions. As for the anticipated references, they include "development needs", which are reflected in the anticipation of increased deforestation in a BAU scenario. In this case, reduction commitments are not absolute but relative reductions: an increase in deforestation less than the anticipated reference is called a "reduction".

Box 19.1: REDD+, a proposal inspired by the theory of rational choices

"REDD+ countries have an incentive to reduce deforestation up to the point where the marginal cost of reductions (i.e. the national supply curve of REDD) is equal to the international compensation, for example, the market price for REDD+ credits" (Angelsen 2008: 59). Such a statement is typical of the "theory of rational choices". The "storyline" of REDD+ as an incentive-based system, inspired by the rational choices theory, can be stated as follows: "::Deforestation in developing countries is a problem of opportunity cost: the governments *decide* to deforest, or not, or the countries choose to deforest as they earn more compared to conservation or SFM. The State is assimilated to any other economic agent, making rational decisions by comparing the relative prices associated with the alternatives offered. Then, the government is acting by adopting the appropriate measures for reducing deforestation and modifying its development pathway" (Karsenty and Ongolo 2012: 39). This storyline is consistent with the position that describes REDD+ as "not encroaching on the sovereign discretion of nations to design acceptable and adequate policies and measures nationally" (Streck 2010: 389).

The references (BAU scenario, but also forest definitions) are proposed by governments. Methodologies are reviewed by experts appointed by the UNFCCC, but a rule adopted by CoP 19 disallows experts from commenting on the public policy assumptions underlying the proposed reference (UNFCCC, 2014).[3] In other words, if a potentially recipient government considers that its BAU development would require a massive conversion of forest areas to agriculture or livestock, and designs its reference scenario accordingly, this is not questionable.

Strategic behaviour of states?

Adopting the assumption of governments acting on the basis of "rational self-interest expectations", the obvious interest of countries wishing to be remunerated is to choose a reference, past or projected, that minimizes the efforts to be made to combat deforestation and maximizes the expectation of being remunerated. In practice, this means choosing a past reference period during which deforestation was high (a strategy adopted by Brazil), or proposing a BAU scenario anticipating a sharp increase in deforestation rates. While reference scenarios are criticized for their rather arbitrary nature and the impossibility of being able to verify them (Obersteiner et al. 2009), adopting a historical reference does not guarantee the additionality of reductions either.

Unlike emissions in the energy or transport sector, emissions from deforestation are much more variable, as they are linked to different economic and political factors, but also to climate irregularities (rainfall/droughts). Since the main source of deforestation is agriculture or livestock, prices at the farm gate are one of the main factors determining the marginal level of deforestation. The higher these prices, the more producers will increase the area devoted to their speculation. Higher marginal profit margins make the conversion of less fertile or more distant land profitable. In Brazil, deforestation rates are correlated with beef producer prices (Chomitz et al. 2007; Verburg et al. 2014), which are themselves linked to the money exchange rate (Richards et al. 2012). In Suriname and Guyana, there is a correlation with the gold price (Dezécache et al. 2018), in Indonesia and Malaysia with the palm oil price (Gaveau et al. 2019), in Côte d'Ivoire and Ghana with the cocoa price, and so on. Carrero and Fearnside (2011) also highlighted the role of land speculation in Brazil (over 30% with speculation than without, according to their calculations), which relativized the "potential income per hectare" factor used in many land use change models.

Other factors are involved, such as road opening, or population density in forest areas. Although population growth can be easily anticipated over mid-term periods, it is not the case for regional conflicts leading refugees to settle in forested areas (Carr 2009).

Is Brazil's "reference level" appropriate?

Brazil was the first country to benefit from "payments for results" for the decrease in deforestation observed in 2014 and 2015 compared to a baseline derived from deforestation for the 1996–2015 period (forest reference level). Brazil uses a historical reference (see Figure 19.1), and therefore "benefits" from the very high levels of past deforestation, before the policy measures implemented by the Lula government. Formally, this reference does not contradict the rules adopted for the REDD+ mechanism, but it does reveal several problems. With the election of Bolsonaro, Brazilian environmental policy changed radically. Ironically, it is his anti-ecology government that will benefit from the "payments for results" of the Green Climate Fund, results partly inherited from the Lula government's environmental policy. In addition, Brazil receives payments while deforestation in the Amazon is resuming sharply, with an increase of 43% over one year (2018–2019 against 2017–2018).

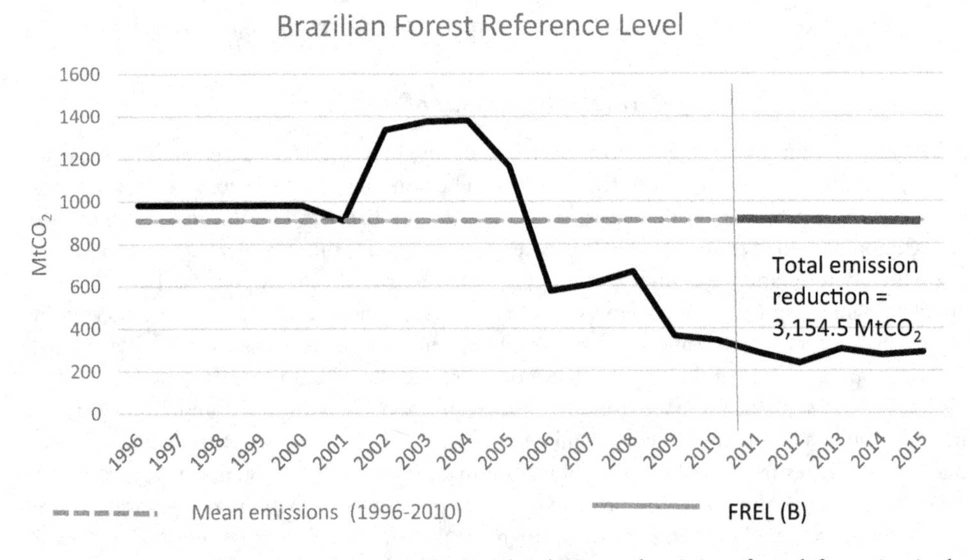

Figure 19.1 Forest reference level proposed by Brazil and estimated emissions from deforestation in the Amazon biome

Source: Data taken from the second biennial update report of Brazil to the United Nations Framework Convention on Climate Change (2017); figure by the author.

"Worst-case policy" or common (but differentiated) responsibility?

Beyond the case of Brazil, questions arise about rules that base results on a comparison with a reference understood as a business-as-usual evolution (the historical reference being implicitly a predictor of the most probable future without new efforts to fight deforestation). The BAU would thus always be a virtual implementation of the "worst-case policy", as if internal and external developments in political contexts were not to evolve with the realization of the urgent need to act against climate change.

To say that, without financial incentives, "my future would have been irresponsible" is to speak of a dubious future that was unlikely to happen once states accepted the international regime of "common but differentiated responsibility" in whose name they would claim remuneration for their performance.

Part III: Rethinking results-based payments

The different policy responses considered to stop deforestation and conserve forest-related ecosystem services have some genuine potential, but their common shortcoming is to overlook the political economy dimension of the various issues they try to address. Responses targeting international trade, be they certification schemes or zero-deforestation commitments, fail to question the implicit assumption by which "the global demands have to be satisfied" and hardly question the WTO rules of international trade.

The growing attraction of economic incentives proposed to governments for reforming and designing more friendly forest-related policies fail to recognize the composite nature of states, where policy decisions are not the result of a rational cost-benefit evaluation but of complex policy decision processes, where vested interests and corruption play central roles. REDD+, with its ambitions to be the largest results-based payment in terms of scale, reflects this naive view of the political economy.

It would be futile to seek a miraculous solution to stop deforestation, whose causes are numerous and some of which are the consequence of growth dynamics in the countries of the South, while being linked to poverty and underdevelopment in a context of growing inequalities in tropical countries. The systemic aspects of the problems and their political economy dimension are obvious, and so must be the solutions. A non-comprehensive set of priority policy interventions can be suggested:

- Collective action through the patrimonialization of services and the institutionalization of local rights;
- Conditioning international remuneration on changes in forest-related public policies;
- Give priority to investments and incentives to local actors;
- Coordinated action on demand, international trade policies and rules.

Patrimonialization and institutionalization

The notion of ecosystem services, sometimes criticized for its utilitarian dimension, nevertheless makes it possible to defend the idea of a common heritage constituted by these global services without calling into question the principle, set out at the Rio conference in 1992, of countries' sovereignty over their natural resources.

Recognizing global ecosystem services as a common heritage whose destruction is an environmental crime would provide arguments and support initiatives aiming to set up international courts for judging this type of crime. The issue is as much cultural as legal, in order to enable civil societies to legitimize their actions to fight deforestation and amend legislation.

Mapping and recognizing customary land and resources rights of communities and families in developing countries should be one of the top priorities in the international policy dialogue. The difficulty is both conceptual and political: which one of these identified rights should be recognized as exclusive rights, to protect the poor against land grabbing, and which one should be recognized as overlapping rights, shared with other stakeholders and to be managed as new commons? In increasingly populated landscapes and diversified economies, this is one of the biggest challenges to be addressed by land-use planning inclusive processes.

Investment, the neglected priority

The fate of forests is largely determined outside the forest sector. The evolution of agriculture and livestock systems is a key issue. Ecological intensification through peasant agroecology, crop-livestock associations and agroforestry should become the priority of public policies. The necessary investment could be channelled through payment programmes for environmental services (PES) to finance changes in producer practices. Such programmes would combine conditionalities and conservation incentives to counteract the "rebound effects" associated with intensification (Karsenty 2011). Part of the financing effort could come from domestic taxation of some goods consumed mainly by urban populations. Earmarked taxes with low rates and a broad base, as implemented by emerging countries such as Costa Rica and Mexico with fuel or water distribution to finance their national PES programmes, should be considered. In the poorest countries, where the cost of fuel is too socially sensitive, other taxation bases might be contemplated, such as phone units, beer and sodas.

It will be necessary to finance policies for the recognition of local rights (participatory mapping, registration of rights, etc.) and appropriate forms of land tenure security to protect rural communities from land grabbing for agribusiness. The recognition of territorialized rights, be they exclusive or not, for the benefit of communities and rural families using forested areas is not only an act of justice but also the necessary, but not sufficient, condition for working in the long term to conserve and restore forest resources. Forest concessions might evolve through the recognition of overlapping tenure rights, which should be mapped and institutionalized to be used for co-management and benefit sharing (Karsenty and Vermeulen 2016).

Investment in education, particularly girls' access to long-term education, is essential to accelerate the demographic transition. The African continent, where this transition is lagging behind in many countries, is particularly affected.

Moving away from the reference trap and rethinking payments for results

The principle of "payments for results" has no chance of achieving its objectives without extensive and sustained support for the investments needed to produce those results, particularly in countries with failing institutions. If developed countries decided to support such large-scale investments, a principle of financial incentives to encourage reform initiatives may be useful, provided that the notion of "results" is meaningfully rethought.

From this point of view, the problem of the "right" reference level for REDD+ is unresolved. On the one hand, no spatial and economic model is able to predict the evolution of major economic and climatic variables that control deforestation rates (agricultural commodity prices, droughts and rainfall, etc.), which leaves the door open to the construction of "optimized" scenarios, with variables chosen according to the strategic interests of the proposing states. On the other hand, the very logic of the business-as-usual projection provides perverse incentives, insofar as it encourages governments to virtually free themselves from the "common but differentiated" liability regime inherited from Rio 1992 through the construction of worst-case scenarios.

The aim might be to keep the principle of payment for results, without tying one's hands with an automatic payment procedure based on an unverifiable level/reference scenario. The only meaningful criterion is the coherence of public policies that potentially have impacts on forests. The effectiveness of measures to contain deforestation (formal adoption of laws and regulations, land use planning, enforcement efforts, etc.), the effectiveness of sanctions imposed on perpetrators of environmental offences, and other public choices affecting ecosystems can be relied upon. Independent collective expertise, under the joint aegis of the "Climate" and "Biodiversity" conventions, should be able to evaluate the efforts made by governments to combat deforestation and degradation.

Reforming international trade

Without profound changes in consumption patterns and strict control of the demand for products involved in deforestation, it would be illusory to seek to stop deforestation. The absolute reduction of certain consumptions (e.g. beef), the selectivity of purchases (guided by information and certification systems), and the abandonment of first generation biofuels (in particular those using palm oil) are major policy priorities.

However, leaving the rules of international trade unchanged would ruin the efforts of committed consumers. Strategies to combat imported deforestation must combine measures banning products involved in illegal deforestation, as the EU is trying to do for timber, and (where deforestation is allowed in third countries) a differentiation of tariffs favouring products certified as "zero deforestation" by internationally recognized standards. The current WTO rules, which do not allow products to be discriminated against based on environmental externalities (in this case, deforestation) related to their production, must be amended. The proceeds of this differentiated taxation, whose revenues are expected to decrease over time, should be fully allocated to support programmes for small producers in the countries of origin. Agribusiness companies will find their own interest in engaging with producers to finance the expected transitions and thus maintain their market share. The "zero deforestation" commitments of major agribusiness firms, particularly soybean buyers, have contributed to reducing deforestation in the Amazon, the most widely publicized biome (Lambin et al. 2018).

Finally, it is the responsibility of states to withdraw from trade agreements with countries that encourage the conversion of forestlands, and to ensure that any new trade agreements contain legally enforceable anti-deforestation clauses.

Notes

1 *Sciences & Vie*, November 2019.
2 "Community-based conservation institutions and local governance regimes have often been effective, at times even more effective, than formally established protected areas, in preventing habitat loss . . . Several studies have highlighted contributions by indigenous peoples and local communities in limiting deforestation" (p. 22 of the "Summary for Policymakers").
3 "The assessment team shall refrain from making any judgment on domestic policies taken into account in the construction of forest reference emission levels and/or forest reference levels" (Decision 13/CP 19, Warsaw 2014).

Bibliography

Angelsen, A. 2008. How Do We Set the Reference Levels for REDD Payments. In *Moving Ahead with REDD: Issues, Options and Implications*, ed. A. Angelsen, 53–64. Bogor: CIFOR.

Angelsen, A. and D. Kaimowitz. 1999. Rethinking the Causes of Deforestation: Lessons from Economic Models. *World Bank Research Observer* 14, n°1: 73–98.

Angelsen, A. and D. Kaimowitz (eds). 2001. *Agricultural Technologies and Tropical Deforestation*. New York: CABI Publishing.

Austin, K. G., González-Roglich, M., Schaffer-Smith, D., Schwantes, A. M., and J. J. Swenson. 2017. Trends in Size of Tropical Deforestation Events Signal Increasing Dominance of Industrial-scale Drivers. *Environmental Research Letters* 12, n°5: 054009.

Bahuchet, S. and P. de Maret (eds.). 2000. *Les peuples des forêts tropicales aujourd'hui: 3. Région Afrique centrale*. Bruxelles (BEL), Bruxelles: APFT, ULB, 456 p.

Barr, C. 1998. Bob Hasan, the Rise of Apkindo, and the Shifting Dynamics of Control in Indonesia's Timber Sector. *Indonesia*, n° 65 (April), Cornell University Southeast Asia Program

Barr, C. 2002. Timber Concession Reform: Questioning the "Sustainable Logging" Paradigm. In *Which Way Forward? People, Forests, and Policymaking in Indonesia*, ed. C. J. P. Colfer, and I. A. P. Resosudarmo, 191–220. Washington, DC: Resources for the Future.,

Bojanic, A. and E. Bulte. 2002. Financial Viability of Natural Forest Management in Bolivia: Environmental Regulation and the Dissipation and Distribution of Profits. *Forest Policy and Economics* 4: 239–250.

Boserup, E. 1965. *The Conditions of Agricultural Growth*. New York: Aldine, 124 p.

Boucher, D. H., Elias, P., Lininger, K., May-Tobin, C., Roquemore, S. and E. Saxon. 2011. *The Root of the Problem: What's Driving Tropical Deforestation Today?* Cambridge, MA: Union of Concerned Scientists.

Brack, D., Saunders, J. and C. House. 2004. *Public Procurement of Timber: EU Member States Initiatives for Sourcing Legal and Sustainable Timber*. London: Royal Institute of International Affairs.

Brown, S. and D. Zarin. 2013. What Does Zero Deforestation Mean? *Science* 342, 15 November n°6160: 805–807.

Brunner, J. and F. Ekoko. 2000. Environmental Adjustment: Cameroon Case Study. In *The Right Conditions: The World Bank, Structural Adjustment, and Forest Policy Reform*, ed. F. Seymour and N. Dubash, 59–82. Washington, DC: World Resources Institute.

Burgess, R., M. Hansen, B. A. Olken, P. Potapov and S. Sieber. 2012. The Political Economy of Deforestation in the Tropics. *The Quarterly Journal of Economics* 127, n°4: 1707–1754.

Carr, D. 2009. Population and Deforestation: Why Rural Migration Matters. *Progress in Human Geography* 33, n°3: 355–378.

Carrero, G. C. and P. M. Fearnside. 2011. Forest Clearing Dynamics and the Expansion of Landholdings in Apuí, a Deforestation Hotspot on Brazil's Transamazon Highway. *Ecology and Society* 16, n°2: 26.

Cashore, B., Auld, G. and D. Newsom. 2004. *Governing Through Markets: Forest Certification and the Emergence of Non-State Authority*. New Haven: Yale University Press.

Casson, A. 2002. The Political Economy of Indonesia's Oil Palm Subsector. In *Which Way Forward? People, Forests, and Policymaking in Indonesia*, ed. C. J. P. Colfer and I. A. P. Resosudarmo, 221–245. Washington, DC: Resources for the Future.

Cheyns, E., Daviron, B., Djama, M., Fouilleux, E. and S. Guéneau. 2016. La normalisation du développement durable par les filières agricoles insérées dans les marchés internationaux. In *Développement durable et filières tropicales*, ed. E. Biénabe, A. Rival and D. Loeillet, 275–294. Paris: Quæ Editions.

Chomitz, K. M., Buys, P., De Luca, G., Thomas, T. S. and S. Wertz-Kanounnikoff. 2007. *At Loggerheads? Agricultural Expansion, Poverty Reduction, and Environment in the Tropical Forests*. World Bank Policy Research Report. The World Bank, Washington DC.

Defourny, P., Delhage, C. and J.-P. Kibambe. 2011. *Analyse Quantitative des Causes de la Déforestation et de la Dégradation des Forêts en République Démocratique du Congo*. Louvain, Université Catholique de Louvain.

Dezécache, C., Salles, J.-M. and B. Hérault. 2018. Questioning Emissions-based Approaches for the Definition of REDD+ Deforestation Baselines in High Forest Cover/low Deforestation countries. *Carbon Balance Management* 13: 21.

EEA. 2011. Opinion of the EEA Scientific Committee on Greenhouse Gas Accounting in Relation to Bioenergy (European Environment Agency Scientific Committee, 15 September 2011). Available via http://go.nature.com/exGbJX

Ellison, D. et al. 2017. Trees, Forests and Water: Cool Insights for a Hot World. *Global Environmental Change* 4: 41–61

European Commission. 2013. The Impact of EU Consumption on Deforestation: Comprehensive Analysis of the Impact of EU Consumption on Deforestation, Prepared by VITO, IIASA, HIVA and IUCN.

FAO. 2012. FRA 2015 Terms and Definitions. Rome. Available via www.fao.org/docrep/017/ap862e/ap862e00.pdf (accessed May 20, 2020)

FAO. 2014. *State of the World's Forests – Enhancing the Socioeconomic Benefits from Forests.* Rome.

Fauroux, E. 2001. Dynamiques migratoires, tensions foncières et déforestation dans l'ouest malgache. In *Sociétés paysannes, transitions agraires et dynamiques écologiques dans le Sud-Ouest de Madagascar*, eds. L. Razanaka et al., 91–105. Antananarivo: Actes de l'Atelier CNRE/IRD du 8–10 novembre 1999.

Feintrenie, L., W. P. Chong, and P. Levang. 2010. Why Do Farmers Prefer Oil Palm? Lessons Learnt from Bungo District, Indonesia. *Small-Scale Forestry* 9, n°3: 379–396.

Garrett, R. D., S. Levy, K. M. Carlson, T. A. Gardner, J. Godar, J. Clapp, . . . and N. Villoria. 2019. Criteria for Effective Zero-Deforestation Commitments. *Global Environmental Change* 54: 135–147.

Gatti, R. C., Liang, J., Velichevskaya, A., and M. Zhou. 2019. Sustainable Palm Oil May Not Be So Sustainable. *Science of the Total Environment* 652: 48–51.

Gaveau, D. et al. 2019. Rise and Fall of Forest Loss and Industrial Plantations In Borneo (2000–2017). *Conservation Letters* e12622(e): 1–8.

Gawel, E. and G. Ludwig. 2011. The iLUC Dilemma: How to Deal with Indirect Land Use Changes When Governing Energy Crops? *Land Use Policy* 28: 846–856.

Geist, H. and E. Lambin. 2002. Proximate Causes and Underlying Driving Forces of Tropical Deforestation. *BioScience* 52: 143–150.

Gibbs, H. K., Ruesch, A. S., Achard, F. et al. 2010. Tropical Forests were the Primary Sources of New Agricultural Land in the 1980s and 1990s. *PNAS* 107, n 38: 16732–16737.

Gibbs, H. K. et al. 2015. Brazil's Soy Moratorium. *Science* 6220, n°347: 377–378

Global Carbon Project. 2019. Carbon Budget and Trends 2019. Available via www.globalcarbonproject.org/carbonbudget

Grut, M., Gray, J. A. and N. Egli. 1991. Forest Pricing and Concession Policies: Managing the High Forests of West and Central Africa. World Bank Technical Paper 143, Africa Technical department. Washington, DC: World Bank.

Haber, S. H. 2002. Introduction: The Political Economy of Crony capitalism. In *Crony Capitalism and Economic Growth in Latin America: Theory and Evidence*, ed. S. H. Haber, xi–xxi. Stanford, CA: Hoover Institution Press.

Hoare, A. 2005. Irrational numbers: Why the FAO's Forest Assessments are Misleading. Rainforest Foundation UK. Available via www.eldis.org/document/A21495 (accessed May 20, 2020)

Hosonuma, N., Herold, M., de Sy, V. et al. 2012. An Assessment of Deforestation and Forest Degradation Drivers in Developing Countries. *Environmental Research Letters* 7, n°4: 4009.

Humphreys, D. 2012. *Logjam: Deforestation and the Crisis of Global Governance.* New York: Routledge, 302 p.

Humphreys, D. 2014. *Forest Politics : The Evolution of International Cooperation.* New York: Routledge.

IPBES, 2019. Summary for Policymakers of the Global Assessment Report on Biodiversity and Ecosystem Services of the Intergovernmental Science-Policy Platform on Biodiversity and Ecosystem Services. Bonn, Germany: IPBES Secretariat. Available via https://ipbes.net/global-assessment

Karsenty, A. 2011. Payments for Environmental Services and Development. Combining Conservation Incentives with Investment. *Perspective* 7. Montpellier: Cirad.

Karsenty, A. 2018. A New "Great Game" over the World's Arable Land? *Soils as a Key Component of the Critical Zone 2: Societal Issues* 2: 13–37. Wiley on-line library. https://onlinelibrary.wiley.com/doi/abs/10.1002/9781119438137.ch2

Karsenty, A. 2019. Certification of Tropical Forests: A Private Instrument of Public Interest? A focus on the Congo Basin. *Forest Policy and Economics* 106: 101974.

Karsenty, A. and S. Assembe. 2011. Les régimes fonciers et la mise en œuvre de la REDD+ en Afrique centrale. *Land Tenure Journal* 2: 105–129.

Karsenty, A. and S. Ongolo. 2012. Can "Fragile States" Decide to Reduce Their Deforestation? The Inappropriate Use of the Theory of Incentives with Respect to the REDD Mechanism. *Forest Policy and Economics* 18: 38–45.

Karsenty, A. and C. Vermeulen. 2016. Toward "Concessions 2.0": Articulating Inclusive and Exclusive Management in Production Forests in Central Africa. *International Forestry Review* 18 (S1): 1–13.

Karstensen, J., Peters, G. P. and R. M. Andrew. 2013. Attribution of CO2 Emissions from Brazilian Deforestation to Consumers between 1990 and 2010. *Environmental Research Letters* 8, n 2.

Kaul, I., Grunberg, I. and M. Stern. 1999. *Global Public Goods: International Cooperation in the 21st Century*. New-York and Oxford: Oxford University Press for the UNDP.

Krasner, S. D., and T. Risse. 2014. External actors, state-building, and service provision in areas of limited statehood: Introduction. *Governance* 27, n°4: 545–567.

Lambin, E. F., Gibbs, H. K., Heilmayr, R., Carlson, K. M., Fleck, L. C., Garrett, R. D. et al. 2018. The Role of Supply-chain Initiatives in Reducing Deforestation. *Nature Climate Change* 8, n°2: 109.

Landell-Mills, N. and T. I. Porras. 2002. *Silver Bullet or Fools' Gold? A Global Review of Markets for Forest Environmental Services and their Impact on the Poor.* Instruments for sustainable private sector forestry series. London: International Institute for Environment and Development. 272 p.

Lemeilleur, S. and Allaire, G. 2018. Système participatif de garantie dans les labels du mouvement de l'agriculture biologique. Une réappropriation des communs intellectuels. *Économie rurale* 365: 7–27.

Marchal, J.-Y. 1985. L'évolution récente du rapport population/ressources au Yatenga (Burkina Faso). In *Des labours de Cluny à la révolution verte. Techniques agricoles et population*, eds. P. Gourou, and E. Gilbert, 223–242. Paris: PUF.

Messerli, P., Giger, M., Dwyer, M.B., Breu, T. and S. Eckert. 2014. The Geography of Large-scale Land Acquisitions: Analysing Socio-ecological Patterns of Target Contexts in the Global South. *Applied Geography* 53: 449–459.

Millennium Ecosystem Assessment. 2005. *Millennium Ecosystem Assessment. Ecosystems and Human Wellbeing: A Framework for Assessment.* Washington, DC: Island Press.

Möller, D. 2010. *Chemistry of the Climate System.* Berlin: De Gruyter, 722 p.

Moonen, P. C. J., Verbist, B., Schaepherders J, Bwama Meyi, M., Van Rompaey, A. and B. Muys. 2016. Actor-based Identification of Deforestation Drivers Paves the Road to Effective REDD+ in DR Congo. *Land Use Policy* 58: 123–132.

Napitupulu, D. M. and R. Rafiq. 2018. RSPO Certification Impacts on Oil Palm Smallholders' Welfare in Jambi Province. Available via http://ejurnal.litbang.pertanian.go.id/index.php/akp/article/view/8544 (accessed May 20, 2020)

Navrud, S. and J. Strand. 2013. Valuing Global Public Goods: A European Delphi Stated Preference Survey of Population Willingness to Pay for Amazon Rainforest Preservation. World Bank Policy Research Working Paper 6637. Available via https://ssrn.com/abstract=2336347 (accessed May 20, 2020)

Niesten, E., and Rice, R., 2004, Gestion durable des forêts et incitations directes à la conservation de la biodiversité. *Revue Tiers Monde* 177: 129–152.

Obersteiner, M., Huettner, M., Kraxner, F., McCallum, I., Aoki, K., Böttcher, et al. 2009. On fair, effective and efficient REDD mechanism design. *Carbon Balance and Management* 4, n°11.

Pacheco, P. 2009. Smallholder Livelihoods, Wealth and Deforestation in the Eastern Amazon. *Human Ecology* 37: 27–41.

Penna, I. 2010, Understanding the FAO's 'Wood Supply from Planted Forests' Projections, ed. A.J.J. Lynch, University of Ballarat Centre for Environmental Management Monograph Series n° 2010/01.

Phalan, B. et al. 2016. Conservation Ecology. How Can Higher-yield Farming Help to Spare Nature? *Science* 351, n°6272: 450–451.

Pirard, R. and K. Belna. 2012. Agriculture and Deforestation: Is REDD+ Rooted in Evidence? *Forest Policy and Economics* 21: 62–70.

Richards, P.D., Myers, R.J., Swinton, S.M. and R.T. Walker. 2012. Exchange Rates, Soybean Supply Response, and Deforestation in South America. *Global Environmental Change* 22: 454–462.

Robinson, B. E., Holland, M.B. and L. Naughton-Treves. 2011. Does Secure Land Tenure Save Forests? A Review of the Relationship between Land Tenure and Tropical Deforestation. CCAFS Working Paper 7. CGIAR, Copenhagen

Rudel, T. K., de Fries, R., Asner, G. P. and W. F. Laurance. 2009. Changing Drivers of Deforestation and New Opportunities for Conservation. *Conservation Biology* 23: 1396–1405.

Ruf, F. and F. Varlet. 2017. The Myth of Zero Deforestation Cocoa in Côte d'Ivoire. *ETFRN News* 58, June.

Ruysschaert, D. and D. Salles. 2014. Towards Global Voluntary Standards: Questioning the Effectiveness in Attaining Conservation Goals: The Case of the Roundtable on Sustainable Palm Oil (RSPO). *Ecological Economics* 107: 438–446.

Saadun, N., Lim, E. A. L., Esa, S. M., et al. 2018. Socio-ecological Perspectives of Engaging Smallholders in Environmental-friendly Palm Oil Certification Schemes. *Land Use Policy* 72: 333–340.

Samuelson, P. 1954. The Pure Theory of Public Expenditure. *The Review of Economics and Statistics*, 36, n° 4: 387–389.

Santilli, M., Moutinho, P., Schwartzman, S., Nepstad, D., Curran, L. and C. Nobre. 2005. Tropical Deforestation and the Kyoto Protocol. *Climate Change* 3, n°71: 267–276.
Seymour, F. and N. Dubash. 2000. *The Right Conditions: The World Bank, Structural Adjustment, and Forest Policy Reform*. Washington, DC: World Resources Institute
Singer, B. 2008. Putting the National Back into Forest-Related Policies: The International Forests Regime and National Policies in Brazil and Indonesia. *International Forestry Review* 10, n°3: 523–537.
Sist, P., Nolan, T., Bertault, J.-G. and D. Dykstra. 1998. Harvesting Intensity versus Sustainability in Indonesia. *Forest Ecology and Management* 108: 251–260.
Smouts, M. 2003. *Tropical Forests, International Jungle: The Underside of Global Ecopolitics*. New York: Springer, 310 p.
Staal, A., Tuinenburg, O. A., Bosmans, J. H. C. et al. 2018. Forest-rainfall Cascades Buffer against Drought across the Amazon. *Nature Climate Change* 8: 539–543.
Stiglitz, J. 1999, Knowledge as a Global Public Good. In *Providing Global Public Goods*, eds. I. Kaul et al., 308–324. Oxford: Oxford University Press.
Streck, C. 2010. Reducing Emissions from Deforestation and Forest Degradation: National Implementation of REDD schemes – An Editorial Comment. *Climatic Change* 100, n 3–4: 389–394.
United Nations Conference on Trade and Development (UNCTAD). 2006. International Tropical Timber Agreement, 2006. Available via https://www.itto.int/direct/topics/topics_pdf_download/topics_id=3363&no=1&disp=inline (accessed May 20, 2020)
United Nations Framework Convention on Climate Change (UNFCCC). 1997. Kyoto Protocol to the United Nations Framework Convention on Climate Change. Available via https://unfccc.int/sites/default/files/resource/docs/cop3/l07a01.pdf (accessed May 20, 2020)
United Nations Framework Convention on Climate Change (UNFCCC). 2014. Report of the Conference of the Parties on its Nineteenth Session, Held in Warsaw from 11 to 23 November 2013. Addendum. Part Two: Action Taken by the Conference of the Parties at Its Nineteenth Session. Available via https://unfccc.int/resource/docs/2013/cop19/eng/10a01.pdf (accessed May 20, 2020)
Verburg, R., Filho, S.R., Lindoso, D., Debortoli, N., Litre, G. and M. Bursztyn. 2014. The Impact of Commodity Price and Conservation Policy Scenarios on Deforestation and Agricultural Land Use in a Frontier Area within the Amazon. *Land Use Policy* 37: 14–26.
Weatherley-Singh, J. and A. Gupta. 2018. "Embodied Deforestation" as a New EU Policy Debate to Tackle Tropical Forest Loss: Assessing Implications for REDD+ Performance'. *Forests* 9, n°12: 751.

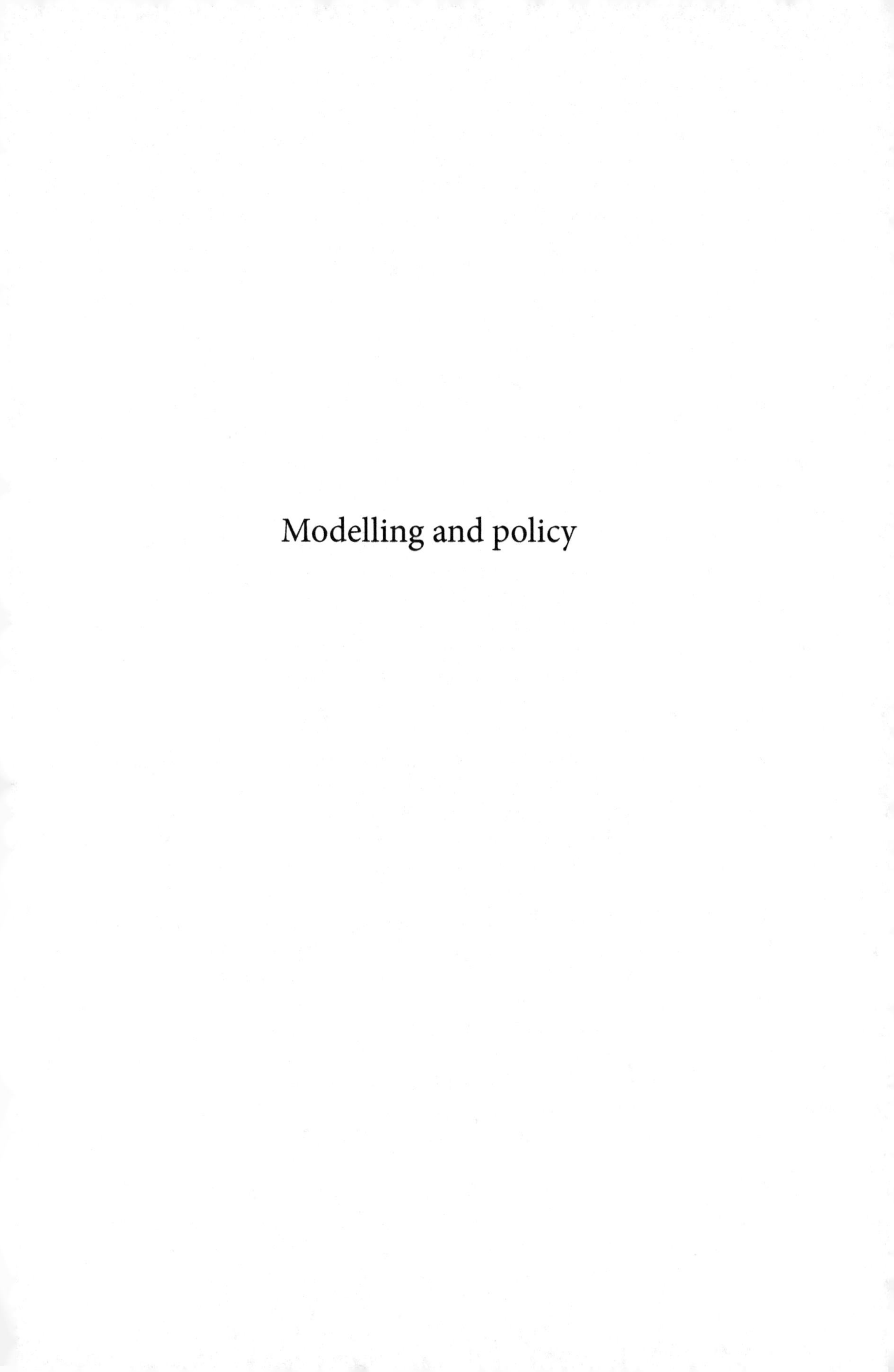

Modelling and policy

20

INFORMING THE POLITICAL ECONOMY OF ENERGY AND CLIMATE TRANSITIONS

Modelling tools, pathways design frameworks and analytical challenges

Patrick Criqui and Henri Waisman

Introduction

Climate change is clearly identified as the major challenge humanity has to face in the next decades.[1] While the scientific basis for the identification of the problem has been known since the end of the 19th century (Arrhenius, 1896), only by the end of the 20th century has the problem been set on the agenda of international politics with the adoption of the United Nations Framework Convention on Climate Change.[2] This framework introduced the successive Conference of the Parties[3] and the creation of the Intergovernmental Panel on Climate Change (IPCC), tasked to regularly synthesize scientific insights that can inform policy decisions.[4]

Climate mitigation policies raise major economic and societal challenges, given the major impact of measures adopted to reduce greenhouse gas emissions on all essential human activities. Mitigation is therefore not only a technical question but also, more fundamentally, a question of political economy.

Internationally, the core issue relates to the implementation of the "common but differentiated responsibilities and respective capabilities" principle, which acknowledges the need to consider the heterogeneity of national circumstances when debating the distribution of emission reduction efforts across countries. This is because these efforts define the magnitude of the national economic and social impacts of the struggle against climate change. Energy and climate policies also raise major political economy challenges inside countries, given their differential impacts on various sectors and activities and the different categories of households. The design of national low carbon strategies therefore involves the resolution of conflicts of interest across the stakeholders that represent the different sectors of society and thus careful attention to the expected redistributive effects of the transition policies and associated instruments.

A massive effort of academic research, of both a theoretical and an applied nature, has been devoted to the task of analyzing the economic and social dimensions of energy and climate policies, with the objective to provide policy-relevant insights for climate mitigation. New economic models have been developed, new research communities have been created and specific analyses have been conducted in order to explore the many dimensions of the problem.

DOI: 10.4324/9780367814533-22

In the first section, we propose a brief history and analysis of the role of economic models in the first phase of international climate negotiations, from 1992 to 2014, ending with the publication of IPCC's Fifth Assessment Report.[5] It presents the main paradigms of economic modelling to assess climate policies during this period. It also analyzes their strengths and limitations, notably regarding the assessment of economic and social dimensions of climate policy.

The subsequent section describes how a new modelling paradigm, based on national decarbonization scenarios, progressively emerged in the lead-up to the Paris Agreement. It discusses how this new paradigm is relevant to inform the political economy dimensions of climate action by capturing the economic and social impacts of mitigation policies in different national circumstances.

We then propose two illustrations of the new paradigm as developed in the scenario-making process of the French National Debate on the Energy Transition in 2013 and in the international Deep Decarbonization Pathways Project in 2014–2016.

The final section presents key insights from recent literature addressing the "carbon neutrality" objective, which is the new horizon of climate policy adopted collectively since the Paris Agreement. This ambitious target imposes a consideration of radical changes in each and every sociotechnical system, with major implications in terms of the political economy of the impacts, efforts and benefits of the transition in each society. This clearly calls for new approaches and new solutions and opens new research avenues for the economic analysis of climate policies.

The chapter concludes with a discussion on both the opportunities and the most controversial issues raised by the extreme degree of ambition of the corresponding new climate policies.

Economic assessment models for climate policies: dealing with global economic effectiveness and equity

Since the creation of the IPCC in 1988 and the rise to the fore of climate concerns, economic analysis and assessment models for energy and climate have played a crucial role in the evolution of the topics of international negotiation and the definition of associated policies. One can even trace a process of coevolution between economic modelling, international climate regime and low carbon transition policies. We examine this question from the historical perspective of the dynamics of interactions between the assessment tools' construction and the identification of negotiated solutions, both internationally and nationally. These negotiations clearly refer to political economy debates on how the efforts necessary for the struggle against climate change will have to be shared among the different stakeholder categories inside each country and across the different nations in the international arena. We draw on the development of modelling capacities that accompanied the evolutions in the climate policy debate in an attempt to bring to light the merits and limits of modelling to assess the main stakes in the struggle against climate change, notably social and economic dimensions.

1972–1992: genesis of the integrated energy-economy-environment models

The MIT Report to the Club of Rome, "The Limits to Growth", in 1972 (Meadows et al., 1972) was certainly the first complete modelling exercise representing the interactions between the economic system and the limits that the planet's natural resources may impose. The thesis is well known: in the first decades of the 21st century, the constraints imposed by the depletion of different natural resources are likely to cause a collapse of the world economy and even population.

Its impact was all the more resounding since, barely a year later, the first oil shock made real the fears of shortage for oil, an essential resource in the world economy. The Club of Rome report sparked intense reactions in the economics community, particularly in the USA, while triggering an intense flow of research on "the economics of sustainability" (Pezzey and Toman, 2002). Formal theoretical models were developed to provide an economic perspective on the questioned conditions for long-term substitutability of natural capital and artificial capital. W. Nordhaus participated in this effort, by pioneering the problem of climate change with models that would be the precursors of his cost-benefit analysis models in the fight against climate change: Dynamic Integrated Climate Economy model (DICE) and Regional Integrated Climate Economy model (RICE).

At the same time, the oil shocks induced an unprecedented effort of economic modelling for energy policies. The development of operational research methods and linear optimization algorithms first allowed the development of supply models exploring, for a given energy demand, supply at the lowest cost (MARKet ALlocation (MARKAL), Energy Flow Optimizati Model (EFOM), etc.). And the question of demand policies also rapidly emerged with the first simulation models (Modele d'evolution de la demande d'energie (MEDEE)) that allowed going beyond simple extrapolation to reveal the determinants of – and levers for action on – energy demand.

It was also in the context of oil shocks that the development of macroeconomic models, then predominantly Keynesian (harmonised econometric research for modelling economic systems (HERMES)), made it possible to explore the consequences of external trade shocks and, correlatively, energy independence policies. The late 1980s saw a true revolution, that of computable general equilibrium models (CGEM). Inspired by Walrasian general equilibrium theories, these models were made "computable" thanks to the progress in resolution algorithms and the computing capacity of modern computers, which were then rapidly increasing. This duality between neoclassical general equilibrium modes and Keynesian models is still relevant today.

1992–2014: the "golden age" of integrated assessment models

From the early 1990s, the European Commission launched two main lines of research in environmental economics. The first relates to the evaluation of the external costs of the different energy sources and technologies. It was organized in the successive External Costs of Electricity (ExternE) programmes, carried out initially in parallel with the work of the Oak Ridge National Laboratory in the United States. The second relates to a line of energy models intended to inform decisions on energy policy under climate constraints: GEM-E3 (E3 for Economy-Energy-Environment) and New Econometric Model of Evaluation by Sectoral Interdependency and Supply (NEMESIS) for the European economy and, for the energy system, Price-Induced Market Equilibrium System (PRIMES) for Europe and Prospective Outlook on Long-term Energy Systems (POLES) for the world.

These models were developed and mobilized in connection with the preparation of international negotiations before and after COP 3, which led to the Kyoto Protocol in 1997. They allowed in particular an assessment of the costs of the protocol for the different countries, even though the figures put forward in Kyoto for emission reduction targets were the pure result of a bargaining process between the main developed countries.

With the ramp-up of the work of the IPCC and in particular of its Working Group 3 in charge of analyzing mitigation options, the number of available assessment models increased quickly (e.g. IMAGE, MESSAGE, REMIND, WITCH, IMACLIM, for the European ones). They featured different paradigms or structures, and international cooperation was gradually being formed among them to explore common scenarios addressing the key political economy questions of the time (Kelly and Kolstad, 1999).

In the 2000s, the climate discussion centred on the extension of the Kyoto Protocol and was dominated by the economic vision of the problem of building an "international climate regime", which formed a stabilized paradigm for international negotiations between Kyoto (1997) and Copenhagen (2009). At the national level, models allowed the identification of least-cost policies by the equalization of marginal abatement costs, through either a carbon tax or an emission trading system. At the international level, and in line with the logics of the Kyoto Protocol, most modelled scenarios supposed the existence of an international emission trading system.

This paradigm considered political economy dimensions essentially with an international justice angle. More specifically, a global emissions cap is first fixed consistently with climate targets, and this global constraint is then shared among the different countries through an allocation of emission permits supported by specific equity principles. The creation of an international emission trading scheme then leads to cost-effectiveness through a single carbon price on the market. According to the permit allocation and actual emission reductions, monetary transfers would happen across countries implementing redistribution.

This scheme was simple and consistent with pure environmental economics; it also lent itself very well to a description by applied models. Multiple studies were then carried out with different models to identify and assess the consequences of different permit allocation schemes, based on various principles and concepts of international equity.

Without going into the analysis of the results, one can however question the "realism" of this approach, which assumes collective ex-ante agreement on each country's efforts and permanent transfers of hundreds of billions of dollars each year, for example from the United States or Europe to China, India or Africa. In 2009, COP 15 in Copenhagen (see later), marked the impossibility of concluding an international agreement based on such an approach.

After Copenhagen, however, the modelling work continued in large intercomparison modelling exercises (in Europe ADAM, AMPERE, ADVANCE), and the international modelling community was still organized in the framework of large consortia associated with the work of the IPCC, such as the IAMC.[6] The emphasis was put on the analysis of economic consequences of different assumptions on technological progress and on the timing of actions, notably with a focus on the impact of delayed climate action. For example, the AMPERE project provided the basis for the adoption by Europe in 2014 of an ambitious 2030 objective, that is, a 40% reduction in GHG emissions, compared to 1990.

After 2014: models, from the capital to the Tarpeian rock?

Paradoxically, in spite of the rise in the descriptive power of large applied global energy-climate models, the redesign of the international negotiation process after Copenhagen questions the fundamental paradigm on which they are based. The Paris Agreement indeed consecrates a significant change of perspectives, which strays away from the pure economic theory schemes and reintroduces the political economy dimension of transformative scenarios:

- Beyond a top-down, global perspective, more emphasis is placed on national climate policies that take into account country conditions to develop an action plan, leading progressively to a bottom-up approach. This was institutionalized in the negotiation with the introduction in 2013 of the concept of INDCs (intended nationally determined contributions) in preparation for the Paris COP 21.

- Beyond a simplified approach to policies focused on carbon price, more attention is given to policy packages. In almost all economic models applied to climate policies, the carbon price or the carbon value is used as a key driver for the desired changes in the system. For some of the modellers, this was understood only as a "proxy variable" for the intensity of the climate policies. But for the more standard economists, the carbon price was simply the signal to be introduced as a tax to all fossil fuel consumers and the models would simply deliver the right level of carbon tax (or the permit price under the vision where an international emission trading system would be introduced) to reach a globally cost-effective solution. After Copenhagen and with the abandonment of the top-down architecture, the fiction of a universal tax or permit price progressively faded away, except for those economists who are more strictly attached to the pure theory (Cooper et al., 2017). Simultaneously, those governments that seriously tried to introduce a real carbon tax encountered, at the least, difficulties due to social acceptability problems. Only some countries characterized by a strong social consensus, like Sweden, succeeded in that direction (Criqui, Jaccard, and Sterner, 2019). More and more the universal carbon price seems to be an aporia, although carbon pricing remains an essential dimension of national or regional policies.
- The international dimension is not absent in the Paris Agreement perspective, but it results from an iterative process aimed at identifying key cooperation instruments. The new paradigm derives international cooperation needs from the diversity of the national priorities in each country, rather than a top-down vision that would envision cooperation as a means to solve the international justice issues according to an ex-ante vision of equity.
- Many modelling exercises either used intertemporal optimization algorithms[7] or focused on the "early or late action" debate (Kriegler, Riahi and Bauer, 2015). But looking at national decarbonization perspectives implies a change in perspective, from pure intertemporal optimization[8] to an approach in terms of lock-ins, stranded assets, infrastructures, learning by doing and endogenous technological change. This requires taking into account elements of path-dependency and irreversibility, which are in many cases ignored in the models.
- Last but not least, the focus set on national perspectives and on sectoral priorities and policies implies taking a step back from the sole consideration of GDP as the measure of the social effectiveness of climate policies. Later on, this will be particularly clear in those projects that combine climate policies with the multidimensional national development strategies of emerging nations (Bataille, Waisman and Colombier, 2016).

Thus, while models have been structuring tools to elaborate an international climate regime from Kyoto to Copenhagen, their status has since been in practice largely reconsidered. Talking of the Tarpeian Rock might be exaggerating, as models and modelled scenarios are still used and useful. This is particularly true in the context of the IPCC reports, where the implementation of large economic assessment models is an essential condition of the dialogue between natural and social sciences (O'neill, Kriegler and Riahi, 2014). But their role in describing the one and best way to define and negotiate both the global climate regime and the corresponding national policies has been significantly weakening.

An innovative pathway design framework for informing the political economy of energy and climate transitions

In the preparation of the Paris Agreement process, it rapidly emerged that a new approach to modelling is needed. It has to combine key elements of the global integrated assessment model

(IAM) approach framing with national bottom-up modelling studies, in order to reflect country circumstances and national political priorities, while ensuring the compatibility with the collective ambition towards the common temperature goal (Waisman, Bataille and Winkler, 2019).

The change in perspective ensures the reintroduction of the political economy dimensions of energy and climate transitions: the taking into account of environmental co-benefits of climate policies, the consideration of social equity, the necessary consistency with national economic development strategies and the heterogeneity of effects among different economic groups (sectors and households). The 17 UN Sustainable Development Goals (SDGs) (Sachs, 2012) adopted by the UN General Assembly three months before the Paris Agreement provide a detailed and consistent framework for this enlargement of perspectives.

This new perspective defines an evolution of the role of models, which are no longer an instrument to provide a normative expert-view of policy solutions, but rather a tool to inform policy discussions among stakeholders. In this new approach, modelling is only one component of a broader pathways design framework, which serves to translate detailed, qualitative narratives expressed in the language of stakeholders into a synthetic set of quantitative indicators needed to characterize the transformations according to key metrics (Figure 20.1).

Country-based narratives, reflecting differing interests, in a context of deep uncertainty

Bottom-up studies, starting from the country perspectives, allow incorporating new dimensions in the foresight exercises and associating different categories of stakeholders to the scenario development process. In that way, they make possible the confrontation of different interests and visions, thus clearly positioning the scenario development process in a political economy framework.

To support the debates around the different options available to decision makers in energy and climate policies, the visions and their narratives must be sufficiently clear, understood and accepted by a working majority of stakeholders, both those responsible for implementation and those affected by the transformation: governments, indigenous peoples' organizations, sector associations, firms, energy utilities, trade unions, experts, households, non-governmental organizations, citizen groups, and so on.

The narratives have to allow an open exploration of a small number of internally consistent strategies and create a structured space for dialogue among them, in order to design and rigorously debate such pathways. This requires the strategies to be formulated in a qualitative or "lightly quantitative" language understandable to all stakeholders and to be structured around the main determinants that critically drive the evolution of the system. This also requires that the narratives explore different strategies in response to various plausible futures and are fit to support an adaptive decision-making process (Mathy, Criqui and Knoop, 2016) that allows policymakers to learn and adjust to evolving information, technology and unexpected events (Lempert, Groves and Popper, 2006; Haasnoot, Kwakkel and Walker, 2013).

Modelling development pathways in conjunction with the development goals

To be useful for policy making, quantitative national scenarios should not only describe emissions trajectories but also be designed within the context of other development goals as defined by national circumstances (Paris Agreement preamble), including the Sustainable Development Goals (SDGs) relating to energy access and security, air quality, poverty alleviation and

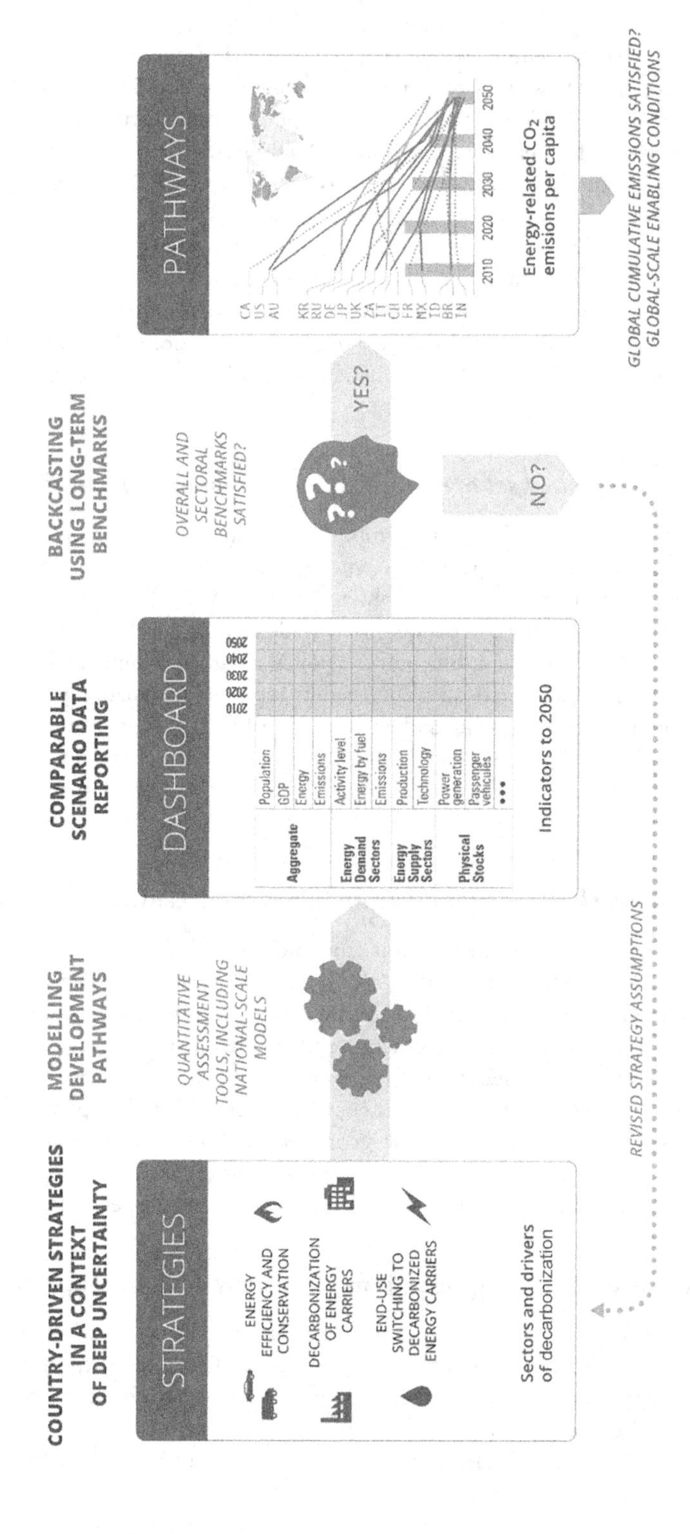

Figure 20.1 Milestones for the development of deep decarbonization pathways

employment creation (Shukla, Dhar and Mahapatra, 2008; Winkler, Boyd and Gunfaus, 2015). To this aim, they must provide transparent sectoral detail of the broader social, economic and technological changes, within which they are founded. Trade-offs and complementarity between objectives should also be clearly identified. So are the conflict of visions and interests among the different stakeholders at the core of political economy discussions.

Modelling is useful for this purpose, but no single model is able to encompass all the sectoral and socio-economic indicators required to characterize development trajectories. Different model types are appropriate for different scales and sectors, and different modelling paradigms or tools should be accommodated, as appropriate, for quantification given the specific focus of the analysis (Pfenninger, Hawkes and Keirstead, 2014). The choice of models and their level of complexity should be made whilst considering their capacity to inform the practical needs of the political dialogue and policy formation (Pye and Bataille, 2016). The modelling approach must also be pragmatic and sensitive to ease of use, data availability, budget and timescales.

A consistent set of quantified and comparable metrics

An adequate organizational architecture combines research teams, government agencies and other stakeholders, working independently with ex-ante guidance, to define the necessary physical transformations to meet the emissions and development objectives. As model outputs vary from one tool to the other, depending on paradigm, research focus and scope, this can lead to stakeholders' confusion and difficulties in policy design and implementation. However, a consistent set of comparable and quantified results reported systematically across modelling tools can facilitate knowledge sharing and enable a global composite to emerge. This requires a systematic quantitative structure identifying key sectoral and development metrics and built to accommodate scenarios from different sources. We refer to this reporting structure as a "dashboard".

A framework put in motion by an iterative backcasting approach

The purpose of the analysis is to help identify the options to reach mid-century development objectives and low emission targets, starting from the present. The design starts from the definition of realistic future benchmark values, for example for 2050, for the key indicators listed in the dashboard. A backcasting approach is then needed to identify the systemic changes required to move these indicators from their present values to ranges in line with these benchmarks. The "backcasted" pathways have to ensure consistency of national near-term planning, investment and policy decisions with long-term social, economic and environmental goals in a context of technical and structural inertia, lock-in risks, stranded assets and technological learning by doing (Sachs and Schmidt-Traub, 2016; Rockström, Gaffney and Rogelj, 2017).

Two illustrations of applications of the new analytical approach

In this section, we describe two examples that illustrate the change in perspective leading to the introduction of a broader economic social perspective. This new perspective goes beyond pure theoretical economics and integrates different considerations of political economy in the definition of climate and energy strategies: the French experience of the National Debate on Energy Transition in 2013 and the international Deep Decarbonization Pathways Project (DDPP, 2014–2015).

Visions of the future in the French National Debate on Energy Transition

Shortly after the 2012 presidential elections, France had to revise the terms of its energy policy. In the fall, a comprehensive scheme was set up, which can be termed as supporting a "deliberative democracy process". It combined a National Council for Energy Transition made up of seven colleges of 16 members, with different expert groups. This system was active during several months in 2013 and produced reports from eight working groups on (i) sobriety and energy efficiency, (ii) scenarios and trajectories, (iii) renewable energies and new technologies, (iv) transition financing, (v) governance, (vi) professional transitions, (vii) industrial competitiveness, and (viii) distribution and networks (Figure 20.2).

Concerning the scenario development, one of the main innovations of this process was undoubtedly that it had been decided *ab initio* to not create new scenarios but to use preexisting ones, as produced by NGOs, think tanks or research centres. Sixteen were identified, but it quickly appeared that from the "jungle of scenarios" it was possible to produce a reduced number of visions of the energy future, called "trajectories". Four of these can be described by the following features:

- *Decarbonization*: the reduction in demand in 2050 is moderate (−20%), the electrical system remains low carbon by maintaining the share of nuclear at 75% of the total.
- *Diversification*: demand reduction remains moderate, but renewable energies – biomass, wind and solar – are mobilized, in order to allow the share of nuclear to drop to 50%.
- *Efficiency*: the mobilization of energy efficiency potentials makes it possible to reduce demand much more significantly (−50%), while the electricity mix remains diversified at around 50% for nuclear power in 2050.
- *Sobriety* (or *sufficiency*): the strong reduction in demand is obtained for a large part by energy sobriety, while there is an exit from nuclear power and renewable energies represent up to 80% of the electric mix.

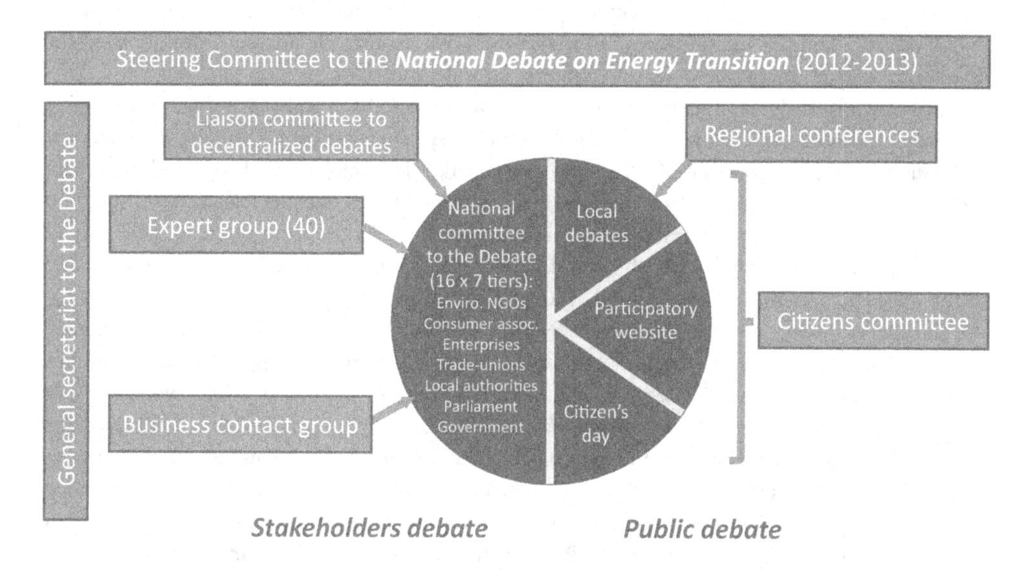

Figure 20.2 The institutional scheme for the National Debate on Energy Transition (2012–2013)

The most important thing in the "Débat National sur la Transition Energétique" (DNTE) approach was that the trajectories stem from contrasting visions of the future, whose characteristics cannot be described in exclusively economic terms. In some cases, models were used to construct the quantitative details, but not to produce an explicit economic evaluation of each scenario. Nevertheless, the four trajectories have been described in a coherent way in tables of standardized results, as a precursor to the aforementioned "dashboards". They constituted, therefore, the necessary common objects, serving as a support for the debates and allowing the political anchoring of the energy transition on the French political scene (Aykut and Nadaï, 2019). Later on, the work of the DNTE provided the basis for developing the Energy Transition for Green Growth Act, adopted by French Parliament in 2015, just ahead of COP 21.

Deep decarbonization pathways in support of international negotiations before COP 21

In the lead-up to COP21, the approach of the DNTE was transposed at the international level to support the efforts performed by the different countries to prepare their nationally determined contributions to the struggle against climate change. The basic underlying idea was that countries would be able to engage in the international discussions only if they can develop their own vision of a decarbonized future. This was done in the Deep Decarbonization Pathways Project (DDPP), co-convened by the Institut du Developpement Durable et des Relations Internationales (IDDRI) and the Sustainable Development Solutions Network (SDSN), which contributed to the preparation of the international agreement at Paris-COP 21.

This collaborative global research initiative operationalized the development of country-driven deep decarbonization pathways (DDPs) and investigated how individual countries can establish and use such pathways to understand how they can reduce emissions in a way that is consistent with the 2°C limit, taking into account the specificities of their socio-economic conditions, development priorities, existing infrastructure, natural resource endowments and other relevant factors.[9] The DDPP comprised 16 country research teams from both industrialized and emerging economies covering 74% of global energy related CO_2 emissions.[10] The teams did not represent the positions of their respective governments, but were all engaged in their domestic policy debates, and the main purpose of the exercise was to provide analytical foundations for in-country debates involving political economy considerations.

Cross-cutting analyses of national DDP scenarios indicate that ambitious mitigation requires simultaneous strong action on "three pillars" of energy system transformation: energy efficiency and conservation; decarbonization of energy carriers like electricity, biofuels and hydrogen; and fuel switching of energy end uses to decarbonized energy carriers like electricity. Because of the interactive effects between the pillars (e.g. using low carbon electricity to electrify vehicles), all the national scenarios consistently showed that deep decarbonization cannot be achieved if any of the pillars are absent or implemented at insufficient scale.

The analysis shows that these common features will be achieved very differently in different country contexts. A clear example is the composition of electricity supply. All DDP studies project a fully or nearly fully decarbonized electricity supply system at 2050, typically with significant increases in total electricity generated given the shift from direct fossil fuel use in, for example, road transport, heating and industrial processes. Some countries rely exclusively on renewable energy for electricity in their core scenarios, while others project a substantial role for nuclear power and/or fossil fuel combustion with carbon capture and storage. The differences are due to different physical and economic characteristics, as well as judgements about societal acceptability of different technologies.

Deep decarbonization pathways accommodate the expansion of energy services needed to meet countries' economic growth targets and social priorities and were designed to ensure that crucial domestic socio-economic objectives are met in each country. The analysis shows that, if enabling conditions are met and adequate policy packages are introduced, the infrastructure transformation required for deep decarbonization can be performed in a way that provides multiple economic, social and environmental benefits and opportunities for raising living standards. These include improved air quality, enhanced energy security, addressing energy poverty, improved employment, reduction in basic poverty and improved income distribution and good macroeconomic performance.

Investigating the political economy of carbon neutrality

The concept of "climate neutrality" has been present in the scientific literature and in different climate policy initiatives since the early 2000s. However, the Paris Agreement is the first international treaty to refer to this concept, in its Article 4.1 that defines the objective to "achieve a balance between anthropogenic emissions by sources and removals by sinks of greenhouse gases in the second half of this century". Since then, the European Commission has presented its detailed carbon neutrality strategy by the end of 2018.[11] Among a first set of countries, France adopted its Climate Plan and National Low Carbon Strategy, respectively in 2017 and 2018. Both explicitly mention carbon neutrality as the official national target for 2050.

The IPCC Report on "Global Warming of 1.5°C" (Masson-Delmotte, Zhai and Pörtner, 2018) confirms the emergence of carbon neutrality as a new focal point for climate policies. It shows notably that the achievement of Paris-compatible climate objective requires as a necessary condition that global carbon neutrality be achieved between 2050 and 2070. This landmark report highlights that this carbon neutrality objective would require transitions in energy, land and urban infrastructure and industrial systems that are rapid, far-reaching and unprecedented in terms of scale and speed. Investigating these accelerated and profound transitions raises specific political economy questions and challenges standard analytical paradigms.

The radical transformation of socio-technical systems

Several studies explore the many dimensions of reaching carbon neutrality by mid-century. Among them, the ZEN-2050 (Zéro Emissions Nettes 2050) performed for a French environmentally oriented business association, Entreprises pour l'Environnement, describes the necessary transformations, considering their sociological and systemic dimensions (Senard, Bonnafe and Bardin, 2019).

This study identifies six major "nexus",[12] or sub-systems, with strongly interconnected variables, each being strategic for the deployment of carbon neutrality:

- *The food–agriculture–forests–bioenergy and land use nexus.* Obviously, carbon neutrality will involve changes in the feeding regimes, with a lesser consumption of meat (particularly beef), hence different trade-offs in land use patterns, with new but competing opportunities for producing materials, producing bioenergies or storing carbon in wood and in soils. Competition for land use dedicated to food, energy and carbon storage will be a major issue in the carbon neutrality transition.
- *The urban planning–housing–transport and accessibility nexus.* The analysis of urban systems clearly identifies the necessity of acting jointly on urban zoning and on transport systems in order to limit and optimize mobility (Lefèvre and Giraud, 2007). Recent social events in

many locations also indicate that the cost of transport is a key element in the accessibility of populations to essential services. It should thus be taken into account when considering options of limiting transport through pure price effects, such as carbon taxation.

- *The industry, materials and circular economy nexus*. The reduction of emissions in productive activities encompasses new production processes and efficiency improvements in traditional activities, but also a careful management of structural and critical materials in new technologies, so as a consideration of the positive and negative environmental impacts of digital industries. The concepts of industrial ecology and circular economy may allow the conception of productive systems compatible with carbon neutrality.
- *The decarbonized energy systems*. The transformation of energy systems away from fossil fuels started almost half a century ago, but the deployment of low or zero CO_2 energy supplies has not met the rise in total energy demand, so fossil fuels still represent more than 80% of total world primary energy supply. The complete decarbonization of energy systems implies consistent action on the moderation of energy consumption, the decarbonization of the main energy carriers (electricity, gas, probably hydrogen) and the penetration of these zero-carbon carriers in final uses for each sector.
- *Behavioural changes and new consumption patterns*. Current household consumption models are clearly not sustainable in the carbon neutrality perspective and the "big is beautiful" paradigm cannot be generalized at world level (Frank, 2001). Mimetic behaviours will have to give place to more emphasis on sufficiency and efficiency in every type of consumption. Given the irreducible sociological diversity of modern societies, one way to describe and stimulate this process is to encourage the shift in lifestyles, the "laggards" progressively adopting the "leaders in transition" behaviours (Senard, Bonnafe and Bardin, 2019).
- *A new macroeconomy*. Carbon neutrality policies basically aim at replacing the intensive use of depletable fossil resources by a controlled and efficient consumption of renewable resources. This will imply a structurally new balance between consumption and investment and the creation of new channels for green investments. Measuring the precise impacts on GDP is probably not feasible and, incidentally, might be a somewhat superficial question. More important should be the task of ensuring a soft transition for the work force, while maximizing everywhere the creation of decent employment in the transition.

Capturing the interconnections at the heart of these systems' change requires going beyond an aggregated description of the different systems to propose instead a detailed representation of the main drivers of their possible changes.

Changing lifestyles

To analyze the aforementioned profound systems' transformation required by the carbon neutrality objective, one must represent as explicitly as possible societal changes consistent with ambitious emission climate goals, as a way to inform about the multiple conditions (economic, technological, social, cultural, institutional, etc.) for implementing them and the policy levers available to drive them. In particular, demand for services (energy, food or mobility, for example) is a "dimensioning" variable that structures the possible transformations of technico-economic systems. It is considered as a natural, intangible factor by conventional foresight analyses, although it is in fact the result of an interaction between societal and technological systems, meant to evolve significantly in the coming decades (Shove, 2004). Radical changes in demand can be an essential component of carbon neutrality, which in turn requires opening up the question of lifestyles and behaviours to explore transformations that would not be considered under a framing focused on technical solutions only (McCollum, Wilson and Pettifor, 2017).

For example, a significant reduction of energy demand through efficiency and sobriety reduces the size of the energy system and enables remaining within a 1.5 °C emission budget without the need for negative emission technologies (Grubler, Wilson and Bento, 2018). Dietary changes are also a key enabler of a transformation towards agro-ecology in Europe, by accommodating a productive model with lower yields (Poux and Aubert, 2018). Carrying out these exercises makes it possible to rely on quantitative analysis in order to enter into highly structured debates based on quantitative evaluations. This is not intended to perpetuate an opposition between technical solutions and societal solutions, but on the contrary to propose other ways of carrying out the transition, consisting in combining technological and societal changes.

Considering policy packages addressing social justice in the ecological transition

The social impact of transitions to very ambitious climate objectives largely depends on the combination of policies and measures adopted to trigger the required transformations. In general, along with other country-specific policies, implementing sectoral and economy-wide systems' transitions will likely require some mix of the following:

- Regulations and information for less price-sensitive sectors, particularly with respect to buildings and transport efficiency (e.g. building codes, performance standards).
- Carbon pricing, in order to incentivize behavioural changes and technology innovation in price-sensitive sectors.
- Policies that support innovation and infrastructure change, pushing the technology frontier forward (e.g. municipal land use, R&D, prototyping and commercialization support such as municipalities decarbonizing their vehicle fleets).
- Institutions to monitor sectoral progress towards decarbonization and to adjust policy if necessary. These institutions may also be responsible for supporting and monitoring key R&D programs (e.g. for Carbon Capture and Storage if applicable), and for being ready to implement alternative plans if technological aspirations do not meet necessary performance levels.

There is no unique or "optimal" policy package – ambitious emission reductions in a given country can be reached through very different policies. The choice of policy instruments depends on societal preferences, political economy concerns and institutional considerations. The policy tools need to be determined by national circumstances, complementary policy objectives (e.g. promoting energy access), policy preferences (e.g. preference for market-based mechanisms over taxation), as well as specific sector needs.

Carbon pricing is an important component of policy packages to carbon neutrality, which can be implemented through economy-wide and sectoral cap-and-trade systems or taxation. The recycling method (e.g. reduced taxes or direct financing of emission reduction programs) can have a significant effect on the performance of price instruments and their distributive effects. The role of carbon pricing must be thought of in light of national circumstances. It is an especially efficient policy instrument in mature market economies, where price signals can naturally be integrated into agents' decision-making processes. But even in these favourable contexts, the carbon price is not the silver bullet able to trigger an efficient and equitable transition, notably because some sectors have a low price sensitivity because of the multiplicity of other drivers affecting decisions of actors. In these cases, complementary measures may be needed to make the low-carbon alternatives available, accessible and affordable.

In the less-mature market systems of developing countries, where markets are incomplete because of the strong role of informal exchanges, institutional instability and poor information availability, carbon prices would probably be a secondary tool limited to the mitigation potentials that can be tapped through market incentives. In these cases, the core of decarbonization should be triggered, at least in a transition period, by a more complex set of policies and measures tailored to the constraints, potentials and development needs of the country in question.

The necessity to adopt a holistic approach to policy packages imposes going beyond conventional economic modelling exercises. They indeed lead to a superficially simple focus on price-oriented mitigation policies based on cost-benefit or cost-effectiveness approaches (Ackerman, Decanio and Howarth, 2009), and thus restrict consideration of a wider range of policy instruments (Scrieciu, Barker and Ackerman, 2013; Van Vuuren, Stehfest and Gernaat, 2018) and objectives.

Revisiting the approach to international equity

The carbon neutrality objective requires abandoning the idea that ambitious international climate transitions can simply be managed by an ex-ante burden-sharing allocation among countries. It requires instead that, in the long term, each country faces the challenge of reaching very low emission levels in its own boundaries. It remains clear, however, that the nature of the domestic transformation – and notably the political economy aspects of the transition – and even possibly the capacity of the national economy to follow a transformation towards carbon neutrality still depend on a number of parameters that are not in the hands of domestic decision makers. These "international boundary conditions" of country trajectories reflect the nature of international cooperation on specific aspects of the global transition and the external efforts in support of a given country's domestic measure.

In the bottom-up framing of the Paris Agreement, the political economy of the global climate transition is considered through the lens of this cooperation agenda. A collective international discussion and negotiation is one of the key purposes of the regular global stock take in the Paris Agreement. Explicit documentation of international enablers for a country to be able to follow a pathway to carbon neutrality can therefore form a core input to the global stocktaking discussion, which is to be viewed "in the light of equity" (Article 14.1). This in turn requires an approach to scenario design that allows an explicit and transparent depiction of international enablers that are critical for the achievement of the transition.

Conclusion

The historical process of development of models and scenarios for climate policies clearly shows that, starting from a perspective grounded on the standard environmental economics, the tools, analytical framework and decision processes had to evolve in parallel with the dynamics of international climate negotiations.

In the first stages, global assessment models, mostly based on price signals, allowed to cast light on the main transformations needed to meet climate targets, in a cost-effective way. But it soon appeared that this approach failed to support effectively actual implementation for two main reasons. First, the implementation of standard economic instruments, taxes or permits, raised significant difficulties at both national and international levels. Second, partly due to the advances and retreat of the globalization movement, it clearly appeared that climate policies had to be defined at the national level and not imposed by any supranational institution. Collective action in the struggle against climate change has to be made compatible with national interests and political will. This clearly calls for a political economy approach at the international and national level.

New analytical processes and tools have emerged to support the process codified in the Paris Agreement, notably the design of national low greenhouse gas emission development strategies that are consistent with global ambition and can support national policy formation and implementation. They should also inform the sectoral and international discussions needed to reveal the key points of global cooperation. Principles and methodologies for such an approach include the definition of multiple country-specific strategies framed by common drivers of decarbonization in a context of deep uncertainty; the use of a variety of national modelling tools to translate narrative strategies into quantified scenarios and indicators reported in a common dashboard; and national and sectoral benchmarks to provide guidance towards collective mid-century mitigation ambition.

These building blocks are combined in an iterative integrated framework for pathways design, encouraging cross-stakeholder communication and learning, enabling the assessment of compliance with national development and global emissions goals, and providing concrete support to policy formation in the context of the Paris Agreement, taking into account political economy dimensions of national and international conversations.

A key frontline for future research under this framework is the deepening of incorporation of social science insights through integration of techno-economic modelling, socio-technical studies and political analysis. This bridge would enable the alignment of conceptual languages, a mutual understanding of key ideas and iteration of alternative visions until a working understanding is achieved amongst stakeholders and decision makers (Fortes et al., 2015). Based on this, flexible and robust policy packages could be designed to meet national development and global emissions goals, taking into account the considerations of political economy dimensions both at the national level and among nations.

Notes

1 This chapter was written while the coronavirus crisis was still under full development.
2 https://unfccc.int/process-and-meetings/the-convention/what-is-the-united-nations-framework-convention-on-climate-change
3 https://unfccc.int/process/bodies/supreme-bodies/conference-of-the-parties-cop
4 www.ipcc.ch/
5 In 2014, www.ipcc.ch/report/ar5/syr/
6 Integrated Assessment Modelling Consortium.
7 The MARKAL-TIMES modelling systems.
8 With the underlying issue of the choice of discount rate.
9 Further information on the DDP initiative, the 2015 Global Synthesis Report, and country level reports can be found at https://ddpinitiative.org/
10 Australia, Brazil, Canada, China, France, Germany, India, Indonesia, Italy, Japan, Mexico, Russia, South Africa, South Korea, the United Kingdom, and the United States.
11 https://ec.europa.eu/clima/news/commission-calls-climate-neutral-europe-2050_en
12 To use the term coined by Ignacy Sachs for the Food and Energy Nexus programme of the UN University.

Bibliography

Ackerman, F., Decanio, S. J. and Howarth, R. B. (2009) 'Limitations of integrated assessment models of climate change', *Climatic Change*, 95(3–4), pp. 297–315.
Arrhenius, S. (1896) 'On the influence of carbonic acid in the air upon the temperature of the ground', *Philosophical Magazine and Journal of Science*, 5(41), pp. 237–241.

Aykut, S. C. and Nadaï, A. (2019) 'Le calcul et le politique Le Débat National sur la Transition Énergétique et la construction des choix énergétiques en France', *Revue d'Anthropologie des Connaissances*, 13(4).

Bataille, C., Waisman, H. and Colombier, M. (2016) 'The need for national deep decarbonization pathways for effective climate policy', *Climate Policy*, 16(Sup1), pp. S7–S26.

Cooper, Richard N., Gollier, C. and Nordhaus, W. D. (2017) *Global Carbon Pricing: The Path to Climate Cooperation*. Cambridge: MIT Press.

Criqui, P., Jaccard, M. and Sterner, T. (2019) 'Carbon taxation: A tale of three countries', *Sustainability*, 11(22), p. 6280.

Fortes, P. et al. (2015) 'Long-term energy scenarios: Bridging the gap between socio-economic storylines and energy modeling', *Technological Forecasting and Social Change*, 91(June), pp. 161–178. doi: 10.1016/j.techfore.2014.02.006.

Frank, R. H. (2001) *Luxury Fever: Why Money Fails to Satisfy in an Era of Excess*. New York: Simon and Schuster.

Grubler, A., Wilson, C. and Bento, N. (2018) 'A low energy demand scenario for meeting the 1.5 C target and sustainable development goals without negative emission technologies', *Nature energy*, 3(6), pp. 515–527.

Haasnoot, M., Kwakkel, J. H. and Walker, W. E. (2013) 'Dynamic adaptive policy pathways: A method for crafting robust decisions for a deeply uncertain world', *Global Environmental Change*, 23(2), pp. 485–498.

Kelly, D. L. and Kolstad, C. D. (1999) 'Integrated assessment models for climate change control', in *International Yearbook of Environmental and Resource Economics*, pp. 171–197, Northampton: Edward Elgar Publishing Ltd.

Kriegler, E., Riahi, K. and Bauer, N. (2015) 'Making or breaking climate targets: The AMPERE study on staged accession scenarios for climate policy', *Technological Forecasting and Social Change*, 90, pp. 24–44.

Lefèvre, B. and Giraud, P.-N. (2007) 'La réduction des consommations énergétiques dans les transports urbains exige une politique foncière active L'utilisation du modèle TRANUS-SETU pour l'aide à la décision', in *Les Annales de la Recherche Urbaine*. Paris: Centre de Recherche d'Urbanisme, pp. 42–53.

Lempert, R. J., Groves, D. G. and Popper, S. W. (2006) 'A general, analytic method for generating robust strategies and narrative scenarios', *Management Science*, 52(4), pp. 514–528.

Masson-Delmotte, V., Zhai, P. and Pörtner, H. O. (2018) *IPCC Special Report Global Warming of 1.5 C*. Summary for Policymakers.

Mathy, S., Criqui, P. and Knoop, K. (2016) 'Uncertainty management and the dynamic adjustment of deep decarbonization pathways', *Climate Policy*, 16(Sup1), pp. S47–S62.

McCollum, D. L., Wilson, C. and Pettifor, H. (2017) 'Improving the behavioral realism of global integrated assessment models: An application to consumers' vehicle choices', *Transportation Research Part D: Transport and Environment*, 55, pp. 322–342.

Meadows, D. H., Meadows, D., Dennis, L. and Randers, J. (1972) *The limits to growth*. New York: Potomac Associates – Universe Books.

O'Neill, B. C., Kriegler, E. and Riahi, K. (2014) 'A new scenario framework for climate change research: The concept of shared socioeconomic pathways', *Climatic Change*, 122(3), pp. 387–400.

Pezzey, J. C. and Toman, M. (2002) The economics of sustainability: A review of journal articles. Discussion Paper 02-03, Resources for the Future, Washington DC, 36pp.

Pfenninger, S., Hawkes, A. and Keirstead, J. (2014) 'Energy systems modeling for twenty-first century energy challenges', *Renewable and Sustainable Energy Reviews*, 33, pp. 74–86.

Poux, X. and Aubert, P.-M. (2018) An agroecological Europe in 2050: Multifunctional agriculture for healthy eating: Findings from the Ten Years For Agroecology (TYFA) Modelling Exercise. Iddri-AScA, Study N°09/18, Paris, France, 74 p.

Pye, S. and Bataille, C. (2016) 'Improving deep decarbonization modelling capacity for developed and developing country contexts', *Climate Policy*, 16(Sup1), pp. S27–S46.

Rockström, J., Gaffney, O. and Rogelj, J. (2017) 'A roadmap for rapid decarbonization', *Science*, 355(6331), pp. 1269–1271.

Sachs, J. D. (2012) 'From millennium development goals to sustainable development goals', *The Lancet*, 379(9832), pp. 2206–2211.

Sachs, J. D. and Schmidt-Traub, G. W. J. (2016) 'Pathways to zero emissions', *Nature Geoscience*, 9(11), p. 799.

Scrieciu, S. Ş., Barker, T. and Ackerman, F. (2013) 'Pushing the boundaries of climate economics: Critical issues to consider in climate policy analysis', *Ecological Economics*, 85, pp. 155–165.

Senard, J.-D., Bonnafe, J.-L. and Bardin, F. (2019) *ZEN 2050 Net Zero-Imagining and Building a Carbon-neutral France*. Paris: Entreprises pour l'Environnement.
Shove, E. (2004) 'Changing human behaviour and lifestyle: A challenge for sustainable consumption?', *The Ecological Economics of Consumption*, p. 1116131.
Shukla, P. R., Dhar, S. and Mahapatra, D. (2008) 'Low-carbon society scenarios for India', *Climate Policy*, 6(Sup1), pp. S156–S176.
Van Vuuren, D. P., Stehfest, E. and Gernaat, D. E. (2018) 'Alternative pathways to the 1.5 C target reduce the need for negative emission technologies', *Nature Climate Change*, 8(5), pp. 391–397.
Waisman, H., Bataille, C. and Winkler, H. (2019) 'A pathway design framework for national low greenhouse gas emission development strategies.', *Nature Climate Change*, 9(4), pp. 261–268.
Winkler, H., Boyd, A. and Gunfaus, M. T. (2015) 'Reconsidering development by reflecting on climate change', *International Environmental Agreements: Politics, Law and Economics*, 15(4), pp. 369–385.

Acknowledgment: This paper was supported by the Agence Nationale de la Recherche of the French government through the Investissements d'avenir [ANR-10-LABX-14-01] programme. The authors gratefully acknowledge the contribution of Ivan Pharabod for the design of Figure 1.

21

DIAGNOSTICS AND POLICY TOOLS TO MEASURE AND MITIGATE ENVIRONMENTAL HEALTH INEQUALITIES

Julien Caudeville

Introduction: the emergence of environmental health inequalities

Environmental conditions are a central foundation for health and well-being and account for at least 15% of mortality in the World Health Organization (WHO) European Region (WHO Regional Office for Europe, 2018).

The adverse health consequences of exposure to environmental contamination are major and growing problems yet receive insufficient attention for the political economy discipline. The full costs of environmental health impacts (including burdens on healthcare services, reduced economic productivity and lost utility associated with premature death, pain and suffering) is rarely integrated by the market. The absence of internalization is a source of environmental inequalities between those who create damage to others and reduce their well-being and those who suffer from it. Many different market failures create a compelling economic rationale for institution intervention in health inequality mitigation and environmental prevention, as a way of improving social welfare.

Environmental health inequalities (EHIs) refer to health hazards disproportionately or unfairly distributed among the most vulnerable social groups or territories, which are generally the most discriminated against, poor populations and minorities affected by environmental risks. They correspond to a declination of the environmental inequality terminology that considers the widest environmental impacts (climate change, biodiversity loss, etc.) with more specific possible configurations:

- differentiated exposure to pollution;
- different capacity of the public to influence decisions affecting the environment;
- different distribution and accessibility of environmental amenities and ecosystem services;
- differentiated vulnerability of individual, groups or territories.

Such EHIs occur in all countries in the WHO European Region, posing a triple challenge: reduction of social inequalities, mitigation of EHIs and prevention of health inequalities. However, the interrelations of these challenges offer opportunities to achieve multiple benefits

DOI: 10.4324/9780367814533-23

through environmental or social interventions (WHO Regional Office for Europe, 2019). Since 2012, WHO assessment findings indicate:

- EHIs occur in all countries, irrespective of the national level of development and the environmental preservation or economic activities;
- the occurrence of EHI has tended to persist or even increase over time, despite the improvement of environmental conditions observed in most countries in the WHO European Region;
- inequalities can often be significant, with some population subgroups exposed or affected five times more than others to certain pollutions (WHO Regional Office for Europe, 2012, 2019).

WHO has identified EHI as a priority issue in need to be addressed by the national governments in Europe. At a regional or continental scale, the characterization of EHI expresses the idea that populations are not equal in the face of pollution in terms of exposure and sensitivity and implies the identification and management of the areas at risk of overexposure where pollutions are suspected to generate a potential risk to human health. Yet the magnitude of health impacts caused by EHI is difficult to quantify, as it requires detailed information on specific population groups as well as identifying and characterizing exposure in order to interpret how risk accumulates across a territory and prioritizes interventions. The development of methods is a prerequisite for implementing public health actions aimed at protecting populations (Ioannidou et al., 2018).

Whatever their source (natural or anthropogenic), contaminants can be transferred and transported between different environmental compartments, and humans can come in contact with them through various media (drinking water, air inhalation, etc.). This plurality of exposure routes during the lifetime, along with the difficulty of assessing the exposure for a population, contributes to the strong need for research in this domain. The recently emerged concept of *exposome* (Wild, 2005) is used to describe these complex exposures, considering all sources, routes and – when possible – the interactions of stressors, that are likely to contribute to the health alteration of individuals. The external contribution to the human exposome is determined by environmental exposure, also termed the eco-exposome (Lioy and Smith, 2013), such as exposure via air, food, water, dust and use of consumer products. Needed for understanding and tackling EHIs and the related health effects is a coherent conceptual framework of exposure assessment that permits estimating or measuring the magnitude, frequency and duration of exposure to chemicals, along with the number and characteristics of the population exposed.

Quantitative exposure assessment for EHI characterization poses specific questions that need to be addressed:

- the identification of contamination source;
- the estimation of the contamination from different environmental media;
- the characterization of exposure mechanisms (pathways and relevant routes) (Asante-Duah, 2002).

However, constructing methods and tools to help orient public policies in order to reduce territorialized EHI requires the evaluation of phenomena not always easy to apprehend and the reliability and representativeness of information that usually demand statistical processing (Ioannidou, 2018).

Exposure assessment to identify and characterize the territorialized EHI depends on the availability of data. However, data on this subject are not always available. The available databases are often assembled for diverse objectives and reprocessed using statistical methods. Exposure assessment is generally considered as complex due to a lack of data and the inherent natural variability in exposure levels, leading to uncertainty in the estimates (Ciffroy et al., 2016). As a result, common spatial support is lacking, and therefore prior spatial analysis in order to homogenize them or increase their resolution is required. In addition, temporal support also differs between the available data (punctual measurements, year averages, etc.), which also requires additional treatment. Spatialization and crossing of available databases pose several methodological difficulties and can introduce uncertainties in achieving the cartography process. For this reason, different methods and techniques are employed to specifically treat environmental databases in order to benefit from all available information and to reduce the uncertainties in the final map (WHO Regional Office for Europe, 2015).

This chapter presents the European institutional and scientific contexts in which EHI characterization operates. The connection between the environment and health in public policies in Europe will be presented to address the need of integrating data from the environmental health tracking information system as a strategy to characterize the exposure of populations living in a territory.

To enhance mitigation of EHI and prevention of health inequalities in the scientific exposome concept emergence context, an environmental health methodological framework will be presented using different examples to identify a common taxonomy for conceptualizing and operationalizing environmental exposures as an important step towards articulating a science of environmental health disparities.

Institutional context: the gradual recognition of environmental health inequalities

Segmented EU policies to mitigate EHI at the national level

By acknowledging the increasing relevance of the unequal distribution of environmental risks, EHIs have been recognized as a cross-cutting challenge for the European Environment and Health Process, as highlighted at the 2010 and 2017 Ministerial Conferences on Environment and Health (WHO Regional Office for Europe, 2017). In the conference declarations, member states stressed the need to address environmental justice, undertaking to

- act on socioeconomic and gender inequalities in environment and health and tackle health risks to children and other vulnerable groups posed by poor environmental, working and living conditions;
- consider equity and social inclusion in environmental and health policies and prevent inequalities related to environmental pollution and degradation (WHO Regional Office for Europe, 2010).

The Health 2020 European policy framework outlines the various dimensions of this goal and is based on the evidence that many health inequalities can be effectively addressed through action on the social and environmental determinants of health. Inequalities in health are also a major challenge for both development and overall progress in achieving the transformation required for the 2030 Agenda for Sustainable Development. By recognizing the intrinsic link between the state of the environment and quality of life, Priority Objective 3 of the Seventh Environment

Action Programme (7th EAP) aims 'to safeguard the Union's citizens from environment-related pressures and risks to health and well-being' (EC, 2013). The profound dependency of human society on supporting ecosystems lies at the very core of the 7th EAP vision that 'in 2050 we live well, within the planet's ecological limits' (EEA, 2019).

By striving to integrate health equity considerations into all national policies, the principal goals of EU initiatives are

- promoting existing EU environmental policy and legislation at the national level;
- developing a framework to improve the methodology for understanding and assessing EHIs;
- providing road maps and menus of policy options for member states working to put health high on the environmental, social, economic and political agenda.

The EU put in place a broad range of segmented environmental legislation to address environmental impacts on health, which have resulted in overall reduced air, water and food contamination.

Some examples from the main environmental policy areas include

- EU air quality legislation;
- The European Water Framework directive;
- The Environmental Noise Directive;
- The regulation concerning the registration, evaluation, authorisation and restriction of chemicals (REACH).

The EU has modernized the chemicals legislation framework but focuses on single substances in isolated sectors (Kienzler et al., 2014, 2016). If mixtures are covered, they are mostly addressed within a specific regulatory sector (e.g., mixtures of pesticides, biocides, food additives, etc.). However, as humans and the environment are in fact exposed to chemicals via many different routes and sources, co-exposure to chemicals regulated under different sectorial legislations does occur, as can be confirmed by human and environmental (bio)monitoring data (Bopp et al., 2019).

Over the last 20 years, the general public's right to information has been at the heart of environmental policies in many countries. Directive 2003/4/EC on access to environmental information (AIE) provides that individuals have the right to access certain environmental information held by public authorities. The INSPIRE Directive, establishing an infrastructure for spatial information in Europe to support Community environmental policies, and policies or activities, which entered into force in May 2007, is based on infrastructures for spatial information established and operated by the 28 member states of the European Union and makes data available to potentially assess EHIs.

The EU also supports research in this field. Since 1998 over 400 environment and health research projects, receiving over EUR 1.2 billion from the EU, have been funded under the Fifth (1998–2002), Sixth (2002–2006) and Seventh Framework Programmes for Research (2007–2013) and the current Framework Programme Horizon 2020 (2014–2020). The increased knowledge base produced by the projects contributes to providing the scientific support for preventive actions and for assessing the efficacy of policy actions. Environment and health research at the European level already gained visibility under the Fifth Framework Programme (FP5–1998–2002) thanks to the introduction of a specific key action referred to as 'Environment and health', allowing over 100 small- to medium-scale collaborative projects to be funded. In general, the Horizon 2020 projects will provide scientific support for the United Nations Sustainable Development Goals (UN, 2015), especially Goal 3 ('Ensure healthy lives and promote well-being for all at all ages'), Goal 10 ('Reduce inequality within and amongst countries') and Goal 13 ('Take urgent action to combat climate change and its impacts'). Refocusing and reinforcing efforts take place to improve stakeholders' understanding of the environmental causes of disease as well as the environmental burden of disease through support for human exposome research (Karjalainen et al., 2017).

Territorialized exposome: from theory to operationality

Risk managers are currently facing two realities: the need to consider the impact of multiple exposures and combined risks, and the substantial growth of environmental data production at the territorial level. The exposome has been proposed as an emergent exposure science paradigm for conceptualizing the cumulative effects of environmental exposures across the lifecycle (from conception to death) and for examining the dynamic, multidimensional interrelationships between environment and health (Buck Louis et al., 2012; Lioy, 2010; Gee and Payne-Sturges, 2004; Brunekreef, 2013). In 2005, Wild (2005) first described the exposome – the totality of one's lifetime exposures – as a conceptual framework for understanding the environmental context of health outcomes. He differentiated between the "eco-exposome" as the point of contact between an external environmental stressor and biological receptor of an individual and the "endo-exposome" viewed as the set of inward effects arising from exposure on those receptors (Juarez et al., 2014).

Ideally, direct measures of exposure (e.g., biomarkers or personal monitoring data) would be available for all key stressors related to a common health effect throughout the critical period of exposure and in the population of interest (Evans et al., 2013). However, the exclusive use of biomarker data for characterizing EHIs is currently not practicable when considering a large number of diverse chemicals due to analytical and resource limitations (US EPA, 2007), especially when the assessment should cover a large territory. The need for risk assessors to identify populations at risk, in the context of substantial data deficiencies that hinder the evaluation of EHIs, places the operational declination of the concept on a territorial scale. The characterization of the territorialized (eco)exposome implies the development of dynamic, multidimensional, temporal approaches, as well as information systems that require the adoption of transdisciplinary data analysis methods. Integrated approaches could bring together all the information required for assessing the source-to-dose continuum using a geographic information system (GIS) and an integrated exposure assessment framework. Different approaches could be considered for the screening-level analysis of spatialized cumulative risks based on toxicology data, or a multivariate approach could be embraced to combine exposure variables at population level. In other cases, finding a common metric for dissimilar risks is not a strictly analytic process, because judgements must be made as to how to link two or more separate scales of risks during a deliberative process, including stakeholders to identify operational actions adapted to policy objectives (Caudeville et al., 2017).

In the context of EHI mapping, the identification of vulnerable individuals and communities at risk in order to target public health interventions relies on additional requirements in the exposure assessment processes as compared with the traditional risk assessment methodology. The design study should be able to

- investigate the processes taking place at the interface between the environment contaminants of interest and the organisms;
- characterize the principal exposure pathways;

- build realistic scenarios that integrate past and present sources;
- describe the phenomena at a fine temporal and spatial resolution.

Since the exposome would require the estimation of exposures over time across the life course of an individual, the territorialized exposome requires possible population class characterization where representative cross-sectional exposure assessments could be made in order to capture the key exposure periods of interest (*in utero*/infancy, childhood; adolescence; adulthood, etc.). According to this design, measurement and modeling approaches both allow the exposure estimation of the populations considered, with their own limitations and uncertainties. These two approaches are complementary and must be conducted jointly and in consultation. Crossing these approaches will improve the effectiveness of decision support tools (orientation and optimization of measurement campaigns, improvement of the spatial representativeness of evaluations, improved integration and understanding of determinants). The coupling also allows the integration of information on the sources of contamination, the quality of environmental media and resident populations on the same geographical referent units. A GIS thus provides the opportunity to cross the estimated exposure with biological impregnation data to provide interpretive elements of the environmental determinants of exposure. The coupling of numerical and statistical models allows for the integration, data processing and assessment of the transfer of contaminants from the environment to the populations.

In order to build a calculation infrastructure able to characterize the eco-exposome at territorial level, several methodological issues have to be solved: (1) define an integrated exposure assessment framework that first requires different scientific limitations to be overcome, such as the linkage of the global source-effect chain, (2) provide statistical methods that would allow spatial and temporal data processing from an existing environmental and populational database, (3) link, adapt or develop transport and transfer models to make practical tools available. This framework could be a layered structure that describes the elements of exposure pathways (Figure 21.1), the relationship between those elements, and how data describing the elements is stored and utilized for selected outputs, such as exposure assessment, exposure prediction, epidemiology and public health decision-making (Teeguarden et al., 2016). Refined aggregate exposure assessment is data intensive, requiring detailed information at every step of the source-to-dose pathway.

Diagnostics tools to prioritize prevention action

WHO assessment findings indicate that the lack of data on inequalities in environmental conditions restricts a broad assessment in many countries and therefore represents a major concern (WHO Regional Office for Europe, 2012). Hence, an environmental public health tracking information system (EPHTIS) may be used as a surveillance or a warning tool providing useful data for the prioritization of prevention actions. The explicit aim of the EPHTIS is to provide the information needed to improve the health of the population and optionally reduce EHIs. An EPHTIS is a data processing strategy for characterizing the exposure of populations living in a given territory. It can combine monitoring networks developed for study, observation or decision support purposes. Data, human and computer resources available via the information systems allow the improvement of the approaches to

- identify areas of population overexposure and detect evolution trends;
- generate etiological hypotheses on the relationships between environmental exposures and population health;

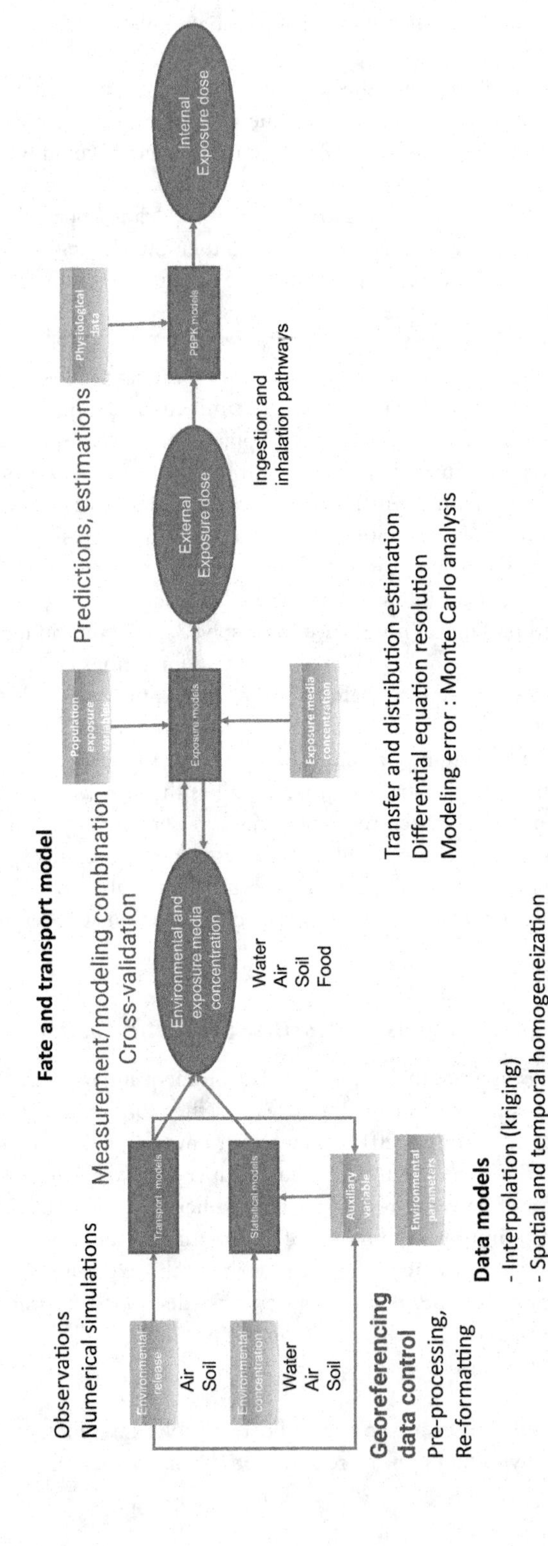

Figure 21.1 Modeling framework example to characterize the territorialized (eco)exposome

- guide prevention actions and strategies;
- provide analytical elements to estimate the efficiency to the general public (Liu et al., 2012).

Various approaches have been developed at an international level, with varying degrees of maturity, to provide devices that require consistency of data production with respect to the monitoring objectives, thus enabling their operationality within a management framework. Figure 21.2 presents the different interactions between the different actors with an elaborate EPHTIS (Caudeville and Habran, 2019). Public participation in decision-making and access to justice in environmental matters coincide with new consideration attributed to the scientific and democratic debate on environmental justice and articulation between social and environmental policies. The first EPHTIS appeared in this context in the 2000s and provided access and ability to integrate data or databases at different administrative levels, promote the interoperability of systems, improve the quality of data produced and better enhance the incorporation of the environmental dimension in all policies.

The approach is based on the systematic collection, integration, analysis, interpretation and dissemination of data relative to environmental exposure, socioeconomic and health effects within the network, allowing for the identification of areas and populations likely to be the most impacted. The information collected also examines the possible relationships between health and the environment and is part of the recent new scientific discipline: "environmental health tracking" described by McKone and Ozkaynak (2009). Unlike epidemiology, which is more concerned with accurate classifications of diseases than with exposure and health risk assessment

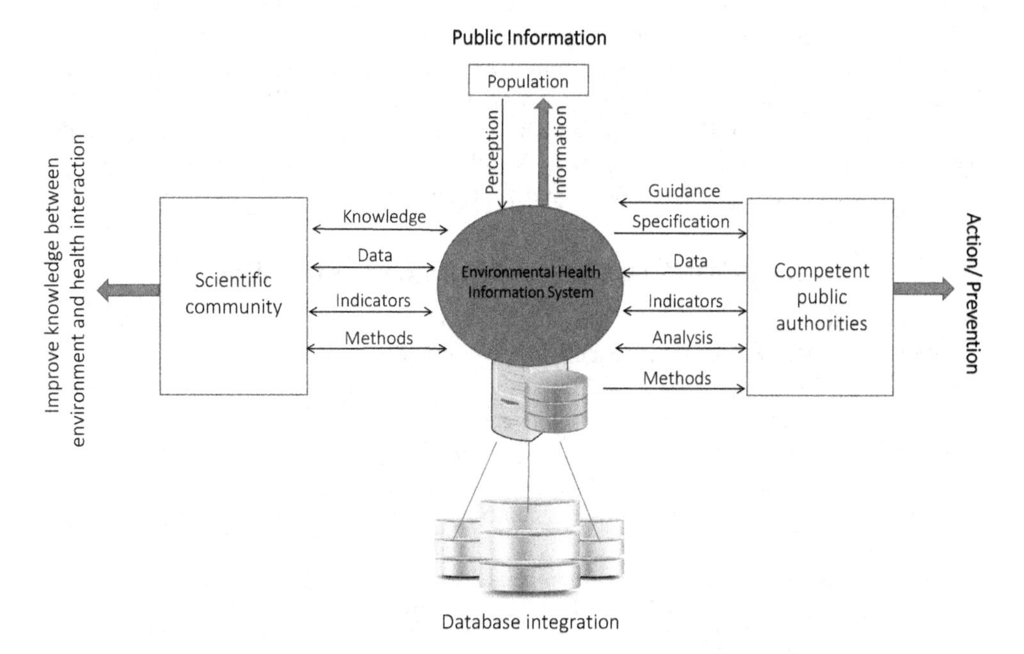

Figure 21.2 Diagram showing the components and interactions between the different actors of an environmental health tracking system

Source: Caudeville and Habran (2019).

that provides broad margins of safety, environmental health tracking aims to characterize the exposure of populations or to link diseases and exposure indicators from spatialized approaches.

In the United States, since the adoption of the Emergency Planning and Community Right-to-Know Act, risk managers must provide to the public the information needed to estimate the health risks to which they are exposed (Pew Environmental Health Commission, 2000). This increasing demand came from communities, represented by associations. In this sense, different systems make it possible to provide this type of information to the population. They provide a foundation on which civil society can rely to advocate for access to a health-friendly environment. Integrated, processed and system-enabled data make it easier to access data also for research.

In 2006, Health Canada started developing a national monitoring and surveillance system, modeled on the US system (Abelsohn et al., 2009). A strategy has been built to develop a coherent national system of scientifically valid indicators applicable to the Canadian context. In Quebec, a joint monitoring plan, including environmental health, occupational health and infectious diseases, has been established within the Ministry of Health and Social Services and centralized within the Public Health Institute. The indicators are selected by consensus of experts in accordance with the objectives of the public health program (CCE, 2014).

In Europe, the Aarhus Convention on "Access to Information, Public Participation in Decision-Making and Access to Justice in Environmental Matters", adopted on 25 June 1998 (Medina et al., 2009), coincides with the new consideration attributed to the scientific and democratic debate on EHIs and on the link between social and environmental policies. Following the WHO Ministerial Conference on Environment and Health in 2004, several European projects such as ENHIS, ENHIS2 (European Environment and Health Information System) (Kim et al., 2005) and ECOEHIS (European Community Health and Environment Information System project) were born. They aimed to propose a global information system(s) to better prioritize risks and improve the comparability of indicators on key themes describing environmental policies and their repercussions in terms of health across Europe. The European project known as "Information Platform for Chemical Monitoring (IPCheM)" promotes the networking of various existing databases and information systems with a view to advancing the environmental data exchange across different disciplines. Although it does not constitute a real EPHTIS, it corresponds to a new integrated data platform permitting to reuse existing data.

The PLAINE platform: from diagnostics to policy

After more than 10 years of actions aimed at the prevention of environmental health risks, the third French National Environmental Health Plan (FNEHP) (2015–2019) proposed a new approach to environmental health, both robust and more connected to the territories while integrating the development of new scientific concepts like that of the exposome. Four actions are specifically dedicated to the environmental inequality theme:

- identify and analyze the methods for constructing spatialized and integrated exposure indicators;
- develop and disseminate, via a common platform, reference methodologies at the national level for the characterization of locally disparate EHIs, taking into account the vulnerable situations of the populations;
- use tools for analyzing EHIs to cross exposure models and population data (biomonitoring, epidemiological data, social and health vulnerabilities);
- implement multi-exposure studies conducted in several territories, based on methodological references.

The GIS-based modeling PLAINE platform (Plateforme d'Analyse Intégrée des Inégalités Environnementales: Environmental Health Inequality Analysis Platform) developed by INERIS has been developed in the context of the FNEHP to pinpoint EHIs.

The PLAINE platform: algorithms and big data

By enabling the systematic collection, integration and analysis of data on emission sources, environmental contamination, exposure to environmental hazards and population and health, the platform makes it possible to deal with different purposes, such as

- mapping exposure disparities at a fine resolution by building environmental indicators related to human health relying on exposure assessment methods;
- identifying vulnerable populations and determinants of exposure to manage and plan remedial actions.

Following the territorialized exposome concept, source-to-exposure continuum approaches are used for connecting the source of the exposure with the target exposure. The exposure assessment integrates a data processing workflow able to connect the global source-effect chain.

Spatial analysis and statistical methods are employed to process and assemble the databases using R (R Core Team, 2013) and QGIS (QGIS Development Team, 2018) software. EHIs operate on different scales (global, regional, local) and could not be gleaned via the examination of a single medium, but could via the integration of varied contamination pathways: air, water, soil and food. In order to respond to the general objective, the integration and combination of various levels of data from different environmental compartments and exposure media are required. To this end, the MODUL'ERS exposure model is employed (Bonnard and McKone, 2010) to calculate the spatialized exposure indicators using georeferenced environmental databases (air, water, soil, foodstuffs) from monitoring networks to estimate ingestion and inhalation pathway contributions.

Characterizing EHIs requires the use of measurement data from existing environmental databases and methodological tools particularly suitable for the construction of exposure indicators. Different examples are shown in order to illustrate platform usages.

The case of spatial exposure of lead in France

Statistical and geostatistical methods were used to spatialize concentration measurements of lead in France in air, water and soil compartments. To address the issue of the limited number of observations and representativeness, appropriate spatial variables are constructed (Table 21.1). Spatial analysis and statistical methods are employed to process (georeferencing, data controlling, pre-processing, reformatting) and assemble the databases for the purposes of the study.

Lastly, the constructed spatial database is integrated in the probabilistic multimedia exposure model to yield the exposure indicators. To assess the exposure and dose, predicted concentrations of Pb from local exposure media contaminated by environmental transfer and national data for commercial foodstuffs are integrated in the model. In the analysis conducted by Caudeville et al., 2012b, the largest contribution to lead exposure is drinking water (38%), followed by soil (15%), vegetation (17%) and milk ingestion (13%, mainly from commercial origin). These modeling results differ from findings in another French exposure modeling analysis (Glorennec et al., 2007), emphasizing that data and modeling assumption can impact the results used

Table 21.1 Spatial variables in the multimedia exposure model

Parameter	*Support and resolution*	*Spatialization mode and source*
Atmospheric deposition and concentration	Centroid (point) of 0.5 × 0.5 ° Grid	Kriging of the Eulerian atmospheric dispersion model Chimere data (Menut et al., 2013)
Soil concentration	Point: sample; surface: commune	Kriging method (Caudeville et al., 2012a) that integrates data from surface and point spatial supports relying on the trace metal surface soil database (BD ETM, INRA & ADEME Program: concentration of trace metals in topsoil) and the French Soil Quality Monitoring Network
Background soil concentration	1 × 1 km grid	Linear mixed model using the French National Soil Quality Monitoring Network (trace metal concentration in soil) and parental material data (Marchant et al., 2010)
Water concentration	Multiple imputation method: district	Multiple imputation method using the SISE-Eaux database (Davezec et al., 2008), the administrative boundary map of France and distribution unit serve map

in estimation. Lead paint or strictly commercial foodstuff pathways was not included since no spatial data are available for determining this contribution in geographical terms.

The mapping of environmental concentrations such as soil compartment (Figure 21.3a) and associated uncertainties are estimated by using geostatistical methods. Figure 21.3b presents the predicted error (kriging variance) of lead concentrations in topsoil. The spatial structure of Pb concentration maps reflects the influence of a complex set of spatial and environmental factors with great variation and operating on different spatial scales. The analysis of these structures allows for the quantification of the combination and contribution of different scales of spatial variability. Uncertainty is propagated via the concentration-exposure-dose models and is depicted in the maps during the various processing steps carried out in the GIS, data used, parametric uncertainties and modeling. This involves estimating the errors contained in each type of information and estimating their cumulation within the entire modeling.

Associated with the location of potentially polluted sites (French BASOL database, crossed area in Figure 21.3b), this map allows data to be prioritized to improve the spatial representativeness of the variables investigated in the context of EHIs.

Work on lead exposure has made it possible to integrate parametric variability associated with (1) environmental and behavioral parameters, and (2) pollutant concentration variances in the calculation of exposure via the Monte Carlo simulation method. The main route of exposure for lead is ingestion of drinking water. The second route of exposure is the ingestion of soil with high contributions in the proximity zones of polluted sites and soils. Similar works have been carried out using this design for other trace metals (nickel, chromium and cadmium).

The case of spatial exposure of benzo[a]pyrene in France

The polycyclic aromatic hydrocarbon (PAH) family has been identified by the US EPA, the US Agency for Toxic Substances and Disease Registry (ATSDR) and the European Food Safety

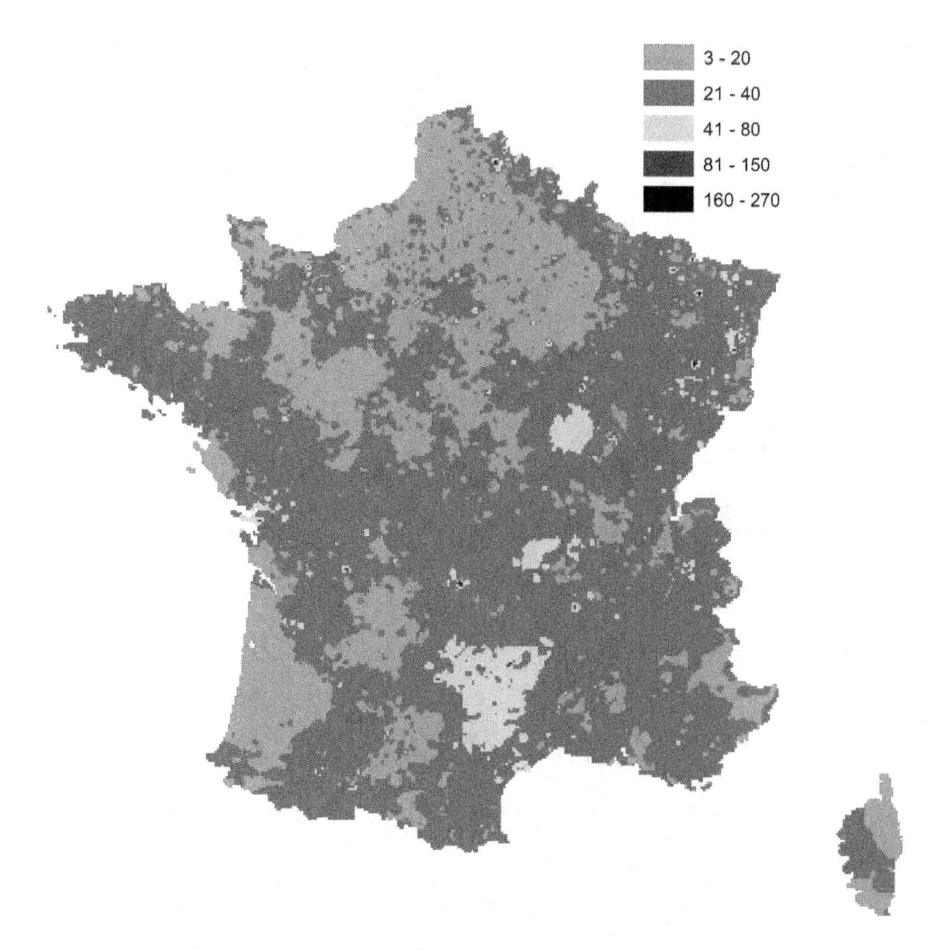

Figure 21.3 Topsoil lead kriging (a) concentrations (mg.kg^{-1}) and (b) variances in France

Authority (EFSA) as a priority, due to their carcinogenicity or genotoxicity and/or ability to be monitored. The PAH congener benzo[a]pyrene (BaP) is widely used as a marker for PAHs in the atmosphere as BaP constitutes a substantial proportion of the total carcinogenic potential of the total PAH burden (Delgado-Saborit et al., 2011).

Data from the different environmental compartments (water, air, soil, food) are available in France in different databases. In order to construct the exposure maps from spatialized databases during the evaluation of EHIs, methods have been developed to process and harmonize the available data, with respect to their specificities (missing values, limited number of observations, etc.) in the same resolution and support (Ioannidou et al., 2018). By way of example, estimation of atmospheric concentrations over France by classical interpolation method could lead to a misrepresentation of the spatial distribution due to the limited number of observations. To address this issue, auxiliary variables in the context of external drift kriging (Goovaerts, 1997) are employed. The best auxiliary variable for defining linear drifts was the one that includes atmospheric emissions as well as population and altitude.

Measurements of PAH topsoil concentrations are available via the French Soil Monitoring Network. Qualitative data on the localization of polluted sites are integrated by processing

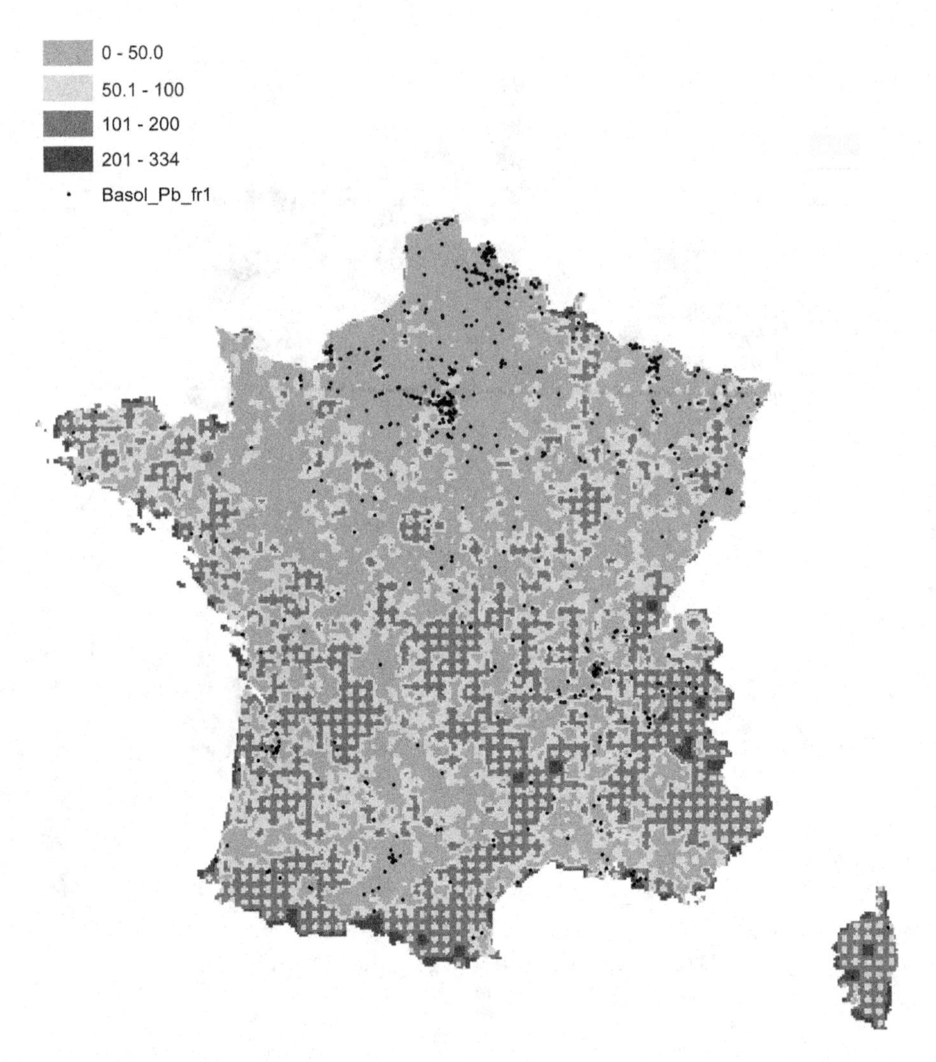

Figure 21.3 (Continued)

distance-to-polluted soil proxy. These along with 14 variables relating to physicochemical soil properties were combined in a hybrid regression-kriging and fitted by using random forest models, which was shown to outperform the traditionally used linear regression. Due to its hydrophobic nature, BaP is globally found in water in small concentrations, therefore the exact measurement cannot always be reported. Several concentrations under the detection limit were observed and required careful handling. A complex multiple imputation method was developed to extract maximum information from the available measurements without introducing too much bias in the results. The latter makes it possible to take advantage of the temporal aspect and the correlations between the substance of interest and other PAH substances. Spatial estimation of water concentrations is carried out by taking into account the multi-annual data and the network water distribution complexity using a bootstrap based on the expectation-maximization algorithm. The methods detailed previously enabled the construction of a representative spatial database in a 9 km^2 reference grid used to perform the integrated exposure assessment. The final results of this work showed that the PAH exposure map is a result of the inhalation and ingestion contribution combination (Figure 21.4).

By examining the map referring to total exposure, the inhalation route proves to be an important contribution. Soil appears to have the smallest contribution for both age groups. Higher intake estimates for children are a result of the higher bodyweight normalized ingestion rate, amount of soil and daily food ingested. In order to understand the dietary intake of food contaminants like PAHs, the relationship between contamination quantities contained in food products as well as nutritional habits need to be explicated (Xia et al., 2010; Duan et al., 2016).

The highest hotpot due to water ingestion is situated near an old gas factory. High values are also found in mountainous areas, since cold winters increase the need for domestic combustion in high-altitude regions and, accordingly, increased PAH emissions are observed.

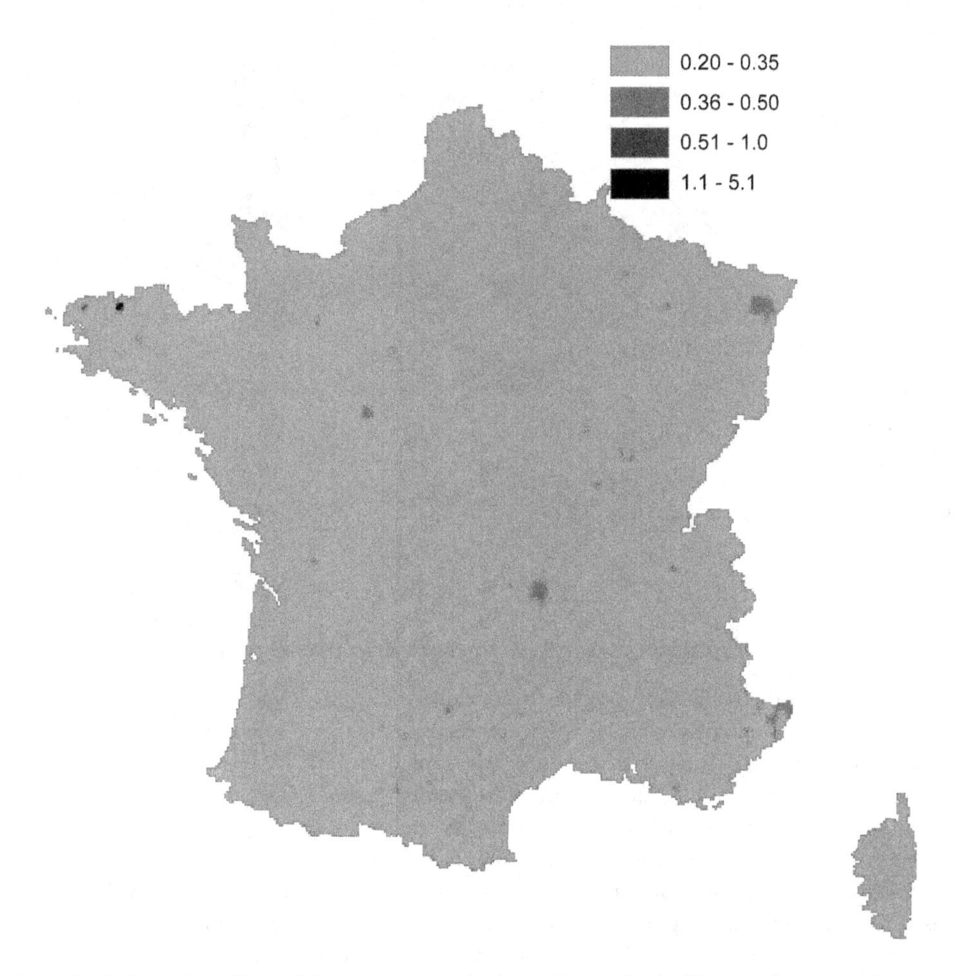

Figure 21.4 Mapping of benzo[a]pyrene exposure in the study area for the lifetime of exposure: (a) exposure due to ingestion, (b) exposure due to inhalation and (c) total exposure indicator

Source: Ioannidou et al. (2018).

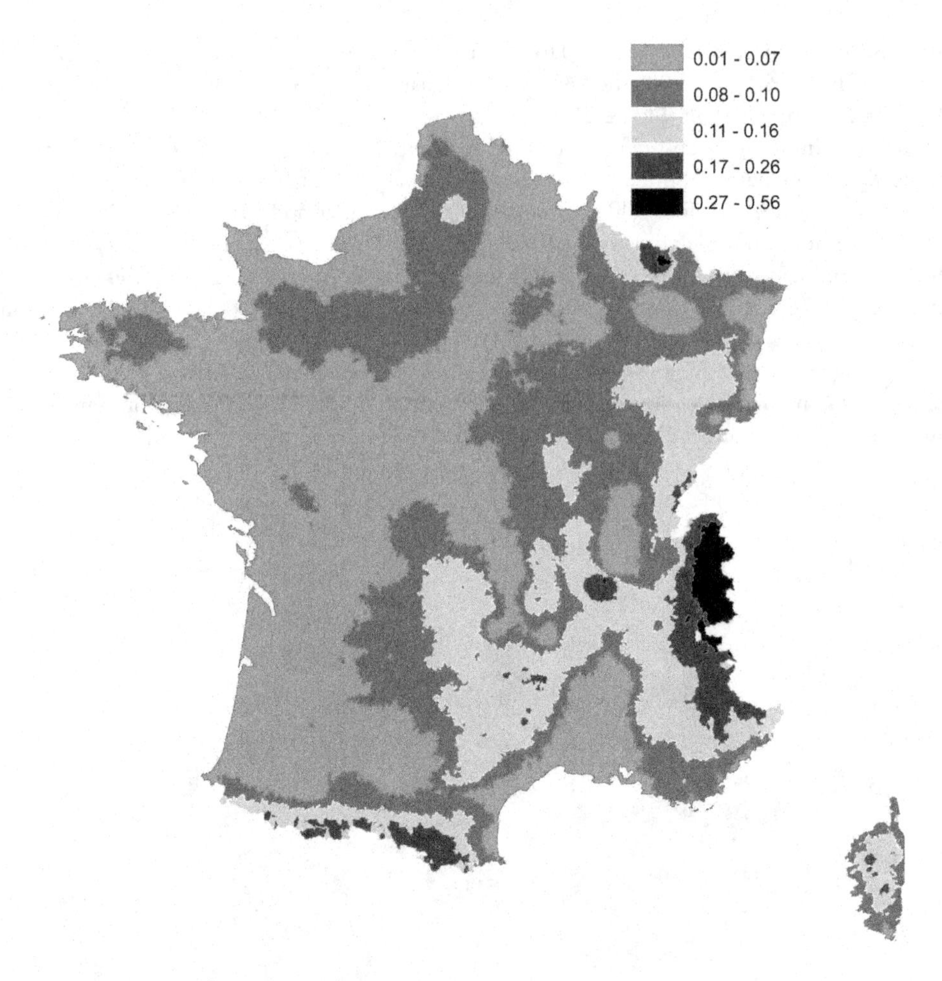

Figure 21.4 (Continued)

The case of spatial exposure of cypermethrin in a region of France

The objective of the CartoExpo project (Caudeville et al., 2019) was to test the feasibility of an integrated methodology to map exposure indicators for fine resolutions (4 km^2 grids) and short intervals (weeks). To illustrate the approach, contamination of the general population was examined for a pyrethroid (cypermethrin) in the Picardy region (Northern France). Phytosanitary assessment was an interesting application considering the recent acquisition of databases, innovative measurement devices and methodological developments (Figure 21.5).

For atmospheric dispersion, an innovative statistical meta-model method was developed, based on a machine learning technique that relied on a large database of simulations on a representative parcel. The meta-model was then applied to cypermethrin application at three-hourly intervals on all agricultural parcels in the Picardy region (Figure 21.6a). Multimedia exposure models were used to (1) estimate emissions from soil and plant volatilization phenomena on agricultural land, (2) quantify the contamination of local food products (excluding home

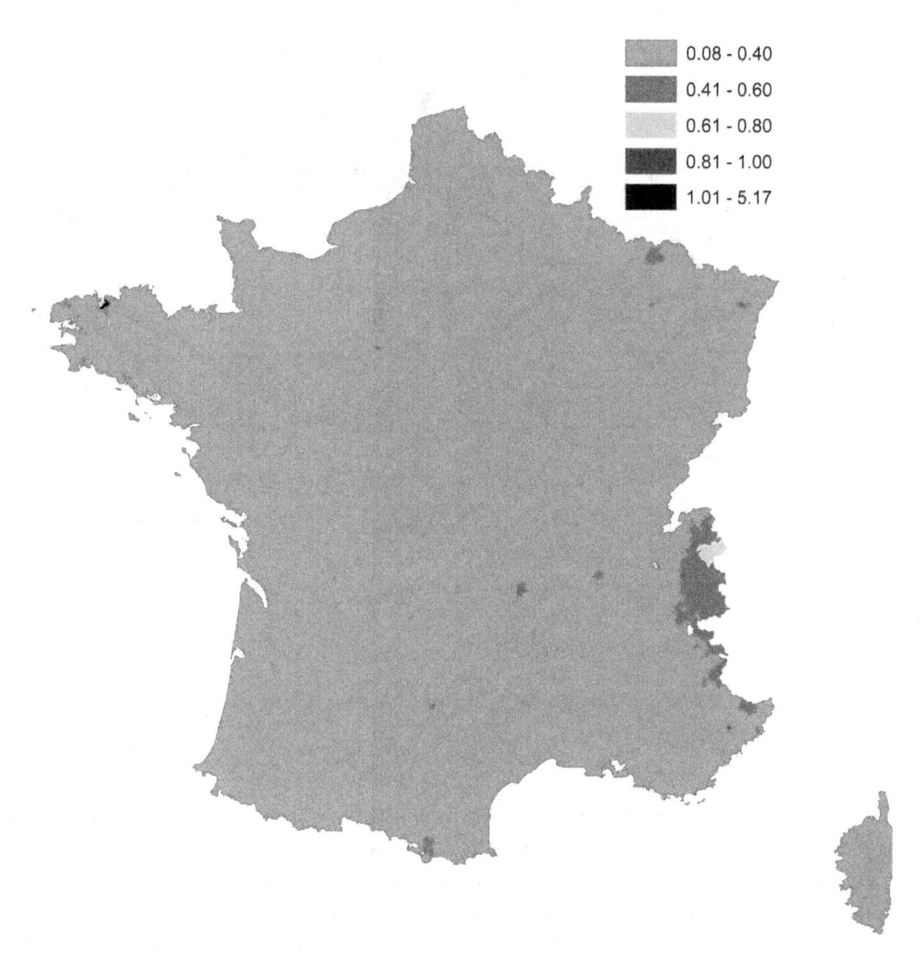

Figure 21.4 (Continued)

vegetable gardens) due to proximity to agricultural land and (3) combine external exposures (Figure 21.6b).

Comparisons of exposure estimations with biomonitoring measures in biological matrices proved feasible since estimations are available at fine spatial and temporal resolutions. Moreover, providing the information needed to evaluate reliability and accuracy of modeling results, the integration and cross-analysis of environmental, exposure and human biomonitoring data would enable researchers to build a more robust portrayal of the exposome, scalable on a larger scale.

From result ownership to action implementation

Each region drew up a regional environment and health action plan to implement the main objectives of the French National Action Plan according to its own specific needs. Different regions in France have included the reduction of EHIs in their planning, which requires

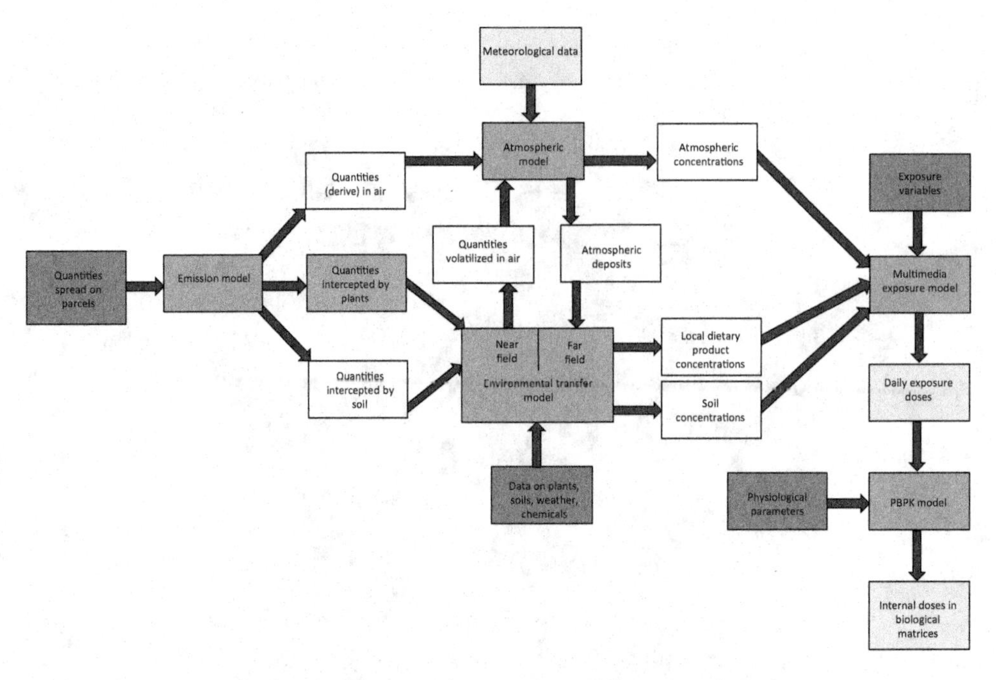

Figure 21.5 Conceptual scheme of the pesticide exposure modeling approach used

Source: Caudeville et al. (2019).

assessment to guide priorities for voluntary action. In the process of characterizing EHIs to help manage the related risk, the construction of continuous variables was required, over a specific area, for which data are not exhaustively monitored, thus introducing uncertainties. It is therefore necessary to provide elements that allow for the evaluation of the representativeness of the input data and their impact to the overall risk calculation. The use of the proposed results in a management framework should therefore be accompanied by additional maps. The way the results from the spatial exposure assessment should be reported varies according to the objective and the audience. Uncertainty should be included in the decision-making process, since ignoring it could result in misinterpretation.

The characterization of the information density and the type of data from which the indicators were constructed allows for interpreting the representativeness of the predictions. Such additional maps will include the kriging variance for air and soil or the emission zone characterization for inhalation and deposits. Those maps will be used for the orientation of additional data collection campaigns in areas where limited data are available or those where overexposure of the population is suspected. Finally, the additional maps will allow

- to estimate the level of confidence of the results;
- to acknowledge the intrinsic biases of the input data;
- to evaluate the potential over- or underestimations in the predictions;
- to determine the areas for which additional data collection is required.

This additional spatial information has important implications for policymakers and could be used to prioritize actions with potentially modifiable health determinants.

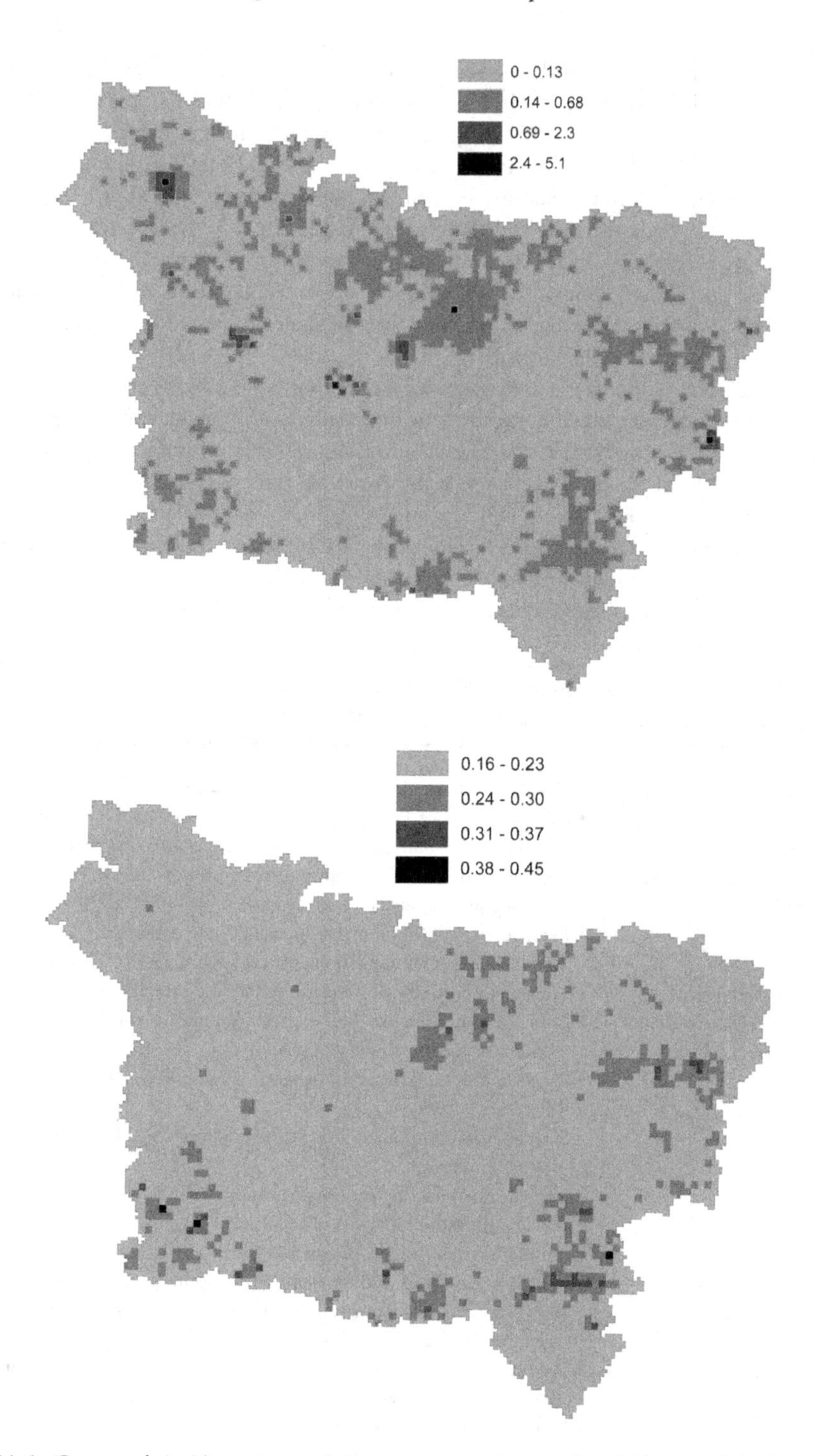

Figure 21.6 Cypermethrin (a) mean annual air concentration in $ng.m^{-3}$ and (b) mean daily dose of total exposure in $ng.kg^{-1}.j^{-1}$ integrating the contributions of soil, water and local food ingestion

Source: Caudeville et al. (2019).

Other approaches have been developed for assessing spatial relationships across health and environmental and socioeconomic data and identifying factors that influence the variability of disease patterns. Saib et al. (2015) propose a methodology for combining environmental, socioeconomic and health indicators in the context of the FNEHP by using a suite of statewide indicators to characterize both pollution burden and population characteristics.

The construction of composite indicators was also needed to provide diagnostics at a territorial level by integrating various environmental, socioeconomic and health dimensions. The definition of a cumulative indicator that combines different dimensions is broad and does not suggest a specific process. Different approaches have been already tested to combine indicators (Morello-Frosch et al., 2011; Sexton, 2012).

Integrating stakeholders in the decision process of defining the subjective conceptual analysis framework or assumptions when uncertainty or knowledge gaps operate contributes to formulating problems, collecting data, accessing data and facilitating action implementation.

Action to tackle inequalities needs to be informed by evidence on the population groups most affected and the sociodemographic features associated with the unequal distribution of risks and opportunities. Hence, better quality evidence and adequate identification of the specific target groups could help to make interventions more effective. Table 21.2 indicates the potential benefits of using inequality evidence for policy action, suggesting that such actions can

Table 21.2 Benefits of environmental inequality data for effective action

Inequality evidence	*Policy actions*
Evidence on societal structures and mechanisms leading to environmental inequalities	• Review and learn from examples of good/equitable societal practices. • Formulate equitable policy options on environmental protection. • Improve public participation in planning and decision-making processes affecting people's local environment. • Incorporate environmental and health equity issues into economic, social and infrastructural regulations, strategies and plans.
Evidence on differential exposure to environmental health risks	• Enforce environmental standards where they are exceeded. • Implement appropriate interventions to improve environmental conditions for the whole population. • Target action on pollution hot spots and population subgroups with the highest exposures. • Shift attention to policies that assure environmental protection and population health. • Support intersectoral action and extend "health in all policies" approaches. • Review the equity impacts of regulations directly or indirectly affecting environmental conditions (such as urban and infrastructure planning, taxation and social welfare) and their implementation.
Evidence on differential vulnerability to environmental health risks	• Ensure that adequate environmental and infrastructural services and conditions are accessible for all. • Provide environmental resources and social benefits to compensate for the influence of environmental risks or social stressors. • Increase targeted protection measures in areas or settings with a high density of vulnerable, sensitive or disproportionately affected populations. • Improve environmental standards in the vicinity of child care centers, schools, hospitals, nursing homes and similar.

Source: WHO Regional Office for Europe (2019) based on WHO Regional Office for Europe, (2012).

be focused on societal structures and mechanisms as well as on resulting disparities in exposure and/or vulnerability.

Conclusion

In a recent report, WHO identified EHIs as a priority issue that needs to be addressed by the national governments in Europe. The reduction of health inequalities requires identifying and characterizing exposure as well as social factors in order to interpret how they accumulate across a territory and prioritize interventions. As the health status of a population is the result of complex interactions taking place across several social, territorial and environmental factors, all related information needs to be studied to assess it. On a regional or continental scale, the characterization of EHIs expresses the idea that populations are not equal in the face of pollution and is based on analysis in order to identify and manage the areas at risk of overexposure where they are suspected of generating potential increased risk to human health. The development of methods is a prerequisite for the implementation of public health actions aimed at the protection of populations. However, constructing methods and tools to help orient public policies to reduce territorialized EHIs requires the evaluation of phenomena not always easy to apprehend and the reliability and representativeness of available information that usually demand statistical processing.

The exposome concept has been proposed as an emergent exposure science paradigm for conceptualizing the cumulative effects of environmental exposures across the entire human life. The need for risk managers to identify populations at risk in the context of substantial data deficiencies that hinder the evaluation of cumulative health risks places the operational interpretation of the concept on a territorial scale. The characterization of the territorialized exposome entails the development of dynamic, multidimensional, longitudinal approaches, as well as information systems that require the adoption of transdisciplinary data analysis methods. For instance, integrated approaches could merge all the information required for appraising the source-to-human-dose continuum by using the geographic information system, multimedia exposure and toxicokinetic model. Beyond the established feasibility of coupling the different models implemented, sensitivity studies will allow for the identification of data and processes of specific interest requiring additional work to reduce uncertainties.

This framework could be used for many purposes, such as

- mapping environmental disparities;
- identifying vulnerable populations and determinants of exposure to manage and plan remedial actions;
- highlighting hot spot areas with significantly elevated exposure indicator values, so as to define environmental monitoring campaigns;
- assessing spatial relationships across health and environmental and socioeconomic data to identify factors that influence the variability of disease patterns.

This approach addresses the need for a methodological tool particularly suitable for the construction of integrated platforms from source to health risk. The data production design also requires optimization for their integration in EHI characterization, extending beyond the original objectives of intended use. The structuring of existing acquisition networks into a coherent network covering the entire territory would allow for the collection, integration, analysis, interpretation and dissemination of environmental, exposure, socioeconomic and health data. The aim would be to establish the prerequisites for the collection and dissemination of "smart" health and environmental data, namely to provide the information needed to characterize EHIs and facilitate actions to improve personal health. To this end, the EPHTIS model developed in the United States is an interesting source of inspiration thanks to the establishment of a politically and technically legitimate coordination allowing for a very real exchange across multidisciplinary organizations able to monitor, identify, organize and improve the quality of data relevant for the characterization of EHIs.

Bibliography

Abelsohn, A., Frank, J. and Eyles, J. Environmental public health tracking/surveillance in Canada: A commentary. *Health Policy* 2009, 4 (3), 37–52.

Asante-Duah, K. *Public health risk assessment for human exposure to chemicals.* Netherlands: Kluwer, 2002.

Bonnard, R. and McKone, T.E. Integration of the predictions of two models with dose measurements in a case study of children exposed to the emissions of a lead smelter. *Human and Ecological Risk Assessment* 2010, 15 (6), 1203–1226.

Bopp, S.K., Kienzler, A., Richarz, A.N., et al. Regulatory assessment and risk management of chemical mixtures: Challenges and ways forward. *Critical Reviews in Toxicology* 2019. https://doi.org/10.1080/10408444.2019

Brunekreef, B. Exposure science, the exposome, and public health. *Environmental and Molecular Mutagenesis* 2013, 54, 596–598.

Buck Louis, G.M. and Sundaram, R. Exposome: Time for transformative research. *Statistics in Medicine* 2012, 31, 2569–2575.

Caudeville, J., Bonnard, R., Boudet, et al. Development of a spatial stochastic multimedia model to assess population exposure at a regional scale. *Science of the Total Environment* 2012b, 432, 297–308.

Caudeville, J., Goovaerts, P., Carré, F., et al. 2012a. Spatial modeling of human exposure to soil contamination – an example of Digital Soil Assessment. Chapter in *Digital soil assessments and beyond.* Eds. Minasny, B., Malone, B.P., McBratney, A.B. Sydney: CRC Press.

Caudeville, J. and Habran, S. Systèmes d'information de surveillance en santé environnement. Environ Risque Santé 2019, 18, 1–10. https://doi.org/10.1684/ers.2019.1307

Caudeville, J., Ioannidou, D., et al. Cumulative risk assessment in the Lorraine region: A framework to characterize environmental health inequalities. *International Journal of Environmental Research and Public Health* 2017, 14 (3), 291.

Caudeville, J., Regrain, C., Bonnard, R., Lemaire, V., Létinois, L., Tognet, F., Brochot, C., Bach, V., Chardon. K. and Zeman, F. Caractérisation de l'exposition environnementale spatialisée à un pyréthrinoïde en Picardie. *Environnement, Risque et Santé* 2019, 18, 1–9. http://doi.org/10.1684/ers.2019.1340

CCE. *Document-cadre sur les facteurs à prendre en considération dans la caractérisation de la vulnérabilité à la pollution de l'environnement en Amérique du Nord.* Montréal (Québec), Canada: Commission de coopération environnementale, 2014 (50 p.).

Ciffroy, P., Alfonso, B., Altenpohl, A., et al. Modelling the exposure to chemicals for risk assessment: A comprehensive library of multimedia and PBPK models for integration, prediction, uncertainty and sensitivity analysis – the MERLIN-expo tool. *Science of the Total Environment* 2016. http://doi.org/10.1016/j.scitotenv.2016.1003.1191.

Davezac, H., Grandguillot, G., Robin, A. and Saout, C. *L'eau Potable en France 2005–2006.* Ministère de la Santé, de la Jeunesse, des Sports et de la Vie associative Report, Paris, France, 2008.

Delgado-Saborit, J.M., Stark, C. and Harrison, R.M. Carcinogenic potential, levels and sources of polycyclic aromatic hydrocarbon mixtures in indoor and outdoor environments and their implications for air quality standards. *Environment International* 2011, 37, 383–392.

Duan, X., Shen, G., Yang, H., Tian, J., Wei, F., Gong, J. and Zhang, J.J. Dietary intake polycyclic aromatic hydrocarbons (PAHs) and associated cancer risk in a cohort of Chinese urban adults: Inter- and intra-individual variability. *Chemosphere* 2016, 144, 2469–2475.

EC. *Living well, within the limits of our planet.* 7th EAP – The new general Union Environment Action Programme to 2020, European Commission, 2013.

EEA. *Unequal exposure and unequal impacts: Social vulnerability to air pollution, noise and extreme temperatures in Europe.* EEA Report No 22/2018, 2019. ISSN 1977-8449.

Evans, A.M., Rice, G., Wright, J.M. and Teuschler, L.K. Exploratory cumulative risk assessment (CRA) approaches using secondary data. *Human and Ecological Risk Assessment* 2013, 20, 704–723.
Gee, G.C. and Payne-Sturges, D.C. Environmental health disparities: A framework integrating psychosocial and environmental concepts. *Environmental Health Perspectives* 2004, 112, 1645–1654.
Glorennec, P., Bemrah, N., Tard, A., et al. Probabilistic modeling of young children's overall lead exposure in France: Integrated approach for various exposure media. *Environment International* 2007, 33, 937–945.
Goovaerts, P. Kriging vs stochastic simulation for risk analysis in soil contamination. *Geostatistics for Environmental Applications* 1997, 9, 247–258.
Ioannidou, D. *Characterization of environmental inequalities due to Polyaromatic Hydrocarbons in France: Developing environmental data processing methods to spatialize exposure indicators for PAH substances.* Santé publique et épidémiologie. Conservatoire national des arts et metiers – CNAM, 2018.
Ioannidou, D., Malherbe, L., Beauchamp, M., et al. Characterization of environmental health inequalities due to Polyaromatic Hydrocarbons exposure in France. *International Journal of Environmental Research and Public Health* 2018, 15, 2680. http://doi.org/10.3390/ijerph15122680.
Juarez, P.D., Matthews-Juarez, P., Hood, D.B., et al. The public health exposome: A population-based, exposure science approach to health disparities research. *International Journal of Environmental Research and Public Health* 2014, 11, 12866–12895.
Karjalainen, T., Hoeveler, A. and Draghia-Akli, R. European Union research in support of environment and health: Building scientific evidence base for policy. *Environment International* 2017, 103, 51–60.
Kienzler, A., Berggren, E., Bessems, J., et al. Assessment of mixtures – review of regulatory requirements and guidance EUR 26675 EN, 2014, 10.2788/84264.
Kienzler, A., Bopp, S.K., van der Linden, S., et al. Regulatory assessment of chemical mixtures: Requirements, current approaches and future perspectives. *Regulatory Toxicology and Pharmacology* 2016. http://doi.org/10.1016/j.yrtph.2016.05.020.
Kim, R., Dalbokova, D. and Krzyzanowski, M. Development of environmental health indicators for European Union countries. *EpiMarker* 2005, 9 (3), 1–4.
Lioy, P.J. Exposure science: A view of the past and milestones for the future. *Environmental Health Perspectives* 2010, 118, 1081–1090.
Lioy, P.J. and Smith, K.R. A discussion of exposure science in the 21st century: A vision and a strategy. *Environmental Health Perspectives* 2013, 121, 405–409.
Liu, H.Y., Bartonova, A., Pascal, M., et al. Approaches to integrated monitoring for environmental health impact assessment. *Environmental Health* 2012, 11, 88.
Marchant, B.P., Saby, N.P.A., Lark, M., Bellamy, P.H., Jolivet, C.C. and Arrouays, D. Robust analysis of soil properties at the national scale: Cadmium content of French soils. *European Journal of Soil Science* 2010, 61, 144–152.
McKone, T.E. and Ozkaynak, R.P. Exposure information in environmental health research: Current opportunities and future directions for particulate matter, ozone, and toxic air pollutants. *Journal of Exposure Science & Environmental Epidemiology* 2009, 19, 30–34.
Medina, S., Lim, T.A., Declercq, C., et al. Les programmes de surveillance en santé environnementale en France: Apports des travaux européens et internationaux. *Bulletin épidémiologique hebdomadaire* 2009, 27–28, 309–312 (numéro thématique).
Menut, L., Bessagnet, B., Khvorostyanov, D., et al. Chimere: A model for regional atmospheric composition modelling. *Geoscientific Model Development* 2013, 6 (4).
Morello-Frosch, R., Zuk, M., Jerrett, M., Shamasunder, B. and Kyle, A.D. Understanding the cumulative impacts of inequalities in environmental health: Implications for policy. *Health Aff (Millwood).* 2011, 30, 879–887.
Pew Environmental Health Commission. *America's environmental health gap: Why the county needs a nationwide health tracking network: Technical report.* Baltimore, MD: Johns Hopkins University School of Public Health, 2000.
QGIS Development Team. *QGIS geographic information system.* Open Source Geospatial Foundation Project, 2018. http://qgis.osgeo.org.
R Core Team. *R: A language and environment for statistical computing.* Vienna, Austria: R Foundation for Statistical Computing, 2013.
Saib, M.S., Caudeville, J., Beauchamp, M., et al. Building spatial composite indicators to analyze environmental health inequalities on a regional scale. *Environmental Health* 2015, 14, 68.

Sexton, K. Cumulative risk assessment: An overview of methodological approaches for evaluating combined health effects from exposure to multiple environmental stressors. *Journal Environment Research Public Health* 2012, 9, 370–390.

Teeguarden, J.G., Tan, Y.M., Edwards, S.W., et al. Completing the link between exposure science and toxicology for improved environmental health decision making: The aggregate exposure pathway framework. *Environmental Science and Technology* 2016, 50 (9), 4579–4586.

UN. *Sustainable development goals*, 2015. www.un.org/sustainabledevelopment/sustainable-development-goals.

US EPA. *Concepts, methods and data sources for cumulative health risk assessment of multiple chemicals, exposures and effects: A resource document (EPA/600/R-8 06/014F)*. Cincinnati, OH: National Center for Environmental Assessment, Office of Research and Development, U.S. Environmental Protection Agency, 2007.

WHO Regional Office for Europe. *Parma declaration on environment and health*. Copenhagen: WHO Regional Office for Europe, 2010.

WHO Regional Office for Europe. *Environmental health inequalities in Europe: Assessment report*. Copenhagen: WHO Regional Office for Europe, 2012.

WHO Regional Office for Europe. *Improving environment and health in Europe: How far have we gotten?* Copenhagen: WHO Regional Office for Europe, 2015.

WHO Regional Office for Europe. *Declaration of the sixth ministerial conference on environment and health*. Copenhagen: WHO Regional Office for Europe, 2017.

WHO Regional Office for Europe. *Healthy environments for healthier people*. Copenhagen: WHO Regional Office for Europe, 2018.

WHO Regional Office for Europe. *Environmental health inequalities in Europe Second assessment report*. Copenhagen: WHO Regional Office for Europe, 2019.

Wild CP. Complementing the genome with an "exposome": The outstanding challenge of environmental exposure measurement in molecular epidemiology. *Cancer Epidemiology, Biomarkers & Prevention* 2005, 14, 1847–1850.

Xia, Z., Duan, X., Qiu, W., Liu, D., Wang, B., Tao, S., Jiang, Q., Lu, B., Song, Y. and Hu, X. Health risk assessment on dietary exposure to polycyclic aromatic hydrocarbons (PAHs) in Taiwan, China. *Science of The Total Environment* 2010, 408 (22), 5331–5337.

22

BUILDING ON THE RIGHT TO KNOW

Data interlinkage and information intermediation for environmental and corporate regulation

Richard Puchalsky, Michael Ash and James K. Boyce

Introduction

Environmental injustice is systematic exclusion from environmental goods and exposure to environmental "bads" based on social difference, often in ascribed categories such as caste or, especially in the US, race and ethnicity. One of the fundamental tasks in researching environmental justice is to link pollution sources to people affected, which then enables systematic analysis of the social correlates and determinants of exposure.

Arguments over what constitutes proximity between polluting facilities and the people potentially exposed haunted the early years of environmental justice research. The method of unit-hazard coincidence – which finds exposure only for the population residing in the same geographic unit as the polluting facility, be it small area geography of the US Census (tracts or block groups, corresponding more or less to neighborhoods), postal codes, counties, states, or fixed-radius circular buffers has substantive weaknesses (Mohai and Saha 2003), as the shared geography unit may be either too large or too small to make a substantive determination regarding exposure.

Innovative US Environmental Protection Agency (EPA) databases combine information about the type, amount, and geographic origin of pollution with modeled transport to receptor locations that can be associated with US Census data. Linkages between databases are technically complex, often involving imperfect matching methods across linkages of facility ownership, function and scale, and geography and proximity. A fundamental problem is that some applications of data were not envisioned by the entities that produce the underlying databases. For instance, linking individual polluting facilities to the parent companies that own them can relate causes of environmental injustice to corporate policies, but EPA typically does not analyze pollution by parent company even though this information is collected as part of several EPA databases.

In this chapter, we describe broadening uses of two emissions databases from the US EPA that have been important for empirical environmental justice research: the Toxic Release Inventory (TRI), which provides annual, chemical-specific, quantified amounts of air, water, and land pollution from industrial facilities, and the Greenhouse Gas Reporting Program (GHGRP)

DOI: 10.4324/9780367814533-24

database, which provides CO_2-equivalent emissions of greenhouse gases from large, fixed sites. The discussion is technical and detailed; it is intended to function as a guide for practitioners to carry out analysis of environmental justice and corporate environmental justice performance using databases and tools. Availability of data may vary across countries and context. For example, the pollutants listed and the industries covered in pollutant release and transfer registers varies across countries, although the European Pollutant Release and Transfer Register (E-PRTR) harmonizes consistent data collection across Europe. The public availability of data on corporate ownership of pollution sources and on economic and employment characteristics of facilities varies substantially. Census and household survey data vary substantially in terms of geographic detail – many countries do not report data at the neighborhood level – and of the social and economic variables reported. France, notably, does not collect data on national origin for descendants of immigrants or on ethnicity.

The TRI additionally supports yearly estimates of the potential chronic human health risk from each facility by running the underlying data through a value-added EPA model, RSEI (Risk-Screening Environmental Indicators), which produces comparative estimates of chronic human health risk at finely distinguished geographic locations of receptors. RSEI was initially developed at the US EPA to assist in setting priorities for investigation and enforcement among TRI reporting facilities. Hence, its initial structure was designed to generate a univariate comparative risk score for each facility, reflecting the potential chronic human health risk from the facility based on the quantity, toxicity, fate and transport, and population exposure of its emissions. The potential of the high-resolution geographic model of fate and transport was soon recognized by its developers as a potential tool for assessing differential subpopulation risk as well as total population risk (Bouwes et al. 2003).

The GHGRP is designed for reporting greenhouse gas emissions, which are themselves a global problem with limited health impact near the point of release. But greenhouse gas emissions are often accompanied by co-pollutants with substantial local and regional effects, such as particulate matter, nitrogen oxides, and sulfur dioxide, and hence greenhouse gases can be used as a proxy for the local effects of fossil fuel combustion and for environmental justice profiles.

We have linked the TRI/RSEI and GHGRP with each other and with sources of information. The data enable us to rank corporations based on airborne potential chronic human health risk generated by all of their TRI-reporting facilities. The rankings, which are produced on a regular basis from annually updated data from the US EPA, are used by socially responsible investors, corporate environmental managers, regulators, and activists interested in assessing environmental performance at the level of the corporation as well as individual facilities. These data are published on a maintained website, toxic100.org, and are also made available for download to researchers. Using the methodology developed in Ash and Boyce (2011), we characterize the distribution across vulnerable environmental justice communities of chronic human health effects caused by toxic air pollution, both for individual facilities and for parent companies that own them. We additionally characterize the distribution across vulnerable environmental justice communities of co-pollutants of fossil fuel combustion by facilities and the companies that own them.

The data also permit estimation of pollution levels at key receptor sites, including schools, neighborhoods, and other socially vulnerable locations, which enables community, regulatory, and corporate response and permits generalizable research and hypothesis testing about the underlying political and economic processes that lead to differential exposure. Interlinkage of parent company information on emissions of toxics and greenhouse gases to company data on the receipt of public subsidies and the assessment of fines and penalties for environmental, labor, financial, and other infractions enables a more comprehensive characterization of corporate environmental and social governance. Interlinkage of pollutant risk scores with facility-level administrative data on employment from the Equal Employment Opportunity Commission has permitted generalizable research on the often-posited trade-off between jobs and the environment, in particular concerning the employment of minorities in relation to their disproportionate exposure to industrial pollution, and could in principle be applied to holding facilities and companies responsible for their broad social and environmental impact.

Developing interlinkages presents substantial technical challenges, from the integrative assessment model that describes spatial exposure to pollutant releases to selecting and joining spatial, organizational, and socioeconomic data to characterize the landscape of risk. We describe the development of the underlying pollutant release and transfer register data and the process and challenges of linkage. We then briefly survey the results of the studies enabled by the linkages.

We first introduce the toxic air pollutant data used with sections titled "The Toxics Release Inventory: The World's First PRTR" and "Risk Screening Environmental Indicators: An Integrative Assessment Model to Estimate Human Health Risk from Industrial Air Pollution," and then go on to describe how to improve these data with sections titled "Parent Assignment: Corporate Research to Assign TRI Facilities to Ultimate Owners" and "Environmental Justice Ratios: Measuring Corporate Environmental Justice Performance." Subsequent sections introduce the greenhouse gas reporting data used with "Linking Local and Global Pollutants: The Greenhouse Gas Reporting Program" and describe how to improve it with "Extending EJ to Greenhouse Gas Emitters," "Assigning Responsibility for Greenhouse Gas Emissions," and "GHGRP EJ Ratio Analysis for Parent Companies." The section "Lessons on Linkages" describes how to link US EPA databases together at the facility level. Finally, the section "Applied and Social Scientific Research" describes previous research that has been done using these data and "Public Intermediation for Policy Impact" describes previous efforts to provide free and useful public access to the data.

The toxics release inventory: the world's first PRTR

Created by the Emergency Preparedness and Community Right-to-Know Act of 1986 (EPCRA), the Toxics Release Inventory became the world's first national pollutant release and transfer register (PRTR), an innovation that has since been adopted in many industrialized countries. Annual TRI data collection began in 1987, and data from 1988 forward are considered to be of high quality. The TRI was an important advance for researchers because it provides yearly reports of the total mass of toxic chemicals released, broken down by facility, chemical, and medium (air, land, and water) – data that were previously impossible to obtain.

Right-to-know regulation envisions policy or market-based changes resulting from better public information, and right-to-know regulation mandates disclosure of information. Key examples of public information mandates include the pollution data we discuss here and other environmental data such as residential water quality, energy efficiency to guide the purchase of cars and consumer durables, disclosure of the risks inherent in financial assets and loans for investors and consumers, lending performance by banks especially in the domain of racial equity, school performance on standardized tests, and health care provider performance on a range of indicators. Right-to-know regulation often emerges as a compromise between public demands for more concrete regulation and industry resistance to outright regulation. Fung et al. (2007) survey and analyze many domains of regulation by disclosure.

The conversion of right-to-know data into concrete fulfillment of the right to a clean and safe environment requires not only that stakeholders have access to the information, but also, critically, that they have the ability to interpret the information and the capacity and incentive to respond to it (Hersh 2006).

TRI requires large industrial facilities in the United States to report on an annual basis what quantity of each of roughly 600 different listed toxic chemicals or chemical categories the facility released into the environment over the course of the past year. A facility must report to the TRI if it operates in a TRI-covered industrial sector based on its North American Industrial Classification System (NAICS, which replaced the Standard Industrial Classification system), employs at least 10 full-time equivalent employees, and manufactures, processes, or otherwise uses quantities of the chemical in excess of published, chemical-specific thresholds. Most facilities engaged in manufacturing, electrical energy generation from fossil fuels and biomass, coal and metal mining, wholesaling of petroleum products and other chemicals, and hazardous waste storage and disposal must report. Noteworthy exclusions include fracking and oil extraction, electrical energy generation from natural gas combustion, mobile sources, ports, and airports.

The facility, rather than the company, is the reporting unit for TRI and more generally is the object of permitting and enforcement by the US EPA. Each facility submits annual Form R reports to the TRI Central Data Exchange signed by the facility's certifying official. The Form R reports the quantity of each toxic chemical released and the release media (i.e., whether the chemical was released to air, land, or water or transferred offsite). Air media include fugitive releases, stack releases, incineration-based releases, and off-gassing of effluents from publicly operated treatment works. While the term "fugitive release" may evoke industrial accidents and spills, which are subject to reporting requirements, most of the data reported to the TRI (including the fugitive-release category) involves business-as-usual releases. Quantities are reported on an annual basis with no indication of the timing of releases. In their reporting, facilities may use direct measurement of the mass of inputs and outputs or alternative methods, including estimates based on engineering specifications for particular industrial processes.

The TRI data are mandatory and standardized across industries and states, but they are self-reported. While penalties for failing to report and for misreporting are, in principle, high, the occasional TRI enforcement that does occur is generally for nonreporting, and there are indications of systematic underreporting.[1]

The Form R reporting instrument includes a parent company field for facilities to report their corporate ownership, which EPA has attempted to standardize. Remarkably, the US does not maintain a single unique public identifier of ultimate corporate parents. The US EPA uses the Dun and Bradstreet DUNS Number, a private and proprietary unique nine-digit identifier for businesses. The frequency of reorganizations of corporate ownership via mergers, acquisitions, and divestments often leaves parent company information out of date, and the quality of this information, in terms of accurate assignment of facilities to final parent companies, is poor.

Early analysis of TRI data generally consisted of adding up the pounds of releases across chemicals and media for the entities under consideration, be they geographical areas, industrial sectors, or individual facilities. This was unsatisfactory for many purposes because some TRI chemicals are far more hazardous to human health than others. Adding them together by pounds meant that the total often was dominated by lower-risk chemicals released in large quantities, rather than identifying high-risk, low-volume chemicals. EPA's RSEI model was developed to address this and similar issues. RSEI uses a peer-reviewed system of toxicity weights that express how dangerous each chemical is on a per-pound basis; the toxicity weights make it easier to understand the importance of obscurely named chemicals for actual human health risk.

Risk Screening Environmental Indicators: an integrative assessment model to estimate human health risk from industrial air pollution

The US EPA's Risk Screening Environmental Indicators model uses various sources of toxicological information to weigh TRI-listed chemicals for both cancer and non-cancer human health effects, which are put into the same scoring system. RSEI then runs a fate-and-transport model for each TRI facility using weather patterns, velocity and altitude of release, and physicochemical properties of the released chemical to estimate where air pollution from the facility goes. The estimates are computed for each 810 m by 810 m cell within 50 km of the releasing facility. (RSEI uses a different exposure model for surface water pollution.) Finally, RSEI multiplies the amount of pollution at each receptor location by the number of people residing in that location. The resulting estimate constitutes a comparative risk score that can be added up over any subset of releases and that takes into account the release quantity, chemical toxicity, fate and transport, and the size of the exposed population. A typical EPA use of RSEI is to add up the risk score for each TRI release from a facility over the entire area for which population exposure is computed to establish a risk score for the facility as a whole.

As part of the production of RSEI, estimates of air exposure to each chemical from each facility are made for each grid cell within 50 km of each releasing facility. This allows coverage of the entire US with a consistent closely spaced, high-resolution geographic grid that can be associated with US Census geographic areas. Because the data are broken out by both chemical identity and the individual facility releasing the pollution, the dataset allows for many kinds of analysis. The data providing the toxicity-weighted concentrations for every 810 m by 810 m receptor site, by source facility and chemical, are referred to as the RSEI Geographic Microdata (RSEI-GM). These data are free and publicly available.

Production and use of the RSEI-GM data present several challenges. First, the production is both data intensive, requiring facility-specific information that may not be included in typical PRTR data collection, and computationally intensive, requiring the estimation of a concentration based on a plume model at roughly 12,000 sites for each release from each facility (for roughly 100,000 air releases in 2017).

Second, the RSEI-GM data are very large, requiring substantial computing facilities simply to maintain and access the data. For example, the 2017 data included roughly 1.1 billion data points, each characterizing the effect of one release on one grid cell.

Third, the RSEI-GM grid cells are labelled with an RSEI-specific X-Y coordinate system, which requires some geographic sophistication to use. These X-Y locations can be converted to or from lat-long coordinates, and the US EPA publishes a full crosswalk between X-Y cells and US Census Bureau blocks, the finest geographic unit for the census. With the crosswalk, RSEI concentrations can be compared with or aggregated to US Census American Community Survey five-year data, which contain demographic information suitable for environmental justice research, at the census block group or tract level.

A unique feature of the RSEI model is its tight coupling of source and receptor in the analysis of risk from industrial toxic pollution. Datasets such as the US EPA's National Air Toxics Assessment (NATA) have some advantages over RSEI with respect to the wider range of included pollution sources, including mobile and so-called area sources in addition to the industrial point sources included in TRI. NATA also reports airborne risk from these activities on a high-resolution geographic basis. But the inclusion of multiple sources comes at the cost of losing the association between specific sources and community receptors. As RSEI fully models each toxic release from each releasing facility and maintains release-specific exposure data, it is possible to attach the community burden at the receptor location to the source facility.

This association enables two scorings of facilities: one based on the total potential chronic human health risk from the facility, called the RSEI score; and another based on the potential chronic human health risk for populations and subpopulations of interest, for example, the Hispanic RSEI score indicating the total potential chronic human risk from the facility for the Hispanic, or Latino, population. The subpopulation-specific scores for the facility sum to the total population score for the facility.

The tight connection between the high-resolution toxicity-weighted concentration estimates of potential chronic human health hazard from industrial pollution and small-area socioeconomic data on residents allows the production of environmental justice (EJ) ratios that expresses the total RSEI exposure for people within a demographic category of interest, for example people with income under the US poverty line, divided by RSEI exposure for the entire population from the same source. Numbers of people affected times exposure concentration times toxicity weight can be aggregated over multiple pollution releases, so these EJ ratios can be created for any summative entity: states, cities, parent companies, and so on (Ash and Boyce 2011).

The RSEI-GM data include four different routes of public exposure to air pollution from industrial facilities: (1) direct releases to the air from point sources at the facility, such as smokestacks; (2) "fugitive" releases from undetermined points at the facility, such as open storage containers or spills; (3) releases of chemicals not destroyed by incineration that occur after transfer of the chemical from the originating facility to an incineration facility; and (4) transfers of chemicals by public or private sewerage to publicly operated treatment works (POTWs) resulting in air emissions from the volatilization of the chemical from the POTW. RSEI tracks the transfer of chemicals from TRI facilities to treatment facilities, that is, incinerators and POTWs, and models the transfer sites as the source of release within the RSEI-GM grid. For purposes of estimating receptor concentrations, this tracking follows chemicals into the landscape by introducing source locations that are not necessarily themselves TRI facilities. Responsibility for these offsite releases is assigned to the TRI facility that originally produced and transferred the chemical.

There are consistency issues in data analysis of RSEI-GM data that researchers should take into account: late revisions to TRI data, regulatory changes in TRI reporting rules, and variation in chemical speciation for TRI chemicals (notably in the case of chromium). Late revisions to TRI data occur because EPA allows TRI reporters to revise past data submissions at any time: these can be corrected in RSEI by multiplying scores by the ratio of the new to old release amount (only downward, since upward corrections often would involve creating a new score where none existed). Variation in the methods used to calculate RSEI scores can occur from both changes in the RSEI estimation methods and changes in the regulations that require TRI reporting. To compare RSEI scores across TRI data years, these changes have to be removed by using a "core chemical data set" or "core industry set" that excludes chemicals or industries whose reporting requirements have changed across the relevant range of years. Lastly, some of the chemicals that are reported to TRI are actually groups of chemicals rather than single chemical entities. For instance, chromium is reported to TRI as either chromium or chromium compounds and is modeled by RSEI as a single chemical category. However, hexavalent chromium Cr(VI) and trivalent chromium Cr(III) have very different human health risks, and the researcher may need to look into RSEI's chromium speciation estimates in detail.

Parent assignment: corporate research to assign TRI facilities to ultimate owners

The process described earlier can be used to compute EJ ratios by geographic area, facility, industry, chemical, and location, but not by parent company. The latter information is important, however, since corporate policy can affect the severity of environmental justice disparities, and finding out a corporation's total and comparative responsibility is one of the tools that communities sometimes use when they try to make political change. An early social-scientific analysis of TRI (Hamilton 1995) demonstrated that financial markets respond with reduced valuations to information about companies with facilities represented in the TRI, for example, because shareholder estimates of legal liability may be higher when EPA publishes new toxics information.

The logic of EPA's facility-level data collection effort is that monitoring, regulation, and enforcement are facility-level responsibilities. However, beginning with the Hamilton (1995) analysis, public and private decision makers have observed the value of relating environmental performance to corporate policy and corporate responsibility. By joining data on facility activity with data on corporate ownership, it becomes possible for socially responsible investors, corporate environmental managers, regulators, and activists to associate corporate policy and environmental activity. Both facility-level regulations, for example the requirement of filter or scrubbers in polluting industrial processes, and systematic regulation of the owner can contribute to improving the environment.

Although EPA collects parent company information and has made some effort to standardize names in reporting of parent companies, its most generally used data distribution method, the TRI National Analysis, does not feature parent company analysis. For instance, the 2017 TRI National Analysis displays data by release and transfer type, geographic location, chemical, and industry, but the only apparent place where it breaks out the data by parent company is in the source reduction and pollution prevention section, which describes generally beneficial activities. Similarly, the EPA Envirofacts TRI Basic Search allows search by facility name, geography, industry, or chemical but not corporate parent. These are data that exist within the TRI database, but they are not generally presented by EPA in the context of responsibility for pollution.

The TRI database, in principle, contains a field for parent company information, but the ownership data are not generally reliable. There are three kinds of problems: (1) the parent company may be left blank, reported inaccurately, reported with variant spelling (as there is no standardized company identification code), or reported as a subsidiary owner rather than the ultimate parent; (2) a facility may be jointly owned by more than one parent company; and (3) a facility that has changed hands may fail to update the parent company.

We regularize parent names to reduce variation. For exchange-traded companies, the nonprofit CorpWatch provides access to a US Securities and Exchange Commission database linking subsidiaries to parents. These automated database methods improve facility-parent matching, but gaps remain. For many facilities, we use Web or library searches or contact the technical contact listed on the Form R to ascertain the parent company. These time-intensive procedures improve the quality of matches. For facilities that are jointly owned by multiple parents, we assign the pollution from the facility to the majority owner. In the case of 50/50 joint ventures, we divide the pollution from the facility between the parent companies.

Many facilities and companies change hands over time. We research mergers, acquisitions, and divestments to update facility ownership data. Ownership of individual facilities can be affected by sales of specific assets or entire companies. We establish a contemporary snapshot of ownership and assign current and historical pollution to the current owner on the principle that ownership includes responsibility for the past pollution.

Environmental justice ratios: measuring corporate environmental justice performance

The Table 22.1 shows results from the Toxic 100 Air project for 2017 (the latest data year available at the time this was written) for five parent companies that rank high for disproportionate

Table 22.1 Five selected records from PERI's Toxic 100 Air Polluters Index, 2017

Company	*EJ: minority share*	*EJ: poor share*	*Toxic 100 Air rank*	*RSEI score*	*Share of score from top facility*
Chevron	76%	20%	40	807,162	68%
Schlumberger	75%	25%	96	189,054	88%
Goodyear Tire & Rubber	74%	22%	62	368,213	65%
TMS International	74%	29%	12	3,285,626	53%
Ecolab	71%	16%	79	265,619	57%

chronic human health risk to minority groups. The Toxic 100 companies are chosen from a list of major companies (as defined by them being on various Forbes, Fortune, or S&P 500 lists); this table has been further limited to those companies with less than 90% of their risk score from a single facility. The Toxic 100 rankings are on the Web, along with the underlying data for all companies in the TRI database. In the following table, "EJ: minority share" is the share of the total population health risk borne by minority racial/ethnic groups, and "EJ: poor share" is the share borne by people living below the poverty line. For comparison, in the US population, approximately 39% were members of minority racial/ethnic groups and approximately 13% lived below the poverty line in 2017.

Examination of the individual facilities for the companies listed in Table 22.1 shows especially high burdens on minority communities, the EJ: minority share, at sites in El Segundo, California; Richmond, California; Houston, Texas; Beaumont, Texas; Gary, Indiana; East Chicago, Indiana; and Fresno, Texas. These data, which integrate pollution releases, the social distribution of pollution releases, and the ultimate corporate responsibility for the exposure can intervene in public and private decision-making in several ways. First, the data connect environmental justice to corporate decisions and show how corporate policy is expressed through siting decisions and the management of facilities. Second, the publication of these data showing both the facilities and parents can connect multiple communities affected by separate facilities with common ownership, with the potential to identify patterns in company relationships with disadvantaged communities. These connections may also be useful to regulators and to socially responsible investors who can use the tool to coordinate environmental and social corporate governance (ESG).

Linking local and global pollutants: the Greenhouse Gas Reporting Program

In 2008, Congress directed the EPA to use its existing authority under the Clean Air Act to develop a mandatory greenhouse gas reporting rule, intended to cover both upstream production of fossil fuels from suppliers and downstream sources of GHGs that were large, fixed facilities (excluding mobile sources, agriculture, residential, etc.) The supplier information is useful but not immediately applicable to EJ studies since it contains locations of production rather than release. The downstream information consists of annual reports of greenhouse gas emissions from facilities in certain industries, primarily large facilities releasing 25,000 metric tons or more of CO_2-equivalent emissions (including CO_2, methane, nitrous oxide, and some fluorinated gases). The first reports were for data collected in 2010: a number of additional industries were added in 2011. Downstream reports include nearly all emissions from electricity generation and most emissions from industrial facilities, accounting altogether for about half of all US GHG emissions.

Stated justifications for the creation of the GHGRP database generally do not include explicit right-to-know language but do include general informational purposes. For instance, EPA's FAQ page (updated September 23, 2019) on GHGRP describes the benefits of the data as follows:

> Information in the database can be used by communities to identify nearby sources of greenhouse gas emissions, help businesses track emissions and identify cost- and fuel-saving opportunities, inform policy at the state and local levels, and provide important information to the finance and investment communities.

While greenhouse gases have global effects on anthropogenic climate change, the research discussed here has to do with local human health effects from breathing co-pollutants from combustion of fossil fuels such as particulate matter, NO_X, and volatile organic compounds. Co-pollutant emissions are not directly reported in the GHGRP database, but GHG emissions can be used either as a proxy or as a link to direct estimates of these emissions from another EPA source such as eGRID (Emissions & Generation Resource Integrated Database), although those databases usually cover only the electric power generation industry.

Extending EJ to greenhouse gas emitters

GHGRP source emissions have associated lat-long points and therefore can be related to US Census American Community Survey five-year data. In calculating EJ ratios from the GHGRP database, we made certain simplifying assumptions.

First, we assumed that co-pollutant severity was proportional to CO_2-equivalent emissions from fossil fuel combustion. This assumption could be improved upon in future work by treating different fossil fuels as having different co-pollutant profiles. We also omitted biogenic CO_2-equivalent emissions from the total because they are excluded from most global climate change reporting on the basis that they are not a net source of CO_2 in the atmosphere over the medium term, yet this does not prevent them from producing local co-pollutants. Second, we assumed that demographic composition of populations affected by co-pollutants could be modeled as those living within a 10-mile radius of the facility releasing them, since there is no equivalent of RSEI for the GHGRP database that does detailed exposure modeling at the facility level.[2] Total populations affected by each facility were taken as those within census blocks whose centroids were within 10 miles of the facility point location. For parent companies, the 10-mile radius population around each facility was weighted by the facility's CO_2-equivalent emissions, and these were aggregated for all facilities it owns.

Some CO_2-equivalent emissions are "non-direct emissions": for example, oil and natural gas producers report their emissions from operations within geologic basins, and distribution companies report emissions that take place over their distribution system within a state. Because these emissions do not come from point sources, they are excluded from this analysis.

Assigning responsibility for greenhouse gas emissions

As with TRI, the GHGRP database contains parent company information, but this information is not displayed by EPA as a summed-up table in its Data Highlights default public data presentation. EPA's Envirofacts and FLIGHT database do allow searches by parent company name. As with TRI, there is no overall parent company ID. Unlike TRI, the GHGRP database allows reporting of multiple parent company owners for individual facilities, instead of a single parent company, and includes percentages of ownership for each.

In general, this permits a parent company assignment procedure similar to that described earlier for TRI: attempting automatic regularization of parent names and SEC filing lookup, final decision informed through Web searches done by a researcher, facility ownership either assigned to the majority owner or to two 50%/50% owners, facility ownership assigned to most recent owner, and so on.

However, determining parent company ownership presents a few challenges that are particular to the electric power generation industry, which is the largest single sector for emissions in the GHGRP database. Ownership determination is also sometimes complicated because the facility may be named in connection with its operating company, rather than with its owner. In addition, in this industry sometimes different owners own different power generating units at the same overall facility, resulting in cases in which a facility has no 50% owner. For this reason, a number of single facilities have large enough CO_2-equivalent emissions to make the top 100 list of "companies," but are treated, in effect, as a parent company that consists of that single facility. These single-facility emissions could be divided up and assigned to other companies by percentage of ownership, but this would be somewhat problematic since percentage of ownership may not equate to percentage of the facility's emissions generated.

GHGRP EJ ratio analysis for parent companies

Table 22.2 shows results from the "Greenhouse 100" project for 2017 (the latest data year available at the time this was written) for the five parent companies that rank highest for disproportionate modeled co-pollutant exposure to minority groups. The Greenhouse 100 project and its underlying data are publicly available on the Web for all companies in the GHGRP database. Again, "EJ: minority share" refers to the share of the total population health risk borne by minority racial/ethnic groups, and "EJ: poor share" is the share borne by people living below the poverty line.

Lessons on linkages

Some EJ analyses are best done not for parent companies but for individual facilities. For these purposes, it is often helpful to link facility data from multiple sources together. In connection with our research, projects have been undertaken linking TRI to GHGRP facilities; GHGRP

Table 22.2 Five selected records from PERI's Greenhouse 100 Index, 2017

Company	*EJ: minority share*	*EJ: poor share*	*Greenhouse 100 rank*	*2017 CO_2-equivalent emissions (metric tons)*	*Share of emissions from top facility*
San Antonio Public Service Board	78%	19%	48	12,839,604	47%
LyondellBasell	77%	19%	91	7,602,442	26%
BP	74%	20%	39	15,185,278	31%
Hilcorp Energy	73%	22%	95	6,928,281	32%
Enterprise Products Partners	73%	21%	80	8,518,255	17%

Examination of individual facilities owned by these companies reveals EJ: minority share scores to be dominated by facilities in San Antonio, Channelview, Houston, Corpus Christi, and Mont Belvieu (all in Texas), and a few other locations including Whiting, Indiana, and Bloomfield, New Mexico.

to the US EPA's eGRID (for comparison of emissions with power generation, co-pollutants, and fuel quantities); and TRI to US Equal Employment Opportunity Commission (EEOC) data to compare employment of members of minority groups at facilities with the share of environmental burdens borne by members of the same minority groups living near these facilities.

These comparisons require addressing a number of difficulties with the design and accuracy of data. In some cases, the unit of data collection may focus on different managerial or engineering concepts. For example, the unit of observations for the EEOC data on employment by race, sex, and occupation considers the "establishment," an economic concept, while the TRI collects data on "facilities"; these entities often coincide, but not always. Other reporting systems, especially those concerned with energy production and industrial processes, can be based on specific activities or processes, with for example each boiler within an electricity generating facility reporting separately.

Addresses of facilities may be recorded differently in different databases, with some facilities, for example, having no set physical address other than a point some miles down a rural road. Mailing addresses may be listed instead of physical addresses, which if uncorrected could lead to pollution being attributed to a corporate headquarters rather than a physical plant. Some of these problems can be mitigated by using GIS or other methods of comparing lat-long coordinates, although these are sometimes missing, incorrect, or poorly defined (as when a facility with a large physical extent must be reduced to a single point). As with parent company assignments, we have found no better way to make these matches than to have them automatically suggested by programs as far as possible, with a researcher making the final decision.

The US EPA constructed an additional database, the EPA Facility Registry Service (FRS) that is intended to assist with linkages across EPA datasets to facilitate comparisons and correlations. FRS succeeded previous EPA internal systems that were intended to solve a fundamental informational and regulatory problem: EPA has different programs authorized under different laws with a host of differing definitions. Although all of the regulations refer to facilities in some sense, definitions of what constitutes a facility may differ. Even in cases where the definitions largely correspond, data are collected by each of these programs independently, without any mandate for any agent or regulator to figure out whether, for example, a facility with an air pollution permit under the Clean Air Act is the same facility as one listed with a hazardous waste permit in the Resource Conservation and Recovery Act (RCRA) Information System.

FRS assigns a single ID to each facility that EPA has identified, and attaches this ID to all of the separate air, water, hazardous waste, and so on IDs that the same facility has under various EPA programs. There may be zero, one, or many IDs for each EPA program that are associated with a single FRS facility ID. The FRS database can be obtained through EPA's Envirofacts website, and we recommend it as the starting point for any kind of database-to-database facility comparison using EPA data.

In the case of the join of TRI/RSEI data with EEOC data (Ash and Boyce 2018), the join that permitted the analysis of jobs and pollution required matching two completely independent sets of identifiers, that of TRI/RSEI and that of EEOC. The set of facilities targeted for the join was limited to the 1,000 highest-impact facilities in terms of RSEI score, out of approximately 20,000 reporting facilities. The join, based on matching name and address, succeeded in joining more than 700 facilities. Walker (2013, online appendix A.4) describes matching rates using name and address to join data between the Census Bureau Standard Statistical Establishment List (SSEL) and US EPA facility lists. A key distinction is that, unlike the TRI right-to-know data that are publicly available and specifically and intentionally identify facilities, the EEOC and Census Bureau datasets, while collected by government agencies with a public mandate, are confidential, and access and use are tightly restricted. Access is limited to research by stringent application, and only summary results and generalized findings may be reported. For example, the access to EEOC data for Ash and Boyce (2018) required formal appointment of the investigator as an (unpaid) employee of the EEOC, demonstration that the research would contribute specifically to EEOC meeting its agency mandate, and strict regulation that no individual facility data be released.

Other datasets on firm and facility activity are proprietary with expensive access, for example, the Compustat dataset on the financial and real activity of firms traded on stock exchanges. In many cases, the right to publish data about specific firms is limited by user agreements for proprietary databases. The interface of right-to-know data with other datasets is of potentially great value, but the usefulness of right-to-know data is curtailed when joined datasets are proprietary or otherwise restricted in access and results are limited to aggregated summaries and general findings.

The web of connections to other databases can also expand value. The Toxic 100 and Greenhouse 100 indexes link to public and non-governmental databases on chemical toxicity, to additional facility-level information maintained by the US EPA, to mapping services provided by private providers, and to several public watchdog databases maintained by the non-governmental organization Good Jobs First.

Applied and social scientific research

Once the data have been assembled from the multiple sources described previously, they can be used in a variety of overlapping research projects. This section describes some of the uses that have been made of these data generated by researchers centered at the University of Massachusetts Amherst.

Bouwes et al. (2003) and Ash and Fetter (2004) pioneered the application of RSEI to environmental justice. In both studies, the unit of observation is the geographic receptor – in the case of Bouwes et al., the RSEI square kilometer cell and in the case of Ash and Fetter, the census block group. The dependent variable is human health risk, and the key explanatory variable is the minority share of the population. Important methodological differences between Bouwes et al. (2003) and Ash and Fetter (2004) include the assessment of all areas in Bouwes et al., as opposed to urbanized areas in Ash and Fetter, and the inclusion of population-weighted risk score in Bouwes et al., as opposed to the analysis of individual unit risk in Ash and Fetter.

Both studies found substantial evidence of environmental inequality on racial and ethnic lines. An enormous advantage to the high-resolution modeling of fate and transport of pollution is that it obviates the need to debate "how close is close" that plagued earlier studies based simply on proximity to a polluting facility (see Mohai and Saha 2006, 2007 for discussion of these problems). The comprehensive receptor-based modeling of TRI data with RSEI enabled analysis that was both national in scope and precise regarding exposure.

The high geographic resolution of the RSEI model enables the analysis of neighborhood-level differences in exposure to industrial toxics. Ash and Fetter focused on within-city risk differences, comparing this to overall (pooled within- and between-city) differences in risk. Given the importance of residential segregation in US cities, local siting decisions by companies, and local regulatory enforcement, the focus on distribution of industrial toxic exposure within urban areas allows Ash and Fetter to pose the question, "Who lives on the wrong side of the environmental tracks?"

The distinction between within-city and between-city differences in exposure provided new information on the disproportionate exposure of Latinos, or Hispanics, to industrial toxics in the United States. Earlier research had focused on and identified disproportionate exposure of

African Americans to industrial toxics, with disproportionate exposure occurring on essentially every geographic scale, both neighborhoods within cities and excess exposure based on population concentration in US industrial cities in the industrial Midwest and the urban South. Latinos were more concentrated in parts of the US with less toxic-intensive heavy industry, and city-level comparisons did not identify disproportionate Latino exposure. However, neighborhood-level analysis within cities demonstrated that Latinos live in parts of cities that have systematic excess exposure (Ash and Fetter 2004). Case studies of specific regions, for example the analysis by Morello-Frosch et al. (2001) of the "riskscape" of the Los Angeles basin, had detected this phenomenon, and RSEI-based national analysis confirmed its systemic character of disparities within place.

The high-resolution RSEI-GM data can support hierarchical models that examine simultaneously the distribution of pollution within a polity, which requires high-resolution distinction among neighborhoods, and the overall level of average pollution in that polity compared to others. Building on Ash and Boyce (2011) and Ash et al. (2009), which developed an empirical measure of the segregation of pollution, Ash et al. (2013) tested a political economy model in which the degree of environmental disparity, that is, the capacity to displace pollution onto a vulnerable social group, affects the overall level of pollution in metropolitan areas. This operationalizes Boyce's work on the theory of inequality and environmental degradation, which hypothesizes that the ability to displace environmental bads onto vulnerable populations (into spaces that effectively become "sacrifice zones") and to appropriate environmental goods into spaces reserved for a privileged few affect the political calculus regarding environmental bads and goods. Ash et al. (2013) find that in high-disparity metropolitan areas, not only do vulnerable social groups, including people of color and low-income people, experience substantially higher pollution exposure, but also the overall level of pollution exposure is higher.

The high geographic resolution of the Geographic Microdata also makes it possible to compare environmental justice gradients, the extent to which vulnerable communities are disproportionately exposed, across locations. For example, the states of the industrial Midwest – Illinois, Indiana, Michigan, Minnesota, Ohio, and Wisconsin (together designated as EPA Region 5), have high exposure of the average resident and also a very steep gradient, in which racial and ethnic minorities are disproportionately exposed (Zwickl et al. 2014). It is also possible to compute vertical inequality measures that describe the variation in exposure between the most exposed and least exposed communities, and to compare these to horizontal inequality measures based on differential exposure by race or class (Boyce et al. 2016). Currie et al. (2015) used variation in RSEI scores to value environmental health risks from changes in housing values induced by plant openings and closings.

More recent research involves the integration of Risk Screening Environmental Indicators data with carbon emissions data from the Greenhouse Gas Reporting Program. Boyce and Pastor (2013) drew attention to the importance of explicitly considering air quality co-benefits and environmental justice in the design of carbon policy. Subsequent work, including Cushing et al. (2018) and Boyce and Ash (2018), have expanded the global-local analysis of greenhouse gas reductions, the potential for co-benefits, and the peril of overlooking co-benefits for environmental justice communities. The combination of RSEI and GHGRP data makes it possible to explore further the ways in which this can be achieved.

Public intermediation for policy impact

In addition to research uses, these data have been used in various information intermediation efforts, with purposes including regulatory compliance and enforcement, socially responsible investment, corporate environmental management, and popular and mass movement awareness, action, and redress. Typically, these projects have taken the form of a website allowing the public to search and display the data for free, in an attempt to empower one or more of these types of uses. This kind of activity has taken place in connection with TRI since 1989, before the World Wide Web was created, on Bulletin Board Systems and through other early means of networked data sharing and display.

The US EPA itself has created search-and-display sites, which generally also include data download and mapping capabilities, for disseminating the TRI and GHGRP data. The major EPA sites at the time of writing are TRI Explorer, EnviroFacts (which contains both TRI and GHGRP as well as many other EPA databases), and FLIGHT (a GHGRP interface). Other governmental sites that distribute these data include international sites focused on PRTRs (pollutant release and transfer registries) that include TRI along with similar data from other countries. This is done by OECD (the Organisation for Economic Co-operation and Development) and UNITAR (United Nations Institute for Training and Research), and has been done by the CEC (the North American Commission for Environmental Cooperation, established under the North American Agreement on Environmental Cooperation).

Nonprofit, journalistic, and academic organizations have operated websites to increase public access to the data and enable analyses that are difficult to undertake on the official sites. One of the earliest efforts at enhancing public access to TRI and other EPA databases was RTK NET (the Right-to-Know Network), a project of the nonprofit Center for Effective Government (previously named OMB Watch). RTK NET has provided access to TRI and other databases since 1989. After the Center for Effective Government closed in 2016, the *Houston Chronicle* newspaper sponsored RTK NET. Another notable site was Scorecard, which provided a value-added interface to TRI and related exposure data prior to the advent of RSEI. Scorecard was initiated by Environmental Defense (formerly the Environmental Defense Fund), a major environmental advocacy nonprofit. GoodGuide.org temporarily operated as a legacy site that maintained but did not update the Scorecard data.

The Toxic 100 and Greenhouse 100 are public data intermediation projects run by the Corporate Toxics Information Project of the Political Economy Research Institute (PERI) at the University of Massachusetts Amherst, a public university, as part of its public mission to engender greater public participation in decision-making about environmental policy. These lists rank US companies by their emissions responsible for global climate change, by chronic human health risk from air toxics exposure, and by chronic human health hazard from water pollution exposure. The PERI analysis also includes environmental justice indicators, examples of which were given earlier, to assess impacts on minorities and low-income people. These indexes are frequently cited in news media, on Wiki pages about individual corporations, and in shareholder resolutions on corporate environmental performance.

Several additional intermediation projects have in turn used the Toxic 100 index to add further value to the pollution information from the Corporate Toxics Information Project. In 2008, the UK-based Business and Human Rights Resource Centre moderated a dialogue between the top 10 firms listed on the Toxic 100 index published that year and the Corporate Toxics Information Project of PERI (Business and Human Rights Resource Centre 2008). Good Jobs First, a non-governmental policy resource center and the Corporate Toxics Information Project reciprocally link company specific data between the Toxic 100 and Greenhouse 100 and the Violation Tracker and Subsidy Tracker sites, which monitor and report fines and penalties that corporations pay for violation of environmental, health, occupational, financial, and fiscal regulations and laws and federal, state, and local public subsidies to corporations.

Conclusion

To affect environmental justice policy and practice, the right-to-know movement and regulation by right to know require consistent intermediation by public, university, and non-governmental organizations to draw out meaningful connections in the data. The Corporate Toxics Information Project has experimented with data interlinkages and the intermediation of results to expand the impact and utility of right-to-know data from pollutant release and transfer registers. Concrete scientific findings include results on the political economy of pollution exposure and environmental racism in the United States and the weak empirical case for a widely assumed jobs-environment trade-off. Public intermediation of the data has affected shareholder intervention in the dimension of socially responsible investment and activist and journalistic interventions. Effective interlinkage and intermediation depend on the availability of data for integrative assessment models, the establishment of corporate ownership of fixed assets, and socioeconomic variation on a geographic basis. Establishment of common identifiers at the facility, corporate, and geographic level is a significant challenge for environmental justice researchers.

Notes

1 Who's Counting? The Systematic Underreporting of Toxic Air Emissions. Environmental Integrity Project. June 2004. www.environmentalintegrity.org/pdf/publications/TRIFINALJune_22.pdf
2 The selection of the 10-mile radius reflects an expert judgment on the most affected area for many TRI releases in the RSEI fate-and-transport model. Criticisms of distance-based buffer models include Mohai and Saha (2003), Mohai and Saha (2006, 2007).

References

Ash, M. and Boyce, J.K., 2011. Measuring corporate environmental justice performance. *Corporate Social Responsibility and Environmental Management*, 18(2), pp. 61–79.
Ash, M. and Boyce, J.K., 2018. Racial disparities in pollution exposure and employment at US industrial facilities. *Proceedings of the National Academy of Sciences*, 115(42), pp. 10636–10641.
Ash, M., Boyce, J.K., Chang, G., Pastor, M., Scoggins, J. and Tran, J., 2009. *Justice in the air: Tracking toxic pollution from America's industries and companies to our states, cities, and neighborhoods*. Amherst, MA: Political Economy Research Institute.
Ash, M., Boyce, J.K., Chang, G. and Scharber, H., 2013. Is Environmental justice good for White folks? Industrial air toxics exposure in Urban America. *Social Science Quarterly*, 94(3), pp. 616–636.
Ash, M. and Fetter, T.R., 2004. Who lives on the wrong side of the environmental tracks? Evidence from the EPA's risk-screening environmental indicators model. *Social Science Quarterly*, 85(2), pp. 441–462.
Bouwes, N.W., Hassur, S.M. and Shapiro, M.D., 2003. Information for empowerment: The EPA's risk-screening environmental indicators project. In *Natural Assets: Democratizing Ownership of Nature*. Eds. J.K. Boyce and B. Shelley. Washington, DC: Island Press, pp. 117–134.
Boyce, J.K. and Ash, M., 2018. Carbon pricing, co-pollutants, and climate policy: Evidence from California. *PLoS Medicine,* 15(7), e1002610.
Boyce, J.K. and Pastor, M., 2013. Clearing the air: Incorporating air quality and environmental justice into climate policy. *Climatic Change*, 120(4), pp. 801–814.
Boyce, J.K., Zwickl, K. and Ash, M., 2016. Measuring Environmental Inequality. *Ecological Economics*, 124, pp. 114–123.
Business and Human Rights Resource Centre, 2008. Toxic 100 & company responses – 2008. www.business-humanrights.org/en/documents/toxic-100-company-responses-2008.
Currie, J., Davis, L., Greenstone, M. and Walker, R., 2015. Environmental health risks and housing values: Evidence from 1,600 toxic plant openings and closings. *American Economic Review*, 105(2), pp. 678–709.
Cushing, L., Blaustein-Rejto, D., Wander, M., Pastor, M., Sadd, J., Zhu, A. and Morello-Frosch, R., 2018. Carbon trading, co-pollutants, and environmental equity: Evidence from California's cap-and-trade program (2011–2015). *PLoS Medicine*, 15(7), p.e1002604.

Fung, A., Graham, M. and Weil, D. 2007. *Full disclosure: The perils and promise of transparency*. New York: Cambridge University Press.
Hamilton, J.T., 1995. Pollution as news: Media and stock market reactions to the toxics release inventory data. *Journal of Environmental Economics and Management*, 28(1), pp. 98–113.
Hersh, A. 2006. Corporate Toxics Information Project Interview with co-directors James Boyce and Michael Ash. Political Economy Research Institute. March. www.peri.umass.edu/images/CTIP_Interview.pdf.
Mohai, P. and Saha, R., 2003, August. Reassessing race and class disparities in environmental justice research using distance-based methods. In *Annual Meeting of the American Sociological Association*, Atlanta, GA.
Mohai, P. and Saha, R., 2006. Reassessing racial and socioeconomic disparities in environmental justice research. *Demography*, 43(2), pp. 383–399.
Mohai, P. and Saha, R., 2007. Racial inequality in the distribution of hazardous waste: A national-level reassessment. *Social Problems*, 54(3), pp. 343–370.
Morello-Frosch, R., Pastor, M. and Sadd, J., 2001. Environmental justice and Southern California's "riskscape" the distribution of air toxics exposures and health risks among diverse communities. *Urban Affairs Review*, 36(4), pp. 551–578.
Walker, W.R., 2013. The transitional costs of sectoral reallocation: Evidence from the clean air act and the workforce. *Quarterly Journal of Economics*, 128(4), pp. 1787–1835.
Zwickl, K., Ash, M. and Boyce, J.K., 2014. Regional variation in environmental inequality: Industrial air toxics exposure in US cities. *Ecological Economics*, 107, pp. 494–509.

23

CONCLUSION

New frontiers in the political economy of the environment

Éloi Laurent and Klara Zwickl

From biophysical limits to social-ecological frontiers

The blind spots identified in the introduction of this handbook have been individually addressed with each chapter's contribution in a powerful and original way. This fills the void left by the lack of collaborative efforts to bridge political economy and environmental economics. The collective picture formed by these chapters leads us to conclude that an important next step will be the development of a robust and consistent framework connecting ecological crises to social systems. This consolidated approach could help strengthen the discipline of political economy of the environment.

One way to start thinking about the task standing before us is to consider the value and limitations of existing frameworks that have recently attempted to represent the magnitude of global environmental change. Certainly, the most influential representation of our unprecedented time is the model of "planetary boundaries". Planetary boundaries are quantitative thresholds "within which humanity can continue to develop and thrive for generations to come" (Steffen et al., 2015: 1). Crossing these boundaries would amplify the risk of generating large-scale abrupt or irreversible environmental and social changes. The authors warn us in unambiguous terms about this peril: "Four of nine planetary boundaries have now been crossed as a result of human activity: climate change, loss of biosphere integrity, land-system change, and altered biogeochemical cycles (phosphorus and nitrogen)". Two of these, climate change and biosphere integrity, are "core boundaries", hence, significantly altering either of these would "drive the Earth System into a new state" (Steffen et al., 2015: 1).

While this is certainly breakthrough science, classical economists from the 18th and the 19th century had the intuition that human development was in fact constrained by the scarcity of nature. More importantly, planetary boundaries seem to be silent on human frontiers: who is responsible for ecological trespassing? Who is vulnerable? Similarly, the Anthropocene theory presents us with a unified human species that has set in motion a geological revolution, to the perverse effects of which all humans are now exposed. However, this is clearly misleading, in terms of both analysis and policy: to borrow from UNFCC language, if humans share a common responsibility in setting global environmental change in motion, this responsibility is highly differentiated (U.N., 1992).

DOI: 10.4324/9780367814533-25

As a result, we would like to argue here, in light of the information presented in each chapter of this volume, that the notion of social-ecological frontiers, rather than planetary boundaries, offers a more meaningful framework for studying the political economy of the environment. In other words, the conceptualization of "boundaries" should be extended to consider limits determined not only objectively by chemistry or biophysics, but also normatively by principles of justice. As the chapters gathered in this handbook make clear, ecological crises are social issues, in which the central question should be to understand which social causes generate ecological damage and crises and which social consequences they in return induce on social systems and human groups.

The case of climate change illustrates the need for such an approach. The 1.5 to 2 degrees limit of warming mentioned in the 2015 Paris Agreement[1] is indeed a chosen threshold and not an imposed limit – a threshold designed by humans that will determine the fate of hundreds of millions of people in decades to come. In addition, the COVID-19 pandemic, which is very likely a zoonosis, shows us that the human-animal limit itself is a moving and porous frontier that depends on the expansion of human systems in the biosphere (WHO, 2020). If the "one health" approach offers insights into the contemporary outburst of pathogens transmitted from animals to humans, it is precisely because it posits that there is no boundary between humans and animals (WHO, 2017). More generally, it could be that planetary boundaries might be better understood and respected by human societies if they are conceived as social-ecological frontiers endowed with social values and justice principles.

However, "new frontiers" has a more precise meaning within this general framework, referring to the new domains that are emerging in relation to the political economy of the environment. Specifically, we see four such domains that are not sufficiently covered in this volume.

Ecological and digital transitions: friends or foes?

The first such domain is the emerging contradiction between the reality of digital transition and the necessity of ecological transition, which could be thought of as the political economy of conflicting transitions. The state of California represents a compelling example of this growingly visible contradiction. California is home to two types of exceptional ecosystems: natural ecosystems and digital ecosystems. While the region has become the center of global tech capitalism, it is also a region where, in the last 20 years, natural ecosystems have entered a structural crisis (water, forests, air quality, drought, etc.). In other words, while digital ecosystems are flourishing (e.g. Apple is now worth more than the entire French stock market and will soon surpass the GDP of France), natural ecosystems are floundering (Bradshaw & McGee, 2020). Is there a contradiction or a convergence between natural and digital ecosystems? One way to answer this question is to look at the extraction of natural resources that has accompanied digital transition.

The world economy currently extracts three times more natural resources than it did in 1970. Furthermore, the early 2000s and the mid-2000s, when the digital transition really took off, marked an acceleration in extraction patterns (40 billion more tons were extracted between 2000 and 2017 in comparison to the 20 billion tons extracted between 1970 and 2000) (IRP, 2017). Even more striking, the relative decoupling between economic production and consumption of natural resources observed throughout the 20th century and up until the early 2000s has since been reversed (Laurent, 2020a: 92). But why? And with which consequences for whom?

Social-ecological urban environmental justice[2]

The beginning of the 21st century is characterized by two large-scale geographical dynamics. The first is the global urban revolution that began in the second half of the 20th century. Whereas in 1700 only 2% of the planet's inhabitants lived in cities (and then 3% in 1800), this proportion rose to 15% in 1900 and then doubled to reach 30% in barely 50 years, with the 50% threshold having been crossed in 2007 (U.N., 2019). In 2018, according to figures published by the United Nations, 55% of the world's population lived in urban areas (U.N., 2019). Interestingly, this trend can also be observed in terms of the size of cities. This is to say, in the 21st century, the growth of urban spaces has been twice as fast as that of the world population. Thus, demographic experts predict that the physical expansion of the world's cities during the first three decades of the 21st century will surpass that of all urban spaces from the origins of the human species until the end of the 20th century (U.N., 2019). In the short term, the urban population, which was roughly 4.2 billion inhabitants in 2020, is expected to reach 5.1 billion by 2030 (or about 60% of the world population), with almost half of that increase occurring in urban spaces that have more than one million inhabitants (U.N., 2019).

This dynamic of urbanization is clearly global, and thus, one can speak without hyperbole of a universal urban revolution. Of course, not all regions of the world are equally urbanized, but they are converging toward urbanization at a rapid rate. Leading the urbanized regions are North America (with 82% of its population living in urban areas in 2018), Latin America and the Caribbean (81%), Europe (74%), and Oceania (68%) (U.N., 2019). Conversely, India has the largest rural population (893 million), followed by China (578 million), while African nations remain predominantly rural, with only 43% of their population living in urban areas (U.N., 2019). However, both Asia and Africa are expected to account for 90% of urban population growth over the next three decades. The case of Nigeria, whose population became predominantly urban in 2018, is a compelling case study of the global convergence towards urbanization. While the country had less than 10% of its population living in urban areas in 1960 (three times less than the global average), it is anticipated that by 2050 this figure will rise to 70%, the same level as the world average (U.N., 2019).

A stylized fact, resulting from the two previous trends, is that environmental sustainability is now an urban issue. As the urban revolution accelerates before our eyes, cities are now recognized as key places for both mitigation of and adaptation to ecological crises, starting with climate change. In particular, cities hold the key to the necessary reduction in the consumption of natural resources, which causes considerable damage to biodiversity and ecosystems and increasingly affects depopulated rural areas. While they only occupy 5% of the planet's surface, cities represent 66% of the energy consumed and 75% of CO_2 emissions (Laurent, 2020a: 173). According to the International Energy Agency (IEA), buildings alone are the largest source of energy consumption in the world (Laurent, 2020a: 173).

As surprising as it may seem given the salience of these dynamics, there is still no formal, universally accepted definition of what a city is. There is therefore, *a fortiori*, no consensus on what a sustainable city would be, nor what a just and sustainable city would entail.

The political economy of cities is of course nothing new. City dwellers are citizens in places: a city is by definition a politicized space, a domain defined by the legal authority under which it is placed by a human community. It is therefore a place of justice between humans. Thus, if Latin distinguishes *urbs* (the physical space) and *civitas* (the community of citizens), the Greek *polis* means both the city and the political community. The legal space thus defined determines in turn the social prospects of the people who occupy it. Social injustice is thus reflected within space (through the spatial inscription of social inequality, such as the striking racial segregation of the city of Detroit); spatial organization generates injustice (the spatial creation of social inequalities, such as the externalization of industrial risk in the poorest departments of the Ile-de-France region around Paris); finally, social relations, in particular inequalities, produce space.

Social sciences took up the question of urban justice in the 1920s and 1930s, with the development of the "urban ecology" approach by the sociological school of Chicago, which was particularly interested in the influence of spatial factors on social phenomena. The first appearance of the concept of "spatial justice" in academic literature dates back to the pioneering work of David Harvey in the early 1970s (Harvey, 1973).

However, we are presented with new and evolving urban realities in the face of ecological crises. At the start of the 21st century, the discussion/issue of urban spatial justice seemed to align with the overall goals of environmental justice, at all levels of governance of cities. Natural elements, deeply altered by humans, give rise to a new, third, type of inequality, which is neither natural nor social but rather socio-ecological. While heat waves linked to climate change are certainly natural phenomena, it seems that in the 21st century these are exacerbated by harmful human activities (via climate change) and thus, their social impact is in turn considerable (e.g. 70,000 dead in Europe and nearly 15,000 in France after the 2003 heat wave) (Robine et al., 2007). These inequalities are thus socio-ecological in nature, both upstream and downstream.

In fact, the major challenge for developed cities at the start of the 21st century is no longer, as it was in the 18th and 19th centuries, the problematic proximity of places of residence and production sites. Rather, it is the problematic distance between employment and residential areas leading to constrained mobility, which generates unequal local and global pollution and the ecological vulnerability of urban spaces (in which almost the entire population is concentrated).

Similarly, while one of the aims of a city is to protect against natural disasters, it might just as well expose its residents to greater risk, as is the case with flooding. "Hydrometeorological events" (such as floods and storms) constitute the social-ecological risk affecting the greatest number of people in the world and represent two-thirds of the costs linked to so-called natural disasters in Europe. By way of illustration, the "natural" shock of the June 2013 floods (the most serious in recent decades on the continent) was considerable: a 10-year flood along the river plain of the Danube, the Elbe, the Saale, and the Vltava, affecting Germany, Austria, Hungary, the Czech Republic, and Slovakia. In fact, it is the human factor that explains the scale of the disaster, since increased urbanization entails the artificialization of soils and thus degradation or even inversion of their flood regulatory capacities. The flood risk is not natural but results from the combination of exposure and sensitivity of populations. Therefore, social-ecological analysis and policy may have a great future in the 21st century.

Toward a comprehensive assessment of the distributional impacts of environmental policy

Another important avenue for further research in the field of political economy of the environment is the development of a broader framework to understand the distributional effects of environmental policies. The field of environmental justice distinguishes between different types of environmental inequalities, for example Laurent (2011) distinguishes between four types: exposure and access inequalities, policy effect inequalities, environmental impact inequalities, and policymaking inequalities. Yet, most empirical studies in the field of environmental justice focus on exposure and access inequalities, while some environmental economists have become increasingly interested in assessing policy effect inequalities, for example the distributional impacts of carbon taxes (see Boyce, chapter 17). Currently, very limited evidence exists on how an environmental policy would affect both or all of the four dimensions of environmental inequality. Theoretically, it is possible that an environmental policy reduces all of the aforementioned categories, or reduces some but increases other dimensions of inequality. Thus, the broader our analysis, the more likely we can assess the overall distributional effects. Such a broad framework could build upon different streams of literature discussed in this handbook with the aim of integrating them into a comprehensive analysis.

Economic theory, policy discourse, and the public perception consider environmental policies, especially price-based policies, to be regressive. A carbon tax, for example, increases the price of energy, which disproportionately affects low-income households since they spend a larger share of their income on energy consumption. Thus, while an analysis limited to the consumption effects of environmental policies will thus likely find regressive effects, the broader and more outcome variables that are included, the more likely it is that the overall effect will be reversed. If only the impact of carbon pricing on a specific carbon intensive sector is analyzed, then the employment or output effects are likely to be negative. This however neglects that job-shifting from carbon intensive to less carbon intensive sectors can occur and that overall employment effects might thus turn positive (see for example Yamazaki (2017) who found this for the carbon tax introduced in British Columbia). Moreover, carbon pricing can increase the cost of capital and therefore may lead to a substitution from capital to labor, which in turn leads to a rising wage share (Metcalf, 2019). Carbon pricing also generates revenues that can be recycled progressively (see Boyce, chapter 17).

These examples suggest that the broader the assessment, the more likely it is that the carbon tax is found to yield overall positive economic and distributional effects. Yet, the current literature is still largely focused on economic variables. However, different strands of literature – the aforementioned literature on the economic and distributional impacts of environmental policies, the literature on environmental justice and climate justice, and the literature on co-benefits of environmental policies – could be combined to develop a framework to conceptualize and measure the full distributional impacts of environmental policies.

The literature on environmental inequality provides strong evidence that environmental goods and bads are distributed unequally across society. Various chapters in this handbook emphasize the fact that poor and minority neighborhoods often face higher pollution burdens and associated adverse health effects, in addition to being more vulnerable to the health effects of pollution due to lower access to health care (for example Laurent in chapter 3, and Pellow in chapter 6). Poor households are also more often affected by natural disasters, and when they are, their assets are disproportionately more affected than those of rich households (see Hallegatte and Walsh in chapter 8). Conversely, policies that reduce these environmental burdens would have positive redistributive effects on health, productivity, and asset values. However, little empirical evidence currently exists on the effects of environmental policy on the distribution of these outcomes. One would expect that if environmental policies improve environmental quality, the health effects of these environmental improvements would disproportionately fall on socioeconomically vulnerable populations that have faced high levels of pollution in the past.

Similarly, the literature on air quality co-benefits emphasizes the existence of substantial and significant positive spillovers of climate policy on public health through air quality improvements (see Zwickl and Sturn, chapter 13). However, public health benefits are typically aggregated across the whole economy and not differentiated by income or socioeconomic status. The only two exceptions in the literature are Boyce and Pastor (2013) and Cushing et al. (2018), who find that incorporating air quality co-benefits into climate policy in the US could also narrow environmental inequality, since co-benefits are highest in socioeconomically disadvantaged neighborhoods. More empirical research linking co-benefits, environmental inequality, and the distributional effects of climate policy would be very beneficial.

Developing a broad concept of the distributional effects of environmental policies becomes even more challenging when future generations are also considered, when intergenerational inequality is added to intragenerational inequality. While current climate policy still focuses on climate change mitigation, the reality is that the longer we take to reduce carbon emissions, the bigger the need for climate change adaptation will be in the future. Although some adaptation measures can be very costly now, they will be crucial for the survival of millions of people in the future. Thus, when assessing the distributional effects of climate policies, we should also consider that they will result in fewer adaptation measures, which in turn will bring along new challenges of allocation among competing needs in society.

Understanding the belief structures of environmental policy

This handbook has emphasized that ecological crises are social issues, while also presenting environmental degradation in the light of power inequality. Many articles explicitly or implicitly deal with the question of which power constellations maintain and reinforce current environmental degrading activities (for instance Sicotte in chapter 11 and Cardenas in chapter 12). Sometimes, however, social norms, attitudes, and beliefs could result in outcomes in which the majority of the population would benefit from an environmental regulation but individuals and groups still harbor strong negative views against the considered measure.

These norms, attitudes, and beliefs vary across countries and time. In fact, sudden changes in public opinions are especially common for environmental issues, since on the one hand environmental conservation is considered desirable by many, but on the other hand environmental policies are often viewed as socially regressive for the reasons discussed in the previous section. Many trade-offs are thus constructed in the political discourse, such as between environmental and social goals, or between different environmental goals.

An interesting case to examine is nuclear energy. Attitudes towards nuclear energy vary substantially across countries and over time. One country with strong negative views on nuclear energy is Austria. In a nuclear referendum in the late 1970s, the (small) majority of voters objected to the launch of the already constructed first and only nuclear power plant in Austria (which to date still exists and now produces solar energy). While nuclear energy was still strongly contested in the early 1980s, the Chernobyl nuclear disaster of 1986 lead to a near consensus against nuclear energy in the Austrian society, with no political party promoting it ever since (though Austria imports nuclear energy from other European countries). In contrast, in the United States, nuclear energy is widely used and even advocated for by some environmental policymakers, who promote nuclear energy as an important alternative to carbon combustion (or at least as a transition technology). Other countries fall somewhere in between Austria and the US. However, public perceptions change, as they have after nuclear disasters (e.g. Germany decided on a nuclear phaseout in 2011, shortly after the nuclear disaster in Fukushima) or when an inevitable trade-off between the threats from nuclear energy and the threats from climate change receives wide public attention. The latter can explain why in Switzerland a popular initiative to phase out nuclear power in 2016 initially expected strong support; however, within only eight weeks, support in favor of the phaseout dropped from 60% to around 46% (Rinscheid & Wüstenhagen, 2018).

An even stronger change in public attitudes in a popular vote can be found in California's rejection of "The California Right to Know Genetically Engineered Food Act" in 2011. The new law would have required the labelling of genetically modified food products, a proposition initially enjoying a large – up to 90% – support by people wanting to know – and thus decide – whether they were eating genetically modified foods. Prior to the election, a large campaign funded by producers of genetically modified organisms (GMOs) successfully reframed the debate (with the help of a budget that was five times as big as that of the supporters of the initiative) (Paull, 2012). In this case, an argument used by the opponents that ultimately shifted the election outcome was that an additional label on every food product would unnecessarily confuse consumers, given that food products already require other labels, including information on nutrients. The vote was lost 47% to 53%.

The previous examples suggest that the framing of an environmental goal and the economic, social, and distributional consequences of achieving it strongly affect political processes and popular views, often at the expense of more stringent policies. Hence, this should be considered in the field of political economy of the environment.

Moreover, the perceptions and actual distributional effects of policies may differ substantially. A large-scale French survey on the attitudes on climate change and policy found that the French generally prefer regulations or green public investments over a carbon tax, because they perceive the latter to have strongly regressive effects (Douenne & Fabre, 2020). In fact, regulations, for example mandatory technology requirements, might lead to price increases similar to a tax but without the advantage of the tax generating a revenue that can be distributed progressively. In the previous section we hypothesized that the broader the economic, social, and environmental outcome variables that are measured, the more likely it is that an environmental policy is considered to have positive distributional effects. This section illustrates that it is not only necessary for the field of political economy of the environment to empirically assess this, but it is also necessary to communicate the findings to a broader public audience. It also illustrates the centrality of narratives in human behaviors and should lead us, finally, to extend the definition put forth in the Introduction of this volume: the political economy of the environment should also be concerned with narrative power inequality.

Notes

1 Article 2a of the Paris Agreement states the objective of "Holding the increase in the global average temperature to well below 2 °C above pre-industrial levels and pursuing efforts to limit the temperature increase to 1.5 °C above pre-industrial levels, recognizing that this would significantly reduce the risks and impacts of climate change".

2 This section is adapted from Laurent (2020b).

Bibliography

Boyce, J. K. & Pastor, M. 2013. Clearing the air: Incorporating air quality and environmental justice into climate policy. *Climate Change*, 120, 801–814.

Bradshaw, T. & McGee, P. 2020. Apple market value hits $2tn. *Financial Times*.

Cushing, L., Blaustein-Rejto, D., Wander, M., Pastor, M., Sadd, J., Zhu, A. & Morello-Frosch, R. 2018. Carbon trading, co-pollutants, and environmental equity: Evidence from California's cap-and-trade program (2011–2015). *PLoS Medicine*, 15.

Douenne, T. & Fabre, A. 2020. French attitudes on climate change, carbon taxation and other climate policies. *Ecological Economics*, 169.

Harvey, D. 1973. *Social justice and the city*, Baltimore: Johns Hopkins University Press.

IRP. 2017. *Assessing global resource use: A systems approach to resource efficiency and pollution reduction.*

Laurent, É. 2011. Issues in environmental justice within the European Union. *Ecological Economics*, 70, 1846–1853.

Laurent, É. 2020a. *The New Environmental Economics: Sustainability and Justice.* Cambridge: Polity Press.

Laurent, É. 2020b. Soutenabilité des systèmes urbains et inégalités environnementales. Le cas français. *Revue de l'OFCE,* 165(1), 145–168.
Metcalf, G. E. 2019. On the economics of a carbon tax for the United States. *Brookings Papers on Economic Activity*, 405–484.
Paull, J. 2012. USA: California rejects mandatory GMO labelling. *Organic News*, 1–2.
Rinscheid, A. & Wüstenhagen, R. 2018. Divesting, fast and slow: Affective and cognitive drivers of fading voter support for a nuclear phase-out. *Ecological Economics*, 152, 51–61.
Robine, J.-M., Cheung, S. L., Le Roy, S., Van Oyen, H., Griffiths, C., Michel, J.-P. & Herrmann, F. R. 2007. Death toll exceeded 70,000 in Europe during the summer of 2003. *C.R. Biologies*, 331.
Steffen, W. et al. 2015. Planetary boundaries: Guiding human development on a changing planet. *Science*, 347.
U.N. 1992. *United Nations framework convention on climate change.* New York: UN General Assembly.
U.N. 2019. *World population prospects 2019.* New York: Population Division of the Department of Economic and Social Affairs of the United Nations Secretariat.
WHO. 2017. *One health* [Online]. World Health Organization. Available: www.who.int/news-room/q-a-detail/one-health [Accessed].
WHO. 2020. *Zoonoses and the environment* [Online]. World Health Organization. Available: www.who.int/foodsafety/areas_work/zoonose/en/ [Accessed].
Yamazaki, A. 2017. Jobs and climate policy: Evidence from British Columbia's revenue-neutral carbon tax. *Journal of Environmental Economics and Management*, 83, 197–216.

INDEX

Page numbers in *italic* indicate a figure and page numbers in **bold** indicate a table on the corresponding page.

Printed in the United States
by Baker & Taylor Publisher Services